Selected Titles in This Series

42 **Jin Hong and Seok-Jin Kang,** Introduction to quantum groups and crystal bases, 2002
41 **Georgi V. Smirnov,** Introduction to the theory of differential inclusions, 2002
40 **Robert E. Greene and Steven G. Krantz,** Function theory of one complex variable, 2002
39 **Larry C. Grove,** Classical groups and geometric algebra, 2002
38 **Elton P. Hsu,** Stochastic analysis on manifolds, 2002
37 **Hershel M. Farkas and Irwin Kra,** Theta constants, Riemann surfaces and the modular group, 2001
36 **Martin Schechter,** Principles of functional analysis, second edition, 2002
35 **James F. Davis and Paul Kirk,** Lecture notes in algebraic topology, 2001
34 **Sigurdur Helgason,** Differential geometry, Lie groups, and symmetric spaces, 2001
33 **Dmitri Burago, Yuri Burago, and Sergei Ivanov,** A course in metric geometry, 2001
32 **Robert G. Bartle,** A modern theory of integration, 2001
31 **Ralf Korn and Elke Korn,** Option pricing and portfolio optimization: Modern methods of financial mathematics, 2001
30 **J. C. McConnell and J. C. Robson,** Noncommutative Noetherian rings, 2001
29 **Javier Duoandikoetxea,** Fourier analysis, 2001
28 **Liviu I. Nicolaescu,** Notes on Seiberg-Witten theory, 2000
27 **Thierry Aubin,** A course in differential geometry, 2001
26 **Rolf Berndt,** An introduction to symplectic geometry, 2001
25 **Thomas Friedrich,** Dirac operators in Riemannian geometry, 2000
24 **Helmut Koch,** Number theory: Algebraic numbers and functions, 2000
23 **Alberto Candel and Lawrence Conlon,** Foliations I, 2000
22 **Günter R. Krause and Thomas H. Lenagan,** Growth of algebras and Gelfand-Kirillov dimension, 2000
21 **John B. Conway,** A course in operator theory, 2000
20 **Robert E. Gompf and András I. Stipsicz,** 4-manifolds and Kirby calculus, 1999
19 **Lawrence C. Evans,** Partial differential equations, 1998
18 **Winfried Just and Martin Weese,** Discovering modern set theory. II: Set-theoretic tools for every mathematician, 1997
17 **Henryk Iwaniec,** Topics in classical automorphic forms, 1997
16 **Richard V. Kadison and John R. Ringrose,** Fundamentals of the theory of operator algebras. Volume II: Advanced theory, 1997
15 **Richard V. Kadison and John R. Ringrose,** Fundamentals of the theory of operator algebras. Volume I: Elementary theory, 1997
14 **Elliott H. Lieb and Michael Loss,** Analysis, 1997
13 **Paul C. Shields,** The ergodic theory of discrete sample paths, 1996
12 **N. V. Krylov,** Lectures on elliptic and parabolic equations in Hölder spaces, 1996
11 **Jacques Dixmier,** Enveloping algebras, 1996 Printing
10 **Barry Simon,** Representations of finite and compact groups, 1996
9 **Dino Lorenzini,** An invitation to arithmetic geometry, 1996
8 **Winfried Just and Martin Weese,** Discovering modern set theory. I: The basics, 1996
7 **Gerald J. Janusz,** Algebraic number fields, second edition, 1996
6 **Jens Carsten Jantzen,** Lectures on quantum groups, 1996

For a complete list of titles in this series, visit the AMS Bookstore at **www.ams.org/bookstore/**.

Introduction to Quantum Groups and Crystal Bases

GRADUATE STUDIES IN MATHEMATICS 42

Introduction to Quantum Groups and Crystal Bases

Jin Hong
Seok-Jin Kang

AMERICAN MATHEMATICAL SOCIETY
Providence, Rhode Island

2000 *Mathematics Subject Classification.* Primary 17B37, 17B65;
Secondary 81R50, 82B23.

This book was written with support from KOSEF Grant #98-0701-01-050-L.
The second author acknowledges support from the
Young Scientist Award, Korean Academy of Science and Technology.

ABSTRACT. This book provides an elementary introduction to the theory of quantum groups and crystal bases. We start with the basic theory of quantum groups and their representations, and then give a detailed exposition of the fundamental features of crystal basis theory. We also discuss its applications to the representation theory of classical Lie algebras and quantum affine algebras, solvable lattice model theory, and combinatorics of Young walls.

Library of Congress Cataloging-in-Publication Data

Hong, Jin, 1972–
Introduction to quantum groups and crystal bases / Jin Hong, Seok-Jin Kang.
p. cm. — (Graduate studies in mathematics, ISSN 1065-7339 ; v. 42)
Includes bibliographical references and index.
ISBN 0-8218-2874-6 (alk. paper) | Softcover ISBN 1-4704-7975-3
1. Quantum groups. 2. Representations of groups. I. Kang, Seok-Jin. II. Title. III. Series.

QA252.3.H66 2002
512′.55—dc21 2001053274

Printed in the United States of America.
Reprinted by the American Mathematical Society, 2025.

♾ The paper used in this book is acid-free and falls within the guidelines
established to ensure permanence and durability.
Visit the AMS home page at URL: http://www.ams.org/

12 11 10 9 8 7 6 5 4 3 30 29 28 27 26 25

Contents

Introduction

The notion of a *quantum group* was introduced by V. G. Drinfel′d and M. Jimbo, independently, in their study of the quantum Yang-Baxter equation arising from two-dimensional solvable lattice models ([**10, 23**]). Quantum groups are certain families of Hopf algebras that are deformations of universal enveloping algebras of Kac-Moody algebras. Over the past 20 years, they turned out to be the fundamental algebraic structure behind many branches of mathematics and mathematical physics such as:

(1) solvable lattice models in statistical mechanics,

(2) topological invariant theory of links and knots,

(3) representation theory of Kac-Moody algebras,

(4) representation theory of algebraic structures, e.g., Hecke algebra,

(5) topological quantum field theory,

(6) geometric representation theory,

(7) C^*-algebras.

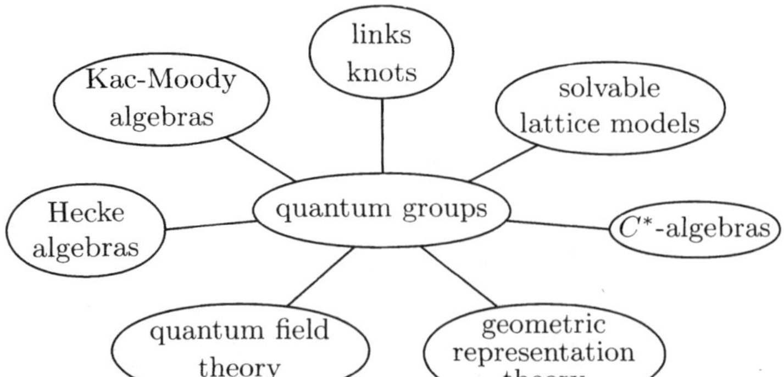

In particular, the theory of *crystal bases* or *canonical bases* developed independently by M. Kashiwara and G. Lusztig provides a powerful combinatorial and geometric tool to study the representations of quantum groups ([**38, 39, 48**]). The purpose of this book is to provide an elementary introduction to the theory of quantum groups and crystal bases focusing on the combinatorial aspects of the theory.

In such an introductory book, the first question to be answered would be: *What are quantum groups?* In his famous lecture given at the International Congress of Mathematicians held at Berkeley in 1986, Drinfel′d gave a *definition* of quantum groups: it was defined to be the *spectrum of a certain Hopf algebra* [**11**]. That is, Drinfel′d noted that any suitable category of groups (algebraic, topological, etc.) is antiequivalent to a suitable category of *commutative* Hopf algebras. In such a situation, one goes from the group to the algebra by considering a suitable algebra of functions, while the group can be reconstructed by taking the *spectrum* in the sense of Grothendieck. Thus, even when one has a noncommutative Hopf algebra, it becomes natural to think of the corresponding object in the opposite category as a *quantum group*, and this is the meaning of Drinfel′d's definition.

In this book, we focus on the quantum groups that appear as certain deformations of universal enveloping algebras of Kac-Moody algebras. For example, let $\mathfrak{g}$ be a finite dimensional simple Lie algebra, and let $U(\mathfrak{g})$ be its universal enveloping algebra. Choose a generic parameter q. Then, for each q, we can associate a Hopf algebra $U_q(\mathfrak{g})$, called the *quantum group* or the *quantized universal enveloping algebra*, whose structure *tends to* that of $U(\mathfrak{g})$ as q approaches 1. Therefore, we get a family of Hopf algebras $U_q(\mathfrak{g})$, and when $q = 1$, it is the same as the Hopf algebra $U(\mathfrak{g})$.

The following example shows how one can understand the above statement in a naive way. This example is not rigorous, not even mathematical, but it gives us a certain intuition. Let $\mathfrak{g} = \mathfrak{sl}_2$ be the complex Lie algebra of 2×2 matrices of trace 0. It is generated by the elements e, f, and h with defining relations

$$[e, f] = h, \qquad [h, e] = 2e, \qquad [h, f] = -2f.$$

Thus its universal enveloping algebra $U(\mathfrak{sl}_2)$ is an associative algebra over $\mathbf{C}$ with 1 generated by the elements e, f, and h with defining relations

$$ef - fe = h, \qquad he - eh = 2e, \qquad hf - fh = -2f.$$

Now, the quantum group $U_q(\mathfrak{g}) = U_q(\mathfrak{sl}_2)$ is defined to be the associative algebra over $\mathbf{C}(q)$ with 1 generated by the elements e, f, and q^h with defining relations

$$ef - fe = \frac{q^h - q^{-h}}{q - q^{-1}}, \qquad q^h e q^{-h} = q^2 e, \qquad q^h f q^{-h} = q^{-2} f.$$

Let us look at the first of these defining relations. As q approaches 1, the left-hand side remains the same as $ef - fe$, but the right-hand side is undetermined. If we apply L'Hospital's rule (however absurd it might be), then the right-hand side is equal to

$$\lim_{q\to 1} \frac{q^h - q^{-h}}{q - q^{-1}} = \lim_{q\to 1} \frac{hq^{h-1} + hq^{-h-1}}{1 + q^{-2}} = \frac{2h}{2} = h,$$

as desired.

For the second relation, if we let $q \to 1$, then we get $e = e$, which gives nothing new. But if we *differentiate* both sides with respect to q (again, however absurd it might be), we get

$$hq^{h-1}eq^{-h} + q^h e(-h)q^{-h-1} = 2qe.$$

Thus, if we take the limit $q \to 1$, we get

$$he - eh = 2e.$$

Similarly, the last relation gives the desired relation as $q \to 1$.

Therefore, one can say that for each generic parameter q, there is a quantum group $U_q(\mathfrak{sl}_2)$ which is a Hopf algebra, so we have a family of Hopf algebras, and the structure of quantum group $U_q(\mathfrak{sl}_2)$ *tends to* that of $U(\mathfrak{sl}_2)$ as $q \to 1$. But of course this cannot be regarded as a mathematical treatment at all. So the first goal of this book is to make the above idea rigorous enough to convince ourselves.

In Chapters 1 and 2, we will give a brief review of the basic theory of Lie algebras, Hopf algebras, and Kac-Moody algebras. The notion of *universal enveloping algebras, highest weight modules,* and the *category* $\mathcal{O}_{\text{int}}$ will be introduced. The *Poincaré-Birkhoff-Witt theorem* and the *Weyl-Kac character formula* will be presented without proof. The readers may refer to [**1, 17, 28, 53**] for more detail and complete proofs.

Let $\mathfrak{g}$ be a symmetrizable Kac-Moody algebra, and let $U(\mathfrak{g})$ be its universal enveloping algebra. In Chapter 3, we will define the *quantum group* $U_q(\mathfrak{g})$ as a certain deformation of $U(\mathfrak{g})$ with a Hopf algebra structure and show that the Hopf algebra structure of $U_q(\mathfrak{g})$ *tends to* that of $U(\mathfrak{g})$ as q approaches 1.

Moreover, we will give a rigorous proof of the statement: *The representation theory of Kac-Moody algebra* $\mathfrak{g}$ *is the same as the representation theory of quantum group* $U_q(\mathfrak{g})$. The essential part of this statement is a theorem proved by G. Lusztig in [**47**]:

The $\mathfrak{g}$*-modules in the category* $\mathcal{O}_{\text{int}}$ (= *integrable modules over* $\mathfrak{g}$ *in the category* $\mathcal{O}$) *can be deformed to* $U_q(\mathfrak{g})$*-modules in the category* $\mathcal{O}^q_{\text{int}}$ *in* *such a way that the dimensions of weight spaces are invariant under the deformation.*

More precisely, let M be a $U(\mathfrak{g})$-module in the category $\mathcal{O}_{\text{int}}$. Then it has a *weight space decomposition* $M = \bigoplus_{\lambda \in P} M_\lambda$, where M_λ is the common eigenspace for the Cartan subalgebra. Now Lusztig's theorem tells that for each generic q, there exists a $U_q(\mathfrak{g})$-module M^q in the category $\mathcal{O}^q_{\text{int}}$ with a weight space decomposition $M^q = \bigoplus_{\lambda \in P} M^q_\lambda$ such that $\dim_{\mathbf{C}(q)} M^q_\lambda = \dim_{\mathbf{C}} M_\lambda$ for all $\lambda \in P$ and that the structure of M^q *tends to* that of M as q approaches 1.

Pictorially, the results obtained in Chapter 3 can be illustrated in the following figure.

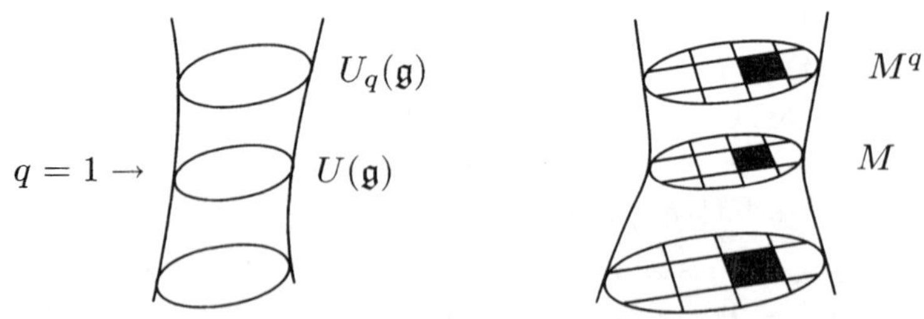

Actually, this is one of the motivations for the theory of *crystal bases.* For an integrable module M over $U(\mathfrak{g})$ in the category $\mathcal{O}_{\text{int}}$, consider the formal power series defined by

$$\operatorname{ch} M = \sum_{\lambda \in P} (\dim_{\mathbf{C}} M_\lambda) e^\lambda.$$

The formal series $\operatorname{ch} M$ is called the *character* of the $U(\mathfrak{g})$-module M. The characters of $U(\mathfrak{g})$-modules in the category $\mathcal{O}_{\text{int}}$ *characterize* the representations in the sense that if $M \cong N$, then $\operatorname{ch} M = \operatorname{ch} N$. The converse is not always true, but will hold if the two modules are both highest weight modules with one of them either a Verma module or an irreducible highest weight module. The characters often represent important and interesting mathematical quantities such as *modular forms* in number theory and *one-point functions* in solvable lattice models.

Similarly, one can define the character of a $U_q(\mathfrak{g})$-module M^q in the category $\mathcal{O}^q_{\text{int}}$ to be

$$\operatorname{ch} M^q = \sum_{\lambda \in P} (\dim_{\mathbf{C}(q)} M^q_\lambda) e^\lambda.$$

Since M^q is a quantum deformation of M, by Lusztig's theorem, $\operatorname{ch} M^q$ is the same for all generic parameter q, and it is just the character of M. So if one can calculate $\operatorname{ch} M^q$ for some special value of q, then it suffices to focus on that special case only. The natural question is: *When is the situation simple?* The crystal basis theory tells that it is so when $q = 0$.

In Chapters 4 and 5, we develop the *crystal basis theory* following the combinatorial approach given by Kashiwara [**38, 39**]. In [**48**], a more geometric approach was developed by Lusztig, and it is called the *canonical basis theory.* In [**43–45**], P. Littelmann introduced a combinatorial theory called the *path model* and obtained a colored oriented graph for irreducible highest weight modules over Kac-Moody algebras. It turned out that Littelmann's graphs coincide with Kashiwara's *crystal graphs* ([**25, 40**]).

A *crystal basis* can be understood as a basis at $q = 0$ and is given a structure of colored oriented graph, called the *crystal graph*, with arrows defined by the *Kashiwara operators.* The crystal graphs have many nice combinatorial features reflecting the internal structure of integrable modules over quantum groups. For instance, one of the major goals in combinatorial representation theory is to find an explicit expression for the characters of representations, and this goal can be achieved by finding an explicit combinatorial description of crystal bases. The following picture is the crystal graph for the adjoint representation of $U_q(\mathfrak{sl}_3)$.

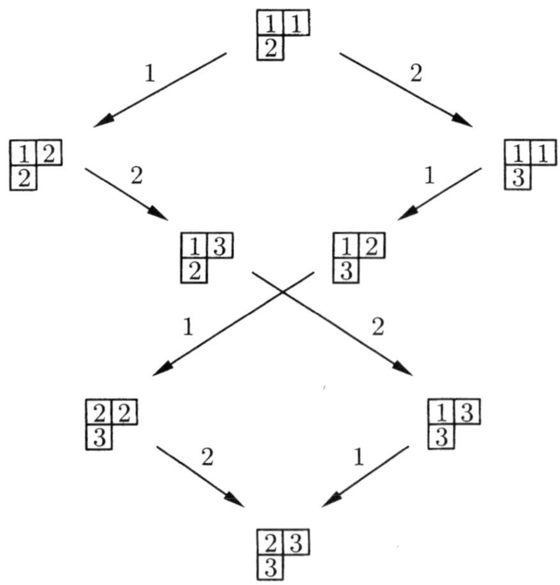

Moreover, crystal bases have extremely nice behavior with respect to taking the tensor product. The action of Kashiwara operators is given by the simple *tensor product rule* and the irreducible decomposition of the tensor product of integrable modules is equivalent to decomposing the tensor product of crystal graphs into a disjoint union of connected components. Thus, the crystal basis theory provides us with a powerful combinatorial method of studying the structure of integrable modules over quantum groups.

Our exposition is based on the combinatorial approach developed by Kashiwara [**39**], and some of our arguments overlap with those given in [**21**]. The existence theorem for crystal bases will be proved using Kashiwara's *grand-loop argument* (Section 5.3). We will simplify the original argument, which consists of 14 interlocking inductive statements, to proving 7 interlocking inductive statements. Still, the spirit of the argument is the same as the original one: the fundamental properties of crystal bases for $U_q^-(\mathfrak{g})$ will play the crucial role in the proof.

The next step is to *globalize* the main idea of crystal bases. More precisely, let M^q be a $U_q(\mathfrak{g})$-module in the category $\mathcal{O}^q_{\text{int}}$ with crystal basis $(\mathcal{L}, \mathcal{B})$. As we mentioned earlier, the crystal basis $\mathcal{B}$ can be regarded as a *local basis* of M^q at $q = 0$. In Chapter 6, we will show that there exists a unique *global basis* $\mathcal{G}(\mathcal{B}) = \{G(b) \mid b \in \mathcal{B}\}$ of M^q satisfying the properties

$$G(b) \equiv b \mod q\mathcal{L}, \qquad \overline{G(b)} = G(b) \quad \text{for all } b \in \mathcal{B},$$

where $\bar{\ }$ denotes the automorphism on M given by (6.5). The existence theorem for global bases will be proved using the notion of a *balanced triple* and the triviality of vector bundles over $\mathbf{P}^1$. Our argument closely follows the original proof given by M. Kashiwara in [**39**].

Over the past 100 years, it has been discovered that there is a close connection between representation theory and combinatorics. We can see this in the classical works by A. Young ([**57–59**]), D. E. Littlewood and A. R. Richardson ([**46**]), D. Robinson ([**52**]), and H. Weyl ([**55**]). In Chapter 7, we study the connection between the crystal basis theory of finite dimensional $U_q(\mathfrak{gl}_n)$-modules and combinatorics of Young diagrams and Young tableaux. The notion of *admissible reading* (e.g., *Far-Eastern reading* and *Middle-Eastern reading*) lies at the heart of this connection. The crystal graph of a finite dimensional irreducible $U_q(\mathfrak{gl}_n)$-module will be realized as the set of semistandard Young tableaux of a given shape. Moreover, using the tensor product rule for Kashiwara operators, we will give a combinatorial rule (*Littlewood-Richardson rule*) for decomposing the tensor product of finite dimensional $U_q(\mathfrak{gl}_n)$-modules into a direct sum of irreducible components. One may refer to [**46**] for the classical approach.

In Chapter 8, we will extend the above idea to the study of crystal graphs for classical Lie algebras. The crystal graph of a finite dimensional irreducible module over a classical Lie algebra will be realized as the set of semistandard Young tableaux satisfying certain additional conditions depending on the type of the Lie algebra. We will also give a combinatorial rule (*generalized Littlewood-Richardson rule*) for decomposing the tensor product of crystal graphs. Most of the results in Chapters 7 and 8 can be found in [**41**] and [**50**].

As the theory of quantum groups originated from the study of the quantum Yang-Baxter equation, the theory of solvable lattice models can be best explained in the language of representation theory of *quantum affine algebras* (which are the quantum groups corresponding to the affine Kac-Moody algebras). In Chapter 9, we will describe the very basic theory of solvable lattice models and discuss its connection with the representation theory of the quantum affine algebra $U_q(\widehat{\mathfrak{sl}}_2)$ (see, for example, [**24, 36**]). In particular, the *one-point function* for the 6-vertex model will be expressed as the quotient of the string function by the character of the basic representation of $U_q(\widehat{\mathfrak{sl}}_2)$.

In Chapter 10, we will develop the theory of *perfect crystals* for quantum affine algebras (see [**36, 37**]), which has a lot of important applications to the representation theory of quantum affine algebras and vertex models (see, for example, [**7, 24**] and the references therein). We will first study the properties of *vertex operators* and then prove a fundamental crystal isomorphism theorem. Using this crystal isomorphism, the crystal graph of an irreducible highest weight module over a quantum affine algebra will be realized as the set of certain *paths*.

The final chapter will be devoted to the study of crystal bases for basic representations of classical quantum affine algebras using some new combinatorial objects which we call the *Young walls* (see [**34**]). The Young walls consist of colored blocks with various shapes that are built on the given *ground-state wall* and can be viewed as generalizations of Young diagrams. The rules for building Young walls and the action of Kashiwara operators will be given explicitly in terms of combinatorics of Young walls. (They are quite similar to playing with LEGO® blocks and the Tetris® game.) The crystal graph of a basic representation will be characterized as the set of all *reduced proper Young walls*. We expect that there exist interesting and important algebraic structures whose irreducible representations (at some specializations) are parameterized by reduced proper Young walls. It still remains to extend the results in this chapter to the quantum affine algebras of type $C_n^{(1)}$ $(n \geq 3)$.

Acknowledgments

This book grew out of the graduate courses given by Seok-Jin Kang at Seoul National University, Massachusetts Institute of Technology, and Yale University during the period of 1998–1999. He is very grateful to all the people who attended his lectures for their keen interest in this subject and stimulating discussions.

We would like to express our deepest thanks to Professors G. Benkart, S. Berman, E. Date, M. Kashiwara, M. Kim, D. J. Melville, T. Miwa, G. B. Seligman, and the reviewers for their careful reading of the manuscript and many invaluable suggestions which improved the exposition of this book in many ways.

Special thanks should be given to the graduate students and postdoctoral fellows at Seoul National University: Kyeonghoon Jeong, Hyeonmi Lee, Jae-Hoon Kwon, Kyu-Hwan Lee, So-Nam Choi, Young-Tak Oh, Jeong-Ah Kim, and Dong-Uy Shin. Without their constant help and criticism, this book would not have materialized.

Finally, we would like to thank our family members and close friends who have provided us with full support and encouragement.

Chapter 1

Lie Algebras and Hopf Algebras

In this chapter, we will briefly review some of the basic definitions and facts about Lie algebras and Hopf algebras. The readers may refer to [**17, 19**] and [**53**] for more detail and complete proofs on these subjects. Throughout this book, $\mathbf{F}$ will denote an arbitrary field of characteristic zero. We denote by $\mathbf{Z}$ the ring of integers and by $\mathbf{Q}$ the field of rational numbers inside $\mathbf{F}$.

1.1. Lie algebras

Lie algebras originally developed as means of studying the local properties of *Lie groups*. Roughly speaking, a Lie group is a manifold with a group structure satisfying certain smoothness and compatibility conditions, and the Lie algebra appears as the tangent space to this manifold at the identity with a bilinear product which is neither commutative nor associative. There exists a good correspondence between the subgroups of a Lie group and the subalgebras of its Lie algebra. That is, the structure of a Lie group G is reflected in the structure of its Lie algebra $L = \text{Lie}(G)$. Moreover, since the Lie algebra L is a vector space, a linear object, it is much easier to deal with a Lie algebra L than with a Lie group G. Thus, to understand a Lie group, a *geometric* object, we study its Lie algebra, an *algebraic* object. We will start with an algebraic definition of Lie algebras.

Definition 1.1.1. A vector space L over $\mathbf{F}$ with a bilinear operation $L \times L \to L$, denoted by $(x, y) \mapsto [x, y]$ and called the ***bracket***, is a ***Lie algebra*** if the following conditions are satisfied:

$$(1.1) \qquad \begin{aligned} &[x, x] = 0 \quad \text{for all } x \in L, \\ &[x, [y, z]] + [y, [z, x]] + [z, [x, y]] = 0 \quad \text{for all } x, y, z \in L. \end{aligned}$$

The first identity implies that the bracket operation is *anticommutative*:

$$[x, y] = -[y, x] \quad \text{for all } x, y \in L,$$

and the second identity in (1.1) is known as the ***Jacobi identity***.

A subspace L' of a Lie algebra L is a ***(Lie) subalgebra*** of L if L' itself is a Lie algebra with the bracket induced from L. A subalgebra I of L is an ***ideal*** of L if $[x, y] \in I$ for all $x \in L$, $y \in I$. If I is an ideal of L, the quotient space L/I becomes a Lie algebra with the bracket defined by

$$[x + I, y + I] = [x, y] + I \quad \text{for all } x, y \in L.$$

A Lie algebra L is ***simple*** if it contains no ideal other than 0 and L. One-dimensional Lie algebras are always simple and they are called *trivial* Lie algebras. We are mainly interested in nontrivial simple Lie algebras.

Let L and L' be Lie algebras. A ***homomorphism*** of L into L' is a linear map $\phi : L \to L'$ satisfying

$$\phi([x, y]) = [\phi(x), \phi(y)] \quad \text{for all } x, y \in L.$$

The ***kernel*** of ϕ, $\ker \phi = \{x \in L \mid \phi(x) = 0\}$, is an ideal of L. Actually, every ideal of L is the kernel of the canonical homomorphism $\pi : L \to L/I$. The notions of ***monomorphisms***, ***epimorphisms***, and ***isomorphisms*** are defined in the usual way. Moreover, the usual isomorphism theorems also hold for Lie algebras. For example, given a Lie algebra homomorphism $\phi : L \to L'$, we have an isomorphism $\operatorname{Im} \phi \cong L/\ker \phi$.

Example 1.1.2.

(1) Let A be an associative algebra over $\mathbf{F}$ and define the bracket on A by

$$[a, b] = ab - ba \quad \text{for all } a, b \in A.$$

Then the bracket operation satisfies the anticommutativity and the Jacobi identity, and hence the pair $(A, [\ ,\ ])$ becomes a Lie algebra. Thus any associative algebra can be made into a Lie algebra.

(2) Let V be a vector space over $\mathbf{F}$. Denote by $\operatorname{End} V$ the set of all linear transformations on V. Define the bracket on $\operatorname{End} V$ by

$$[x, y] = xy - yx \quad \text{for all } x, y \in \operatorname{End} V.$$

Then, $\operatorname{End} V$ becomes a Lie algebra called the ***general linear Lie algebra*** and is denoted by $\mathfrak{gl}(V)$. If $V = \mathbf{F}^n$, the general linear Lie algebra is denoted by $\mathfrak{gl}_n(\mathbf{F})$ or by $\mathfrak{gl}(n, \mathbf{F})$. In this case, we identify the linear endomorphisms of $\mathbf{F}^n$ with $n \times n$ matrices over $\mathbf{F}$.

(3) Let $L = \{x \in \mathfrak{gl}(n, \mathbf{F}) \mid \operatorname{tr} x = 0\}$ be the subspace of $\mathfrak{gl}(n, \mathbf{F})$ consisting of matrices with trace zero. Then L is a Lie subalgebra of $\mathfrak{gl}(n, \mathbf{F})$ called the ***special linear Lie algebra***. This Lie algebra is simple and will be denoted by $\mathfrak{sl}_n(\mathbf{F})$ or $\mathfrak{sl}(n, \mathbf{F})$ (Exercise 1.1).

(4) The three-dimensional simple Lie algebra $\mathfrak{sl}(2, \mathbf{F})$ plays a central role in the theory of Kac-Moody algebras and their representations. It is easy to see that the Lie algebra $\mathfrak{sl}(2, \mathbf{F})$ has a basis consisting of

$$h = \begin{pmatrix} 1 & 0 \\ 0 & -1 \end{pmatrix}, \quad e = \begin{pmatrix} 0 & 1 \\ 0 & 0 \end{pmatrix}, \quad f = \begin{pmatrix} 0 & 0 \\ 1 & 0 \end{pmatrix},$$

and that they satisfy the relations

$$[h, e] = 2e, \quad [h, f] = -2f, \quad [e, f] = h.$$

1.2. Representations of Lie algebras

Definition 1.2.1. Let L be a Lie algebra and let V be a vector space over $\mathbf{F}$.

(1) A ***representation*** of L on V is a Lie algebra homomorphism $\varphi : L \to \mathfrak{gl}(V)$.

(2) A vector space V is called an ***L-module*** if there is a bilinear map $L \times V \to V$, denoted by $(x, v) \mapsto x \cdot v$, satisfying

$$[x, y] \cdot v = x \cdot (y \cdot v) - y \cdot (x \cdot v) \quad \text{for } x, y \in L,\ v \in V.$$

Note that a representation $\varphi : L \to \mathfrak{gl}(V)$ of a Lie algebra L on a vector space V defines an L-module structure on V by

$$x \cdot v = \varphi(x)(v) \quad \text{for } x \in L,\ v \in V. \tag{1.2}$$

Conversely, if V is an L-module, we obtain a representation $\varphi : L \to \mathfrak{gl}(V)$ defined by (1.2). Hence we often say that V is a *representation* of L when it is an L-module. The dot signifying the Lie algebra action will frequently be omitted.

A subspace W of an L-module V is called a ***submodule*** of V if

$$x \cdot W \subset W \quad \text{for all } x \in L.$$

In this case, the quotient space V/W becomes an L-module with the action of L given by

$$x \cdot (v + W) = x \cdot v + W \quad \text{for } x \in L,\ v \in V.$$

An L-module V is called ***irreducible*** if it has no submodule other than 0 and V.

Example 1.2.2.

(1) Let $L = \mathfrak{gl}(n, \mathbf{F})$ be the general linear Lie algebra and $V = \mathbf{F}^n$. Define a map $L \times V \to V$, $(x, v) \mapsto xv$, by matrix multiplication. Then V is given an L-module structure called the ***vector representation*** or the ***natural representation*** of $\mathfrak{gl}(n, \mathbf{F})$.

(2) Let L be a Lie algebra. Define a map $\mathrm{ad} : L \to \mathfrak{gl}(L)$ by

$$\mathrm{ad}\, x(y) = [x, y] \quad \text{for } x, y \in L.$$

Then ad is a Lie algebra homomorphism called the ***adjoint representation*** of L.

Definition 1.2.3. Let $U(L)$ be an associative algebra over $\mathbf{F}$ with unity and let $\iota : L \to U(L)$ be a linear map satisfying

$$\iota([x, y]) = \iota(x)\iota(y) - \iota(y)\iota(x) \quad \text{for all } x, y \in L.$$

The pair $(U(L), \iota)$ is called the ***universal enveloping algebra*** of the Lie algebra L if it has the following universal property. For any associative algebra $\mathcal{A}$ and any linear map $j : L \to \mathcal{A}$ satisfying

$$j([x, y]) = j(x)j(y) - j(y)j(x) \quad \text{for all } x, y \in L,$$

there exists a unique homomorphism of associative algebras $\phi : U(L) \to \mathcal{A}$ such that $\phi \circ \iota = j$. Pictorially, the universal property of $U(L)$ can be illustrated as follows.

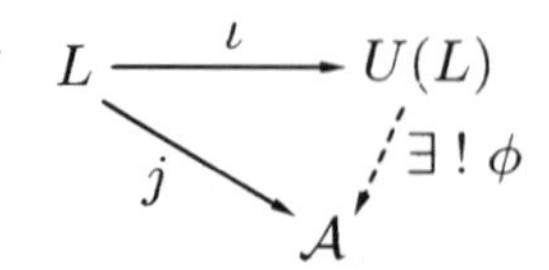

The uniqueness of the universal enveloping algebra can be proved in the usual way (see, for example, [**18, 20**]). As for its existence, first set $\mathcal{T}(L) = \bigoplus_{k=0}^{\infty} L^{\otimes k}$ to be the tensor algebra of L and let $\mathcal{I}$ be the two-sided ideal of $\mathcal{T}(L)$ generated by the elements of the form $x \otimes y - y \otimes x - [x, y]$ $(x, y \in L)$. Using the universal property of tensor algebras, it can be shown that $U = \mathcal{T}(L)/\mathcal{I}$ together with the composition of natural maps $\iota : L \to \mathcal{T} \to U$ satisfies all the conditions for the universal enveloping algebra of L. Hence the universal enveloping algebra $U(L)$ of L can be viewed as the maximal associative algebra over $\mathbf{F}$ with unity generated by L satisfying the relation

$$xy - yx = [x, y] \quad \text{for } x, y \in L.$$

The natural question would be whether the canonical map $\iota : L \to \mathcal{T} \to \mathcal{U}$ is injective so that we can view L as a subspace of $U(L)$. The answer is affirmative, and, as a by-product, we obtain a linear basis of $U(L)$ as is shown in the following theorem known as the ***Poincaré-Birkhoff-Witt Theorem*** (see, for example, [**17**, Ch.V, Thm.17.3], [**19**, Ch.V, Thm.3]).

Theorem 1.2.4.

(1) *The map $\iota : L \to U(L)$ is injective.*

(2) *Let $\{x_\alpha \mid \alpha \in \Omega\}$ be an ordered basis of L. Then, all the elements of the form $x_{\alpha_1} x_{\alpha_2} \cdots x_{\alpha_n}$ satisfying $\alpha_1 \le \alpha_2 \le \cdots \le \alpha_n$ together with 1 form a basis of $U(L)$.*

Example 1.2.5.

(1) Let $L = \mathbf{F}x$ be the one-dimensional trivial Lie algebra. Then the universal enveloping algebra $U(L)$ is isomorphic to the polynomial algebra $\mathbf{F}[x]$.

(2) Let $L = \mathfrak{sl}(2, \mathbf{F}) = \mathbf{F}f \oplus \mathbf{F}h \oplus \mathbf{F}e$ be the Lie algebra of 2×2 matrices of trace zero. Then the elements of the form $f^k h^l e^m$ with $k, l, m \in \mathbf{Z}_{\ge 0}$ form a basis of the universal enveloping algebra $U(L)$ of L.

Let $\mathcal{A}$ be an associative algebra over $\mathbf{F}$ with unity. A ***representation*** of $\mathcal{A}$ on a vector space V is an algebra homomorphism $\varphi : \mathcal{A} \to \operatorname{End} V$. As we have seen in the case with Lie algebras, a representation ψ of $\mathcal{A}$ on V defines an $\mathcal{A}$-module structure on V, and vice versa.

Let L be a Lie algebra and V be an L-module. Then the universal property of $U(L)$ implies that the L-module structure on V can be extended naturally to the $U(L)$-module structure on V. More precisely, since the universal enveloping algebra $U(L)$ of L is generated by L, we can easily check that we may define the action of $U(L)$ on V inductively by setting

$$(x_1 x_2 \cdots x_r) \cdot v = x_1 \cdot ((x_2 \cdots x_r) \cdot v) = x_1 \cdot (x_2 \cdots (x_r \cdot v))$$

for all $x_1, \ldots, x_r \in L$, $v \in V$. Hence, a representation of L naturally extends to that of $U(L)$. Conversely, since we know from the Poincaré-Birkhoff-Witt Theorem that $U(L)$ contains L, a representation of $U(L)$ is also a representation of L. Hence, the representation theory of a Lie algebra is essentially the same as that of its universal enveloping algebra.

1.3. The Lie algebra $\mathfrak{sl}(2, \mathbf{F})$

In this and the next section, we will consider two examples which may be considered prototypes for the Lie algebras we will be dealing with in this book—*Kac-Moody algebras.*

The first example is the three-dimensional simple Lie algebra $\mathfrak{g} = \mathfrak{sl}_2 = \mathfrak{sl}(2, \mathbf{F})$. As we have seen before, it has a basis consisting of

$$h = \begin{pmatrix} 1 & 0 \\ 0 & -1 \end{pmatrix}, \qquad e = \begin{pmatrix} 0 & 1 \\ 0 & 0 \end{pmatrix}, \qquad f = \begin{pmatrix} 0 & 0 \\ 1 & 0 \end{pmatrix},$$

and they satisfy the relations

$$[h, e] = 2e, \qquad [h, f] = -2f, \qquad [e, f] = h.$$

Thus its universal enveloping algebra $U(\mathfrak{sl}_2)$ is the associative algebra over $\mathbf{F}$ with 1 generated by the elements e, f, and h subject to the defining relations

$$he - eh = 2e, \qquad hf - fh = -2f, \qquad ef - fe = h. \tag{1.3}$$

Define a linear functional $\alpha \in (\mathbf{F}h)^*$ by $\alpha(h) = 2$. Then we have

$$\begin{aligned} \mathfrak{g}_\alpha &= \{x \in \mathfrak{g} \mid [h, x] = \alpha(h)x\} = \mathbf{F}e, \\ \mathfrak{g}_{-\alpha} &= \{x \in \mathfrak{g} \mid [h, x] = -\alpha(h)x\} = \mathbf{F}f, \\ \mathfrak{g}_0 &= \{x \in \mathfrak{g} \mid [h, x] = 0\} = \mathbf{F}h, \end{aligned}$$

which yields the ***root space decomposition*** and ***triangular decomposition*** of the Lie algebra $\mathfrak{g} = \mathfrak{sl}_2$:

$$\mathfrak{g} = \mathfrak{sl}_2(\mathbf{F}) = \mathbf{F}f \oplus \mathbf{F}h \oplus \mathbf{F}e.$$

Let U^+, U^0, and U^- be the subalgebra of $U(\mathfrak{sl}_2)$ generated by e, h, and f, respectively. Then the triangular decomposition of $\mathfrak{sl}_2$ implies

$$U(\mathfrak{sl}_2) \cong U^- \otimes U^0 \otimes U^+,$$

which is referred to as the ***triangular decomposition*** of $U(\mathfrak{sl}_2)$.

We recall the classification of finite dimensional irreducible modules over $\mathfrak{sl}(2, \mathbf{F})$, as in [**17, 19**]. Let V be a finite dimensional irreducible $\mathfrak{sl}(2, \mathbf{F})$-module. It decomposes into a direct sum of eigenspaces for h:

$$V = \bigoplus_{\lambda \in \mathbf{F}} V_\lambda, \quad \text{where } V_\lambda = \{v \in V \mid h \cdot v = \lambda v\}. \tag{1.4}$$

We may easily verify (Exercise 1.3)

$$\begin{aligned} e \cdot V_\lambda &\subset V_{\lambda+2}, \\ f \cdot V_\lambda &\subset V_{\lambda-2}. \end{aligned} \tag{1.5}$$

Since V is finite dimensional, there exists a nonzero vector $v_0 \in V_\lambda$ for some $\lambda \in \mathbf{F}$ such that $e \cdot v_0 = 0$. Let $v_j = \frac{1}{j} f^j \cdot v_0$ for $j \geq 0$. Then we have (Exercise 1.3)

$$\begin{aligned} h \cdot v_j &= (\lambda - 2j)v_j, \\ f \cdot v_j &= (j+1)v_{j+1}, \\ e \cdot v_j &= (\lambda - j + 1)v_{j-1}. \end{aligned} \tag{1.6}$$

Let W be the subspace of V spanned by the vectors v_j $(j = 0, 1, 2, \ldots)$. Then the equation (1.6) shows that W is a submodule of V. Since V is irreducible, we must have $V = W$. Since the vectors v_j have different eigenvalues, they are linearly independent. Moreover, since V is finite dimensional, there exists a nonnegative integer m such that $v_m \neq 0$, $v_{m+1} = 0$. Thus we have

$$0 = e \cdot v_{m+1} = (\lambda - m)v_m,$$

which implies $\lambda = m$. We denote this $(m+1)$-dimensional irreducible $\mathfrak{sl}_2$-module by $V(m)$. To summarize, we obtain:

Theorem 1.3.1. [**17**, Ch.II, Thm.7.2] *For each $m \in \mathbf{Z}_{\geq 0}$, there exists a unique irreducible $\mathfrak{sl}_2$-module $V(m)$ of dimension $m+1$ as described above. The $\mathfrak{sl}_2$-module $V(m)$ has the eigenspace decomposition*

$$V(m) = \bigoplus_{j=0}^{m} V_{m-2j},$$

where each of these eigenspaces is one-dimensional. Moreover, every finite dimensional irreducible $\mathfrak{sl}_2$-module has the form $V(m)$ for some $m \in \mathbf{Z}_{\geq 0}$.

The finite dimensional irreducible $\mathfrak{sl}_2$-module $V(m)$ is an example of ***highest weight modules***. In general, an $\mathfrak{sl}_2$-module V, which is not necessarily finite dimensional, is said to be a ***weight module*** if V is decomposed into a direct sum of eigenspaces for h:

$$V = \bigoplus_{\lambda \in \mathbf{F}} V_\lambda, \quad \text{where } V_\lambda = \{v \in V \mid hv = \lambda v\}.$$

If $V_\lambda \neq 0$ for some $\lambda \in \mathbf{F}$, then λ is called a ***weight*** of V, V_λ is the λ-***weight space***, and $\dim V_\lambda$ is called the ***weight multiplicity*** of λ.

A weight module V is called a ***highest weight module*** with ***highest weight*** λ if there exists a nonzero $v_0 \in V_\lambda$, called a ***highest weight vector***, such that $ev_0 = 0$ and $V = U(\mathfrak{sl}_2)v_0$. For such a module, the triangular decomposition of $U(\mathfrak{sl}_2)$ implies $V = U^- v_0$. Note that the same argument preceding Theorem 1.3.1 shows $V = \bigoplus_{k \geq 0} V_{\lambda - 2k}$ with each $\dim V_{\lambda-2k} = 1$ unless trivial.

Fix $\lambda \in \mathbf{F}$. Let $J(\lambda)$ be the left ideal of $U(\mathfrak{sl}_2)$ generated by e and $h - \lambda 1$, and set

$$M(\lambda) = U(\mathfrak{sl}_2)/J(\lambda).$$

Then $M(\lambda)$ is given a $U(\mathfrak{sl}_2)$-module structure by left multiplication. We call $M(\lambda)$ the ***Verma module***. Its properties are summarized in the next proposition.

Proposition 1.3.2. [**17**, Ch.VI, Thm.20.2]

(1) *$M(\lambda)$ is a highest weight $U(\mathfrak{sl}_2)$-module with highest weight λ and highest weight vector $v_0 = 1 + J(\lambda)$.*

(2) *Every highest weight $U(\mathfrak{sl}_2)$-module with highest weight λ is a homomorphic image of $M(\lambda)$.*

(3) *As a U^--module, $M(\lambda)$ is free of rank 1 generated by the highest weight vector $v_0 = 1 + J(\lambda)$.*

(4) *$M(\lambda)$ contains a unique maximal submodule $N(\lambda)$.*

The quotient $V(\lambda) = M(\lambda)/N(\lambda)$ is an irreducible $U(\mathfrak{sl}_2)$-module. Note that $\{f^k v_0 \mid k \geq 0\}$ is a basis of $M(\lambda)$. For $\lambda = m \in \mathbf{Z}_{\geq 0}$, $N(\lambda)$ is the span of $\{f^k v_0 \mid k \geq m+1\}$, so indeed, the definition of $V(\lambda)$ agrees with the previous definition of $V(m)$. For $\lambda \notin \mathbf{Z}_{\geq 0}$, one can show $N(\lambda) = 0$, so $M(\lambda) = V(\lambda)$ is irreducible. The $(m+1)$-dimensional irreducible $U(\mathfrak{sl}_2)$-module $V(m)$ can be characterized as follows.

Proposition 1.3.3. *Let V be a highest weight $U(\mathfrak{sl}_2)$-module with highest weight $m \in \mathbf{Z}_{\geq 0}$ and highest weight vector v_0. If $f^{m+1} v_0 = 0$, then $V \cong V(m)$.*

Proof. We leave it to the readers as an exercise (Exercise 1.4). □

1.4. The special linear Lie algebra $\mathfrak{sl}(n, \mathbf{F})$

Our next example is the *special linear Lie algebra*

$$\mathfrak{g} = \mathfrak{sl}_n(\mathbf{F}) = \{x \in M_{n \times n}(\mathbf{F}) \mid \operatorname{tr} x = 0\}.$$

By definition, it is easy to see that the matrices

$$E_{ij} \quad (i \neq j), \qquad E_{i,i} - E_{i+1,i+1} \quad (i = 1, \ldots, n-1)$$

form a basis of $\mathfrak{sl}_n(\mathbf{F})$, where E_{ij} denotes the $n \times n$ elementary matrix whose (i, j)-entry is one and all other entries are zero.

Let $e_i = E_{i,i+1}$, $f_i = E_{i+1,i}$, $h_i = E_{i,i} - E_{i+1,i+1}$ $(i = 1, \ldots, n-1)$. Then, as a Lie algebra, $\mathfrak{sl}_n(\mathbf{F})$ is generated by the elements e_i, f_i, h_i $(i = 1, 2, \ldots, n-1)$, and they satisfy the relations

$$
\begin{aligned}
&[e_i, f_j] = \delta_{ij} h_i, \\
&[h_i, e_j] = \begin{cases} 2e_j & \text{if } i = j, \\ -e_j & \text{if } |i-j| = 1, \\ 0 & \text{if } |i-j| > 1, \end{cases} \\
&[h_i, f_j] = \begin{cases} -2f_j & \text{if } i = j, \\ f_j & \text{if } |i-j| = 1, \\ 0 & \text{if } |i-j| > 1. \end{cases}
\end{aligned} \tag{1.7}
$$

Let $A = (a_{ij})_{i,j=1,\ldots,n-1}$ be a square matrix whose entries are given by

$$a_{ii} = 2, \qquad a_{ij} = -1 \quad \text{if } |i-j| = 1, \qquad a_{ij} = 0 \quad \text{otherwise.} \tag{1.8}$$

The matrix A is called the ***Cartan matrix*** of the Lie algebra $\mathfrak{sl}_n(\mathbf{F})$. Then the generators e_i, f_i, h_i satisfy the following additional relations called the ***Serre relations*** (Exercise 1.5):

$$(\operatorname{ad} e_i)^{1-a_{ij}}(e_j) = (\operatorname{ad} f_i)^{1-a_{ij}}(f_j) = 0 \quad \text{for } i \neq j. \tag{1.9}$$

For $j = 1, \ldots, n$, define the linear functional $\epsilon_j : M_{n\times n}(\mathbf{F}) \to \mathbf{F}$ by

$$\epsilon_j(T) = t_{jj} \quad \text{where } T = (t_{ij}) \in M_{n\times n}(\mathbf{F}), \tag{1.10}$$

and set

$$\alpha_i = \epsilon_i - \epsilon_{i+1} \quad \text{for } i = 1, \ldots, n-1. \tag{1.11}$$

Note that the entries of the Cartan matrix A are given by

$$a_{ij} = \alpha_j(h_i) \quad \text{for } i, j = 1, \ldots, n-1.$$

Consider the abelian subalgebra $\mathfrak{h} = \mathbf{F}h_1 \oplus \cdots \oplus \mathbf{F}h_{n-1}$ and let

$$h = \begin{pmatrix} \lambda_1 & & 0 \\ & \ddots & \\ 0 & & \lambda_n \end{pmatrix} \in \mathfrak{h}.$$

(Hence, we have $\lambda_1 + \cdots + \lambda_n = 0$.) Then, for $i < j$, we have

$$[h, E_{ij}] = (\lambda_i - \lambda_j)E_{ij} \quad \text{and} \quad [h, E_{ji}] = (\lambda_j - \lambda_i)E_{ji}.$$

Note that $\alpha_i(h) = \lambda_i - \lambda_{i+1}$. Hence the above relation can be written as

$$
\begin{aligned}
[h, E_{ij}] &= (\alpha_i + \cdots + \alpha_{j-1})(h)E_{ij}, \\
[h, E_{ji}] &= -(\alpha_i + \cdots + \alpha_{j-1})(h)E_{ji}.
\end{aligned}
$$

Let $\Pi = \{\alpha_1, \ldots, \alpha_{n-1}\}$ and set

$$\Phi_+ = \{\alpha = \alpha_i + \cdots + \alpha_{j-1} \,|\, i < j\} = \{\epsilon_i - \epsilon_j \,|\, i < j\}, \tag{1.12}$$

and $\Phi_- = -\Phi_+$. The elements of Φ_+ (respectively, Φ_-) are called the ***positive roots*** (respectively, ***negative roots***). We denote by

$$\Phi = \Phi_+ \cup \Phi_- = \{\epsilon_i - \epsilon_j \,|\, i \neq j\}$$

the set of all roots of $\mathfrak{g} = \mathfrak{sl}_n(\mathbf{F})$. For any root $\alpha = \alpha_i + \cdots + \alpha_{j-1} = \epsilon_i - \epsilon_j$ $(i < j)$, we have

$$\mathfrak{g}_\alpha = \{x \in \mathfrak{sl}_n(\mathbf{F}) \,|\, [h, x] = (\epsilon_i - \epsilon_j)(h)x \quad \text{for all } h \in \mathfrak{h}\} = \mathbf{F}E_{ij}$$

and

$$\mathfrak{g}_{-\alpha} = \{x \in \mathfrak{sl}_n(\mathbf{F}) \,|\, [h, x] = (\epsilon_j - \epsilon_i)(h)x \quad \text{for all } h \in \mathfrak{h}\} = \mathbf{F}E_{ji},$$

which yields the ***root space decomposition*** of $\mathfrak{g} = \mathfrak{sl}_n(\mathbf{F})$:

$$\begin{aligned} \mathfrak{g} = \mathfrak{sl}_n(\mathbf{F}) &= \left(\bigoplus_{\alpha \in \Phi_-} \mathfrak{g}_\alpha\right) \oplus \mathfrak{h} \oplus \left(\bigoplus_{\alpha \in \Phi_+} \mathfrak{g}_\alpha\right) \\ &= \left(\bigoplus_{i<j} \mathbf{F}E_{ji}\right) \oplus \left(\bigoplus_{i=1}^{n-1} \mathbf{F}(E_{ii} - E_{i+1,i+1})\right) \oplus \left(\bigoplus_{i<j} \mathbf{F}E_{ij}\right). \end{aligned}$$

Let

$$\begin{aligned} \mathfrak{g}_+ &= \mathfrak{sl}_n(\mathbf{F})^+ = \bigoplus_{\alpha \in \Phi_+} \mathfrak{g}_\alpha = \bigoplus_{i<j} \mathbf{F}E_{ij}, \\ \mathfrak{g}_- &= \mathfrak{sl}_n(\mathbf{F})^- = \bigoplus_{\alpha \in \Phi_-} \mathfrak{g}_\alpha = \bigoplus_{i<j} \mathbf{F}E_{ji}, \end{aligned}$$

and denote by $U^+ = U(\mathfrak{g}_+)$ (respectively, $U^- = U(\mathfrak{g}_-)$ and $U^0 = U(\mathfrak{h})$) the subalgebra of $U(\mathfrak{g})$ generated by the elements e_i (respectively, f_i and $\mathfrak{h}$) for $i \in I$. Then by the Poincaré-Birkhoff-Witt Theorem, we get the ***triangular decomposition*** for $\mathfrak{sl}_n(\mathbf{F})$ and $U(\mathfrak{sl}_n(\mathbf{F}))$:

$$\begin{gathered} \mathfrak{g} = \mathfrak{sl}_n(\mathbf{F}) = \mathfrak{g}_- \oplus \mathfrak{h} \oplus \mathfrak{g}_+, \\ U(\mathfrak{g}) = U(\mathfrak{sl}_n(\mathbf{F})) \cong U(\mathfrak{g}_-) \otimes U(\mathfrak{h}) \otimes U(\mathfrak{g}_+). \end{gathered}$$

For each $i = 1, 2, \ldots, n-1$, we define the ***simple reflection*** $r_i : \mathfrak{h}^* \to \mathfrak{h}^*$ by

$$r_i(\lambda) = \lambda - \lambda(h_i)\alpha_i.$$

The subgroup W of $GL(\mathfrak{h}^*)$ generated by r_i $(i = 1, 2, \ldots, n-1)$ is called the ***Weyl group*** of $\mathfrak{sl}_n(\mathbf{F})$. Note that

$$r_i(\epsilon_j) = \begin{cases} \epsilon_{i+1} & \text{if } j = i, \\ \epsilon_i & \text{if } j = i+1, \\ \epsilon_j & \text{if } j \neq i,\ j \neq i+1. \end{cases} \tag{1.13}$$

Hence the simple reflection r_i can be identified with the transposition $(i, i+1)$ on the set $\{\epsilon_1, \dots, \epsilon_n\}$, and the Weyl group W is isomorphic to the symmetric group S_n of n letters. For an element $w \in W$, the expression $w = r_{i_1} \cdots r_{i_t}$ is called ***reduced*** if t is minimal among all such expressions. The minimal t is called the ***length*** of w and is denoted by $l(w)$.

For the representations of $\mathfrak{sl}_n(\mathbf{F})$, let us consider the n-dimensional natural representation $V = \mathbf{F}^n = \mathbf{F}v_1 \oplus \mathbf{F}v_2 \oplus \cdots \oplus \mathbf{F}v_n$, where v_i's are elements of the standard basis. The Lie algebra $\mathfrak{sl}_n(\mathbf{F})$ acts on V by matrix multiplication (Exercise 1.6):

$$
e_i v_j = \begin{cases} v_i & \text{if } j = i+1, \\ 0 & \text{otherwise}, \end{cases}
$$

$$
f_i v_j = \begin{cases} v_{i+1} & \text{if } j = i, \\ 0 & \text{otherwise}, \end{cases} \tag{1.14}
$$

$$
h_i v_j = \begin{cases} v_i & \text{if } j = i, \\ -v_{i+1} & \text{if } j = i+1, \\ 0 & \text{otherwise}. \end{cases}
$$

In particular, we have

$$
e_i v_1 = 0, \qquad h_i v_1 = \epsilon_1(h_i) v_1, \qquad f_i^{\epsilon_1(h_i)+1} v_1 = 0.
$$

Moreover, $V = U(\mathfrak{sl}_n(\mathbf{F}))v_1 = U(\mathfrak{sl}_n(\mathbf{F})_-)v_1$. We denote $V = \mathbf{F}^n$ by $V(\epsilon_1)$ and call it the ***natural representation*** or ***vector representation*** of $\mathfrak{sl}_n(\mathbf{F})$.

As in the case with $\mathfrak{sl}_2(\mathbf{F})$, the vector representation is an example of a ***highest weight module***. An $\mathfrak{sl}_n(\mathbf{F})$-module V, which is not necessarily finite dimensional, is called a ***weight module*** if V is decomposed into a direct sum of eigenspaces for $\mathfrak{h}$: $V = \bigoplus_{\lambda \in \mathfrak{h}^*} V_\lambda$, where

$$
V_\lambda = \{ v \in V \mid hv = \lambda(h) v \quad \text{for all } h \in \mathfrak{h} \}.
$$

If $V_\lambda \neq 0$ for some $\lambda \in \mathfrak{h}^*$, then λ is called a ***weight*** of V, V_λ is the ***λ-weight space***, and $\dim V_\lambda$ is called the ***weight multiplicity*** of λ.

Suppose that $V = \bigoplus_{\lambda \in \mathfrak{h}^*} V_\lambda$ is a weight module over $\mathfrak{sl}_n(\mathbf{F})$ on which all e_i and f_i $(i = 1, 2, \dots, n-1)$ act locally nilpotently. For each $i = 1, 2, \dots, n-1$, define an $\mathfrak{sl}_n(\mathbf{F})$-module automorphism τ_i on V by

$$
\tau_i = (\exp f_i)(\exp(-e_i))(\exp f_i). \tag{1.15}
$$

Then one can prove that

$$
\tau_i(V_\lambda) = V_{r_i \lambda} \quad \text{for all } i = 1, 2, \dots, n-1 \text{ and } \lambda \in \mathfrak{h}^*,
$$

which implies

$$\dim V_{w\lambda} = \dim V_\lambda \quad \text{for all } w \in W,\ \lambda \in \mathfrak{h}^*.$$

In particular, by taking the adjoint representation, we obtain

$$\dim \mathfrak{g}_{w\alpha} = \dim \mathfrak{g}_\alpha \quad \text{for all } w \in W,\ \alpha \in \Phi$$

(see, for example, [**28**, Ch.3] for more detail).

A weight module V is called a ***highest weight module*** with ***highest weight*** λ if there exists a nonzero $v_\lambda \in V_\lambda$, called a ***highest weight vector***, such that $e_i v_\lambda = 0$ for all $i = 1, 2, \ldots, n-1$ and $V = U(\mathfrak{sl}_n(\mathbf{F}))v_\lambda$. For such a module, the triangular decomposition of $U(\mathfrak{sl}_n(\mathbf{F}))$ implies $V = U^- v_\lambda$.

Fix $\lambda \in \mathfrak{h}^*$. Let $J(\lambda)$ be the left ideal of $U(\mathfrak{sl}_n(\mathbf{F}))$ generated by e_i and $h_i - \lambda(h_i)1$ $(i = 1, 2, \ldots, n-1)$, and set

$$M(\lambda) = U(\mathfrak{sl}_n(\mathbf{F}))/J(\lambda).$$

Then $M(\lambda)$ is given a $U(\mathfrak{sl}_n(\mathbf{F}))$-module structure by left multiplication. We call $M(\lambda)$ the ***Verma module*** and its properties are summarized in the next proposition.

Proposition 1.4.1. [**17**, Ch.VI, Thm.20.2]

(1) *$M(\lambda)$ is a highest weight $U(\mathfrak{sl}_n(\mathbf{F}))$-module with highest weight λ and highest weight vector $1 + J(\lambda)$.*

(2) *Every highest weight $U(\mathfrak{sl}_n(\mathbf{F}))$-module with highest weight λ is a homomorphic image of $M(\lambda)$.*

(3) *As a U^--module, $M(\lambda)$ is free of rank 1.*

(4) *$M(\lambda)$ contains a unique maximal submodule $N(\lambda)$.*

The quotient $V(\lambda) = M(\lambda)/N(\lambda)$ is an irreducible $U(\mathfrak{sl}_n(\mathbf{F}))$-module. It is a well known fact that every finite dimensional irreducible representation of $\mathfrak{sl}_n(\mathbf{F})$ has the form $V(\lambda)$ for some $\lambda \in \mathfrak{h}^*$ such that $\lambda(h_i) \in \mathbf{Z}_{\geq 0}$ for all $i = 1, 2, \ldots, n-1$. In this case, we have $f_i^{\lambda(h_i)+1} v_\lambda = 0$ for all $i = 1, 2, \ldots, n-1$ (see [**17**, Ch.VI]). Actually, the finite dimensional irreducible $\mathfrak{sl}_n(\mathbf{F})$-module $V(\lambda)$ is characterized by these properties.

Proposition 1.4.2. [**17**, Ch.VI] *Let V be a highest weight module over $U(\mathfrak{sl}_n(\mathbf{F}))$ of highest weight $\lambda \in \mathfrak{h}^*$ such that $\lambda(h_i) \in \mathbf{Z}_{\geq 0}$ for all $i = 1, 2, \ldots, n-1$ and let v_λ be its highest weight vector. Suppose we have $f_i^{\lambda(h_i)+1} v_\lambda = 0$ for all $i = 1, 2, \ldots, n-1$, then $V \cong V(\lambda)$.*

1.5. Hopf algebras

Let L be a Lie algebra and let V, W be L-modules. The tensor product $V \otimes W$ is given an L-module structure with the action of L defined by

$$x \cdot (v \otimes w) = x \cdot v \otimes w + v \otimes x \cdot w \quad \text{for all } x \in L,\ v \in V,\ w \in W.$$

Thus, $V \otimes W$ becomes a $U(L)$-module, where the action of $U(L)$ is defined inductively by

$$(x_1 x_2 \cdots x_k) \cdot (v \otimes w) = x_1 \cdot ((x_2 \cdots x_k) \cdot (v \otimes w))$$

for $x_i \in L$, $v \in V$, $w \in W$, $i = 1, \ldots, k$.

For example, if $x, y \in L$, $v \in V$, and $w \in W$, we have

$$\begin{aligned}(xy) \cdot (v \otimes w) &= x \cdot (y \cdot (v \otimes w)) = x \cdot (y \cdot v \otimes w + v \otimes y \cdot w) \\ &= x \cdot y \cdot v \otimes w + y \cdot v \otimes x \cdot w + x \cdot v \otimes y \cdot w + v \otimes x \cdot y \cdot w.\end{aligned}$$

Hence the $U(L)$-module structure on $V \otimes W$ can be explained as follows. Let $\Delta : U(L) \to U(L) \otimes U(L)$ be an algebra homomorphism defined by

$$\Delta(x) = x \otimes 1 + 1 \otimes x \quad \text{for all } x \in L. \tag{1.16}$$

Then $V \otimes W$ is given a $U(L)$-module structure with the $U(L)$-action defined by

$$u \cdot (v \otimes w) = \Delta(u)(v \otimes w) \quad \text{for all } u \in U(L),\ v \in V,\ w \in W.$$

The algebra homomorphism $\Delta : U(L) \to U(L) \otimes U(L)$ is called a ***comultiplication*** on $U(L)$. Thus we can define a $U(L)$-module structure on the tensor product of $U(L)$-modules due to the existence of comultiplication Δ.

Now, let V be a graded L-module with finite dimensional homogeneous spaces, and let V^* denote the finite dual of V. Then the L-module structure on V^* is given by

$$(x \cdot f)(v) = -f(x \cdot v) = f((-x) \cdot v) \quad \text{for all } x \in L,\ f \in V^*,\ v \in V.$$

Again, V^* becomes a $U(L)$-module, where the $U(L)$-action is defined by

$$((x_1 x_2 \cdots x_k) \cdot f)(v) = (x_1 \cdot ((x_2 \cdots x_k) \cdot f))(v)$$

for $x_i \in L$, $f \in V^*$, $v \in V$. However, in this case, the situation is different from the tensor product. If $x, y \in L$, $f \in V^*$, and $v \in V$, we have

$$\begin{aligned}((xy) \cdot f)(v) &= (x \cdot (y \cdot f))(v) = (y \cdot f)((-x) \cdot v) \\ &= f((-y) \cdot ((-x) \cdot v)) = f(((-y)(-x)) \cdot v).\end{aligned}$$

Therefore, the $U(L)$-module structure on V^* can be described in the following way. Let $S : U(L) \to U(L)$ be an antihomomorphism (of algebras) defined by

$$S(x) = -x \quad \text{for all } x \in L. \tag{1.17}$$

Then the finite dual space V^* is given a $U(L)$-module structure by defining

$$(u \cdot f)(v) = f(S(u) \cdot v) \quad \text{for all } u \in U(L),\ f \in V^*,\ v \in V.$$

The antihomomorphism $S : U(L) \to U(L)$ is called an ***antipode*** of $U(L)$. Thus, the finite dual space V^* of a $U(L)$-module V can be made into a $U(L)$-module thanks to the existence of the antipode S on $U(L)$.

Actually, the comultiplication Δ and the antipode S satisfy more conditions, which make $U(L)$ a so-called ***Hopf algebra***. Roughly speaking, a Hopf algebra is an associative algebra that gives module structures on the tensor products and the duals of its modules. We will continue our discussion in a more general setting.

Let $\mathcal{A}$ be an (associative) algebra with unity over $\mathbf{F}$. Recall the axioms for an algebra over $\mathbf{F}$. An *algebra* is a vector space over $\mathbf{F}$ equipped with a bilinear map $\mathcal{A} \times \mathcal{A} \to \mathcal{A}$, denoted by $(a, b) \mapsto ab$, and a unit element $1 \in \mathcal{A}$ satisfying the *associativity* condition and the *unity* condition:

$$\begin{aligned} (ab)c = a(bc) \quad & \text{for all } a, b, c \in \mathcal{A}, \\ a1 = 1a = a \quad & \text{for all } a \in \mathcal{A}. \end{aligned}$$

This may be restated as follows.

Definition 1.5.1. An (associative) ***algebra*** $\mathcal{A}$ over a field $\mathbf{F}$ is a vector space over $\mathbf{F}$ with linear maps $\mu : \mathcal{A} \otimes \mathcal{A} \to \mathcal{A}$, the ***multiplication***, and $\iota : \mathbf{F} \to \mathcal{A}$, the ***unit***, such that the following diagrams are commutative.

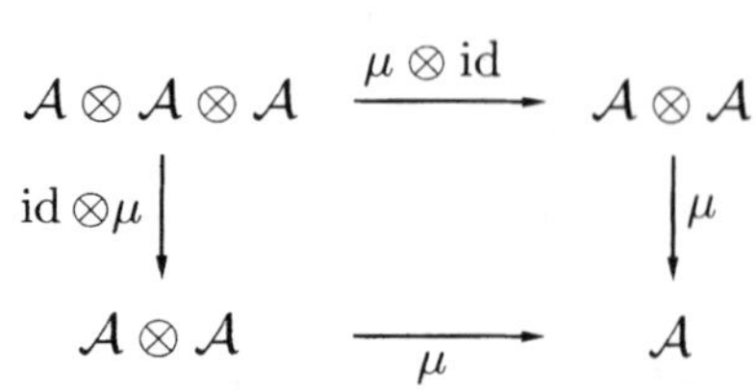

An algebra $\mathcal{A}$ is said to be *commutative* if $ab = ba$ for all $a, b \in \mathcal{A}$. This can be redefined using the diagrams as follows. For any two vector spaces V and W, we define the *transposition map* $\sigma : V \otimes W \to W \otimes V$ by setting $\sigma(x \otimes y) = y \otimes x$ on homogeneous elements and extending this by linearity. Then, a ***commutative algebra*** is an algebra satisfying $\mu \circ \sigma = \mu$.

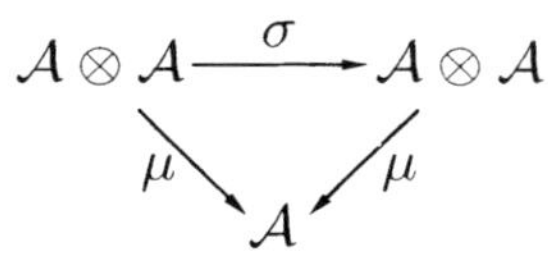

Similarly, we can define the standard notions related to associative algebras using commutative diagrams. For example, an ***algebra homomorphism*** is a linear map $\phi : \mathcal{A} \to \mathcal{A}'$ satisfying $\phi \circ \mu_{\mathcal{A}} = \mu_{\mathcal{A}'} \circ (\phi \otimes \phi)$ and $\phi \circ \iota_{\mathcal{A}} = \iota_{\mathcal{A}'}$.

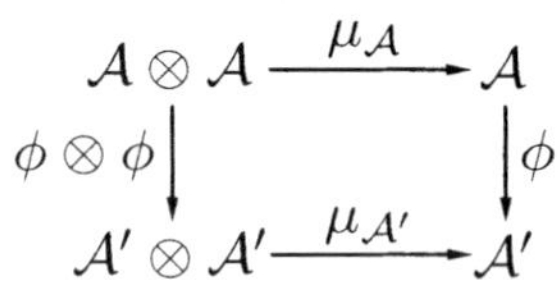

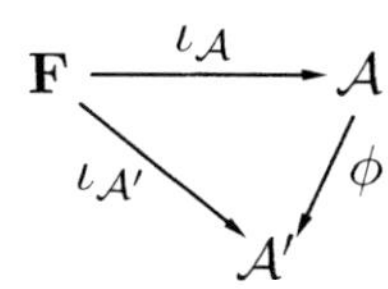

Let $\mathcal{A}$ and $\mathcal{A}'$ be (associative) algebras over $\mathbf{F}$. Then the tensor product $\mathcal{A} \otimes \mathcal{A}'$ has a natural algebra structure given by

$$\begin{aligned} \mu_{\mathcal{A}\otimes\mathcal{A}'} &= (\mu_{\mathcal{A}} \otimes \mu_{\mathcal{A}'}) \circ (\mathrm{id} \otimes \sigma \otimes \mathrm{id}), \\ \iota_{\mathcal{A}\otimes\mathcal{A}'} &= \iota_{\mathcal{A}} \otimes \iota_{\mathcal{A}'}. \end{aligned} \tag{1.18}$$

Using commutative diagrams, these conditions can be expressed as follows.

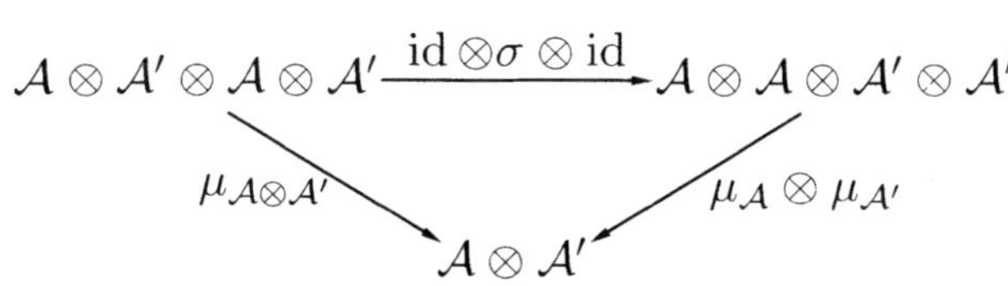

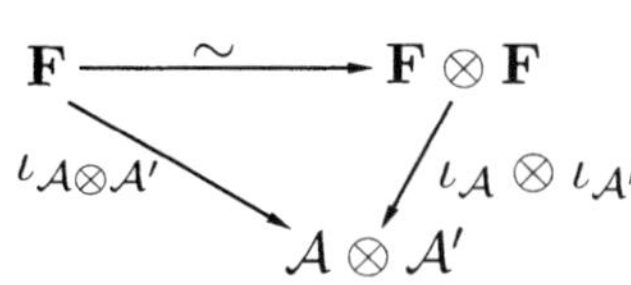

We now define the notion of *coalgebras* by reversing the arrows in the diagrams defining algebras.

Definition 1.5.2. A ***coalgebra*** $\mathcal{C}$ over a field $\mathbf{F}$ is a vector space over $\mathbf{F}$ with linear maps $\Delta : \mathcal{C} \to \mathcal{C} \otimes \mathcal{C}$, the ***comultiplication***, and $\varepsilon : \mathcal{C} \to \mathbf{F}$, the ***counit***, such that the following diagrams are commutative.

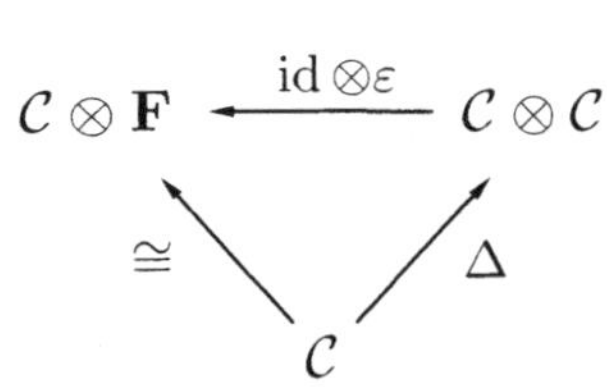

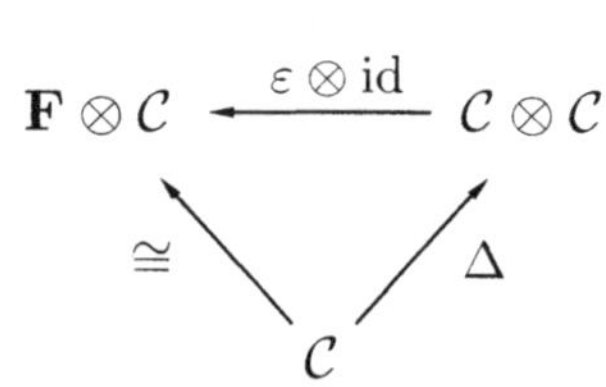

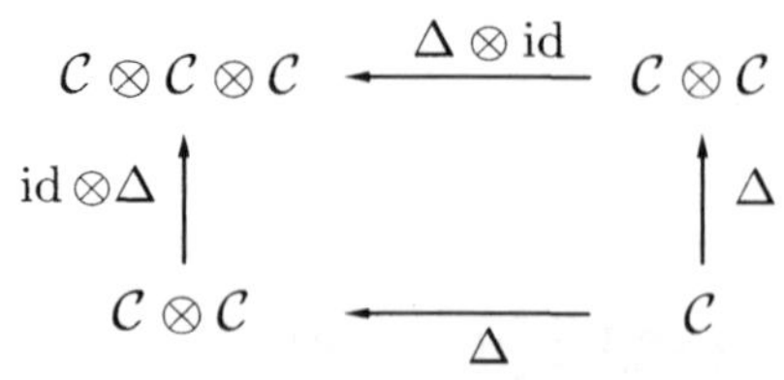

The last diagram is called the ***coassociativity condition***.

A coalgebra $(\mathcal{C}, \Delta, \varepsilon)$ is said to be ***cocommutative*** if $\sigma \circ \Delta = \Delta$.

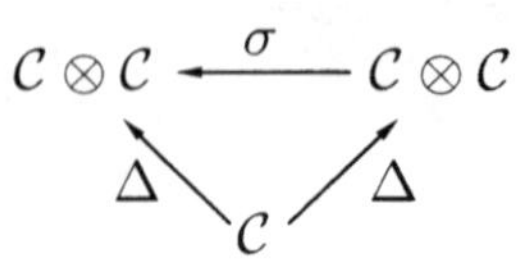

For an element x of a coalgebra $\mathcal{C}$, we often write

$$\Delta(x) = \sum_{(x)} x_{(0)} \otimes x_{(1)} \in \mathcal{C} \otimes \mathcal{C}. \tag{1.19}$$

This is called the ***Sweedler notation***. For example, using the Sweedler notation, a coalgebra $\mathcal{C}$ is cocommutative if $\sum_{(x)} x_{(0)} \otimes x_{(1)} = \sum_{(x)} x_{(1)} \otimes x_{(0)}$ for all $x \in \mathcal{C}$.

Let $(\mathcal{C}, \Delta_{\mathcal{C}}, \varepsilon_{\mathcal{C}})$ and $(\mathcal{C}', \Delta_{\mathcal{C}'}, \varepsilon_{\mathcal{C}'})$ be coalgebras. A ***coalgebra homomorphism*** ϕ of $\mathcal{C}'$ into $\mathcal{C}$ is a linear map $\phi : \mathcal{C}' \to \mathcal{C}$ satisfying the conditions

$$\begin{aligned} (\phi \otimes \phi) \circ \Delta_{\mathcal{C}'} &= \Delta_{\mathcal{C}} \circ \phi, \\ \varepsilon_{\mathcal{C}} \circ \phi &= \varepsilon_{\mathcal{C}'}. \end{aligned} \tag{1.20}$$

Using commutative diagrams, these conditions can be expressed as follows.

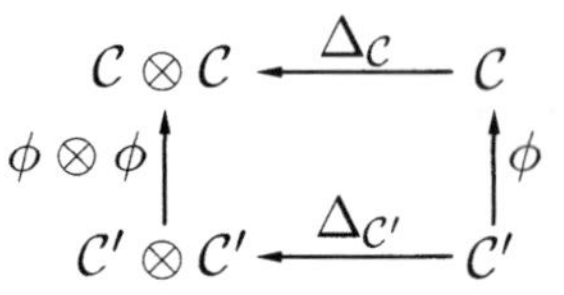

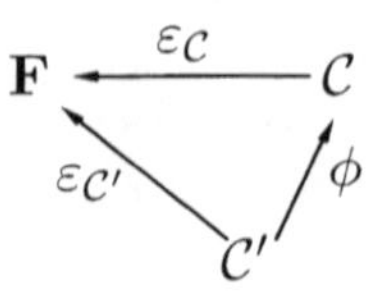

Let $\mathcal{C}$ and $\mathcal{C}'$ be coalgebras over $\mathbf{F}$. Then the tensor product $\mathcal{C} \otimes \mathcal{C}'$ has a natural coalgebra structure given by

$$\begin{aligned} &\Delta_{\mathcal{C}\otimes\mathcal{C}'} = (1 \otimes \sigma \otimes 1) \circ (\Delta_{\mathcal{C}} \otimes \Delta_{\mathcal{C}'}), \\ &\varepsilon_{\mathcal{C}\otimes\mathcal{C}'}(c \otimes c') = \varepsilon_{\mathcal{C}}(c)\varepsilon_{\mathcal{C}'}(c'). \end{aligned} \tag{1.21}$$

Using commutative diagrams, these conditions can be expressed as follows.

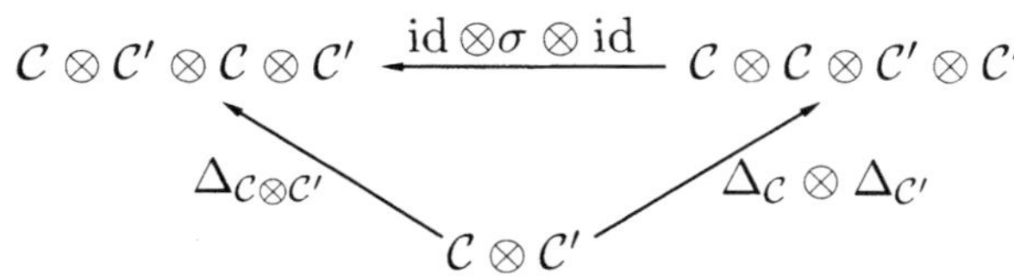

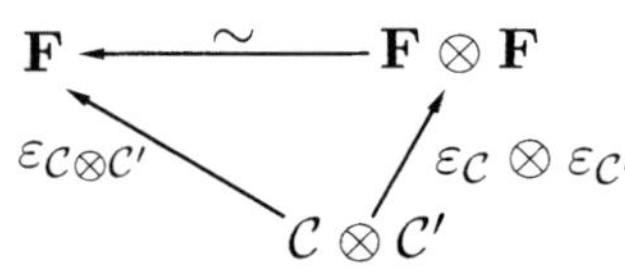

Now we are ready to define the notion of *Hopf algebras.*

Definition 1.5.3. A ***Hopf algebra*** $\mathcal{H}$ over a field $\mathbf{F}$ is a vector space over $\mathbf{F}$ together with the linear maps $\mu : \mathcal{H}\otimes\mathcal{H} \to \mathcal{H}$, the ***multiplication***, $\Delta : \mathcal{H} \to \mathcal{H}\otimes\mathcal{H}$, the ***comultiplication***, $\iota : \mathbf{F} \to \mathcal{H}$, the ***unit***, $\varepsilon : \mathcal{H} \to \mathbf{F}$, the ***counit***, $S : \mathcal{H} \to \mathcal{H}$, the ***antipode*** satisfying the following conditions:

(1) $(\mathcal{H}, \mu, \iota)$ is an algebra over $\mathbf{F}$,

(2) $(\mathcal{H}, \Delta, \varepsilon)$ is a coalgebra over $\mathbf{F}$,

(3) the multiplication and the unit are coalgebra homomorphisms,

(4) the comultiplication and the counit are algebra homomorphisms,

(5) the antipode S satisfies the following commutative diagrams.

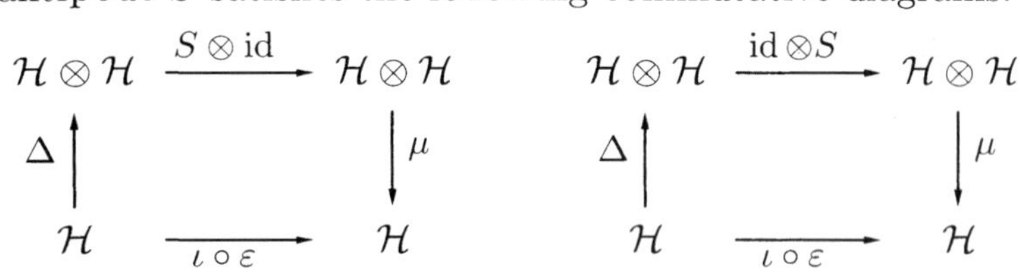

It can be shown that the antipode is always an antihomomorphism of algebras (Exercise 1.8). If $\mathcal{A}$ and $\mathcal{B}$ are Hopf algebras, a linear map $\phi : \mathcal{A} \to \mathcal{B}$ is a ***Hopf algebra homomorphism*** if it is both an algebra homomorphism and a coalgebra homomorphism and if it commutes with the antipode $S_B \circ \phi = \phi \circ S_A$. A ***Hopf ideal*** of a Hopf algebra $\mathcal{H}$ is a two-sided ideal I of the algebra $\mathcal{H}$ such that

$$\Delta(I) \subset I\otimes\mathcal{H} + \mathcal{H}\otimes I, \qquad \varepsilon(I) = 0, \qquad S(I) \subset I. \tag{1.22}$$

The notion of Hopf ideals allows us to define the ***quotient*** of Hopf algebras. Notice that if $\phi : \mathcal{A} \to \mathcal{B}$ is a Hopf algebra homomorphism, then $\ker\phi$ is a Hopf ideal of $\mathcal{A}$ and $\operatorname{Im}\phi \cong \mathcal{A}/\ker\phi$ (Exercise 1.9). The tensor product of Hopf algebras is again a Hopf algebra. The multiplication and comultiplication are given as before and the antipode is given by $S_{\mathcal{H}\otimes\mathcal{H}'} = S_{\mathcal{H}} \otimes S_{\mathcal{H}'}$ (Exercise 1.10).

A ***representation*** of a Hopf algebra $\mathcal{H}$ is a representation of $\mathcal{H}$ as an associative algebra. If V and W are representations of a Hopf algebra $\mathcal{H}$, their tensor product becomes a representation of $\mathcal{H}$ by defining

$$x \cdot (v \otimes w) = \Delta(x)(v \otimes w) = \sum_{(x)} (x_{(0)} \cdot v) \otimes (x_{(1)} \cdot w) \tag{1.23}$$

for $x \in \mathcal{H}$, $v \in V$, $w \in W$.

Let V be a representation of $\mathcal{H}$ and V^* be the finite dual of V. Then V^* becomes a representation of $\mathcal{H}$ by defining

$$(x \cdot f)(v) = f(S(x) \cdot v) \quad \text{for all } x \in \mathcal{H},\ f \in V^*,\ v \in V. \tag{1.24}$$

Moreover, if U, V, and W are representations of $\mathcal{H}$, then we have the following canonical linear isomorphisms (Exercise 1.11):

$$\begin{aligned} \operatorname{Hom}_{\mathcal{H}}(U, V \otimes W) &\xrightarrow{\sim} \operatorname{Hom}_{\mathcal{H}}(V^* \otimes U, W), \\ \operatorname{Hom}_{\mathcal{H}}(U \otimes V, W) &\xrightarrow{\sim} \operatorname{Hom}_{\mathcal{H}}(U, W \otimes V^*). \end{aligned} \tag{1.25}$$

We end this section with various examples of Hopf algebras. We will just give the space $\mathcal{H}$ and define the linear maps $\Delta : \mathcal{H} \to \mathcal{H} \otimes \mathcal{H}$, $\varepsilon : \mathcal{H} \to \mathbf{F}$, and $S : \mathcal{H} \to \mathcal{H}$. Verifications are left to the readers as exercises (Exercise 1.12).

Example 1.5.4.

(1) Let $\mathcal{H} = \mathbf{F}[x, x^{-1}]$ be the algebra of Laurent polynomials in x. Set

$$\Delta(x^{\pm 1}) = x^{\pm 1} \otimes x^{\pm 1}, \qquad \varepsilon(x^{\pm 1}) = 1, \qquad S(x^{\pm 1}) = x^{\mp 1}.$$

Then $\mathcal{H}$ is a commutative, cocommutative Hopf algebra.

(2) Let G be a nonabelian group and $\mathcal{H} = \mathbf{F}[G]$ be the group algebra over $\mathbf{F}$. Set

$$\Delta(g) = g \otimes g, \qquad \varepsilon(g) = 1, \qquad S(g) = g^{-1}$$

for all $g \in G$. Then $\mathcal{H}$ is a noncommutative, cocommutative Hopf algebra.

(3) Let L be a Lie algebra and $\mathcal{H} = U(L)$ be its universal enveloping algebra. Set

$$\Delta(x) = x \otimes 1 + 1 \otimes x, \qquad \varepsilon(x) = 0, \qquad S(x) = -x$$

for all $x \in L$. Then $\mathcal{H}$ is a noncommutative, cocommutative Hopf algebra.

(4) Let G be a real Lie group and let

$$\mathcal{H} = \mathbf{F}(G) = \{f : G \to \mathbf{R} \mid f \text{ is a smooth function on } G\}.$$

The algebra structure on $\mathcal{H}$ is defined by pointwise multiplication:

$$(fg)(x) = f(x)g(x) \quad \text{for all } x \in G.$$

Hence $\mathcal{H}$ is a commutative algebra. Set

$$\Delta(f)(x,y) = f(xy), \qquad \varepsilon(f) = 1, \qquad S(f)(x) = f(x^{-1})$$

for all $f \in \mathcal{H}$, $x, y \in G$. Here we use the embedding $\mathbf{F}(G) \otimes \mathbf{F}(G) \hookrightarrow \mathbf{F}(G \times G)$. Then $\mathcal{H}$ is a commutative, noncocommutative Hopf algebra.

(5) Let $\mathcal{H} = \mathbf{F}[t_{ij}, d^{-1}]$ be the coordinate algebra of $GL_n(\mathbf{F})$, where $1 \le i, j \le n$ and $d = \det(t_{ij})$. Define

$$\begin{aligned} \Delta(t_{ij}) &= \sum_{k=1}^{n} t_{ik} \otimes t_{kj}, & \Delta(d^{-1}) &= d^{-1} \otimes d^{-1}, \\ \varepsilon(t_{ij}) &= \delta_{ij}, & \varepsilon(d^{-1}) &= 1, \\ S(t_{ij}) &= \left((t_{ij})^{-1}\right)_{ij}, & S(d^{-1}) &= d. \end{aligned}$$

Then $\mathcal{H}$ is a commutative and noncocommutative Hopf algebra for any $n > 1$.

(6) Let $\mathbf{F}\{t, x\}$ be the free algebra over $\mathbf{F}$ and let I be the two-sided ideal of $\mathbf{F}\{t, x\}$ generated by $t^2 - 1$, x^2, $xt + tx$. Consider the four-dimensional $\mathbf{F}$-algebra $\mathcal{H} = \mathbf{F}\{x, y\}/I$. Set

$$\begin{aligned} \Delta(t) &= t \otimes t, & \Delta(x) &= x \otimes 1 + 1 \otimes x, \\ \varepsilon(t) &= 1, & \varepsilon(x) &= 0, \\ S(t) &= t, & S(x) &= tx. \end{aligned}$$

Then $\mathcal{H}$ is a noncommutative, noncocommutative Hopf algebra.

Remark 1.5.5. As we will see in Chapter 3, quantum groups are noncommutative and noncocommutative Hopf algebras.

Exercises

1.1. Show that the special linear Lie algebra $\mathfrak{sl}(n, \mathbf{F})$ is simple.

1.2. (a) Find a linear basis of the Lie algebra $\mathfrak{sl}(3, \mathbf{F})$.
(b) Use (a) to construct a Poincaré-Birkhoff-Witt basis of the universal enveloping algebra $U(\mathfrak{sl}(3, \mathbf{F}))$.

1.3. Let V be a finite dimensional $\mathfrak{sl}(2, \mathbf{F})$-module with the eigenspace decomposition $V = \bigoplus_{\lambda \in \mathbf{F}} V_\lambda$.
(a) Show that $eV_\lambda \subset V_{\lambda+2}$, $fV_\lambda \subset V_{\lambda-2}$.

(b) If $v_0 \in V_\lambda$ satisfies $ev_0 = 0$, show that

$$hf^{(j)}v_0 = (\lambda - 2j)f^{(j)}v_0,$$
$$ff^{(j)}v_0 = (j+1)f^{(j+1)}v_0,$$
$$ef^{(j)}v_0 = (\lambda - j + 1)f^{(j-1)}v_0,$$

where $f^{(j)} = f^j/j!$.

(c) Show that

$$e^{(k)}f^{(l)} = \sum_{t=0}^{\min(k,l)} \frac{1}{t!} f^{(l-t)} \prod_{s=1}^{t} \big(h + (t+s) - (k+l)\big)e^{(k-t)}.$$

1.4. Prove Proposition 1.3.3.

1.5. Let $A = (a_{ij})_{i,j=1}^{n-1}$ be the Cartan matrix of the Lie algebra $\mathfrak{sl}(n, \mathbf{F})$ and let $e_i = E_{i,i+1}$ and $f_i = E_{i+1,i}$ $(i = 1, \ldots, n-1)$ be the generators of $\mathfrak{sl}(n, \mathbf{F})$. Verify that the Serre relations hold:

$$(\operatorname{ad} e_i)^{1-a_{ij}}(e_j) = 0, \quad (\operatorname{ad} f_i)^{1-a_{ij}}(f_j) = 0 \quad \text{for } i \neq j.$$

1.6. Verify that the relations in (1.14) define an $\mathfrak{sl}(n, \mathbf{F})$-module structure on $\mathbf{F}^n$.

1.7. Verify the properties of the Verma module stated in Proposition 1.3.2 and 1.4.1.

1.8. Show that the antipode of a Hopf algebra is an antihomomorphism of algebras.

1.9. Let $\mathcal{A}$ and $\mathcal{B}$ be Hopf algebras and $\phi : \mathcal{A} \to \mathcal{B}$ be a Hopf algebra homomorphism. Show that $\ker \phi$ is a Hopf ideal of $\mathcal{A}$ and that there is a Hopf algebra isomorphism $\operatorname{Im}\phi \cong \mathcal{A}/\ker\phi$.

1.10. Verify that the tensor product of Hopf algebras is again a Hopf algebra.

1.11. Let $\mathcal{H}$ be a Hopf algebra and let U, V, W be $\mathcal{H}$-modules. Show that there exists canonical linear isomorphisms

$$\operatorname{Hom}_{\mathcal{H}}(U, V \otimes W) \xrightarrow{\sim} \operatorname{Hom}_{\mathcal{H}}(V^* \otimes U, W),$$
$$\operatorname{Hom}_{\mathcal{H}}(U \otimes V, W) \xrightarrow{\sim} \operatorname{Hom}_{\mathcal{H}}(U, W \otimes V^*).$$

1.12. Verify the statements in Example 1.5.4.

Chapter 2

Kac-Moody Algebras

In this chapter, we review the basic theory of Kac-Moody algebras. Our exposition is based on Kac's book ([**28**]). We will omit most of the proofs. Interested readers may refer to [**28, 49**] for more detail.

2.1. Kac-Moody algebras

Let I be a finite index set. A square matrix $A = (a_{ij})_{i,j \in I}$ with entries in $\mathbf{Z}$ is called a ***generalized Cartan matrix*** if it satisfies

$$\begin{aligned} & a_{ii} = 2 \quad \text{for all } i \in I, \\ & a_{ij} \le 0 \quad \text{if } i \neq j, \\ & a_{ij} = 0 \quad \text{if and only if} \quad a_{ji} = 0. \end{aligned} \tag{2.1}$$

If there exists a diagonal matrix $D = \operatorname{diag}(s_i \mid i \in I)$ with all $s_i \in \mathbf{Z}_{>0}$ such that DA is symmetric, then A is said to be ***symmetrizable***. In this book, we will assume that the generalized Cartan matrix A is symmetrizable. The matrix A is ***indecomposable*** if for every pair of nonempty subsets $I_1, I_2 \subset I$ with $I_1 \cup I_2 = I$, there exists some $i \in I_1$ and $j \in I_2$ such that $a_{ij} \neq 0$.

Let $P^\vee$ be a free abelian group of rank $2|I| - \operatorname{rank} A$ with a $\mathbf{Z}$-basis $\{h_i | i \in I\} \cup \{d_s | s = 1, \ldots, |I| - \operatorname{rank} A\}$ and let $\mathfrak{h} = \mathbf{F} \otimes_{\mathbf{Z}} P^\vee$ be the $\mathbf{F}$-linear space spanned by $P^\vee$. We call $P^\vee$ the ***dual weight lattice*** and $\mathfrak{h}$ the ***Cartan subalgebra***. We also define the ***weight lattice*** to be

$$P = \{\lambda \in \mathfrak{h}^* | \lambda(P^\vee) \subset \mathbf{Z}\}.$$

Set $\Pi^\vee = \{h_i | i \in I\}$ and choose a linearly independent subset $\Pi = \{\alpha_i | i \in I\} \subset \mathfrak{h}^*$ satisfying

$$\alpha_j(h_i) = a_{ij}, \qquad \alpha_j(d_s) = 0 \text{ or } 1 \tag{2.2}$$

for $i, j \in I$, $s = 1, \ldots, |I| - \operatorname{rank} A$. The elements of Π are called ***simple roots***, and the elements of $\Pi^\vee$ are called ***simple coroots***. We also define the ***fundamental weights*** $\Lambda_i \in \mathfrak{h}^*$ $(i \in I)$ to be the linear functionals on $\mathfrak{h}$ given by

$$(2.3) \qquad \Lambda_i(h_j) = \delta_{ij}, \qquad \Lambda_i(d_s) = 0 \quad \text{for } j \in I,\ s = 1, \ldots, |I| - \operatorname{rank} A.$$

Definition 2.1.1. The quintuple $(A, \Pi, \Pi^\vee, P, P^\vee)$ defined as above is said to form a ***Cartan datum*** associated with the generalized Cartan matrix $A = (a_{ij})_{i,j \in I}$.

The free abelian group $Q = \bigoplus_{i \in I} \mathbf{Z}\alpha_i$ is called the ***root lattice*** and $Q_+ = \sum_{i \in I} \mathbf{Z}_{\geq 0}\alpha_i$ is called the ***positive root lattice***. The notation $Q_- = -Q_+$ will also be used. There is a partial ordering on $\mathfrak{h}^*$ defined by $\lambda \geq \mu$ if and only if $\lambda - \mu \in Q_+$ for $\lambda, \mu \in \mathfrak{h}^*$.

For each $i \in I$, we define the ***simple reflection*** r_i on $\mathfrak{h}^*$ by

$$(2.4) \qquad r_i(\lambda) = \lambda - \lambda(h_i)\alpha_i.$$

The subgroup W of $\mathrm{GL}(\mathfrak{h}^*)$ generated by all simple reflections is called the ***Weyl group***. For an element w of the Weyl group, the expression $w = r_{i_1} r_{i_2} \cdots r_{i_t}$ is called a ***reduced expression*** if t is minimal among all such expressions. Note that an element of the Weyl group can have many different reduced expressions. The minimal number t is called the ***length*** of w and is denoted by $l(w)$.

Remark 2.1.2. In [**28**], the triple $(\mathfrak{h}, \Pi, \Pi^\vee)$ is called the *realization* of A. This is sometimes called the *minimal realization*, for $2|I| - \operatorname{rank} A$ is the minimal dimension for the Cartan subalgebra $\mathfrak{h}$ such that α_i $(i \in I)$ are linearly independent. On the other hand, we could use the *maximal realization*. Set $\mathfrak{h}_{\max} = \left(\bigoplus_{i \in I} \mathbf{F} h_i\right) \oplus \left(\bigoplus_{i \in I} \mathbf{F} d_i\right)$ with formal basis elements h_i and d_i $(i \in I)$, and define $\alpha_i \in \mathfrak{h}_{\max}^*$ by setting $\alpha_j(h_i) = a_{ij}$ and $\alpha_j(d_i) = \delta_{ij}$ $(i, j \in I)$. Then the linear functionals α_i are always linearly independent and we can develop the theory of Kac-Moody algebras with the maximal realization in the same way.

We now define the notion of Kac-Moody algebras.

Definition 2.1.3. The ***Kac-Moody algebra*** $\mathfrak{g}$ associated with a Cartan datum $(A, \Pi, \Pi^\vee, P, P^\vee)$ is the Lie algebra generated by the elements e_i, f_i $(i \in I)$ and $h \in P^\vee$ subject to the following defining relations:

(1) $[h, h'] = 0$ for $h, h' \in P^\vee$,

(2) $[e_i, f_j] = \delta_{ij} h_i$,

(3) $[h, e_i] = \alpha_i(h) e_i$ for $h \in P^\vee$,

(4) $[h, f_i] = -\alpha_i(h) f_i$ for $h \in P^\vee$,

(5) $(\operatorname{ad} e_i)^{1-a_{i,j}} e_j = 0$ for $i \neq j$,

(6) $(\operatorname{ad} f_i)^{1-a_{i,j}} f_j = 0$ for $i \neq j$.

The relations (1)–(4) are called the *Weyl relations* and the relations (5)–(6) are called the *Serre relations*. We define $\mathfrak{g}_+$ (respectively, $\mathfrak{g}_-$) to be the subalgebra of $\mathfrak{g}$ generated by the elements e_i (respectively, f_i) with $i \in I$, and for each $\alpha \in Q$, let

$$\mathfrak{g}_\alpha = \{x \in \mathfrak{g} \mid [h, x] = \alpha(h)x \quad \text{for all } h \in \mathfrak{h}\}.$$

The basic properties of Kac-Moody algebras are given in the following proposition.

Proposition 2.1.4. [**28**, Ch.1]

(1) *We have the* ***triangular decomposition***

$$\mathfrak{g} = \mathfrak{g}_- \oplus \mathfrak{h} \oplus \mathfrak{g}_+.$$

(2) *We have the* ***root space decomposition***

$$\mathfrak{g} = \bigoplus_{\alpha \in Q} \mathfrak{g}_\alpha \quad \textit{with } \dim \mathfrak{g}_\alpha < \infty \textit{ for all } \alpha \in Q.$$

(3) *The subalgebra $\mathfrak{g}_+$ (respectively, $\mathfrak{g}_-$) is the Lie algebra generated by the elements e_i $(i \in I)$ (respectively, f_i $(i \in I)$) with the defining relations* (5) *(respectively,* (6)*) in Definition* 2.1.3.

(4) *There exists an involution $\omega : \mathfrak{g} \to \mathfrak{g}$, called the* ***Chevalley involution****, sending $e_i \mapsto -f_i$, $f_i \mapsto -e_i$, and $h \mapsto -h$.*

If $\alpha \neq 0$ and $\mathfrak{g}_\alpha \neq 0$, then α is called a ***root*** of $\mathfrak{g}$ and $\mathfrak{g}_\alpha$ is called the ***root space*** attached to α. The dimension of $\mathfrak{g}_\alpha$ is called the ***root multiplicity*** of α. The set of all roots is denoted by Φ. Denote by $\Phi_\pm = \Phi \cap Q_\pm$ the set of positive and negative roots. Every root can be seen to be either positive or negative. The subalgebra $\mathfrak{g}' = [\mathfrak{g}, \mathfrak{g}]$ is called the ***derived subalgebra***.

The center and the ideals of Kac-Moody algebras are described in the following proposition.

Proposition 2.1.5. [**28**, Ch.1]

(1) *The center of the Kac-Moody algebra $\mathfrak{g}$ is given by*

$$Z(\mathfrak{g}) = \{h \in \mathfrak{h} \mid \alpha_i(h) = 0 \quad \textit{for all } i \in I\}.$$

Hence the dimension of the center is $\dim \mathfrak{h} - |I| = \operatorname{corank} A$.

(2) *Suppose that the generalized Cartan matrix A is indecomposable. Then every ideal of the Kac-Moody algebra $\mathfrak{g}$ either contains its derived subalgebra $\mathfrak{g}'$ or is contained in its center $Z(\mathfrak{g})$.*

We now turn to the *universal enveloping algebra* $U(\mathfrak{g})$ of the Kac-Moody algebra $\mathfrak{g}$. First, note that (Exercise 2.1)

$$(\operatorname{ad} x)^N(y) = \sum_{k=0}^{N} (-1)^k \binom{N}{k} x^{N-k} y x^k \quad \text{for } x, y \in U(\mathfrak{g}),\ N \in \mathbf{Z}_{\geq 0}.$$

Hence we obtain the presentation of $U(\mathfrak{g})$ with generators and relations.

Proposition 2.1.6. *The universal enveloping algebra $U(\mathfrak{g})$ of $\mathfrak{g}$ is the associative algebra over $\mathbf{F}$ with unity generated by e_i, f_i $(i \in I)$ and $\mathfrak{h}$ subject to the following defining relations*:

(1) $hh' = h'h$ *for* $h, h' \in \mathfrak{h}$,

(2) $e_i f_j - f_j e_i = \delta_{ij} h_i$ *for* $i, j \in I$,

(3) $he_i - e_i h = \alpha_i(h) e_i$ *for* $h \in \mathfrak{h}$, $i \in I$,

(4) $hf_i - f_i h = -\alpha_i(h) f_i$ *for* $h \in \mathfrak{h}$, $i \in I$,

(5) $\sum_{k=0}^{1-a_{ij}} (-1)^k \binom{1-a_{ij}}{k} e_i^{1-a_{ij}-k} e_j e_i^k = 0$ *for* $i \neq j$,

(6) $\sum_{k=0}^{1-a_{ij}} (-1)^k \binom{1-a_{ij}}{k} f_i^{1-a_{ij}-k} f_j f_i^k = 0$ *for* $i \neq j$.

As we have seen in Chapter 1, the universal enveloping algebra $U(\mathfrak{g})$ has a Hopf algebra structure given by

$$\begin{aligned} \Delta(x) &= x \otimes 1 + 1 \otimes x, \\ \varepsilon(x) &= 0, \\ S(x) &= -x \quad \text{for } x \in \mathfrak{g}. \end{aligned} \tag{2.5}$$

Let U^+ (respectively, U^0 and U^-) be the subalgebra of $U = U(\mathfrak{g})$ generated by the elements e_i (respectively, $\mathfrak{h}$ and f_i) for $i \in I$. We also define the ***root spaces*** to be

$$\begin{aligned} U_\beta &= \{u \in U \mid hu - uh = \beta(h)u \quad \text{for all } h \in \mathfrak{h}\} \quad \text{for } \beta \in Q, \\ U_\beta^\pm &= \{u \in U^\pm \mid hu - uh = \beta(h)u \quad \text{for all } h \in \mathfrak{h}\} \quad \text{for } \beta \in Q_\pm. \end{aligned}$$

By the Poincaré-Birkhoff-Witt Theorem, the universal enveloping algebra also has the ***triangular decomposition*** and the ***root space decomposition***.

Proposition 2.1.7.

(1) $U(\mathfrak{g}) \cong U^- \otimes U^0 \otimes U^+$.

(2) $U(\mathfrak{g}) = \bigoplus_{\beta \in Q} U_\beta$.

(3) $U^\pm = \bigoplus_{\beta \in Q_\pm} U_\beta^\pm$.

2.2. Classification of generalized Cartan matrices

In this section, we will discuss the classification of generalized Cartan matrices and corresponding Kac-Moody algebras. Let $u = (u_1, \dots, u_n)^t$ be a column vector in $\mathbf{R}^n$. We say that $u > 0$ (respectively, $u \geq 0$) if $u_i > 0$ (respectively, $u_i \geq 0$) for all $i = 1, \dots, n$.

Theorem 2.2.1. [**28**, Ch.4] *Let $A = (a_{ij})_{i,j \in I}$ be an indecomposable generalized Cartan matrix. Then one and only one of the following three possibilities hold for both A and A^t.*

(Fin) $\det A \neq 0$; *there exists $u > 0$ such that $Au > 0$; $Av \geq 0$ implies $v > 0$ or $v = 0$.*

(Aff) $\operatorname{corank} A = 1$; *there exists $u > 0$ such that $Au = 0$; $Av \geq 0$ implies $Av = 0$.*

(Ind) *There exists $u > 0$ such that $Au < 0$; $Av \geq 0$ and $v \geq 0$ imply $v = 0$.*

Definition 2.2.2. A generalized Cartan matrix A is said to be of ***finite*** (respectively, ***affine*** or ***indefinite***) ***type*** if A satisfies the condition (Fin) (respectively, (Aff) or (Ind)) in Theorem 2.2.1.

Corollary 2.2.3. [**28**, Ch.4] *An indecomposable generalized Cartan matrix A is of finite (respectively, affine or indefinite) type if there exists $u > 0$ such that $Au > 0$ (respectively, $Au = 0$ or $Au < 0$).*

To each generalized Cartan matrix $A = (a_{ij})_{i,j \in I}$, we associate an oriented graph, called the ***Dynkin diagram*** of A. The Dynkin diagram of A consists of vertices indexed by I and edges with arrows defined as follows: If $a_{ij}a_{ji} \leq 4$ and $|a_{ij}| \geq |a_{ji}|$, then the vertices i and j are connected with $|a_{ij}|$ edges equipped with an arrow pointing toward i if $|a_{ij}| > 1$. If $a_{ij}a_{ji} > 4$, then the vertices i and j are connected with a bold-faced edge equipped with an ordered pair of integers $(|a_{ij}|, |a_{ji}|)$.

Conversely, from each Dynkin diagram, we can recover the corresponding generalized Cartan matrix, up to the order of indices.

Let us give some examples of 2×2 generalized Cartan matrices and their corresponding Dynkin diagrams.

Example 2.2.4.

(1) $A = \begin{pmatrix} 2 & -1 \\ -1 & 2 \end{pmatrix}$ 

(2) $A = \begin{pmatrix} 2 & -2 \\ -1 & 2 \end{pmatrix}$

(3) $A = \begin{pmatrix} 2 & -2 \\ -2 & 2 \end{pmatrix}$ 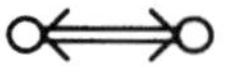

(4) $A = \begin{pmatrix} 2 & -2 \\ -3 & 2 \end{pmatrix}$ 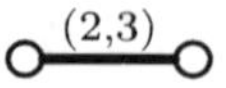

A ***subdiagram*** of a Dynkin diagram consists of a subset of the vertices of the original diagram and all edges of the original diagram joining the chosen vertices.

Proposition 2.2.5. [**28**, Ch.4] *Let A be an indecomposable generalized Cartan matrix.*

(1) *A is of finite type if and only if all its principal minors are positive. Equivalently, A is of finite type if and only if all the subdiagrams of the Dynkin diagram of A are of finite type.*

(2) *A is of affine type if and only if $\det A = 0$ and all its proper principal minors are positive. Equivalently, A is of affine type if and only if $\det A = 0$ and all the proper subdiagrams of the Dynkin diagram of A are of finite type.*

Definition 2.2.6. An indecomposable generalized Cartan matrix A is said to be of ***hyperbolic type*** if A is of indefinite type and every proper subdiagram of the Dynkin diagram of A is of either finite or affine type.

The complete classification of generalized Cartan matrices of finite type and affine type are given in [**28, 54**]. The generalized Cartan matrices of hyperbolic type are classified in [**42, 54**].

Some examples of generalized Cartan matrices of hyperbolic type and their corresponding Dynkin diagrams are listed below.

Example 2.2.7.

(1) $A = \begin{pmatrix} 2 & -a \\ -b & 2 \end{pmatrix}$ with $ab \geq 5$ 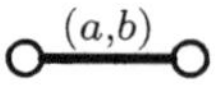

(2) $A = \begin{pmatrix} 2 & -1 & 0 \\ -1 & 2 & -2 \\ 0 & -2 & 2 \end{pmatrix}$ 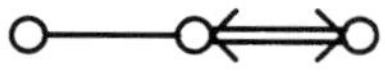

$$(3)\ A = \begin{pmatrix} 2 & -1 & 0 & 0 & 0 & 0 & 0 & 0 & 0 & 0 \\ -1 & 2 & -1 & 0 & 0 & 0 & 0 & 0 & 0 & 0 \\ 0 & -1 & 2 & -1 & 0 & 0 & 0 & 0 & 0 & 0 \\ 0 & 0 & -1 & 2 & -1 & 0 & 0 & 0 & 0 & 0 \\ 0 & 0 & 0 & -1 & 2 & -1 & 0 & 0 & 0 & 0 \\ 0 & 0 & 0 & 0 & -1 & 2 & -1 & 0 & 0 & 0 \\ 0 & 0 & 0 & 0 & 0 & -1 & 2 & -1 & 0 & -1 \\ 0 & 0 & 0 & 0 & 0 & 0 & -1 & 2 & -1 & 0 \\ 0 & 0 & 0 & 0 & 0 & 0 & 0 & -1 & 2 & 0 \\ 0 & 0 & 0 & 0 & 0 & 0 & -1 & 0 & 0 & 2 \end{pmatrix}$$

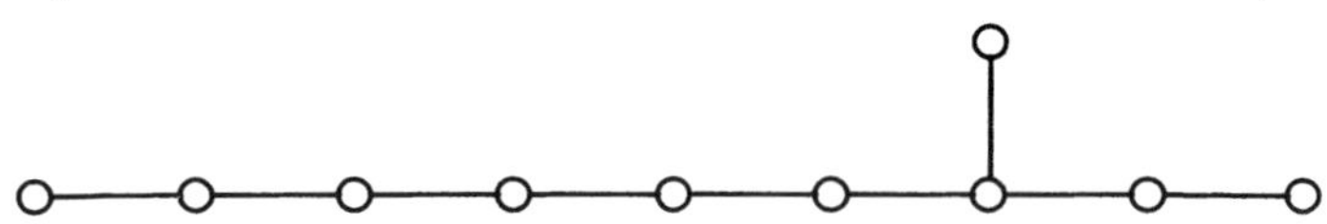

One of the fundamental problems in the theory of Kac-Moody algebras is to find an explicit formula for the root multiplicities. For the Kac-Moody algebras of finite type, the root multiplicities are all one. The root multiplicities of affine Kac-Moody algebras are also well known (see, for example, [**28**]). For the Kac-Moody algebras beyond affine type, only limited information is available (see, for example, [**4, 12, 29, 30, 32**]). There do exist formulas for the root multiplicities of Kac-Moody algebras associated with symmetrizable generalized Cartan matrices. In [**5**], S. Berman and R. V. Moody derived a closed form root multiplicity formula and in [**51**] (see also [**28**]), D. Peterson derived a root multiplicity formula in recursive form. In [**32**], using the Euler-Poincaré principle and Kostant's formula for the homology of Kac-Moody algebras, the general root multiplicity formulas were derived both in closed form and in recursive form. However, the behavior of the root multiplicities is still a mystery and no satisfactory description has yet been discovered.

2.3. Representation theory of Kac-Moody algebras

A $\mathfrak{g}$-module V is called a ***weight module*** if it admits a ***weight space decomposition***

$$V = \bigoplus_{\mu \in \mathfrak{h}^*} V_\mu, \quad \text{where } V_\mu = \{v \in V \mid hv = \mu(h)v \quad \text{for all } h \in \mathfrak{h}\}.$$

A vector $v \in V_\mu$ is called a ***weight vector*** of weight μ. If $e_i v = 0$ for all $i \in I$, v is called a ***maximal vector*** of weight μ. If $V_\mu \neq 0$, μ is called a ***weight*** of V and V_μ is the ***weight space*** attached to μ. Its dimension $\dim V_\mu$ is called the ***weight multiplicity*** of μ. The set of weights of the

$\mathfrak{g}$-module V is denoted by $\mathrm{wt}(V)$. When $\dim V_\mu < \infty$ for all weights μ, the ***character*** of V is defined to be

$$\operatorname{ch} V = \sum_{\mu} \dim V_\mu e^\mu,$$

where e^μ are formal basis elements of the group algebra $\mathbf{F}[\mathfrak{h}^*]$ with multiplication defined by $e^\lambda e^\mu = e^{\lambda+\mu}$.

We leave verification of the following proposition to the readers (Exercise 2.4).

Proposition 2.3.1. [**28**, Ch.1] *Every submodule of a weight module is a weight module.*

For $\lambda \in \mathfrak{h}^*$, set $D(\lambda) = \{\mu \in \mathfrak{h}^* \mid \mu \leq \lambda\}$. Let us define the ***category*** $\mathcal{O}$. Its objects consist of weight modules V over $\mathfrak{g}$ with finite dimensional weight spaces for which there exists a finite number of elements $\lambda_1, \lambda_2, \ldots, \lambda_s \in \mathfrak{h}^*$ such that

$$\mathrm{wt}(V) \subset D(\lambda_1) \cup \cdots \cup D(\lambda_s).$$

The morphisms are $\mathfrak{g}$-module homomorphisms. Note that the category $\mathcal{O}$ is closed under taking the finite direct sum or finite tensor product of objects from the category $\mathcal{O}$. Also, the quotients of $\mathfrak{g}$-modules from the category $\mathcal{O}$ are again in the category $\mathcal{O}$.

The most interesting examples of $\mathfrak{g}$-modules in the category $\mathcal{O}$ may be *highest weight modules* given in the following definition.

Definition 2.3.2. A weight module V is a ***highest weight module*** of ***highest weight*** $\lambda \in \mathfrak{h}^*$ if there exists a nonzero vector $v_\lambda \in V$, called a ***highest weight vector***, such that

$$\begin{aligned} &e_i v_\lambda = 0 \quad \text{for all } i \in I, \\ &h v_\lambda = \lambda(h) v_\lambda \quad \text{for all } h \in \mathfrak{h}, \\ &V = U(\mathfrak{g}) v_\lambda. \end{aligned} \tag{2.6}$$

For a highest weight module V, the triangular decomposition of $U = U(\mathfrak{g})$ (Proposition 2.1.7) yields $V = U^- v_\lambda$. Note also that $\dim V_\lambda = 1$, $\dim V_\mu < \infty$ for all $\mu \in \mathrm{wt}(V)$, and $V = \bigoplus_{\mu \leq \lambda} V_\mu$. Thus any highest weight module belongs to the category $\mathcal{O}$ and the name *highest weight module* is justified.

Fix $\lambda \in \mathfrak{h}^*$ and let $J(\lambda)$ be the left ideal of $U(\mathfrak{g})$ generated by all e_i and $h - \lambda(h)1$ ($i \in I$, $h \in \mathfrak{h}$). Set

$$M(\lambda) = U(\mathfrak{g})/J(\lambda).$$

Then $M(\lambda)$ is given a $U(\mathfrak{g})$-module structure by left multiplication. We call $M(\lambda)$ the ***Verma module***. As we have seen in the case of $\mathfrak{sl}_n(\mathbf{F})$-modules, the properties of Verma modules can be summarized in the following proposition.

Proposition 2.3.3. [**28**, Ch.9]

(1) *$M(\lambda)$ is a highest weight $\mathfrak{g}$-module with highest weight λ and highest weight vector $v_\lambda = 1 + J(\lambda)$.*

(2) *Every highest weight $\mathfrak{g}$-module with highest weight λ is a homomorphic image of $M(\lambda)$.*

(3) *As U^--module, $M(\lambda)$ is free of rank 1, generated by the highest weight vector $v_\lambda = 1 + J(\lambda)$.*

(4) *$M(\lambda)$ has a unique maximal submodule.*

Let us denote by $N(\lambda)$ the unique maximal submodule of $M(\lambda)$. The ***irreducible highest weight module*** $M(\lambda)/N(\lambda)$ is denoted by $V(\lambda)$. The importance of irreducible highest weight modules is reflected in the following proposition.

Proposition 2.3.4. [**28**, Ch.9] *Every irreducible $\mathfrak{g}$-module in the category $\mathcal{O}$ is isomorphic to $V(\lambda)$ for some $\lambda \in \mathfrak{h}^*$.*

Let A be a symmetrizable generalized Cartan matrix with a symmetrizing matrix $D = \operatorname{diag}(s_i \mid i \in I)$. Define a symmetric **F**-valued bilinear form $(\ \mid\ )$ on $\mathfrak{h}$ by

$$(2.7)\qquad \begin{aligned} &(h_i \mid h) = \alpha_i(h)/s_i && \text{for } h \in \mathfrak{h}, \\ &(d_s \mid d_t) = 0 && \text{for } s, t = 1, \dots, |I| - \operatorname{rank} A. \end{aligned}$$

The next lemma may be checked easily (Exercise 2.9).

Lemma 2.3.5. [**28**, Ch.2] *The symmetric bilinear form $(\ \mid\ )$ on $\mathfrak{h}$ is nondegenerate.*

Define a linear map $\nu : \mathfrak{h} \to \mathfrak{h}^*$ by $\nu(h)(h') = (h \mid h')$. The above lemma shows that this map is a vector space isomorphism. Thus we can identify $\mathfrak{h}$ and $\mathfrak{h}^*$ through this map and there is a nondegenerate symmetric bilinear form on $\mathfrak{h}^*$ induced by $(\ \mid\ )$. We will denote this bilinear form by the same notation $(\ \mid\ )$. It satisfies, in particular,

$$(\alpha_i \mid \alpha_j) = s_i a_{ij} \quad \text{for all } i, j \in I.$$

Moreover, it can be easily checked that the symmetric bilinear form $(\ \mid\ )$ is W-invariant; that is, we have (Exercise 2.9)

$$(w\lambda | w\mu) = (\lambda | \mu) \quad \text{for all } w \in W,\ \lambda, \mu \in \mathfrak{h}^*.$$

The nondegenerate symmetric bilinear form on $\mathfrak{h}$ can be extended to a nondegenerate symmetric invariant bilinear form as can be seen in the next proposition.

Proposition 2.3.6. [**28**, Ch.2] *There exists a symmetric bilinear form* $(\ |\ )$ *on* $\mathfrak{g}$ *such that*

(1) $(\ |\ )$ *is given by equations* (2.7) *when restricted to* $\mathfrak{h}$,
(2) $([x,y]\,|\,z) = (x\,|\,[y,z])$ *for all* $x,y,z \in \mathfrak{g}$,
(3) $(\mathfrak{g}_\alpha\,|\,\mathfrak{g}_\beta) = 0$ *if* $\alpha+\beta \neq 0$,
(4) $(\ |\ )$ *is nondegenerate on* $\mathfrak{g}_\alpha \times \mathfrak{g}_{-\alpha}$,
(5) $[x,y] = (x\,|\,y)\nu^{-1}(\alpha)$ *for* $x \in \mathfrak{g}_\alpha$, $y \in \mathfrak{g}_{-\alpha}$.

Choose a linear functional $\rho \in \mathfrak{h}^*$ such that $\rho(h_i) = 1$ for all $i \in I$. Let $\{u_i\}$ and $\{u^i\}$ be two bases of $\mathfrak{h}$, dual to each other with respect to $(\ |\ )$. Also, for each positive root α, fix a basis $\{e_\alpha^{(i)}\}$ of $\mathfrak{g}_\alpha$ and $\{f_\alpha^{(i)}\}$ of $\mathfrak{g}_{-\alpha}$ which are dual to each other with respect to $(\ |\ )$. We define the ***Casimir operator*** to be the formal sum

$$\Omega = 2\nu^{-1}(\rho) + \sum_i u_i u^i + 2\sum_{\alpha\in\Phi_+}\sum_i f_\alpha^{(i)} e_\alpha^{(i)}. \tag{2.8}$$

For now, this may be understood as just a formal sum, but it will be a well defined operator on *restricted* $\mathfrak{g}$-modules defined below.

A $\mathfrak{g}$-module V is ***restricted*** if for every $v \in V$, $\mathfrak{g}_\alpha v = 0$ for all but finitely many positive roots α. Thus the action of Casimir operator is well defined on any restricted $\mathfrak{g}$-module.

Proposition 2.3.7. [**28**, Ch.2]

(1) *The action of Casimir operator* Ω *commutes with the action of* $\mathfrak{g}$ *on any restricted* $\mathfrak{g}$*-module* V.
(2) *If* $v \in V$ *is a maximal vector of weight* λ; *i.e., if* $e_i v = 0$ *for every* $i \in I$ *and* $hv = \lambda(h)v$ *for* $h \in \mathfrak{h}$, *then* $\Omega(v) = (\lambda + 2\rho\,|\,\lambda)v$.

Hence, the Casimir operator acts on any highest weight module of highest weight λ by the constant $(\lambda+2\rho\,|\,\lambda)$.

2.4. The category $\mathcal{O}_{\mathrm{int}}$

Let L be a Lie algebra and V an L-module. We say that $x \in L$ is ***locally nilpotent*** on V if for any $v \in V$ there exists a positive integer N such that $x^N \cdot v = 0$.

Lemma 2.4.1. *Let* L *be a Lie algebra and* V *an* L*-module.*

(1) *Let* $\{y_i|\ i \in \Lambda\}$ *be a set of generators of* L *and let* $x \in L$. *If for each* $i \in \Lambda$ *there exists a positive integer* N_i *such that* $(\operatorname{ad} x)^{N_i}(y_i) = 0$, *then* $\operatorname{ad} x$ *is locally nilpotent on* L.

(2) *Let $\{v_i \mid i \in \Lambda\}$ be a set of generators of V and let $x \in L$. If for each $i \in \Lambda$ there exists a positive integer N_i such that $x^{N_i} \cdot v_i = 0$ and $\operatorname{ad} x$ is locally nilpotent on L, then x is locally nilpotent on V.*

Proof. For a positive integer N and $x, y, z \in L$, we have

$$(\operatorname{ad} x)^N([y, z]) = \sum_{k=0}^{N} \tbinom{N}{k}[(\operatorname{ad} x)^k(y), (\operatorname{ad} x)^{N-k}(z)],$$

$$x^N y = \sum_{k=0}^{N} \tbinom{N}{k}((\operatorname{ad} x)^k(y))x^{N-k}.$$

Here the second equation should be understood as an equation in the universal enveloping algebra with $(\operatorname{ad} x)(y) = xy - yx$. Our assertions follow from the above identities by induction. □

A weight module V over a Kac-Moody algebra $\mathfrak{g}$ is called ***integrable*** if all e_i and f_i $(i \in I)$ are locally nilpotent on V.

Definition 2.4.2. The ***category*** $\mathcal{O}_{\text{int}}$ consists of integrable $\mathfrak{g}$-modules in the category $\mathcal{O}$ such that $\operatorname{wt}(V) \subset P$.

By this definition, any $\mathfrak{g}$-module V in the category $\mathcal{O}_{\text{int}}$ has a weight space decomposition

$$V = \bigoplus_{\lambda \in P} V_\lambda, \quad \text{where } V_\lambda = \{v \in V \mid hv = \lambda(h)v \quad \text{for all } h \in P^\vee\}.$$

Fix $i \in I$. We denote by $\mathfrak{g}_{(i)}$ (respectively, $U_{(i)}$) the subalgebra of $\mathfrak{g}$ (respectively, $U(\mathfrak{g})$) generated by e_i, f_i, h_i. Then we have $\mathfrak{g}_{(i)} \cong \mathfrak{sl}_2$ and $U_{(i)} \cong U(\mathfrak{sl}_2)$. Let V be a $\mathfrak{g}$-module in the category $\mathcal{O}_{\text{int}}$. Since e_i and f_i are locally nilpotent on V, there is a well defined $\mathfrak{g}$-module automorphism of V given by

$$\tau_i = (\exp f_i)(\exp(-e_i))(\exp f_i). \tag{2.9}$$

Moreover we can prove:

Proposition 2.4.3. [**28**, Ch.3] *Let V be a $\mathfrak{g}$-module in the category $\mathcal{O}_{\text{int}}$.*

(1) *For each $i \in I$, V decomposes into a direct sum of finite dimensional irreducible $\mathfrak{h}$-invariant $\mathfrak{g}_{(i)}$-submodules.*

(2) *We have*

$$\tau_i V_\lambda = V_{r_i \lambda} \quad \textit{for all } i \in I,\ \lambda \in \operatorname{wt}(V).$$

Hence $\dim V_\lambda = \dim V_{w\lambda}$ *for all* $w \in W$, $\lambda \in \operatorname{wt}(V)$.

By Lemma 2.4.1, a highest weight $\mathfrak{g}$-module with highest weight λ and highest weight vector v_λ is integrable if and only if for every $i \in I$, there exists $N_i \in \mathbf{Z}_{\geq 0}$ such that $f_i^{N_i} v_\lambda = 0$.

Define the set of ***dominant integral weights*** to be

$$P^+ = \{\lambda \in P \mid \lambda(h_i) \in \mathbf{Z}_{\geq 0} \quad \text{for all } i \in I\}.$$

Lemma 2.4.4. [**28**, Ch.10]

(1) *Let $V(\lambda)$ be the irreducible highest weight $\mathfrak{g}$-module with highest weight $\lambda \in \mathfrak{h}^*$. Then $V(\lambda)$ lies in the category $\mathcal{O}_{\text{int}}$ if and only if $\lambda \in P^+$.*

(2) *Every irreducible $\mathfrak{g}$-module in the category $\mathcal{O}_{\text{int}}$ is isomorphic to $V(\lambda)$ for some $\lambda \in P^+$.*

Proof. (1) Suppose $V(\lambda)$ lies in the category $\mathcal{O}_{\text{int}}$ and let v_λ be a highest weight vector of $V(\lambda)$. Then, by definition, $\lambda \in P$, and for each $i \in I$, there exists a nonnegative integer N_i such that $f_i^{N_i} \cdot v_\lambda \neq 0$ and $f_i^{N_i+1} v_\lambda = 0$. Thus we have

$$0 = e_i f_i^{N_i+1} v_\lambda = (N_i + 1)(\lambda(h_i) - (N_i + 1) + 1) f_i^{N_i} v_\lambda,$$

which implies $\lambda(h_i) = N_i \in \mathbf{Z}_{\geq 0}$. Hence $\lambda \in P^+$.

Conversely, if $\lambda \in P^+$, consider the vector $f_i^{\lambda(h_i)+1} v_\lambda$. If $j \neq i$, then $e_j f_i^{\lambda(h_i)+1} v_\lambda = 0$. Moreover, we have

$$e_i f_i^{\lambda(h_i)+1} v_\lambda = (\lambda(h_i) + 1)(\lambda(h_i) - (\lambda(h_i) + 1) + 1) f_i^{\lambda(h_i)} v_\lambda = 0.$$

Hence if $f_i^{\lambda(h_i)+1} v_\lambda \neq 0$, since its weight is $\lambda - (\lambda(h_i) + 1)\alpha_i < \lambda$, it would generate a nontrivial proper submodule of V, which contradicts the irreducibility of $V(\lambda)$. Therefore, $f_i^{\lambda(h_i)+1} v_\lambda = 0$ for all $i \in I$ and hence $V(\lambda)$ is integrable. Clearly, $\text{wt}(V) \subset P$, and $V(\lambda)$ lies in the category $\mathcal{O}_{\text{int}}$.

(2) By Proposition 2.3.4, every irreducible $\mathfrak{g}$-module V in the category $\mathcal{O}$ is isomorphic to $V(\lambda)$ for some $\lambda \in \mathfrak{h}^*$. If V lies in $\mathcal{O}_{\text{int}}$, by (1), we must have $\lambda \in P^+$. $\square$

Remark 2.4.5. We have just seen that if $V(\lambda)$ is an irreducible highest weight $\mathfrak{g}$-module with highest weight $\lambda \in P^+$ and a highest weight vector v_λ, then we have $f_i^{\lambda(h_i)+1} v_\lambda = 0$ for all $i \in I$. Actually, as we can see in the following theorem, the converse is also true: if V is a highest weight $\mathfrak{g}$-module with highest weight $\lambda \in P^+$ and highest weight vector v_λ such that $f_i^{\lambda(h_i)+1} v_\lambda = 0$ for all $i \in I$, then V is isomorphic to the irreducible highest weight $\mathfrak{g}$-module $V(\lambda)$.

Theorem 2.4.6. [**28**, Ch.10], [**49**, Ch.6] *Let $\mathfrak{g}$ be a Kac-Moody algebra associated with a Cartan datum $(A, \Pi, \Pi^\vee, P, P^\vee)$, and let V be a highest weight $\mathfrak{g}$-module with highest weight $\lambda \in P^+$ and highest weight vector v_λ.*

If $f_i^{\lambda(h_i)+1} v_\lambda = 0$ for all $i \in I$, then the character of V is given by

$$\operatorname{ch} V = \frac{\sum_{w \in W} (-1)^{l(w)} e^{w(\lambda+\rho)-\rho}}{\prod_{\alpha \in \Phi_+} (1-e^{-\alpha})^{\dim \mathfrak{g}_\alpha}}. \tag{2.10}$$

In particular, V is isomorphic to the irreducible highest weight $\mathfrak{g}$-module $V(\lambda)$.

The formula (2.10) is called the ***Weyl-Kac character formula***.

Corollary 2.4.7. [**28**, Ch.10] *Every highest weight $\mathfrak{g}$-module in the category $\mathcal{O}_{\text{int}}$ is isomorphic to some $V(\lambda)$ with $\lambda \in P^+$.*

Proof. Let V be a highest weight $\mathfrak{g}$-module in the category $\mathcal{O}_{\text{int}}$ with highest weight λ and highest weight vector v_λ. From the first part of the proof for Lemma 2.4.4 (1), we find that the nonnegative integer N_i satisfying $f_i^{N_i} \cdot v_\lambda \neq 0$ and $f_i^{N_i+1} v_\lambda = 0$ is actually $\lambda(h_i)$. So we have $\lambda \in P^+$ and $f_i^{\lambda(h_i)+1} v_\lambda = 0$ for all $i \in I$. Hence, by the Weyl-Kac character formula, we obtain $V \cong V(\lambda)$. □

Letting $\lambda = 0$ in (2.10), we obtain the ***denominator identity***

$$\prod_{\alpha \in \Phi_+} (1-e^{-\alpha})^{\dim \mathfrak{g}_\alpha} = \sum_{w \in W} (-1)^{l(w)} e^{w\rho-\rho}. \tag{2.11}$$

The denominator identity is a rich source of interesting mathematical research activity. For instance, the root multiplicity formulas for Kac-Moody algebras mentioned in Section 2.2 were all derived from the denominator identity. Moreover, when it is applied to the affine Kac-Moody algebra of type $A_1^{(1)}$ associated with the generalized Cartan matrix $A = \begin{pmatrix} 2 & -2 \\ -2 & 2 \end{pmatrix}$, it yields the famous *Jacobi triple product identity* ([**26, 28**]):

$$\prod_{n=1}^{\infty} (1-p^n q^n)(1-p^{n-1} q^n)(1-p^n q^{n-1}) = \sum_{k \in \mathbf{Z}} (-1)^k p^{\frac{k(k-1)}{2}} q^{\frac{k(k+1)}{2}}.$$

The denominator identity can also be interpreted as the Euler-Poincaré principle for the Kac-Moody algebras. (See [**31**], [**32**] and [**33**] for more detail and further developments in this direction.)

We conclude this section with a complete reducibility theorem for $\mathfrak{g}$-modules in the category $\mathcal{O}_{\text{int}}$.

Theorem 2.4.8. [**28**, Ch.10] *Let $\mathfrak{g}$ be a Kac-Moody algebra associated with a Cartan datum $(A, \Pi, \Pi^\vee, P, P^\vee)$. Then every $\mathfrak{g}$-module in the category $\mathcal{O}_{\text{int}}$ is isomorphic to a direct sum of irreducible highest weight modules $V(\lambda)$ with $\lambda \in P^+$.*

We will not give a proof for this theorem. But a quantum version of this theorem will be proved in Section 3.5. The original proof given by Kac [**27**] for the nonquantum case uses properties of the Casimir operator. The proof for the quantum case, which does not use Casimir operator, may easily be adopted to the nonquantum case.

Corollary 2.4.9. [**28**, Ch.10] *The tensor product of a finite number of $\mathfrak{g}$-modules in the category $\mathcal{O}_{\text{int}}$ is completely reducible.*

Exercises

2.1. Let L be a Lie algebra and $U(L)$ be its universal enveloping algebra. Verify that for any $x, y \in U(L)$ and $N \in \mathbf{Z}_{\geq 0}$, we have

$$(\operatorname{ad} x)^N(y) = \sum_{k=0}^{k} (-1)^k \binom{N}{k} x^{N-k} y x^k.$$

2.2. Classify all the Dynkin diagrams of affine type with n vertices containing the Dynkin diagram A_{n-1} as a subdiagram.

2.3. Let W be the Weyl group of a Cartan datum. Show that

$$l(w) = |\{\alpha \in \Phi_+ | \, w\alpha < 0\}|.$$

2.4. Prove that every submodule of a weight module is also a weight module.

2.5. (a) Show that the center of a Kac-Moody algebra $\mathfrak{g}$ is

$$Z(\mathfrak{g}) = \{h \in \mathfrak{h} | \, \alpha_i(h) = 0 \quad \text{for all } i \in I\}.$$

(b) Show that $Z(\mathfrak{g}) \subset \mathfrak{h}' = \bigoplus_{i \in I} \mathbf{F} h_i$ and that $\dim Z(\mathfrak{g}) = \operatorname{corank} A$.

2.6. Let $\mathfrak{g}$ be a Kac-Moody algebra associated with an indecomposable generalized Cartan matrix. Prove that every ideal of $\mathfrak{g}$ either contains $\mathfrak{g}'$ or is contained in $Z(\mathfrak{g})$.

2.7. Verify the properties of the Verma module stated in Proposition 2.3.3.

2.8. Show that every irreducible $\mathfrak{g}$-module in the category $\mathcal{O}$ is isomorphic to $V(\lambda)$ for some $\lambda \in \mathfrak{h}^*$.

2.9. Verify that the symmetric bilinear form $(\ |\ )$ on $\mathfrak{h}$ defined by (2.7) is nondegenerate on $\mathfrak{h}$ and is W-invariant.

2.10. Let V be a $\mathfrak{g}$-module in the category $\mathcal{O}_{\text{int}}$ and let Ω be the Casimir operator on V. Show that if v is a maximal vector of weight λ, then

$$\Omega(v) = (\lambda + 2\rho|\lambda)v.$$

2.11. Verify the properties of the automorphism τ_i of $\mathfrak{g}$ given in Proposition 2.4.3.

2.12. Let $\mathfrak{g}$ be the affine Kac-Moody algebra of type $A_2^{(2)}$ associated with the generalized Cartan matrix $A = \begin{pmatrix} 2 & -4 \\ -1 & 2 \end{pmatrix}$. Show that the denominator identity yields the *quintuple product identity*

$$\prod_{n=1}^{\infty}(1-p^{2n}q^n)(1-p^{2n-1}q^{n-1})(1-p^{2n-1}q^n)(1-p^{4n-4}q^{2n-1})(1-p^{4n}q^{2n-1})$$
$$= \sum_{k\in\mathbf{Z}} \left(p^{3k^2-2k}q^{(3k^2+k)/2} - p^{3k^2-4k+1}q^{(3k^2-k)/2}\right).$$

Hint: The root system of $\mathfrak{g}$ is given by

$$\Phi^{\text{re}} = \{(2n \pm 1)\alpha_0 + n\alpha_1,\ 4n\alpha_0 + (2n \pm 1)\alpha_1 \mid n \in \mathbf{Z}\},$$
$$\Phi^{\text{im}} = \{2n\alpha_0 + n\alpha_1 \mid n \in \mathbf{Z}, n \neq 0\},$$

where $\dim \mathfrak{g}_\alpha = 1$ for all $\alpha \in \Phi$. (See [**28**, Ch.8].)

Chapter 3

Quantum Groups

In this chapter, we introduce the *quantum deformations* of the universal enveloping algebras of Kac-Moody algebras, or in more popular terms, the *quantum groups* $U_q(\mathfrak{g})$. We will show that many of the features of the universal enveloping algebras of Kac-Moody algebras carry over to the quantum groups and that the quantum groups are true *deformations* of the universal enveloping algebras. We will also show that the representation theory of Kac-Moody algebras can be *deformed* to the representation theory of quantum groups.

3.1. Quantum groups

In this section, we construct the *quantum deformation* $U_q(\mathfrak{g})$ of the universal enveloping algebra $U(\mathfrak{g})$ of a Kac-Moody algebra $\mathfrak{g}$. It will be given a noncommutative, noncocommutative Hopf algebra structure and we will show that it admits a triangular decomposition. The base field $\mathbf{F}$ will be, as before, an arbitrary field of characteristic zero.

Given $n \in \mathbf{Z}$ and any symbol x, we define the notation

$$[n]_x = \frac{x^n - x^{-n}}{x - x^{-1}}. \tag{3.1}$$

We define $[0]_x! = 1$ and $[n]_x! = [n]_x[n-1]_x \cdots [1]_x$ for $n \in \mathbf{Z}_{>0}$.

For nonnegative integers $m \geq n \geq 0$, the analogues of binomial coefficients are given by

$$\begin{bmatrix} m \\ n \end{bmatrix}_x = \frac{[m]_x!}{[n]_x![m-n]_x!}. \tag{3.2}$$

Fix an indeterminate q. Then, $[n]_q$ and $\begin{bmatrix} m \\ n \end{bmatrix}_q$ are elements of the field $\mathbf{F}(q)$, which are called ***q-integers*** and ***q-binomial coefficients***, respectively. We may show inductively, using the identity

$$\begin{bmatrix} m+1 \\ n \end{bmatrix}_q = q^n \begin{bmatrix} m \\ n \end{bmatrix}_q + q^{-m+n-1} \begin{bmatrix} m \\ n-1 \end{bmatrix}_q, \tag{3.3}$$

that these elements actually belong to $\mathbf{Z}[q, q^{-1}]$ (Exercise 3.1). Note that we have

$$[n]_q \to n \quad \text{and} \quad \begin{bmatrix} m \\ n \end{bmatrix}_q \to \binom{m}{n} \qquad \text{as} \quad q \to 1.$$

Let $A = (a_{ij})_{i,j \in I}$ be a symmetrizable generalized Cartan matrix with a symmetrizing matrix $D = \operatorname{diag}(s_i \in \mathbf{Z}_{>0} \mid i \in I)$ and let $(A, \Pi, \Pi^\vee, P, P^\vee)$ be a Cartan datum associated with A.

Definition 3.1.1. The ***quantum group*** or the ***quantized universal enveloping algebra*** $U_q(\mathfrak{g})$ associated with a Cartan datum $(A, \Pi, \Pi^\vee, P, P^\vee)$ is the associative algebra over $\mathbf{F}(q)$ with 1 generated by the elements e_i, f_i $(i \in I)$ and q^h $(h \in P^\vee)$ with the following defining relations:

(1) $q^0 = 1$, $q^h q^{h'} = q^{h+h'}$ for $h, h' \in P^\vee$,

(2) $q^h e_i q^{-h} = q^{\alpha_i(h)} e_i$ for $h \in P^\vee$,

(3) $q^h f_i q^{-h} = q^{-\alpha_i(h)} f_i$ for $h \in P^\vee$,

(4) $e_i f_j - f_j e_i = \delta_{ij} \dfrac{K_i - K_i^{-1}}{q_i - q_i^{-1}}$ for $i, j \in I$,

(5) $\sum_{k=0}^{1-a_{ij}} (-1)^k \begin{bmatrix} 1-a_{ij} \\ k \end{bmatrix}_{q_i} e_i^{1-a_{ij}-k} e_j e_i^k = 0$ for $i \neq j$,

(6) $\sum_{k=0}^{1-a_{ij}} (-1)^k \begin{bmatrix} 1-a_{ij} \\ k \end{bmatrix}_{q_i} f_i^{1-a_{ij}-k} f_j f_i^k = 0$ for $i \neq j$.

Here, $q_i = q^{s_i}$ and $K_i = q^{s_i h_i}$. For $\alpha = \sum_i n_i \alpha_i \in Q$, the notation $K_\alpha = \prod_i K_i^{n_i}$ will also be used.

Set $\deg f_i = -\alpha_i$, $\deg q^h = 0$, and $\deg e_i = \alpha_i$. Since all the defining relations of the quantum group $U_q(\mathfrak{g})$ are homogeneous, it has a ***root space decomposition***

$$U_q(\mathfrak{g}) = \bigoplus_{\alpha \in Q} (U_q)_\alpha, \tag{3.4}$$

where $(U_q)_\alpha = \{u \in U_q(\mathfrak{g}) \mid q^h u q^{-h} = q^{\alpha(h)} u \text{ for all } h \in P^\vee\}$.

The last two defining relations above are called the *quantum Serre relations*. Define

$$(\operatorname{ad}_q x)(y) = xy - q^{(\alpha|\beta)}yx \quad \text{for} \quad x \in (U_q)_\alpha, y \in (U_q)_\beta \quad (\alpha, \beta \in Q),$$

and extend it to all of $U_q(\mathfrak{g})$ by linearity. These are called *quantum adjoint operators*. Then, we get (Exercise 3.2)

$$(\operatorname{ad}_q e_i)^N(e_j) = \sum_{k=0}^{N}(-1)^k q_i^{k(N+a_{ij}-1)} \begin{bmatrix} N \\ k \end{bmatrix}_{q_i} e_i^{N-k} e_j e_i^k.$$

Hence the quantum Serre relations may be written in the form

$$(3.5) \qquad (\operatorname{ad}_q e_i)^{1-a_{ij}}(e_j) = 0, \qquad (\operatorname{ad}_q f_i)^{1-a_{ij}}(f_j) = 0 \quad \text{for } i \neq j.$$

We will now show that the quantum group $U_q(\mathfrak{g})$ has a Hopf algebra structure.

Proposition 3.1.2. *The quantum group $U_q(\mathfrak{g})$ has a Hopf algebra structure with the comultiplication Δ, counit ε, and antipode S defined by*

(1) $\Delta(q^h) = q^h \otimes q^h,$

(2) $\Delta(e_i) = e_i \otimes K_i^{-1} + 1 \otimes e_i, \quad \Delta(f_i) = f_i \otimes 1 + K_i \otimes f_i,$

(3) $\varepsilon(q^h) = 1, \quad \varepsilon(e_i) = \varepsilon(f_i) = 0,$

(4) $S(q^h) = q^{-h}, \quad S(e_i) = -e_i K_i, \quad S(f_i) = -K_i^{-1} f_i$

for $h \in P^\vee$ and $i \in I$.

Proof. The maps have been defined only on the generators. So we first extend them to the free associative algebra on the given generators by requiring Δ and ε to be algebra homomorphisms and by requiring S to be an antihomomorphism of algebras. To show that these maps are well defined, it suffices to show that all the defining relations are preserved under these maps. The first four relations in Definition 3.1.1 can be easily verified.

To prove that the antipode preserves the quantum Serre relations, we use

$$S(e_i^{N-k} e_j e_i^k) = (-1)^{N+1} q_i^{N(N+a_{ij}-1)} e_i^k e_j e_i^{N-k} K_i^N K_j,$$
$$S(f_i^{N-k} f_j f_i^k) = (-1)^{N+1} q_i^{-N(N+a_{ij}-1)} K_i^{-N} K_j^{-1} f_i^k f_j f_i^{N-k},$$

both of which may be obtained by using the first three defining relations.

We will now prove that comultiplication preserves the quantum Serre relations. By induction, we can show

$$\begin{aligned}\Delta((\operatorname{ad}_q e_i)^N(e_j)) &= (\operatorname{ad}_q e_i)^N(e_j)\otimes K_i^{-N}K_j^{-1}\\ &\quad+\sum_{k=0}^{N-1}\tau_k^{(N)}q_i^{k(N-k)}\begin{bmatrix}N\\k\end{bmatrix}_{q_i} e_i^{N-k}\otimes K_i^{-N+k}(\operatorname{ad}_q e_i)^k(e_j)\\ &\quad+1\otimes(\operatorname{ad}_q e_i)^N(e_j),\end{aligned}$$

where $\tau_k^{(N)}=\prod_{t=k}^{N-1}(1-q_i^{2(t+a_{ij})})$ (Exercise 3.2). Setting $N=1-a_{ij}$, the middle term vanishes and we see that comultiplication preserves the quantum Serre relations.

It remains to check if these maps actually satisfy the conditions for Hopf algebras given in Definition 1.5.3. We have only to verify that these conditions are satisfied on the generators of $U_q(\mathfrak{g})$, which is straightforward (Exercise 3.3). □

Let U_q^+ (respectively, U_q^-) be the subalgebra of $U_q(\mathfrak{g})$ generated by the elements e_i (respectively, f_i) for $i\in I$, and let U_q^0 be the subalgebra of $U_q(\mathfrak{g})$ generated by q^h ($h\in P^\vee$). In addition, let $U_q^{\geq 0}$ (respectively, $U_q^{\leq 0}$) be the subalgebra of $U_q(\mathfrak{g})$ generated by e_i ($i\in I$) and q^h ($h\in P^\vee$) (resp. f_i ($i\in I$) and q^h ($h\in P^\vee$)). We would like to show that the quantum group $U_q(\mathfrak{g})$ has the ***triangular decomposition***

$$U_q(\mathfrak{g})\cong U_q^-\otimes U_q^0\otimes U_q^+.$$

To do this, we first introduce an involution on $U_q(\mathfrak{g})$. Define a linear map $T: U_q(\mathfrak{g})\to U_q(\mathfrak{g})$ by

$$T(q^h)=q^{-h},\quad T(e_i)=f_i,\quad T(f_i)=e_i\quad (h\in P^\vee, i\in I).$$

It is easy to verify that T defines an algebra endomorphism on $U_q(\mathfrak{g})$. Let $\sigma: U_q(\mathfrak{g})\otimes U_q(\mathfrak{g})\to U_q(\mathfrak{g})\otimes U_q(\mathfrak{g})$ be the transposition map defined by

$$\sigma(a\otimes b)=b\otimes a\quad \text{for } a,b\in U_q(\mathfrak{g}).$$

As the following proposition shows, T is actually an involution.

Proposition 3.1.3.

(1) $T^2=\mathrm{id}$.

(2) $\Delta\circ T=\sigma\circ(T\otimes T)\circ\Delta$.

(3) *T restricted to U_q^+ gives an algebra isomorphism between U_q^+ and U_q^-.*

Proof. It suffices to check them on the generators, which is quite straightforward. □

The following lemma is the key step in proving triangular decomposition.

Lemma 3.1.4.

(1) $U_q^{\geq 0} \cong U_q^0 \otimes U_q^+$.

(2) $U_q^{\leq 0} \cong U_q^- \otimes U_q^0$.

Proof. We will just prove the second part. Let $\{f_\zeta\}_{\zeta\in\Omega}$ be a basis of U_q^- consisting of monomials in f_i's $(i \in I)$. Consider the map $\varphi : U_q^- \otimes U_q^0 \to U_q^{\leq 0}$ given by $\varphi(f_\zeta \otimes q^h) = f_\zeta q^h$. Since

$$q^h f_\zeta = q^{-\beta(h)} f_\zeta q^h \quad \text{for } f_\zeta \in (U_q^-)_{-\beta},\ \beta \in Q_+,$$

φ is surjective. Thus it is enough to show that $\{f_\zeta q^h \mid \zeta \in \Omega, h \in P^\vee\}$ is linearly independent over $\mathbf{F}(q)$.

Suppose

$$\sum_{\substack{\zeta\in\Omega \\ h\in P^\vee}} C_{\zeta,h} f_\zeta q^h = 0 \quad \text{for some } C_{\zeta,h} \in \mathbf{F}(q).$$

We may write

$$\sum_{\beta\in Q_+} \Big(\sum_{\substack{\deg f_\zeta=-\beta \\ h\in P^\vee}} C_{\zeta,h} f_\zeta q^h \Big) = 0.$$

(Here, we denote $\deg u = \beta \in Q$ if $u \in (U_q)_\beta$.)

Since $U_q = \bigoplus_{\beta\in Q} (U_q)_\beta$, we have

$$\sum_{\substack{\deg f_\zeta=-\beta \\ h\in P^\vee}} C_{\zeta,h} f_\zeta q^h = 0 \quad \text{for each } \beta \in Q_+. \tag{3.6}$$

Since each f_ζ is a monomial in f_i's $(i \in I)$, if it is of degree $-\beta \in Q_-$, we have

$$\Delta(f_\zeta) = f_\zeta \otimes 1 + (\text{intermediate terms}) + K_\beta \otimes f_\zeta.$$

Applying the comultiplication Δ to (3.6) yields

$$\sum_{\substack{\deg f_\zeta=-\beta \\ h\in P^\vee}} C_{\zeta,h} \big(f_\zeta q^h \otimes q^h + \cdots + K_\beta q^h \otimes f_\zeta q^h \big) = 0.$$

Collecting the terms of degree $(-\beta,0)$, we obtain

$$\sum_{\substack{\deg f_\zeta=-\beta \\ h\in P^\vee}} C_{\zeta,h} \big(f_\zeta q^h \otimes q^h \big) = 0.$$

Since the set $\{q^h\}_{h\in P^\vee}$ is linearly independent, we have

$$\sum_{\deg f_\zeta=-\beta} C_{\zeta,h} f_\zeta q^h = 0 \quad \text{for all } h \in P^\vee.$$

Multiplying by q^{-h} from the right and using linear independence of f_ζ, we conclude all $C_{\zeta,h} = 0$ as desired. □

We are now ready to prove the *triangular decomposition* for $U_q(\mathfrak{g})$.

Theorem 3.1.5. $U_q(\mathfrak{g}) \cong U_q^- \otimes U_q^0 \otimes U_q^+$.

Proof. Let $\{f_\zeta\}_{\zeta\in\Omega}$ and $\{e_\zeta\}_{\zeta\in\Omega}$ be monomial bases of U_q^- and U_q^+, respectively. As in the proof for Lemma 3.1.4, it suffices to show that the set $\{f_\zeta q^h e_\eta \,|\, \zeta,\eta\in\Omega, h\in P^\vee\}$ is linearly independent over $\mathbf{F}(q)$.

Suppose

$$\sum_{\zeta,h,\eta} C_{\zeta,h,\eta} f_\zeta q^h e_\eta = 0 \quad \text{for some } C_{\zeta,h,\eta}\in\mathbf{F}(q).$$

The root space decomposition of $U_q(\mathfrak{g})$ shows that

$$\sum_{\substack{h\in P^\vee \\ \deg f_\zeta + \deg e_\eta = \gamma}} C_{\zeta,h,\eta} f_\zeta q^h e_\eta = 0 \quad \text{for all } \gamma\in Q.$$

We know

$$\Delta(e_\eta) = e_\eta \otimes K_{\deg e_\eta}^{-1} + \cdots + 1\otimes e_\eta,$$
$$\Delta(f_\zeta) = f_\zeta\otimes 1 + \cdots + K_{-\deg f_\zeta}\otimes f_\zeta.$$

Therefore,

$$\begin{aligned} 0 &= \sum_{\substack{h\in P^\vee \\ \deg f_\zeta + \deg e_\eta = \gamma}} C_{\zeta,h,\eta}\Delta\left(f_\zeta q^h e_\eta\right) \\ &= \sum_{\substack{h\in P^\vee \\ \deg f_\zeta + \deg e_\eta = \gamma}} C_{\zeta,h,\eta}(f_\zeta\otimes 1 + \cdots)(q^h\otimes q^h)(\cdots + 1\otimes e_\eta). \end{aligned} \tag{3.7}$$

Recall the partial ordering on $\mathfrak{h}^*$ defined in Section 2.1 and choose $\alpha = \deg f_\zeta$ and $\beta = \deg e_\eta$, which are minimal and maximal, respectively, among those for which $C_{\zeta,h,\eta}$ is nonzero. The terms in (3.7) of degree (α,β) must sum to zero. Hence,

$$\sum_{\substack{h\in P^\vee, \\ \deg f_\zeta=\alpha, \deg e_\eta=\beta}} C_{\zeta,h,\eta}(f_\zeta q^h\otimes q^h e_\eta) = 0.$$

Since the vectors $f_\zeta q^h$ are linearly independent by Lemma 3.1.4, we have

$$\sum_{\deg e_\eta=\beta} C_{\zeta,h,\eta} q^h e_\eta = 0 \quad \text{for all } \zeta \text{ and } h.$$

From this, we may conclude that $C_{\zeta,h,\eta} = 0$, as desired. □

3.2. Representation theory of quantum groups

In this section, we study representations of the quantum group. The theory is quite parallel to that of Kac-Moody algebras.

A $U_q(\mathfrak{g})$-module V^q is called a ***weight module*** if it admits a ***weight space decomposition***

$$V^q = \bigoplus_{\mu \in P} V^q_\mu, \quad \text{where } V^q_\mu = \{v \in V^q \mid q^h v = q^{\mu(h)} v \quad \text{for all } h \in P^\vee\}.$$

A vector $v \in V^q_\mu$ is called a ***weight vector*** of weight μ. If $e_i v = 0$ for all $i \in I$, it is called a ***maximal vector***. If $V^q_\mu \neq 0$, μ is called a ***weight*** of V^q and V^q_μ is the ***weight space*** attached to $\mu \in P$. Its dimension $\dim V^q_\mu$ is called the ***weight multiplicity*** of μ. We will denote by $\mathrm{wt}(V^q)$ the set of weights of the $U_q(\mathfrak{g})$-module V^q. If $\dim V^q_\mu < \infty$ for all $\mu \in \mathrm{wt}(V^q)$, the ***character*** of V^q is defined by

$$\operatorname{ch} V^q = \sum_\mu \dim V^q_\mu e^\mu,$$

where e^μ are formal basis elements of the group algebra $\mathbf{F}[P]$ with multiplication defined by $e^\lambda e^\mu = e^{\lambda+\mu}$.

Proposition 3.2.1. *Every submodule of a weight module over $U_q(\mathfrak{g})$ is also a weight module.*

Proof. Let V^q be a weight module. Suppose there exists some submodule W^q which is not a weight module. Choose $v = v_1 + \cdots + v_p \in W^q$, where $v_k \in V^q_{\mu_k}$, μ_k are distinct, and $v_k \notin W^q$ for some k. We may assume further that every element of W^q with fewer summands has all its summands belonging to W^q. This forces $v_k \notin W^q$ for all k. Choose any $h \in P^\vee$ such that $\mu_1(h) \neq \mu_k(h)$ for at least one k. Then $q^h v - q^{\mu_1(h)} v$ is a nonzero element of W^q with a strictly smaller number of summands for which all its summands do not belong to W^q, which is a contradiction. □

For $\lambda \in P$, set $D(\lambda) = \{\mu \in P \mid \mu \leq \lambda\}$. The ***category*** $\mathcal{O}^q$ consists of weight modules V^q over $U_q(\mathfrak{g})$ with finite dimensional weight spaces for which there exist a finite number of elements $\lambda_1, \lambda_2, \ldots, \lambda_s \in P$ such that

$$\mathrm{wt}(V^q) \subset D(\lambda_1) \cup \cdots \cup D(\lambda_s).$$

As is the case with Kac-Moody algebras, the most important examples among the $U_q(\mathfrak{g})$-modules in the category $\mathcal{O}^q$ may be *highest weight modules*. A weight module V^q is called a ***highest weight module*** with ***highest***

weight $\lambda \in P$ if there exists a nonzero $v_\lambda \in V^q$ such that

$$\begin{aligned} e_i v_\lambda &= 0 \quad \text{for all } i \in I, \\ q^h v_\lambda &= q^{\lambda(h)} v_\lambda \quad \text{for all } h \in P^\vee, \\ V^q &= U_q(\mathfrak{g}) v_\lambda. \end{aligned} \tag{3.8}$$

The vector v_λ, which is unique up to constant multiple, is called the ***highest weight vector***. The triangular decomposition for $U_q(\mathfrak{g})$ (Theorem 3.1.5) shows $V^q = U_q^- v_\lambda$ for any highest weight module. It can be easily verified that $\dim V_\lambda^q = 1$, $\dim V_\mu^q < \infty$ for all $\mu \in \mathrm{wt}(V^q)$, and $V^q = \bigoplus_{\mu \le \lambda} V_\mu^q$. The last property justifies the name *highest weight modules.*

Fix $\lambda \in P$ and let $J^q(\lambda)$ be the left ideal of $U_q(\mathfrak{g})$ generated by e_i $(i \in I)$ and $q^h - q^{\lambda(h)}1$ $(h \in P^\vee)$. Define the ***Verma module*** $M^q(\lambda) = U_q(\mathfrak{g})/J^q(\lambda)$. This is a $U_q(\mathfrak{g})$-module by left multiplication. Set $v_\lambda = 1 + J^q(\lambda)$. Then we have

$$\begin{aligned} q^h v_\lambda &= q^h + J^q(\lambda) = q^{\lambda(h)}1 + J^q(\lambda) = q^{\lambda(h)} v_\lambda, \\ e_i v_\lambda &= e_i + J^q(\lambda) = J^q(\lambda) = 0, \\ U_q(\mathfrak{g}) v_\lambda &= U_q(\mathfrak{g})/J^q(\lambda) = M^q(\lambda). \end{aligned}$$

Thus $M^q(\lambda)$ is a highest weight module with highest weight λ and highest weight vector $v_\lambda = 1 + J^q(\lambda)$.

Proposition 3.2.2.

(1) *As a U_q^--module, $M^q(\lambda)$ is free of rank 1, generated by the highest weight vector $v_\lambda = 1 + J^q(\lambda)$.*

(2) *Every highest weight $U_q(\mathfrak{g})$-module with highest weight λ is a homomorphic image of $M^q(\lambda)$.*

(3) *The Verma module $M^q(\lambda)$ has a unique maximal submodule.*

Proof. (1) Any highest weight module is generated by its highest weight vector as a U_q^--module, so it only remains to prove that it is free. We need to prove that $uv_\lambda = 0$ for $u \in U_q^-$ implies $u = 0$ or, equivalently, $U_q^- \cap J^q(\lambda) = 0$. Combining Lemma 3.1.4 and the triangular decomposition (Theorem 3.1.5), we may write $U_q(\mathfrak{g}) \cong U_q^- \otimes U_q^{\ge 0}$. Thus $J^q(\lambda)$, the left ideal of $U_q(\mathfrak{g})$ generated by e_i $(i \in I)$ and $q^h - q^{\lambda(h)}1$ $(h \in P^\vee)$, cannot have elements that lie in U_q^-.

(2) Let W^q be an arbitrary highest weight module with highest weight λ and highest weight vector w_λ. Then (1) allows us to define a map $\phi : M^q(\lambda) \to W^q$ by $u \cdot (1 + J^q(\lambda)) \mapsto u \cdot w_\lambda$ for $u \in U_q^-$. Since $W^q = U_q^- w_\lambda$, the map ϕ is surjective. It remains to check if ϕ is a $U_q(\mathfrak{g})$-module homomorphism.

The action of an arbitrary $x \in U_q(\mathfrak{g})$ on an arbitrary $uv_\lambda \in M^q(\lambda)$ with $u \in U_q(\mathfrak{g})^-$ may be computed as follows. First, write $xu \in U_q(\mathfrak{g})$ in the form given by the triangular decomposition, say, $xu = \sum u^- u^0 u^+$ with $u^\pm \in U_q^\pm$ and $u^0 \in U_q^0$. We want to see if xuv_λ is sent to xuw_λ by the map ϕ. This is equivalent to checking if $\sum u^- u^0 u^+ v_\lambda$ is sent to $\sum u^- u^0 u^+ w_\lambda$. Since the actions of q^h and e_i on v_λ and w_λ are identical and result in only constant multiples of v_λ, each $u^0 u^+ v_\lambda$ is sent to the corresponding $u^0 u^+ w_\lambda$. The remaining action of u^- is preserved by construction of the map. Hence the map ϕ defined above does preserve $U_q(\mathfrak{g})$-action.

(3) Note that any proper submodule of $M^q(\lambda)$ cannot contain the highest weight vector $v_\lambda = 1 + J^q(\lambda)$; that is, it must lie inside $\bigoplus_{\mu \lneq \lambda} M^q(\lambda)_\mu$. Hence the sum of two proper submodules is again a proper submodule of $M^q(\lambda)$. Therefore, the sum of all proper submodules of $M^q(\lambda)$ is the unique maximal submodule of $M^q(\lambda)$. □

We denote this unique maximal submodule of $M^q(\lambda)$ by $N^q(\lambda)$. Then the quotient $M^q(\lambda)/N^q(\lambda)$ is an ***irreducible highest weight module*** with highest weight λ, which will be denoted by $V^q(\lambda)$.

We now define the main object of our study in this chapter—the *category* $\mathcal{O}^q_{\text{int}}$ of $U_q(\mathfrak{g})$-modules. A weight module V^q over the quantum group $U_q(\mathfrak{g})$ is ***integrable*** if all e_i and f_i $(i \in I)$ are locally nilpotent on V^q.

Definition 3.2.3. The ***category*** $\mathcal{O}^q_{\text{int}}$ consists of $U_q(\mathfrak{g})$-modules V^q satisfying the following conditions:

(1) V^q has a weight space decomposition $V^q = \bigoplus_{\lambda \in P} V^q_\lambda$, where

$$V^q_\lambda = \{v \in V^q \mid q^h v = q^{\lambda(h)} v \ \text{ for all } h \in P^\vee\}$$

and $\dim V^q_\lambda < \infty$ for all $\lambda \in P$,

(2) there exist a finite number of elements $\lambda_1, \dots, \lambda_s \in P$ such that

$$\operatorname{wt}(V^q) \subset D(\lambda_1) \cup \cdots \cup D(\lambda_s),$$

(3) all e_i and f_i $(i \in I)$ are locally nilpotent on V^q.

The morphisms are taken to be usual $U_q(\mathfrak{g})$-module homomorphisms.

Hence the category $\mathcal{O}^q_{\text{int}}$ consists of integrable $U_q(\mathfrak{g})$-modules in the category $\mathcal{O}^q$. Note that the category $\mathcal{O}^q_{\text{int}}$ is closed under taking direct sums or tensor products of finitely many $U_q(\mathfrak{g})$-modules.

Fix $i \in I$. We denote by $U_q(\mathfrak{g}_{(i)})$ the subalgebra of $U_q(\mathfrak{g})$ generated by e_i, f_i, $K_i^{\pm 1}$. Then we have $U_q(\mathfrak{g}_{(i)}) \cong U_{q_i}(\mathfrak{sl}_2)$, and, as for the $\mathfrak{g}$-modules in the category $\mathcal{O}_{\text{int}}$, we can prove (Exercise 3.5):

Proposition 3.2.4 ([**21, 28**]). *Let V^q be a $U_q(\mathfrak{g})$-module in the category $\mathcal{O}^q_{\text{int}}$. Then, for each $i \in I$, V^q decomposes into a direct sum of $U_q(\mathfrak{h})$-invariant finite dimensional irreducible $U_q(\mathfrak{g}_{(i)})$-submodules.*

Set $e_i^{(k)} = e_i^k/[k]_{q_i}!$ and $f_i^{(k)} = f_i^k/[k]_{q_i}!$. They are called the ***divided powers*** of e_i and f_i, respectively. We have the following commutation relations for the divided powers, which can be proved by straightforward induction (Exercise 3.6).

Lemma 3.2.5. *For all $i \in I$ and $k \in \mathbf{Z}_{\geq 0}$, we have*

$$e_i f_i^{(k)} = f_i^{(k)} e_i + f_i^{(k-1)} \frac{K_i q_i^{-k+1} - K_i^{-1} q_i^{k-1}}{q_i - q_i^{-1}}.$$

Proposition 3.2.6. *Let $\lambda \in P^+$ and let $V^q(\lambda)$ be the irreducible highest weight module of highest weight λ and highest weight vector v_λ. Then $f_i^{\lambda(h_i)+1} v_\lambda = 0$ for all $i \in I$.*

Proof. The above lemma shows

$$e_i f_i^{(k)} v_\lambda = [\lambda(h_i) - k + 1]_{q_i} f_i^{(k-1)} v_\lambda. \tag{3.9}$$

Substituting $k = \lambda(h_i) + 1$, we see that $e_i f_i^{\lambda(h_i)+1} v_\lambda = 0$. Moreover, for $j \neq i$, we already know $e_j f_i^{\lambda(h_i)+1} v_\lambda = 0$. Hence if $f_i^{\lambda(h_i)+1} v_\lambda \neq 0$, it would generate a nontrivial proper submodule of $V^q(\lambda)$, contradicting the irreducibility of $V^q(\lambda)$. □

Proposition 3.2.7. *A highest weight $U_q(\mathfrak{g})$-module V^q with highest weight $\lambda \in P$ and highest weight vector v_λ is integrable if and only if for every $i \in I$, there exists some N_i such that $f_i^{N_i} v_\lambda = 0$.*

Proof. We have only to prove the *if* part. Note that $e_i \cdot V_\mu \subset V_{\mu+\alpha_i}$. Since all the weights of a highest weight module are less than or equal to its highest weight, the e_i ($i \in I$) are always locally nilpotent on any highest weight module. So we restrict our attention to the f_i's only.

For homogeneous $u \in U_q^-$ of degree $-\alpha \in Q_-$, we have

$$f_i^n u = \sum_{k=0}^{n} q_i^{(\alpha(h_i)+k)(n-k)} \begin{bmatrix} n \\ k \end{bmatrix}_{q_i} \left((\mathrm{ad}_q f_i)^k (u) \right) f_i^{n-k}.$$

With the help of (3.3), we may check this by induction (Exercise 3.7). Recall that $(\mathrm{ad}_q f_i)^k (f_j) = 0$ if $j \neq i$ and $k > -a_{i,j}$. So if $u = f_j u'$ with $j \neq i$, we have

$$f_i^n u = \sum_{k=0}^{-a_{i,j}} q_i^{(\alpha(h_i)+k)(n-k)} \begin{bmatrix} n \\ k \end{bmatrix}_{q_i} \left((\mathrm{ad}_q f_i)^k (f_j) \right) f_i^{n-k} u'.$$

If $u = f_i u'$, we may set $f_i^n u = f_i^{n+1} u'$. Given $u \in U_q^-$, this shows how we may inductively prove $f_i^n u \in U_q^- \cdot f_i^{N_i}$ for all sufficiently large n. Now, an arbitrary element of V^q may be written in the form $u \cdot v_\lambda$ with $u \in U_q^-$, which completes the proof. □

Proposition 3.2.8. *Let $V^q(\lambda)$ be the irreducible highest weight $U_q(\mathfrak{g})$-module with highest weight $\lambda \in P$. Then $V^q(\lambda)$ belongs to the category $\mathcal{O}_{\text{int}}^q$ if and only if $\lambda \in P^+$.*

Proof. The *if* part is taken care of by Propositions 3.2.6 and 3.2.7. Let us prove the *only if* part. Fix $i \in I$ and let N_i be the smallest nonnegative integer such that $f_i^{N_i} v_\lambda \neq 0$ and $f_i^{N_i+1} v_\lambda = 0$. Then we have

$$0 = e_i f_i^{(N_i+1)} v_\lambda = [\lambda(h_i) - N_i]_{q_i} f_i^{(N_i)} v_\lambda,$$

which implies

$$[\lambda(h_i) - N_i]_{q_i} = \frac{q_i^{\lambda(h_i)-N_i} - q_i^{-\lambda(h_i)+N_i}}{q_i - q_i^{-1}} = 0.$$

It follows that $q_i^{2(\lambda(h_i)-N_i)} = 1$. Since q is an indeterminate, we must have $\lambda(h_i) = N_i \in \mathbf{Z}_{\geq 0}$. □

Remark 3.2.9. As in Lemma 2.4.4 (2), we would like to claim that every irreducible $U_q(\mathfrak{g})$-module in the category $\mathcal{O}_{\text{int}}^q$ is isomorphic to the irreducible highest weight module $V^q(\lambda)$ for some $\lambda \in P^+$. For this, we need to wait until the end of the first half of Section 3.4. In Section 3.4, we will also show that every highest weight $U_q(\mathfrak{g})$-module in the category $\mathcal{O}_{\text{int}}^q$ is isomorphic to $V^q(\lambda)$ with $\lambda \in P^+$.

3.3. $\mathbf{A}_1$-forms

In the previous section, we have seen that the representation theory of $U_q(\mathfrak{g})$ is very similar to that of $\mathfrak{g}$. Hence it is natural to expect that the quantum group $U_q(\mathfrak{g})$ may be regarded as some sort of *deformation* of $U(\mathfrak{g})$ in such a way that the representations of $U_q(\mathfrak{g})$ can also be regarded as the *deformations* of those of $U(\mathfrak{g})$. Moreover, from the defining relations, we can expect the structures of the quantum group $U_q(\mathfrak{g})$ and its representations *tend to* those of $U(\mathfrak{g})$ and its representations as q approaches 1. This observation is one of the most fundamental properties of quantum groups and their representations which was first proved in [**47**].

In this and the next section, we make precise and prove these statements. By somewhat modifying Lusztig's approach, we show that the quantum group $U_q(\mathfrak{g})$ is a *deformation* of $U(\mathfrak{g})$ as a Hopf algebra and show that a highest weight $U(\mathfrak{g})$-module admits a *deformation* to a highest weight $U_q(\mathfrak{g})$-module in such a way that the dimensions of the weight spaces remain the same under the deformation.

We consider the localization of $\mathbf{F}[q]$ at the ideal $(q-1)$:

$$\begin{aligned}\mathbf{A}_1 &= \{f(q) \in \mathbf{F}(q) \mid f \text{ is regular at } q=1\} \\ &= \{g/h \mid g, h \in \mathbf{F}[q],\ h(1) \neq 0\}.\end{aligned} \tag{3.10}$$

Notice that $[n]_{q_i} \in \mathbf{A}_1$ and $\begin{bmatrix} m \\ n \end{bmatrix}_{q_i} \in \mathbf{A}_1$, being elements of $\mathbf{Z}[q, q^{-1}]$. For an integer $n \in \mathbf{Z}$, we formally define

$$[y;n]_x = \frac{yx^n - y^{-1}x^{-n}}{x - x^{-1}} \quad \text{and} \quad (y;n)_x = \frac{yx^n - 1}{x - 1}. \tag{3.11}$$

For example, we have

$$\begin{aligned} [q_i^m; n]_{q_i} &= \frac{q_i^{m+n} - q_i^{-m-n}}{q_i - q_i^{-1}} \in \mathbf{A}_1, \\ (q_i^m; n)_{q_i} &= \frac{q_i^{m+n} - 1}{q_i - 1} \in \mathbf{A}_1, \\ [q^h; n]_q &= \frac{q^h q^n - q^{-h} q^{-n}}{q - q^{-1}} \in U_q^0, \\ (q^h; n)_q &= \frac{q^h q^n - 1}{q - 1} \in U_q^0. \end{aligned}$$

Definition 3.3.1. We define the $\mathbf{A}_1$***-form***, denoted by $U_{\mathbf{A}_1}$, of the quantum group $U_q(\mathfrak{g})$ to be the $\mathbf{A}_1$-subalgebra of $U_q(\mathfrak{g})$ generated by the elements e_i, f_i, q^h, and $(q^h;0)_q$ $(i \in I,\ h \in P^\vee)$.

Let $U_{\mathbf{A}_1}^+$ (respectively, $U_{\mathbf{A}_1}^-$) be the $\mathbf{A}_1$-subalgebra of $U_{\mathbf{A}_1}$ generated by the elements e_i (respectively, f_i) for $i \in I$, and let $U_{\mathbf{A}_1}^0$ be the $\mathbf{A}_1$-subalgebra of $U_{\mathbf{A}_1}$ generated by q^h and $(q^h;0)_q$ for $h \in P^\vee$. The next lemma shows that $U_{\mathbf{A}_1}^0$ contains all of the more frequently appearing elements of U_q^0.

Lemma 3.3.2.

(1) $(q^h;n)_q \in U_{\mathbf{A}_1}^0$ *for all* $n \in \mathbf{Z}$ *and* $h \in P^\vee$.

(2) $[K_i;n]_{q_i} \in U_{\mathbf{A}_1}^0$ *for all* $n \in \mathbf{Z}$ *and* $i \in I$.

Proof. It suffices to check the following identities:

$$(q^h; n)_q = q^n (q^h; 0)_q + \frac{q^n - 1}{q - 1},$$
$$[K_i; 0]_{q_i} = q_i \frac{q - 1}{q_i^2 - 1}(1 + K_i^{-1})(K_i; 0)_q,$$
$$[K_i; n]_{q_i} = q_i^n [K_i; 0]_{q_i} + [n]_{q_i} K_i^{-1},$$

all of which can be verified by straightforward calculations (Exercise 3.8). □

We next show that the triangular decomposition of $U_q(\mathfrak{g})$ carries over to its $\mathbf{A}_1$-form.

Proposition 3.3.3. *We have a natural isomorphism of $\mathbf{A}_1$-modules*

$$U_{\mathbf{A}_1} \cong U_{\mathbf{A}_1}^- \otimes U_{\mathbf{A}_1}^0 \otimes U_{\mathbf{A}_1}^+$$

induced from the triangular decomposition of $U_q(\mathfrak{g})$.

Proof. Consider the canonical isomorphism $\varphi : U_q(\mathfrak{g}) \xrightarrow{\sim} U_q^- \otimes U_q^0 \otimes U_q^+$ given by Theorem 3.1.5. The commutation relations

$$e_i (q^h; 0)_q = (q^h; -\alpha_i(h))_q e_i,$$
$$(q^h; 0)_q f_i = f_i (q^h; -\alpha_i(h))_q,$$
$$e_i f_j = f_j e_i + \delta_{i,j} [K_i; 0]_{q_i},$$

together with Lemma 3.3.2 show that the image of φ lies inside $U_{\mathbf{A}_1}^- \otimes U_{\mathbf{A}_1}^0 \otimes U_{\mathbf{A}_1}^+$ when restricted to $U_{\mathbf{A}_1}$. Its inverse map is given by multiplication. Hence the two spaces are isomorphic as $\mathbf{A}_1$-modules. □

Fix $\lambda \in P$. Throughout this and the next section, V^q will denote a highest weight $U_q(\mathfrak{g})$-module with highest weight λ and highest weight vector v_λ.

Definition 3.3.4. The $\mathbf{A}_1$***-form*** of the highest weight module V^q with highest weight $\lambda \in P$ and highest weight vector v_λ is defined to be the $U_{\mathbf{A}_1}$-module $V_{\mathbf{A}_1} = U_{\mathbf{A}_1} v_\lambda$.

First, observe that we have:

Proposition 3.3.5. $V_{\mathbf{A}_1} = U_{\mathbf{A}_1}^- v_\lambda$.

Proof. In view of Proposition 3.3.3, it suffices to show that $U_{\mathbf{A}_1}^+ v_\lambda = \mathbf{A}_1 v_\lambda$ and $U_{\mathbf{A}_1}^0 v_\lambda = \mathbf{A}_1 v_\lambda$. The first assertion is clear by the definition of highest weight modules. For the second assertion, we observe that

$$q^h v_\lambda = q^{\lambda(h)} v_\lambda,$$
$$(q^h;0)_q v_\lambda = \frac{q^{\lambda(h)}-1}{q-1} v_\lambda.$$

Hence we get $U_{\mathbf{A}_1} v_\lambda = U^-_{\mathbf{A}_1} v_\lambda$. □

Recall that the highest weight $U_q(\mathfrak{g})$-module V^q has the weight space decomposition

$$V^q = \bigoplus_{\mu\le\lambda} V^q_\mu, \quad \text{where } V^q_\mu = \{v \in V^q \mid q^h v = q^{\mu(h)} v \quad \text{for all } h \in P^\vee\}.$$

For each $\mu \in P$, define the ***weight space*** $(V_{\mathbf{A}_1})_\mu = V_{\mathbf{A}_1} \cap V^q_\mu$. The next proposition shows that the weight space decomposition of V^q also carries over to its $\mathbf{A}_1$-form.

Proposition 3.3.6. $V_{\mathbf{A}_1} = \bigoplus_{\mu\le\lambda} (V_{\mathbf{A}_1})_\mu$.

Proof. Assume $v = v_1 + \cdots + v_p \in V_{\mathbf{A}_1}$, where $v_j \in V^q_{\mu_j}$ and $\mu_j \in P$. We may take μ_j to be distinct. Fix an index j. It suffices to show $v_j \in V_{\mathbf{A}_1}$.

For each $k \neq j$, we may choose $H_k \in P^\vee$ such that $\mu_j(H_k) \neq \mu_k(H_k)$. Note that

$$(q^{\mu_j(H_k)}; -\mu_k(H_k))_q = \frac{q^{\mu_j(H_k)-\mu_k(H_k)}-1}{q-1}$$

is invertible in $\mathbf{A}_1$ for each $k \neq j$. Imitating *Lagrange's interpolation formula*, define $u \in U_{\mathbf{A}_1}$ to be

$$u = \prod_{k\neq j} \frac{(q^{H_k}; -\mu_k(H_k))_q}{(q^{\mu_j(H_k)}; -\mu_k(H_k))_q}.$$

Then $uv_j = v_j$ and $uv_k = 0$ for $k \neq j$. Hence $uv = v_j \in V_{\mathbf{A}_1}$. □

An approach to proving the above proposition that mimics the proof of Proposition 3.2.1 fails because the scalars we are dealing with do not form a field. But the proof for the next proposition relies heavily on the fact that $\mathbf{A}_1$ is close enough to a field.

Proposition 3.3.7. *For each $\mu \in P$, the weight space $(V_{\mathbf{A}_1})_\mu$ is a free $\mathbf{A}_1$-module with* $\operatorname{rank}_{\mathbf{A}_1} (V_{\mathbf{A}_1})_\mu = \dim_{\mathbf{F}(q)} V^q_\mu$.

Proof. Notice that $(V_{\mathbf{A}_1})_\mu$ is finitely generated as an $\mathbf{A}_1$-module. Let $\{v_k\}_{k=1}^p$ be an $\mathbf{A}_1$-spanning set of $(V_{\mathbf{A}_1})_\mu$. We will show that this spanning set can be reduced to an $\mathbf{A}_1$-linearly independent set. Then we would have an $\mathbf{A}_1$-basis of $(V_{\mathbf{A}_1})_\mu$, which would imply $(V_{\mathbf{A}_1})_\mu$ is a free $\mathbf{A}_1$-module. Consider an arbitrary $\mathbf{A}_1$-linear dependence relation

$$c_1(q)v_1 + \cdots + c_p(q)v_p = 0 \tag{3.12}$$

with each $c_k(q) \in \mathbf{A}_1$. Dividing out by $(q-1)$ if necessary, we may assume that at least one of the coefficients satisfies $c_k(1) \neq 0$. For example, suppose $c_1(1) \neq 0$. Then $c_1(q)^{-1} \in \mathbf{A}_1$ and

$$v_1 = \frac{-1}{c_1(q)}\{c_2(q)v_2 + \cdots + c_p(q)v_p\}.$$

Repeating this process, we get an $\mathbf{A}_1$-linearly independent spanning set of $(V_{\mathbf{A}_1})_\mu$.

As for its rank, let $\{f_\zeta v_\lambda\}$ be a basis of V_μ^q, where f_ζ are monomials in f_i. The set certainly belongs to $(V_{\mathbf{A}_1})_\mu$ and is also $\mathbf{A}_1$-linearly independent, so $\operatorname{rank}_{\mathbf{A}_1}(V_{\mathbf{A}_1})_\mu \geq \dim_{\mathbf{F}(q)} V_\mu^q$. To show the converse inequality, let $\{v_k\}_{k=1}^p$ be an $\mathbf{A}_1$-linearly independent subset of $(V_{\mathbf{A}_1})_\mu$. Consider an $\mathbf{F}(q)$-linear dependence relation

$$b_1(q)v_1 + \cdots + b_p(q)v_p = 0,$$

where $b_k(q) \in \mathbf{F}(q)$ for $k = 1, \ldots, p$. Multiplying by $(q-1)$ if needed, we may assume that all $b_k(q) \in \mathbf{A}_1$. Since $v_1, \ldots, v_p$ are linearly independent over $\mathbf{A}_1$, we must have $b_k(q) = 0$ for all $k = 1, \ldots, p$. Hence $v_1, \ldots, v_p$ are linearly independent over $\mathbf{F}(q)$ and $\operatorname{rank}_{\mathbf{A}_1}(V_{\mathbf{A}_1})_\mu \leq \dim_{\mathbf{F}(q)} V_\mu^q$, which completes the proof. □

Proposition 3.3.8. *The $\mathbf{F}(q)$-linear map $\varphi : \mathbf{F}(q) \otimes_{\mathbf{A}_1} V_{\mathbf{A}_1} \to V^q$ given by $c \otimes v \mapsto cv$ $(c \in \mathbf{F}(q),\ v \in V_{\mathbf{A}_1})$ is an isomorphism.*

Proof. Combining Propositions 3.3.6 and 3.3.7, we get the desired linear isomorphism. □

Remark 3.3.9.

(1) We see from this proposition that the $\mathbf{A}_1$-form $V_{\mathbf{A}_1}$ of a highest weight module V^q is an *integral form* of V^q over $\mathbf{A}_1$; i.e., it can be viewed as an $\mathbf{A}_1$-lattice in V^q.

(2) In Exercise 3.10, we give an alternative proof of Proposition 3.3.7 which works for more general setting.

3.4. Classical limit

We now proceed to take the limit $q \to 1$ of highest weight $U_q(\mathfrak{g})$-modules. The notation from the previous section will be retained. In particular, V^q

will denote a highest weight $U_q(\mathfrak{g})$-module of highest weight $\lambda \in P$ and highest weight vector v_λ. Let $\mathbf{J}_1$ be the unique maximal ideal of the local ring $\mathbf{A}_1$ generated by $q-1$. There exists an isomorphism of fields

$$\mathbf{A}_1/\mathbf{J}_1 \xrightarrow{\sim} \mathbf{F} \quad \text{given by} \quad f(q) + \mathbf{J}_1 \mapsto f(1).$$

(In particular, q is mapped onto 1.) Define the $\mathbf{F}$-linear vector spaces

$$\begin{aligned} U_1 &= (\mathbf{A}_1/\mathbf{J}_1) \otimes_{\mathbf{A}_1} U_{\mathbf{A}_1}, \\ V^1 &= (\mathbf{A}_1/\mathbf{J}_1) \otimes_{\mathbf{A}_1} V_{\mathbf{A}_1}. \end{aligned} \tag{3.13}$$

Then V^1 is naturally a U_1-module. We would like to show that U_1 is isomorphic to the universal enveloping algebra $U(\mathfrak{g})$ and that V^1 is a highest weight $U(\mathfrak{g})$-module of highest weight λ.

Note that

$$U_1 \cong U_{\mathbf{A}_1}/\mathbf{J}_1 U_{\mathbf{A}_1} \quad \text{and} \quad V^1 \cong V_{\mathbf{A}_1}/\mathbf{J}_1 V_{\mathbf{A}_1}.$$

Consider the natural maps

$$\begin{aligned} U_{\mathbf{A}_1} &\to U_{\mathbf{A}_1}/\mathbf{J}_1 U_{\mathbf{A}_1} \cong U_1, \\ V_{\mathbf{A}_1} &\to V_{\mathbf{A}_1}/\mathbf{J}_1 V_{\mathbf{A}_1} \cong V^1. \end{aligned} \tag{3.14}$$

We use the bar notation for the image under these maps. The passage under these maps is referred to as taking the ***classical limit***. Notice that q is mapped to 1 under these maps. The notation U_1 has been used to call to mind "$U_q(\mathfrak{g})$ at $q = 1$".

For each $\mu \in P$, define $V^1_\mu = (\mathbf{A}_1/\mathbf{J}_1) \otimes_{\mathbf{A}_1} (V_{\mathbf{A}_1})_\mu$. Then we have:

Lemma 3.4.1.

(1) *For each $\mu \in P$, if $\{v_i\}$ is a basis of the free $\mathbf{A}_1$-module $(V_{\mathbf{A}_1})_\mu$, then $\{\bar{v}_i\}$ is a basis of the $\mathbf{F}$-linear space V^1_μ.*

(2) *For each $\mu \in P$, a set $\{v_i\} \subset (V_{\mathbf{A}_1})_\mu$ is $\mathbf{A}_1$-linearly independent if the set $\{\bar{v}_i\} \subset V^1_\mu$ is $\mathbf{F}$-linearly independent.*

Proof. (1) Using [**18**, Thm.5.11, Ch.4], we may show that $\{1 \otimes v_i\}$ is a basis of the $(\mathbf{A}_1/\mathbf{J}_1)$-linear space V^1_μ (Exercise 3.11).

(2) Suppose $\sum c_i(q) v_i = 0$ for $c_i(q) \in \mathbf{A}_1$, not all zero. Dividing out by $q-1$, if necessary, we may assume at least one $c_i(1) \neq 0$. Then, $\sum c_i(1)\bar{v}_i = 0$ is a nontrivial $\mathbf{F}$-linear dependence relation. □

Proposition 3.4.2.

(1) $V^1 = \bigoplus_{\mu \leq \lambda} V^1_\mu$.

(2) *For each $\mu \in P$,* $\dim_{\mathbf{F}} V^1_\mu = \operatorname{rank}_{\mathbf{A}_1} (V_{\mathbf{A}_1})_\mu$.

Proof. The first follows from Proposition 3.3.6. And the second follows from Lemma 3.4.1. □

We now know

$$\dim_{\mathbf{F}} V^1_\mu = \operatorname{rank}_{\mathbf{A}_1}(V_{\mathbf{A}_1})_\mu = \dim_{\mathbf{F}(q)} V^q_\mu \tag{3.15}$$

for all $\mu \in P$.

Let $\bar{h} \in U_1$ denote the classical limit of the element

$$(q^h;0)_q = \frac{q^h - 1}{q-1} \in U_{\mathbf{A}_1}.$$

We first show that the image of $U^0_{\mathbf{A}_1}$ under the classical limit is quite close to $U^0 = U(\mathfrak{h})$.

Lemma 3.4.3.

(1) *For all $h \in P^\vee$, we have $\overline{q^h} = 1$.*

(2) *For any $h, h' \in P^\vee$, $\overline{h+h'} = \bar{h} + \bar{h}'$. Hence, $\overline{nh} = n\bar{h}$ for $n \in \mathbf{Z}$.*

Proof. (1) Note that

$$q^h - 1 = (q-1)(q^h;0)_q.$$

Hence the classical limit of the right-hand side, being a multiple of $q-1$, is zero.

(2) We may easily calculate

$$(q^{h+h'};0)_q = q^{h'}(q^h;0)_q + (q^{h'};0)_q.$$

We take the classical limit of both sides using $\overline{q^{h'}} = 1$ to obtain the desired result. □

Define the subalgebras $U^0_1 = \mathbf{F} \otimes U^0_{\mathbf{A}_1}$ and $U^\pm_1 = \mathbf{F} \otimes U^\pm_{\mathbf{A}_1}$. The next theorem shows that the classical limit of $U_q(\mathfrak{g})$ is almost the same as $U(\mathfrak{g})$.

Theorem 3.4.4.

(1) *The elements $\bar{e}_i$, $\bar{f}_i$, $(i \in I)$ and $\bar{h}$ $(h \in P^\vee)$ satisfy the defining relations of $U(\mathfrak{g})$ given by Proposition 2.1.6. Hence, there exists a surjective $\mathbf{F}$-algebra homomorphism $\psi : U(\mathfrak{g}) \to U_1$ and the U_1-module V^1 has a $U(\mathfrak{g})$-module structure.*

(2) *For each $\mu \in P$ and $h \in P^\vee$, the element $\bar{h}$ acts on V^1_μ as scalar multiplication by $\mu(h)$. So V^1_μ is the μ-weight space of the $U(\mathfrak{g})$-module V^1.*

(3) *As a $U(\mathfrak{g})$-module, V^1 is a highest weight module with highest weight $\lambda \in P$ and highest weight vector $\bar{v}_\lambda$.*

Proof. (1) The first relation for $U(\mathfrak{g})$ is trivial. Let us check the second defining relation. By definition of $U_q(\mathfrak{g})$, we have

$$e_i f_i - f_i e_i = [K_i; 0]_{q_i} = \frac{q_i}{q_i + 1} \frac{q - 1}{q_i - 1} (1 + K_i^{-1})(K_i; 0)_q.$$

Taking Lemma 3.4.3 into account, the classical limit of the right-hand side is

$$\frac{1}{2} \cdot \frac{1}{s_i} \cdot 2 \cdot s_i \bar{h}_i = \bar{h}_i,$$

which yields the second defining relation. As for the third defining relation, note that

$$\begin{aligned}(q^h; 0)_q e_i - e_i (q^h; 0)_q &= e_i (q^h; \alpha_i(h))_q - e_i (q^h; 0)_q \\ &= \frac{q^{\alpha_i(h)} - 1}{q - 1} e_i q^h.\end{aligned}$$

We take the classical limit of both sides to obtain $\bar{h}\bar{e}_i - \bar{e}_i \bar{h} = \alpha_i(h)\bar{e}_i$. Since $\overline{[n]_{q_i}} = n$ and $\overline{\begin{bmatrix} n \\ m \end{bmatrix}_{q_i}} = \binom{n}{m}$ we get the remaining Serre relations.

(2) For $v \in (V_{\mathbf{A}_1})_\mu$ and $h \in P^\vee$, we have

$$(q^h; 0)_q v = \frac{q^{\mu(h)} - 1}{q - 1} v.$$

Taking the classical limit of both sides yields our assertion.

(3) By (2), we have $\bar{h}\bar{v}_\lambda = \lambda(h)\bar{v}_\lambda$ for all $h \in P^\vee$. For each $i \in I$, $\bar{e}_i \bar{v}_\lambda$ is trivially zero. By Proposition 3.3.5 and (1), we get $V^1 = U_1^- \bar{v}_\lambda = U^- \bar{v}_\lambda$, and hence V^1 is a highest weight $U(\mathfrak{g})$-module with highest weight λ and highest weight vector $\bar{v}_\lambda$. □

Summarizing the discussions in Propositions 3.3.7 and 3.4.2, and Theorem 3.4.4, we obtain the following identity between the characters of a $U(\mathfrak{g})$-module and a $U_q(\mathfrak{g})$-module.

Proposition 3.4.5. $\operatorname{ch} V^1 = \operatorname{ch} V^q$.

This shows that the $U_q(\mathfrak{g})$-module V^q can be viewed as a *deformation* of the $U(\mathfrak{g})$-module V^1. The next theorem shows that highest weight $U_q(\mathfrak{g})$-modules in the category $\mathcal{O}_{\text{int}}^q$ and highest weight $U(\mathfrak{g})$-modules in the category $\mathcal{O}_{\text{int}}$ are in good correspondence.

Theorem 3.4.6. *If $\lambda \in P^+$ and V^q is the irreducible highest weight $U_q(\mathfrak{g})$-module $V^q(\lambda)$ with highest weight λ, then V^1 is isomorphic to the irreducible highest weight module $V(\lambda)$ over $U(\mathfrak{g})$ with highest weight λ. Hence, the character of $V^q(\lambda)$ is the same as the character of $V(\lambda)$, which is given by the Weyl-Kac character formula in Theorem* 2.4.6.

Proof. Let v_λ be the highest weight vector of V^q. By Proposition 3.2.6 and Theorem 3.4.4 (3), V^1 is a highest weight $U(\mathfrak{g})$-module with highest weight λ and highest weight vector $\bar{v}_\lambda$ satisfying $f_i^{\lambda(h_i)+1}\bar{v}_\lambda = \bar{f}_i^{\lambda(h_i)+1}\bar{v}_\lambda = 0$ for all $i \in I$. Hence Theorem 2.4.6 shows $V^1 \cong V(\lambda)$. The second assertion follows from Proposition 3.4.5. □

Corollary 3.4.7. *Let $\lambda \in P^+$ and let V^q be a highest weight module over $U_q(\mathfrak{g})$ with highest weight λ and highest weight vector v_λ. If $f_i^{\lambda(h_i)+1}v_\lambda = 0$ for all $i \in I$, then V^q is isomorphic to the irreducible highest weight module $V^q(\lambda)$.*

Proof. As in the proof of Theorem 3.4.6, we have $V^1 \cong V(\lambda)$ as $U(\mathfrak{g})$-modules. Hence, $\operatorname{ch} V^q = \operatorname{ch} V^1 = \operatorname{ch} V(\lambda) = \operatorname{ch} V^q(\lambda)$. Note that there exists a (weight preserving) surjective $U_q(\mathfrak{g})$-module homomorphism $V^q \to V^q(\lambda)$. Since the characters are the same, this must be an isomorphism. □

Corollary 3.4.8.

(1) *If V^q is a highest weight $U_q(\mathfrak{g})$-module in the category $\mathcal{O}^q_{\text{int}}$ with highest weight $\lambda \in P$, then $\lambda \in P^+$ and $V^q \cong V^q(\lambda)$.*

(2) *Every irreducible $U_q(\mathfrak{g})$-module in the category $\mathcal{O}^q_{\text{int}}$ is isomorphic to $V^q(\lambda)$ for some $\lambda \in P^+$.*

Proof. (1) Under the conditions given, V^1 is a highest weight module in category $\mathcal{O}_{\text{int}}$. Hence by Corollary 2.4.7 we have $V^1 \cong V(\lambda)$ with $\lambda \in P^+$ as $U(\mathfrak{g})$-modules with $\lambda \in P^+$. The rest follows as in the proof for Corollary 3.4.7.

(2) Let V^q be an irreducible $U_q(\mathfrak{g})$-module in the category $\mathcal{O}^q_{\text{int}}$. Since $\operatorname{wt}(V^q) \subset \bigcup_{j=1}^s D(\lambda_j)$ for some $\lambda_1, \ldots, \lambda_s \in P$, there exists a maximal vector v_λ of weight λ for some $\lambda \in P$. Then v_λ generates a highest weight $U_q(\mathfrak{g})$-module W^q with highest weight λ. By (1), we must have $\lambda \in P^+$ and $W^q \cong V^q(\lambda)$. Since V^q is irreducible, we conclude that $V^q = W^q$. □

Theorem 3.4.9. *The classical limit U_1 of $U_q(\mathfrak{g})$ inherits a Hopf algebra structure from that of $U_q(\mathfrak{g})$, and it is isomorphic to the universal enveloping algebra $U(\mathfrak{g})$ as a Hopf algebra over* **F**.

Proof. By Theorem 3.4.4 (1), there exists a surjective **F**-algebra homomorphism $\psi : U(\mathfrak{g}) \to U_1$ defined by $e_i \mapsto \bar{e}_i$, $f_i \mapsto \bar{f}_i$, and $h \mapsto \bar{h}$ for $i \in I$ and $h \in P^\vee$. Recall from Proposition 2.1.7 that $U(\mathfrak{g}) \cong U^- \otimes U^0 \otimes U^+$.

We first show that the restriction ψ_0 of ψ to U^0 is an isomorphism of U^0 onto U_1^0. The restricted map ψ_0 is certainly surjective. Choose any **Z**-basis $\{x_i\}$ of the free **Z**-lattice $P^\vee$. It is also a basis of the Cartan subalgebra $\mathfrak{h}$.

Thus any element of U^0 may be written as a polynomial in $\{x_i\}$. Suppose $g \in \ker \psi_0$. Then, for each $\lambda \in P$, we have

$$0 = \psi_0(g) \cdot \bar{v}_\lambda = \lambda(g)\bar{v}_\lambda,$$

where v_λ is a highest weight vector of a highest weight $U_q(\mathfrak{g})$-module of highest weight λ and where $\lambda(g)$ denotes the polynomial in $\lambda(x_i)$ corresponding to g. Hence, we have $\lambda(g) = 0$ for every $\lambda \in P$. Since we may take any integer value for $\lambda(x_i)$, g must be zero, which implies that ψ_0 is injective.

Next, we show that the restriction of ψ to U^-, which we denote by ψ_-, is an isomorphism of U^- onto U_1^-. Suppose $\ker \psi_- \neq 0$ and $u = \sum a_\zeta f_\zeta \in \ker \psi_-$, where $a_\zeta \in \mathbf{F}$ and f_ζ are monomials in f_i's ($i \in I$). Let N be the maximal length of the monomials f_ζ in the expression of u, and choose a dominant integral weight $\lambda \in P^+$ such that $\lambda(h_i) > N$ for all $i \in I$. If $V^q = V^q(\lambda)$ is the irreducible $U_q(\mathfrak{g})$-module of highest weight λ, then by Theorem 3.4.6, the representation V^1 is isomorphic to the irreducible $U(\mathfrak{g})$-module $V(\lambda)$ of highest weight λ. By Theorem 2.4.6 and Remark 2.4.5, the kernel of the map $\varphi : U^- \to V^1$, given by $x \mapsto \psi(x) \cdot v_\lambda$, is the left ideal of U^- generated by the elements $f_i^{\lambda(h_i)+1}$ for $i \in I$. Therefore, $u = \sum a_\zeta f_\zeta$ cannot belong to $\ker \varphi$. That is, $\psi_-(u) \cdot v_\lambda = \psi(u) \cdot v_\lambda \neq 0$, which is a contradiction. Therefore, $\ker \psi_- = 0$ and U^- is isomorphic to U_1^-.

Similarly, we have $U^+ \cong U_1^+$. Hence, by the triangular decomposition, we have the linear isomorphisms

$$U(\mathfrak{g}) \cong U^- \otimes U^0 \otimes U^+ \cong U_1^- \otimes U_1^0 \otimes U_1^+ \cong U_1,$$

where the last one follows from Proposition 3.3.3. It is easy to show that this isomorphism is actually an algebra isomorphism (Exercise 3.14).

For the Hopf algebra structure, we first show that U_1 inherits a Hopf algebra structure from that of $U_q(\mathfrak{g})$. It suffices to show that $U_{\mathbf{A}_1}$ inherits the Hopf algebra structure of $U_q(\mathfrak{g})$. This is accomplished by observing that

$$\begin{aligned} \Delta((q^h;0)_q) &= \frac{q^h \otimes q^h - 1 \otimes 1}{q-1} = (q^h;0)_q \otimes 1 + q^h \otimes (q^h;0)_q, \\ \varepsilon((q^h;0)_q) &= 0, \\ S((q^h;0)_q) &= (q^{-h};0)_q. \end{aligned} \tag{3.16}$$

Hence the maps $\Delta : U_{\mathbf{A}_1} \to U_{\mathbf{A}_1} \otimes U_{\mathbf{A}_1}$, $\varepsilon : U_{\mathbf{A}_1} \to \mathbf{A}_1$, and $S : U_{\mathbf{A}_1} \to U_{\mathbf{A}_1}$ are all well defined and U_1 inherits a Hopf algebra structure from $U_q(\mathfrak{g})$.

Let us now show that the Hopf algebra structure of $U_q(\mathfrak{g})$ coincides with that of $U(\mathfrak{g})$ under the isomorphism we have been considering. Taking the classical limit of the equations in Proposition 3.1.2 and (3.16), we have

$$\Delta(\bar{h}) = \bar{h} \otimes 1 + 1 \otimes \bar{h},$$
$$\Delta(\bar{e}_i) = \bar{e}_i \otimes 1 + 1 \otimes \bar{e}_i,$$
$$\Delta(\bar{f}_i) = \bar{f}_i \otimes 1 + 1 \otimes \bar{f}_i.$$

This coincides with the comultiplication given in (2.5). The classical limit of other maps may also be checked to coincide with the maps for $U(\mathfrak{g})$. □

Since $U^- \cong U_1^-$, it is natural to expect that the classical limit of a Verma module over $U_q(\mathfrak{g})$ is isomorphic to the Verma module over $U(\mathfrak{g})$ with the same highest weight. This is proved in the next theorem.

Theorem 3.4.10. *Let $\lambda \in P$. If V^q is the Verma module $M^q(\lambda)$ over $U_q(\mathfrak{g})$ with highest weight λ, then its classical limit V^1 is isomorphic to the Verma module $M(\lambda)$ over $U(\mathfrak{g})$ with highest weight λ.*

Proof. Let v_λ be the highest weight vector of V^q. Since $U^- \cong U_1^-$, it suffices to show that V^1 is a free U_1^--module of rank one generated by the highest weight vector $\bar{v}_\lambda$.

Recall from Proposition 3.2.2 that $V^q = M^q(\lambda)$ is a free U_q^--module of rank one generated by the highest weight vector v_λ. Noting the fact that $V_{\mathbf{A}_1}$ is a subspace of V^q and taking Proposition 3.3.5 into account, we see that $V_{\mathbf{A}_1}$ is a free $U_{\mathbf{A}_1}^-$-module generated by v_λ. Taking the classical limit, we see that $V^1 = U_1^- \cdot \bar{v}_\lambda$.

It remains to show that $V^1 \cong V_{\mathbf{A}_1}/\mathbf{J}_1 V_{\mathbf{A}_1}$ is a free U_1^--module. Suppose $\bar{u} \cdot \bar{v}_\lambda = 0$ for some $u \in U_{\mathbf{A}_1}^-$. Then $u \cdot v_\lambda \in \mathbf{J}_1 V_{\mathbf{A}_1} = \mathbf{J}_1 U_{\mathbf{A}_1}^- v_\lambda$. But since $V_{\mathbf{A}_1}$ is a free $U_{\mathbf{A}_1}^-$-module generated by v_λ, we must have $u \in \mathbf{J}_1 U_{\mathbf{A}_1}^-$, which implies $\bar{u} = 0$ in $U_1^- \cong U_{\mathbf{A}_1}^-/\mathbf{J}_1 U_{\mathbf{A}_1}^-$ (see, for example, [**18**, Lemma IV.2.10]). Therefore, V^1 is a free U_1^--module of rank one generated by the highest weight vector $\bar{v}_\lambda$. □

3.5. Complete reducibility of the category $\mathcal{O}^q_{\text{int}}$

In this subsection, we will prove the complete reducibility of $U_q(\mathfrak{g})$-modules in the category $\mathcal{O}^q_{\text{int}}$. We first define the notion of *finite dual* (or *restricted dual*) of a $U_q(\mathfrak{g})$-module. Let V be a $U_q(\mathfrak{g})$-module belonging to the category $\mathcal{O}^q$. It is graded by the weight lattice P with each weight space of finite dimension:

$$V = \bigoplus_{\mu \in P} V_\mu \quad \text{with } \dim V_\mu < \infty. \tag{3.17}$$

We define the ***finite dual*** of V to be the vector space

$$V^* = \bigoplus_{\mu} V_\mu^*, \quad \text{where } V_\mu^* = \operatorname{Hom}_{\mathbf{F}(q)}(V_\mu, \mathbf{F}(q)) \tag{3.18}$$

with the action of $U_q(\mathfrak{g})$ on V^* defined by

$$\langle x \cdot \phi, v \rangle = \langle \phi, S(x) \cdot v \rangle \tag{3.19}$$

for each $x \in U_q(\mathfrak{g})$, $\phi \in V^*$, and $v \in V$. From the property

$$x \cdot V_\mu \subset V_{\mathrm{wt}(x)+\mu},$$

we may show that $x \cdot \phi$ actually belongs to V^*. Although we will not have chances to use the real dual of highest weight modules $V(\lambda)$ or $V^q(\lambda)$, to reduce confusion, we shall write $V^*(\lambda)$ and $V^{q*}(\lambda)$ to denote their finite duals.

Since the antipode S is an anti*auto*morphism, we could have defined the dual space using S^{-1} in place of S. The dual of V thus defined will be denoted by V'.

The following lemma is an immediate consequence of the definitions.

Lemma 3.5.1. *Suppose that V is a $U_q(\mathfrak{g})$-module in the category $\mathcal{O}^q_{\mathrm{int}}$ such that* $\mathrm{wt}(V) \subset \bigcup_{j=1}^{s}(\lambda_j - Q_+)$ *for some $\lambda_j \in P$ $(j = 1, \ldots, s)$.*

(1) *There exist canonical isomorphisms $(V^*)' \cong V \cong (V')^*$.*

(2) *The space V_μ^* is a weight space of weight $-\mu$.*

(3) *The finite dual V^* is also integrable and we have*

$$\mathrm{wt}(V^*) \subset \bigcup_{j=1}^{s}(-\lambda_j + Q_+).$$

Proof. We leave it to the readers as an exercise (Exercise 3.16). □

Suppose that V is a $U_q(\mathfrak{g})$-module in the category $\mathcal{O}^q_{\mathrm{int}}$. Recalling the definition of category $\mathcal{O}^q_{\mathrm{int}}$, we may choose a *maximal* weight $\lambda \in \mathrm{wt}(V)$ with the property that $\lambda + \alpha_i$ is not a weight for any $i \in I$. Fix any $v_\lambda \in V_\lambda$ and set $L = U_q(\mathfrak{g})v_\lambda$. Then from Corollary 3.4.8 (1) we know $L \cong V^q(\lambda)$ with $\lambda \in P^+$.

Let v_λ^* denote an element in V_λ^* satisfying $v_\lambda^*(v_\lambda) = 1$, $v_\lambda^*(V_\mu) = 0$ for $\lambda \neq \mu$, and set

$$\overline{L} = U_q(\mathfrak{g})v_\lambda^* \subset V^*.$$

Lemma 3.5.2. *The $U_q(\mathfrak{g})$-module $\overline{L}$ is isomorphic to the irreducible lowest weight module $V^{q*}(\lambda)$ with lowest weight $-\lambda$ and lowest weight vector v_λ^*.*

Proof. Lemma 3.5.1 shows that $\overline{L}$ is integrable. Moreover, from the choice of λ, we know v_λ^* is a *lowest* weight vector of weight $-\lambda$. That is, it satisfies

$$f_i v_\lambda^* = 0 \quad \text{for all } i \in I,$$
$$q^h v_\lambda^* = q^{-\lambda(h)} v_\lambda.$$

Hence $\overline{L}$ is an integrable lowest weight module of lowest weight $-\lambda$. Translating the theory of the category $\mathcal{O}_{\text{int}}^q$ to the case of modules with weights bounded below, we know that it is an irreducible lowest weight module of lowest weight $-\lambda$. Since $V^{q*}(\lambda)$ is one such module, the translation of Corollary 3.4.8 (1) tells us that these two modules must be isomorphic. □

We may now single out at least one irreducible component from V.

Lemma 3.5.3. *Let V be a $U_q(\mathfrak{g})$-module in the category $\mathcal{O}_{\text{int}}^q$ and let L be the submodule of V generated by a maximal vector v_λ of weight λ. Then we have*

$$V \cong L \oplus V/L.$$

Proof. We will show that in the short exact sequence

$$0 \to L \xrightarrow{\iota} V \to V/L \to 0,$$

the map ι has a left inverse. Let us take the dual with respect to S^{-1} of the injection $\overline{L} \to V^*$ to obtain a map $(V^*)' \to (\overline{L})'$. With the help of Lemma 3.5.1, we may consider the following sequence of maps:

$$L \hookrightarrow V \xrightarrow{\varphi} (\overline{L})'.$$

We may easily check that the image of v_λ is nonzero under the composition of these maps. Using Lemma 3.5.2, we also know that both L and $(\overline{L})'$ are isomorphic to the irreducible highest weight module $V^q(\lambda)$. Hence, by Schur's Lemma, the above composition of maps must be an isomorphism. By composing the inverse of this isomorphism with the map φ, we obtain the left inverse of ι. Hence the above short exact sequence splits and we have

$$V \cong (\overline{L})' \oplus (\ker \varphi) \cong L \oplus V/L.$$

□

We may now use this lemma to show the complete reducibility theorem.

Theorem 3.5.4. *Let $U_q(\mathfrak{g})$ be the quantum group associated with the Cartan datum $(A, \Pi, \Pi^\vee, P, P^\vee)$. Then every $U_q(\mathfrak{g})$-module in the category $\mathcal{O}_{\text{int}}^q$ is isomorphic to a direct sum of irreducible highest weight modules $V^q(\lambda)$ with $\lambda \in P^+$.*

Proof. Let $F \subset V$ be a finite dimensional $U_q^{\geq 0}$-submodule and set $V_F = U_q(\mathfrak{g})F \subset V$. We may choose a maximal weight vector of $F \subset V_F$ and apply Lemma 3.5.3 to obtain

$$V_F = L \oplus L_1 \cong L \oplus V_F/L$$

for some irreducible highest weight module L with dominant integral highest weight and its complementary submodule L_1. Note that, as a $U_q(\mathfrak{g})$-module, L_1 is isomorphic to V_F/L which is generated by the $U_q^{\geq 0}$-module $F/(F \cap L)$. Since the dimension of $F/(F \cap L)$ is strictly less than that of F, using induction, we may write the submodule V_F as a direct sum of irreducible highest weight modules with dominant integral highest weights.

Now, for any $v \in V$, by definition of $\mathcal{O}_{\text{int}}^q$, the $U_q^{\geq 0}$-module $F(v) = U_q^{\geq 0} v$ is finite dimensional. Hence, using previous notation, we obtain

$$V = \sum_{v \in V} V_{F(v)},$$

where each $V_{F(v)}$ is a (direct) sum of irreducible highest weight modules with dominant integral highest weights. Thus V can be expressed as a sum of irreducible highest weight modules with dominant integral highest weights. Therefore, by the general argument for semisimplicity ([**8**, Proposition 3.12]), we can deduce that this sum is actually a direct sum, which proves our claim. □

Corollary 3.5.5. *The tensor product of a finite number of $U_q(\mathfrak{g})$-modules in the category $\mathcal{O}_{\text{int}}^q$ is completely reducible.*

Exercises

3.1. Show that the q-integers $[n]_q$ and the q-binomial coefficients $\begin{bmatrix} m \\ n \end{bmatrix}_q$ are elements of $\mathbf{Z}[q, q^{-1}]$ for all nonnegative integers $m \geq n \geq 0$.

3.2. (a) Show that the quantum adjoint operator satisfies

$$(\operatorname{ad}_q e_i)^N(e_j) = \sum_{k=0}^{N} (-1)^k q_i^{k(N+a_{ij}-1)} \begin{bmatrix} N \\ k \end{bmatrix}_{q_i} e_i^{N-k} e_j e_i^k.$$

(b) Verify that the algebra homomorphism Δ defined in Proposition 3.1.2 satisfies

$$\begin{aligned}\Delta((\mathrm{ad}_q\, e_i)^N(e_j)) &= (\mathrm{ad}_q\, e_i)^N(e_j)\otimes K_i^{-N}K_j^{-1}\\ &\quad+\sum_{k=0}^{N-1}\tau_k^{(N)}q_i^{k(N-k)}\begin{bmatrix}N\\k\end{bmatrix}_{q_i} e_i^{N-k}\otimes K_i^{-N+k}(\mathrm{ad}_q\, e_i)^k(e_j)\\ &\quad+1\otimes(\mathrm{ad}_q\, e_i)^N(e_j),\end{aligned}$$

where $\tau_k^{(N)}=\prod_{t=k}^{N-1}(1-q_i^{2(t+a_{ij})})$.

3.3. Verify that the maps Δ, ε, and S defined in Proposition 3.1.2 satisfy all the conditions for Hopf algebras.

3.4. Prove $U_q^{\geq 0}\cong U_q^0\otimes U_q^+$.

3.5. Show that, for each $i\in I$, every $U_q(\mathfrak{g})$-module V^q in the category $\mathcal{O}_{\mathrm{int}}^q$ decomposes into a direct sum of finite dimensional irreducible $U_q(\mathfrak{g}_{(i)})$-submodules, where $U_q(\mathfrak{g}_{(i)})$ is the subalgebra of $U_q(\mathfrak{g})$ generated by e_i, f_i, $K_i^{\pm 1}$.

3.6. Prove the following commutation relation for $k,l\in\mathbf{Z}_{\geq 0}$:

$$e_i^{(k)}f_i^{(l)} = \sum_{t=0}^{\min(k,l)}\frac{1}{[t]_{q_i}!}f_i^{(l-t)}\left(\prod_{s=1}^{t}[K_i;(t+s)-(k+l)]_{q_i}\right)e_i^{(k-t)}.$$

3.7. For $u\in(U_q^-)_{-\alpha}$ with $\alpha\in Q_+$, verify that

$$f_i^n u=\sum_{k=0}^{n}q_i^{(\alpha(h_i)+k)(n-k)}\begin{bmatrix}n\\k\end{bmatrix}_{q_i}\left((\mathrm{ad}_q\, f_i)^k(u)\right)f_i^{n-k}.$$

3.8. Show that the identities in the proof of Lemma 3.3.2 hold.

3.9. Verify the three commutation relations stated in the proof of Proposition 3.3.3.

3.10. (a) Show that $\mathbf{A}_1$ is a principal ideal domain. Deduce that, for each $\mu\in P$, the weight space $(V_{\mathbf{A}_1})_\mu$ is a free $\mathbf{A}_1$-module.

(b) Let $\mathbf{A}$ be an integral domain and let $\mathbf{F}$ be its field of quotients. Consider a vector space V over $\mathbf{F}$. Show that a set of vectors $\{v_1,\dots,v_n\}$ is $\mathbf{F}$-linearly independent if and only if it is $\mathbf{A}$-linearly independent.

(c) Deduce that $\mathrm{rank}_{\mathbf{A}_1}(V_{\mathbf{A}_1})_\mu=\dim_{\mathbf{F}(q)}V_\mu^q$.

3.11. Let $\mathbf{A}$ be a commutative ring with 1 and V be a free $\mathbf{A}$-module with basis $\{v_j \mid j\in J\}$. Show that, for any $\mathbf{A}$-module W, every element u of $W\otimes V$ can be expressed uniquely as

$$u=\sum_{j\in J}w_j\otimes v_j \quad \text{with } w_j\in W.$$

3.12. (a) Show that for each $\mu \in P$, a set $\{v_i\} \subset (V_{\mathbf{A}_1})_\mu$ is a basis of V_μ^q over $\mathbf{F}(q)$ if the set $\{\bar{v}_i\}$ is a basis of V_μ^1 over $\mathbf{F}$.
(b) Verify that the map $V^1 \to V_{\mathbf{A}_1}$ given by $(c(q) + \mathbf{J}_1) \otimes v \mapsto c(1)v$ is injective.
(c) View V^1 as being embedded in $V_{\mathbf{A}_1}$ and hence in V^q. For each $\mu \in P$, any basis of V_μ^1 over $\mathbf{F}$ is a basis of V_μ^q over $\mathbf{F}(q)$.

3.13. As with the modules, we may view $U^\pm \cong U_1^\pm$ as a subset of $U_q^\pm$. Define $(U_{\mathbf{A}_1}^\pm)_{\pm\mu} = U_{\mathbf{A}_1}^\pm \cap (U_q^\pm)_{\pm\mu}$ and $(U_1^\pm)_{\pm\mu} = \mathbf{F} \otimes_{\mathbf{A}_1} (U_{\mathbf{A}_1}^\pm)_{\pm\mu}$ for each $\mu \in Q_+$.
(a) For each $\mu \in Q_+$, show that a set $\{x_i\} \subset (U_{\mathbf{A}_1}^\pm)_{\pm\mu}$ is a basis of $(U_q^\pm)_{\pm\mu}$ if the set $\{\bar{x}_i\}$ is a basis of $(U_1^\pm)_\mu$.
(b) Deduce that, for each $\mu \in Q_+$, any basis of $U_{\pm\mu}^\pm$ is a basis of $(U_q^\pm)_{\pm\mu}$.

3.14. Show that the composition of linear isomorphisms

$$U(\mathfrak{g}) \cong U^- \otimes U^0 \otimes U^+ \cong U_1^- \otimes U_1^0 \otimes U_1^+ \cong U_1$$

is an isomorphism of algebras.

3.15. Let V^q be a $U_q(\mathfrak{g})$-module in the category $\mathcal{O}_{\text{int}}^q$. Show that $\dim V_{w\lambda}^q = \dim V_\lambda^q$ for all $w \in W$, $\lambda \in \text{wt}(V^q)$.

3.16. Prove the statements in Lemma 3.5.1.

Chapter 4

Crystal Bases

The *crystal basis theory* for quantum groups was introduced by Kashiwara ([**38, 39**]). Crystal bases can be viewed as bases at $q = 0$, and they have many nice combinatorial features reflecting the internal structure of integrable representations of quantum groups in the category $\mathcal{O}^q_{\text{int}}$. For instance, one of the major goals in representation theory is to find a nice expression for the characters of representations, and this goal can be achieved by finding an explicit combinatorial description of crystal bases. Furthermore, the crystal bases have extremely simple behavior with respect to the tensor product. Hence the crystal basis theory provides us with a very powerful combinatorial method for studying the structure of integrable representations of quantum groups in the category $\mathcal{O}^q_{\text{int}}$. In this chapter, we develop the crystal basis theory following the framework given in [**39**].

4.1. Kashiwara operators

The *crystal basis theory* is developed for $U_q(\mathfrak{g})$-modules in the category $\mathcal{O}^q_{\text{int}}$. Recall that the ***category*** $\mathcal{O}^q_{\text{int}}$ consists of the $U_q(\mathfrak{g})$-modules M^q satisfying the following conditions:

(1) M^q has a weight space decomposition $M^q = \bigoplus_{\lambda \in P} M^q_\lambda$ such that

$$\dim_{\mathbf{F}(q)} M^q_\lambda < \infty \quad \text{for all } \lambda \in P,$$

(2) there exist finitely many elements $\lambda_1, \dots, \lambda_s \in P$ such that

$$\text{wt}(M^q) \subset D(\lambda_1) \cup \cdots \cup D(\lambda_s),$$

where $D(\lambda) = \{\mu \in P \mid \mu \leq \lambda\}$,

(3) the operators e_i and f_i are locally nilpotent on M^q for all $i \in I$.

Thus the category $\mathcal{O}^q_{\mathrm{int}}$ consists of integrable $U_q(\mathfrak{g})$-modules in the category $\mathcal{O}^q$. Also, recall that the category $\mathcal{O}^q_{\mathrm{int}}$ is semisimple (Theorem 3.5.4) and that for each $i \in I$, every $U_q(\mathfrak{g})$-module M^q in the category $\mathcal{O}^q_{\mathrm{int}}$ is decomposed into a direct sum of finite dimensional irreducible $U_q(\mathfrak{g}_{(i)})$-modules (Proposition 3.2.4). For simplicity, we will omit the superscript q from the notation M^q whenever there is no danger of confusion.

Now we define the modified root operators $\tilde{e}_i$ and $\tilde{f}_i$ on M $(i \in I)$ called the *Kashiwara operators* which will play an essential role in the crystal basis theory. We first prove:

Lemma 4.1.1. *Let $M = \bigoplus_{\lambda \in P} M_\lambda$ be a $U_q(\mathfrak{g})$-module in the category $\mathcal{O}^q_{\mathrm{int}}$. For each $i \in I$, every weight vector $u \in M_\lambda$ $(\lambda \in \mathrm{wt}(M))$ may be written in the form*

$$u = u_0 + f_i u_1 + \cdots + f_i^{(N)} u_N, \tag{4.1}$$

where $N \in \mathbf{Z}_{\geq 0}$ and $u_k \in M_{\lambda + k\alpha_i} \cap \ker e_i$ for all $k = 0, 1, \ldots, N$.

Here, each u_k in the expression is uniquely determined by u and $u_k \neq 0$ only if $\lambda(h_i) + k \geq 0$.

Proof. The existence of the expression (4.1) follows easily from the decomposition of M into a direct sum of finite dimensional irreducible $U_q(\mathfrak{g}_{(i)})$-modules.

For uniqueness, suppose we have an expression

$$\sum_{k=0}^{N} f_i^{(k)} u_k = u_0 + f_i u_1 + \cdots + f_i^{(N)} u_N = 0, \tag{4.2}$$

where $N \in \mathbf{Z}_{\geq 0}$ and $u_k \in M_{\lambda + k\alpha_i} \cap \ker e_i$ for all $k = 0, 1, \ldots, N$. We will show that all $u_k = 0$ by induction on N.

If $N = 0$, there is nothing to prove. If $N > 0$, by applying e_i on (4.2), we get

$$\begin{aligned} 0 &= e_i(u_0 + f_i u_1 + \cdots + f_i^{(N)} u_N) \\ &= \frac{K_i - K_i^{-1}}{q_i - q_i^{-1}} u_1 + \cdots + f_i^{(N-1)} \frac{K_i q_i^{-N+1} - K_i^{-1} q_i^{N-1}}{q_i - q_i^{-1}} u_N \\ &= [\lambda(h_i) + 2]_{q_i} u_1 + \cdots + [\lambda(h_i) + N + 1]_{q_i} f_i^{(N-1)} u_N. \end{aligned}$$

By the induction hypothesis, we have

$$[\lambda(h_i) + 2]_{q_i} u_1 = 0, \ldots, [\lambda(h_i) + N + 1]_{q_i} u_N = 0.$$

Since the length of the i-string through any nonzero u_k is $\lambda(h_i) + 2k$, we have $\lambda(h_i) + 2k \geq k$, which implies $\lambda(h_i) + k + 1 > 0$. Since q is an indeterminate,

$[\lambda(h_i)+k+1]_{q_i} \neq 0$ and hence $u_k = 0$. Thus we get $u_k = 0$ for $k = 1, \ldots, N$, which implies $u_0 = 0$. This proves the uniqueness of the expression (4.1) and the condition $\lambda(h_i) + k \geq 0$ for all nonzero u_k. □

Definition 4.1.2. The ***Kashiwara operators*** $\tilde{e}_i$ and $\tilde{f}_i$ $(i \in I)$ on M are defined by

$$\tilde{e}_i u = \sum_{k=1}^{N} f_i^{(k-1)} u_k, \qquad \tilde{f}_i u = \sum_{k=0}^{N} f_i^{(k+1)} u_k. \tag{4.3}$$

The basic properties of Kashiwara operators are given in the following proposition.

Proposition 4.1.3.

(1) *Let $M = \bigoplus_{\lambda \in P} M_\lambda$ be a $U_q(\mathfrak{g})$-module in the category $\mathcal{O}_{\text{int}}^q$. Then we have*

$$\tilde{e}_i M_\lambda = e_i M_\lambda \subset M_{\lambda+\alpha_i}, \qquad \tilde{f}_i M_\lambda = f_i M_\lambda \subset M_{\lambda-\alpha_i}$$

for all $i \in I$ and $\lambda \in P$.

(2) *The Kashiwara operators $\tilde{e}_i$ and $\tilde{f}_i$ commute with $U_q(\mathfrak{g})$-module homomorphisms.*

Proof. (1) Let $u \in M_\lambda$ and consider the unique expression (4.1) of u. Then

$$\begin{aligned}\tilde{f}_i u &= \sum_{k=0}^{N} f_i^{(k+1)} u_k = \sum_{k=0}^{N} \frac{1}{[k+1]_{q_i}} f_i(f_i^{(k)} u_k) \\ &= f_i\left(\sum_{k=0}^{N} \frac{1}{[k+1]_{q_i}} f_i^{(k)} u_k\right) \in f_i M_\lambda.\end{aligned}$$

On the other hand, we have

$$\begin{aligned}f_i u &= \sum_{k=0}^{N} f_i f_i^{(k)} u_k = \sum_{k=0}^{N} [k+1]_{q_i} f_i^{(k+1)} u_k \\ &= \sum_{k=0}^{N} f_i^{(k+1)}([k+1]_{q_i} u_k) \\ &= \tilde{f}_i\left(\sum_{k=0}^{N} f_i^{(k)}([k+1]_{q_i} u_k)\right) \in \tilde{f}_i M_\lambda.\end{aligned}$$

Similarly, $\tilde{e}_i M_\lambda = e_i M_\lambda$ for all $i \in I$ and $\lambda \in P$.

(2) Let $M = \bigoplus_{\lambda \in P} M_\lambda$ and $M' = \bigoplus_{\mu \in P} M'_\mu$ be $U_q(\mathfrak{g})$-modules in the category $\mathcal{O}_{\text{int}}^q$ and let $\psi : M \to M'$ be a $U_q(\mathfrak{g})$-module homomorphism. For $i \in I$ and $u \in M_\lambda$, consider the unique expression $u = \sum_{k=0}^{N} f_i^{(k)} u_k$ with

$u_k \in \ker e_i \cap M_{\lambda+k\alpha_i}$. Since ψ preserves the weight spaces and commutes with the action of $U_q(\mathfrak{g})$, we have $\psi(u) = \sum_{k=0}^{N} f_i^{(k)} \psi(u_k)$, where $\psi(u_k) \in \ker e_i \cap M'_{\lambda+k\alpha_i}$. It follows that

$$\psi(\tilde{e}_i u) = \psi\left(\sum_{k=1}^{N} f_i^{(k-1)} u_k\right) = \sum_{k=1}^{N} f_i^{(k-1)} \psi(u_k) = \tilde{e}_i \psi(u).$$

Similarly, we get $\psi(\tilde{f}_i u) = \tilde{f}_i \psi(u)$ as desired. □

4.2. Crystal bases and crystal graphs

In this section, we define the notion of *crystal bases* for $U_q(\mathfrak{g})$-modules in the category $\mathcal{O}^q_{\text{int}}$. As a motivation for developing the crystal basis theory, we consider the following example.

Example 4.2.1. Consider the quantum group $U_q(\mathfrak{sl}_2) = \langle e, f, K^{\pm 1} \rangle$ and its two-dimensional natural representation $V = \mathbf{F}(q)v_+ \oplus \mathbf{F}(q)v_-$ with $U_q(\mathfrak{sl}_2)$-module action defined by

$$\begin{aligned} ev_+ &= 0, & fv_+ &= v_-, & Kv_+ &= qv_+, \\ ev_- &= v_+, & fv_- &= 0, & Kv_- &= q^{-1}v_-. \end{aligned}$$

Consider the tensor product $V \otimes V$. It has an obvious basis consisting of

$$v_+ \otimes v_+, \qquad v_+ \otimes v_-, \qquad v_- \otimes v_+, \qquad v_- \otimes v_-.$$

But, when $q \neq 0$, it is not compatible with the irreducible decomposition of $V \otimes V$:

$$V \otimes V \cong V(2) \oplus V(0).$$

The desired (but more complicated) basis of $V \otimes V$ is given as follows. The vectors

$$v_+ \otimes v_+, \qquad v_- \otimes v_+ + q\, v_+ \otimes v_-, \qquad v_- \otimes v_-$$

form a basis of the submodule of $V \otimes V$ which is isomorphic to $V(2)$, and the vector

$$v_+ \otimes v_- - q\, v_- \otimes v_+$$

generates the one-dimensional trivial submodule $V(0)$.

Observe that the two bases coincide when $q = 0$. Hence we *may expect* that there exist particular bases for $U_q(\mathfrak{g})$-modules in the category $\mathcal{O}^q_{\text{int}}$ that have simple and nice behavior at $q = 0$. This is one of the motivations for developing the crystal basis theory.

The *crystal bases* for $U_q(\mathfrak{g})$-modules can be viewed as bases at $q = 0$. To be more precise, consider the localization $\mathbf{A}_0$ of the polynomial ring $\mathbf{F}[q]$ at the ideal (q):

$$\begin{aligned} \mathbf{A}_0 &= \{f(q) \in \mathbf{F}(q) \mid f \text{ is regular at } q = 0\} \\ &= \{g/h \mid g, h \in \mathbf{F}[q],\ h(0) \neq 0\}. \end{aligned} \tag{4.4}$$

Thus the local ring $\mathbf{A}_0$ is a principal ideal domain with $\mathbf{F}(q)$ as its field of quotients.

Definition 4.2.2. Let M be a $U_q(\mathfrak{g})$-module in the category $\mathcal{O}^q_{\text{int}}$. A free $\mathbf{A}_0$-submodule $\mathcal{L}$ of M is called a ***crystal lattice*** if

(1) $\mathcal{L}$ generates M as a vector space over $\mathbf{F}(q)$,
(2) $\mathcal{L} = \bigoplus_{\lambda \in P} \mathcal{L}_\lambda$, where $\mathcal{L}_\lambda = \mathcal{L} \cap M_\lambda$ for all $\lambda \in P$,
(3) $\tilde{e}_i \mathcal{L} \subset \mathcal{L}$, $\tilde{f}_i \mathcal{L} \subset \mathcal{L}$ for all $i \in I$.

We remark that the first condition is equivalent to saying $\mathbf{F}(q) \otimes_{\mathbf{A}_0} \mathcal{L}_\mu \cong M_\mu$ for each $\mu \in \operatorname{wt}(M)$.

Let $\mathbf{J}_0$ be the unique maximal ideal of the local ring $\mathbf{A}_0$ generated by q. There exists an isomorphism of fields

$$\mathbf{A}_0/\mathbf{J}_0 \xrightarrow{\sim} \mathbf{F} \quad \text{given by} \quad f(q) + \mathbf{J}_0 \mapsto f(0)$$

(hence q is mapped onto 0) and we have

$$\mathbf{F} \otimes_{\mathbf{A}_0} \mathcal{L} \xrightarrow{\sim} \mathcal{L}/\mathbf{J}_0\mathcal{L} = \mathcal{L}/q\mathcal{L}.$$

The passage from $\mathcal{L}$ to the quotient $\mathcal{L}/q\mathcal{L}$ is referred to as taking the ***crystal limit***. We will denote by $\overline{v}$ the image of $v \in \mathcal{L}$ under the crystal limit. Notice that since the operators $\tilde{e}_i$ and $\tilde{f}_i$ preserve the lattice $\mathcal{L}$, they also define operators on $\mathcal{L}/q\mathcal{L}$, which we shall denote by the same symbols.

Definition 4.2.3. A ***crystal basis*** of a $U_q(\mathfrak{g})$-module M in the category $\mathcal{O}^q_{\text{int}}$ is a pair $(\mathcal{L}, \mathcal{B})$ such that

(1) $\mathcal{L}$ is a crystal lattice of M,
(2) $\mathcal{B}$ is an $\mathbf{F}$-basis of $\mathcal{L}/q\mathcal{L} \cong \mathbf{F} \otimes_{\mathbf{A}_0} \mathcal{L}$,
(3) $\mathcal{B} = \bigsqcup_{\lambda \in P} \mathcal{B}_\lambda$, where $\mathcal{B}_\lambda = \mathcal{B} \cap (\mathcal{L}_\lambda/q\mathcal{L}_\lambda)$,
(4) $\tilde{e}_i \mathcal{B} \subset \mathcal{B} \cup \{0\}$, $\tilde{f}_i \mathcal{B} \subset \mathcal{B} \cup \{0\}$ for all $i \in I$,
(5) for any $b, b' \in \mathcal{B}$ and $i \in I$, we have $\tilde{f}_i b = b'$ if and only if $b = \tilde{e}_i b'$.

Remark 4.2.4. We would like to point out that the term *crystal base* is more common in the literature.

Take $\mathcal{B}$ as the set of vertices and define the I-colored arrows on $\mathcal{B}$ by

$$b \xrightarrow{i} b' \quad \text{if and only if} \quad \tilde{f}_i b = b' \quad (i \in I).$$

Then $\mathcal{B}$ is given an I-colored oriented graph structure called the ***crystal graph*** of M.

The crystal graph $\mathcal{B}$ reflects the internal structure of M in many ways. For instance, as the following theorem shows, the *character* of M is determined by its crystal graph. Hence, to understand the combinatorial structure of a $U_q(\mathfrak{g})$-module M in the category $\mathcal{O}^q_{\text{int}}$, it often suffices to find a concrete realization of its crystal graph $\mathcal{B}$.

Theorem 4.2.5. *Let M be a $U_q(\mathfrak{g})$-module in the category $\mathcal{O}^q_{\text{int}}$ and let $(\mathcal{L}, \mathcal{B})$ be a crystal basis of M. Then, for all $\lambda \in P$, we have*

$$\dim_{\mathbf{F}(q)} M_\lambda = \operatorname{rank}_{\mathbf{A}_0} \mathcal{L}_\lambda = \#\mathcal{B}_\lambda.$$

In particular,

$$\operatorname{ch} M = \sum_{\lambda \in P} (\#\mathcal{B}_\lambda)\, e^\lambda.$$

Proof. Since $\mathbf{A}_0$ is a localization of $\mathbf{F}[q]$ at $q = 0$, it is a principal ideal domain with $\mathbf{F}(q)$ as its field of quotients. Hence by Exercise 3.10, a set of vectors in $\mathcal{L}_\lambda$ is linearly independent over $\mathbf{A}_0$ if and only if it is linearly independent over $\mathbf{F}(q)$, which implies $\dim_{\mathbf{F}(q)} M_\lambda = \operatorname{rank}_{\mathbf{A}_0} \mathcal{L}_\lambda$. Furthermore, recall that there is an $\mathbf{F}$-linear isomorphism

$$\mathbf{F} \otimes_{\mathbf{A}_0} \mathcal{L}_\lambda \xrightarrow{\sim} \mathcal{L}_\lambda / q\mathcal{L}_\lambda$$

induced by the field isomorphism

$$\mathbf{A}_0/\mathbf{J}_0 \xrightarrow{\sim} \mathbf{F}, \qquad f + \mathbf{J}_0 \mapsto f(0).$$

By Exercise 3.11, if $\{v_j \mid j = 1, \dots, p\}$ is an $\mathbf{A}_0$-basis of $\mathcal{L}_\lambda$, then $\{\overline{v_j} = 1 \otimes v_j \mid j = 1, \dots, p\}$ is an $\mathbf{F}$-basis of $\mathcal{L}_\lambda / q\mathcal{L}_\lambda$. Therefore, we have

$$\operatorname{rank}_{\mathbf{A}_0} \mathcal{L}_\lambda = \dim_{\mathbf{F}} \mathcal{L}_\lambda / q\mathcal{L}_\lambda = \#\mathcal{B}_\lambda.$$

□

Example 4.2.6. Recall that $U_q(\mathfrak{sl}_2)$ is the quantum group generated by the elements e, f, $K^{\pm 1}$ with the defining relations

$$KeK^{-1} = q^2 e, \qquad KfK^{-1} = q^{-2} f, \qquad ef - fe = \frac{K - K^{-1}}{q - q^{-1}}.$$

For $m \in \mathbf{Z}_{\geq 0}$, let $V(m)$ denote the $(m+1)$-dimensional irreducible $U_q(\mathfrak{sl}_2)$-module with basis $\{u, fu, \dots, f^{(m)}u\}$, where

$$eu = 0, \qquad Ku = q^m u,$$

$$f^{(k)} u = \frac{1}{[k]_q!} f^k u \quad (k = 0, 1, \dots, m).$$

Define

$$\mathcal{L}(m) = \bigoplus_{k=0}^{m} \mathbf{A}_0 f^{(k)}u, \qquad \mathcal{B}(m) = \{\overline{u}, \overline{fu}, \dots, \overline{f^{(m)}u}\}, \tag{4.5}$$

where $\overline{f^{(k)}u}$ denotes the image of $f^{(k)}u$ under the crystal limit. By the definition of Kashiwara operators, we have

$$\tilde{e}f^{(k)}u = f^{(k-1)}u \quad \text{and} \quad \tilde{f}f^{(k)}u = f^{(k+1)}u. \tag{4.6}$$

Hence $\mathcal{L}(m)$ is a crystal lattice of $V(m)$. Moreover, by Exercise 3.11, $\mathcal{B}(m)$ is an $\mathbf{F}$-basis of

$$\mathcal{L}(m)/q\mathcal{L}(m) \cong \mathbf{F} \otimes_{\mathbf{A}_0} \mathcal{L}(m).$$

It is now straightforward to verify that $(\mathcal{L}(m), \mathcal{B}(m))$ is a crystal basis of $V(m)$ (Exercise 4.1). The crystal graph $\mathcal{B}(m)$ is

$$\mathcal{B}(m): \quad \overline{u} \longrightarrow \overline{fu} \longrightarrow \overline{f^{(2)}u} \longrightarrow \cdots \longrightarrow \overline{f^{(m)}u}.$$

Example 4.2.7. Let $U_q(\mathfrak{sl}_n)$ be the quantum group associated with the special linear Lie algebra $\mathfrak{sl}_n(\mathbf{F})$. Thus it is an associative algebra over $\mathbf{F}(q)$ with 1 generated by the elements e_i, f_i, $K_i^{\pm 1}$ $(i = 1, \dots, n-1)$ and defining relations

$$\begin{aligned}
&K_iK_i^{-1} = K_i^{-1}K_i = 1, \qquad K_iK_j = K_jK_i,\\
&K_ie_iK_i^{-1} = q^2e_i, \qquad K_if_iK_i^{-1} = q^{-2}f_i,\\
&K_ie_jK_i^{-1} = q^{-1}e_j, \quad K_if_jK_i^{-1} = qf_j \quad \text{if } |i-j| = 1,\\
&K_ie_jK_i^{-1} = e_j, \qquad K_if_jK_i^{-1} = f_j \quad \text{if } |i-j| > 1,\\
&e_if_j - f_je_i = \delta_{ij}\frac{K_i - K_i^{-1}}{q - q^{-1}},\\
&e_i^2e_j - (q+q^{-1})e_ie_je_i + e_je_i^2 = 0 \quad \text{if } |i-j| = 1,\\
&f_i^2f_j - (q+q^{-1})f_if_jf_i + f_jf_i^2 = 0 \quad \text{if } |i-j| = 1,\\
&e_ie_j - e_je_i = 0, \quad f_if_j - f_jf_i = 0 \quad \text{if } |i-j| > 1.
\end{aligned} \tag{4.7}$$

Let $V = \bigoplus_{j=1}^{n} \mathbf{F}(q)v_j$ be the n-dimensional vector space over $\mathbf{F}(q)$ with a basis $\{v_1, \dots, v_n\}$. We define the $U_q(\mathfrak{sl}_n)$-action on V by

$$\begin{aligned}
e_iv_j &= \begin{cases} v_i & \text{if } j = i+1,\\ 0 & \text{otherwise,} \end{cases}\\
f_iv_j &= \begin{cases} v_{i+1} & \text{if } j = i,\\ 0 & \text{otherwise} \end{cases},\\
K_i^{\pm 1}v_j &= \begin{cases} q^{\pm 1}v_i & \text{if } j = i,\\ q^{\mp 1}v_{i+1} & \text{if } j = i+1,\\ v_j & \text{otherwise.} \end{cases}
\end{aligned} \tag{4.8}$$

Then V becomes an irreducible highest weight module over $U_q(\mathfrak{sl}_n)$ with highest weight ϵ_1 (see (1.10)) and highest weight vector v_1. We call V the ***vector representation*** of $U_q(\mathfrak{sl}_n)$.

Let $\mathcal{L} = \bigoplus_{j=1}^n \mathbf{A}_0 v_j$ and $\mathcal{B} = \{\overline{v_1}, \dots, \overline{v_n}\}$, where $\overline{v_j}$ denotes the image of v_j under the crystal limit. Then it is straightforward to verify that $(\mathcal{L}, \mathcal{B})$ is a crystal basis of V (Exercise 4.2). The crystal graph $\mathcal{B}$ is

$$\mathcal{B}: \quad \overline{v_1} \xrightarrow{1} \overline{v_2} \xrightarrow{2} \overline{v_3} \xrightarrow{3} \cdots \xrightarrow{n-1} \overline{v_n}.$$

Proposition 4.2.8. *Let $M = \bigoplus_{\lambda \in P} M_\lambda$ and $M' = \bigoplus_{\mu \in P} M'_\mu$ be $U_q(\mathfrak{g})$-modules in the category $\mathcal{O}^q_{\text{int}}$, and suppose that there exists a $U_q(\mathfrak{g})$-module isomorphism $\psi : M \to M'$. For any crystal basis $(\mathcal{L}, \mathcal{B})$ of M, the pair $(\psi(\mathcal{L}), \overline{\psi}(\mathcal{B}))$ is a crystal basis of M', where $\overline{\psi} : \mathcal{L}/q\mathcal{L} \to \psi(\mathcal{L})/q\psi(\mathcal{L})$ is the $\mathbf{F}$-linear isomorphism induced by ψ.*

Proof. Since the Kashiwara operators $\tilde{e}_i$ and $\tilde{f}_i$ $(i \in I)$ commute with ψ, our assertion follows immediately (Exercise 4.3). □

Definition 4.2.9. Let M be a $U_q(\mathfrak{g})$-module in the category $\mathcal{O}^q_{\text{int}}$ with crystal bases $(\mathcal{L}_j, \mathcal{B}_j)$ $(j = 1, 2)$. We say that two crystal bases $(\mathcal{L}_1, \mathcal{B}_1)$ and $(\mathcal{L}_2, \mathcal{B}_2)$ of M are ***isomorphic*** if there is an $\mathbf{A}_0$-linear isomorphism $\psi : \mathcal{L}_1 \to \mathcal{L}_2$ such that

(1) ψ commutes with all $\tilde{e}_i$ and $\tilde{f}_i$ $(i \in I)$,

(2) the induced $\mathbf{F}$-linear isomorphism $\overline{\psi} : \mathcal{L}_1/q\mathcal{L}_1 \to \mathcal{L}_2/q\mathcal{L}_2$ defines a bijection $\overline{\psi} : \mathcal{B}_1 \cup \{0\} \longrightarrow \mathcal{B}_2 \cup \{0\}$ that commutes with all $\tilde{e}_i$ and $\tilde{f}_i$ $(i \in I)$.

Theorem 4.2.10.

(1) *Let M_j be $U_q(\mathfrak{g})$-modules in the category $\mathcal{O}^q_{\text{int}}$ and let $(\mathcal{L}_j, \mathcal{B}_j)$ be crystal bases for M_j $(j = 1, 2)$. Then $(\mathcal{L} = \mathcal{L}_1 \oplus \mathcal{L}_2, \mathcal{B} = \mathcal{B}_1 \sqcup \mathcal{B}_2)$ is a crystal basis of $M_1 \oplus M_2$. The analogue for a countable direct sum of modules is also true.*

(2) *Conversely, let $M = M_1 \oplus M_2$ be a direct sum of $U_q(\mathfrak{g})$-modules in the category $\mathcal{O}^q_{\text{int}}$ and let $(\mathcal{L}, \mathcal{B})$ be a crystal basis of M. Suppose that there exist free $\mathbf{A}_0$-submodules $\mathcal{L}_j \subset M_j$ and some subsets $\mathcal{B}_j \subset \mathcal{L}_j/q\mathcal{L}_j$ $(j = 1, 2)$ such that $\mathcal{L} = \mathcal{L}_1 \oplus \mathcal{L}_2$ and $\mathcal{B} = \mathcal{B}_1 \sqcup \mathcal{B}_2$. Then $(\mathcal{L}_j, \mathcal{B}_j)$ is a crystal basis of M_j $(j = 1, 2)$.*

Proof. The first of our assertions can be proved in a straightforward manner.

For the second assertion, note that the canonical maps $\mathbf{F}(q)\otimes_{\mathbf{A}_0}(\mathcal{L}_j)_\mu \longrightarrow (M_j)_\mu$ given by

$$f(q) \otimes u_j \longmapsto f(q)u_j \quad (j = 1, 2,\ \mu \in P)$$

are linear isomorphisms; otherwise, we would have

$$\begin{aligned} M_\mu &= (M_1)_\mu \oplus (M_2)_\mu \supsetneqq \mathbf{F}(q) \otimes_{\mathbf{A}_0} (\mathcal{L}_1)_\mu \oplus \mathbf{F}(q) \otimes_{\mathbf{A}_0} (\mathcal{L}_2)_\mu \\ &= \mathbf{F}(q) \otimes_{\mathbf{A}_0} (\mathcal{L}_1 \oplus \mathcal{L}_2)_\mu = \mathbf{F}(q) \otimes_{\mathbf{A}_0} \mathcal{L}_\mu = M_\mu, \end{aligned}$$

which is a contradiction. (Here, we use the equality sign "=" to denote the canonical isomorphism.)

Moreover, it is easy to see that $\mathcal{L}_j = \mathcal{L} \cap M_j$ and $\mathcal{B}_j = \mathcal{B} \cap (\mathcal{L}_j/q\mathcal{L}_j)$. Indeed, if $x \in \mathcal{L} \cap M_1 \subset \mathcal{L} = \mathcal{L}_1 \oplus \mathcal{L}_2$, write $x = x_1 + x_2$ with $x_1 \in \mathcal{L}_1$, $x_2 \in \mathcal{L}_2$. Then $x_2 = x - x_1 \in M_1 \cap M_2 = \{0\}$. Hence $x = x_1 \in \mathcal{L}_1$, which implies $\mathcal{L}_1 = \mathcal{L} \cap M_1$. Also, if $b \in \mathcal{B} \cap (\mathcal{L}_1/q\mathcal{L}_1)$, then either $b \in \mathcal{B}_1$ or $b \in \mathcal{B}_2$. If $b \in \mathcal{B}_2$, choose $v \in \mathcal{L}_2$ such that $\overline{v} = b$. Then we would have $b = \overline{v} \in (\mathcal{L}_1/q\mathcal{L}_1) \cap (\mathcal{L}_2/q\mathcal{L}_2) = \{0\}$ (we take the intersection of their images in $\mathcal{L}/q\mathcal{L}$), which is a contradiction. Hence $b \in \mathcal{B}_1$ and $\mathcal{B}_1 = \mathcal{B} \cap (\mathcal{L}_1/q\mathcal{L}_1)$. Similarly, $\mathcal{L}_2 = \mathcal{L} \cap M_2$ and $\mathcal{B}_2 = \mathcal{B} \cap (\mathcal{L}_2/q\mathcal{L}_2)$.

Now it is straightforward to verify that $(\mathcal{L}_j, \mathcal{B}_j)$ $(j = 1, 2)$ satisfies all the axioms for the crystal bases given in Definition 4.2.3 (Exercise 4.4). □

Proposition 4.2.11. *Let $M = \bigoplus_{\lambda \in P} M_\lambda$ be a $U_q(\mathfrak{g})$-module in the category $\mathcal{O}^q_{\text{int}}$ and let $(\mathcal{L}, \mathcal{B})$ be a crystal basis of M. For $i \in I$ and $u \in \mathcal{L}_\lambda$, consider the unique expression $u = \sum_{k=0}^N f_i^{(k)} u_k$, where $u_k \in M_{\lambda + k\alpha_i} \cap \ker e_i$ and $\langle h_i, \lambda + k\alpha_i \rangle \geq k \geq 0$. Then the following statements are true.*

(1) *$u_k \in \mathcal{L}$ for all $k = 0, 1, \ldots, N$.*

(2) *If $\tilde{e}_i u \in q\mathcal{L}$, then $u_k \in q\mathcal{L}$ for all $k = 1, 2, \ldots, N$.*

(3) *If $u + q\mathcal{L} \in \mathcal{B}$, then there exists a nonnegative integer k_0 such that*
  (a) *$u_k \in q\mathcal{L}$ for all $k \neq k_0$,*
  (b) *$u_{k_0} + q\mathcal{L} \in \mathcal{B}$,*
  (c) *$u \equiv f_i^{(k_0)} u_{k_0} \pmod{q\mathcal{L}}$.*

  In particular, we have

$$\tilde{e}_i u \equiv f_i^{(k_0 - 1)} u_{k_0}, \qquad \tilde{f}_i u \equiv f_i^{(k_0+1)} u_{k_0} \qquad \pmod{q\mathcal{L}}.$$

Proof. We will prove our assertions by induction on N. For the first assertion, if $N = 0$, there is nothing to prove. If $N > 0$, then $\tilde{e}_i u = \sum_{k=1}^N f_i^{(k-1)} u_k \in \mathcal{L}$. By the induction hypothesis, $u_k \in \mathcal{L}$ for all $k = 1, 2, \ldots, N$. Moreover, $f_i^{(k)} u_k = \tilde{f}_i^k u_k \in \mathcal{L}$ for all $k = 1, 2, \ldots, N$. Hence $u_0 = u - \sum_{k=1}^N f_i^{(k)} u_k \in \mathcal{L}$.

For the second assertion, the case when $N = 0$ is trivial. If $N > 0$, consider $\tilde{e}_i u = \sum_{k=1}^{N} f_i^{(k-1)} u_k \in \mathcal{L}$. If $\tilde{e}_i u \in q\mathcal{L}$, then $\tilde{e}_i^k u \in q\mathcal{L}$ for all $k = 1, \ldots, N$. Therefore, we get

$$\begin{aligned} u_N &= \tilde{e}_i^N u \in q\mathcal{L}, \\ u_{N-1} &= \tilde{e}_i^{N-1} u - f_i u_N = \tilde{e}_i^{N-1} u - \tilde{f}_i u_N \in q\mathcal{L}, \end{aligned}$$

and so on. Hence $u_k \in q\mathcal{L}$ for all $k = 1, 2, \ldots, N$ and $u \equiv u_0 \pmod{q\mathcal{L}}$.

For the final assertion, if $N = 0$, our assertion is trivial, so assume that $N > 0$. If $\tilde{e}_i u \in q\mathcal{L}$, we just proved that $k_0 = 0$. If $\tilde{e}_i u \notin q\mathcal{L}$, then since $u + q\mathcal{L} \in \mathcal{B}$, we have $\tilde{e}_i u + q\mathcal{L} \in \mathcal{B}$. By the induction hypothesis, there exists a nonnegative integer $k_0 \geq 1$ such that

(i) $u_k \in q\mathcal{L}$ for all $k \neq k_0$,

(ii) $u_{k_0} + q\mathcal{L} \in \mathcal{B}$,

(iii) $\tilde{e}_i u \equiv f_i^{(k_0-1)} u_{k_0} \pmod{q\mathcal{L}}$.

Therefore, we obtain $u \equiv \tilde{f}_i \tilde{e}_i u \equiv f_i^{(k_0)} u_{k_0} \pmod{q\mathcal{L}}$. Moreover,

$$u_0 = u - f_i^{(k_0)} u_{k_0} - \sum_{k \neq k_0} f_i^{(k)} u_k \in q\mathcal{L},$$

which completes the proof. □

Let $M = \bigoplus_{\lambda \in P} M_\lambda$ be a $U_q(\mathfrak{g})$-module in the category $\mathcal{O}_{\text{int}}^q$ with a crystal basis $(\mathcal{L}, \mathcal{B})$. For $i \in I$ and $b \in \mathcal{B}_\lambda$ ($\lambda \in P$), we define the maps $\varepsilon_i, \varphi_i : \mathcal{B} \longrightarrow \mathbf{Z}$ by

$$\begin{aligned} \varepsilon_i(b) &= \max\{k \geq 0 \,|\, \tilde{e}_i^k b \in \mathcal{B}\}, \\ \varphi_i(b) &= \max\{k \geq 0 \,|\, \tilde{f}_i^k b \in \mathcal{B}\}. \end{aligned} \tag{4.9}$$

Then we have $\varphi_i(b) - \varepsilon_i(b) = \lambda(h_i)$. Moreover, Proposition 4.2.11 shows that the maps ε_i and φ_i satisfy the following properties:

$$\begin{aligned} &\varepsilon_i(\tilde{e}_i b) = \varepsilon_i(b) - 1, \qquad \varphi_i(\tilde{e}_i b) = \varphi_i(b) + 1 \quad \text{if } \tilde{e}_i b \in \mathcal{B}, \\ &\varepsilon_i(\tilde{f}_i b) = \varphi_i(b) + 1, \qquad \varphi_i(\tilde{f}_i b) = \varphi_i(b) - 1 \quad \text{if } \tilde{f}_i b \in \mathcal{B}. \end{aligned} \tag{4.10}$$

It is easy to see that $\tilde{e}_i \mathcal{B}_\lambda \subset \mathcal{B}_{\lambda+\alpha_i} \cup \{0\}$ and $\tilde{f}_i \mathcal{B}_\lambda \subset \mathcal{B}_{\lambda-\alpha_i} \cup \{0\}$.

These properties of crystal bases lead us to define the abstract notion of *crystals* associated with a Cartan datum (Section 4.5).

4.3. Crystal bases for $U_q(\mathfrak{sl}_2)$-modules

In this section, we will investigate the structure of crystal bases for finite dimensional modules over the quantum group $U_q(\mathfrak{sl}_2)$. Since every finite dimensional $U_q(\mathfrak{sl}_2)$-module is completely reducible, it suffices to consider the crystal bases for finite dimensional irreducible modules over $U_q(\mathfrak{sl}_2)$.

Let $m \in \mathbf{Z}_{\geq 0}$ be a nonnegative integer and let $V(m)$ be the $(m+1)$-dimensional irreducible $U_q(\mathfrak{sl}_2)$-module. As we have seen in Example 4.2.6, it has a crystal basis $(\mathcal{L}(m), \mathcal{B}(m))$, where

$$\mathcal{L}(m) = \bigoplus_{k=0}^{m} \mathbf{A}_0 f^{(k)} u, \qquad \mathcal{B}(m) = \{\overline{u}, \overline{fu}, \dots, \overline{f^{(m)}u}\},$$

and u denotes the highest weight vector of weight m.

The Kashiwara operators are given by

$$\tilde{e} f^{(k)} u = f^{(k-1)} u, \qquad \tilde{f} f^{(k)} u = f^{(k+1)} u,$$

and we have the maps $\mathrm{wt}, \varepsilon, \varphi : \mathcal{B}(m) \longrightarrow \mathbf{Z}$ defined by

$$\mathrm{wt}(\overline{f^{(k)}u}) = m - 2k, \qquad \varepsilon(\overline{f^{(k)}u}) = k, \qquad \varphi(\overline{f^{(k)}u}) = m - k.$$

Recall that the crystal graph $\mathcal{B}(m)$ looks like

$$\overline{u} \longrightarrow \overline{fu} \longrightarrow \overline{f^{(2)}u} \longrightarrow \cdots \longrightarrow \overline{f^{(m)}u}\,.$$

Therefore, by Theorem 4.2.10 (1), we have the *existence theorem for crystal bases* of finite dimensional $U_q(\mathfrak{sl}_2)$-modules.

Theorem 4.3.1. *Let M be a finite dimensional $U_q(\mathfrak{sl}_2)$-module such that $M \cong \bigoplus_{m\geq 0} V(m)^{\oplus \nu(m)}$. Then there exists a crystal basis $(\mathcal{L}, \mathcal{B})$ of M which is isomorphic to $(\bigoplus_{m\geq 0} \mathcal{L}(m)^{\oplus \nu(m)}, \bigsqcup_{m\geq 0} \mathcal{B}(m)^{\oplus \nu(m)})$, where $(\mathcal{L}(m), \mathcal{B}(m))$ denotes the crystal basis of $V(m)$ defined in Example* 4.2.6 *and*

$$\mathcal{B}(m)^{\oplus \nu(m)} = \underbrace{\mathcal{B}(m) \sqcup \cdots \sqcup \mathcal{B}(m)}_{\nu(m)}.$$

Our goal in this section is to prove the *uniqueness theorem for crystal bases* for finite dimensional $U_q(\mathfrak{sl}_2)$-modules. More precisely, we would like to prove:

Theorem 4.3.2. *Let M be a finite dimensional $U_q(\mathfrak{sl}_2)$-module with a crystal basis $(\mathcal{L}, \mathcal{B})$. If $M \cong \bigoplus_{m\geq 0} V(m)^{\oplus \nu(m)}$ $(\nu(m) \in \mathbf{Z}_{\geq 0})$, then there exists an isomorphism of crystal bases*

$$\Psi : (\mathcal{L}, \mathcal{B}) \xrightarrow{\sim} \left(\bigoplus_{m\geq 0} \mathcal{L}(m)^{\oplus \nu(m)}, \bigsqcup_{m\geq 0} \mathcal{B}(m)^{\oplus \nu(m)} \right).$$

The proof of Theorem 4.3.2 consists of several lemmas. In this section, we will use the notation $M_{(r)}$ for the *r-weight space* of M; i.e.,

$$M_{(r)} = \{v \in M \mid Kv = q^r v\} \quad (r \in \mathbf{Z}).$$

We may likewise define $\mathcal{L}_{(r)} = \mathcal{L} \cap M_{(r)}$ and $\mathcal{B}_{(r)} = \mathcal{B} \cap \left(\mathcal{L}_{(r)}/q\mathcal{L}_{(r)}\right)$.

Let us first prove:

Lemma 4.3.3. *Let $V(m)$ be the $(m+1)$-dimensional irreducible $U_q(\mathfrak{sl}_2)$-module with highest weight m and highest weight vector u of weight m, and let $(\mathcal{L}(m), \mathcal{B}(m))$ be the crystal basis of $V(m)$ defined in Example* 4.2.6.

(1) *If*

$$U(m) = \{\overline{v} \in \mathcal{L}(m)/q\mathcal{L}(m) \mid \tilde{e}\,\overline{v} = 0\},$$
$$W(m) = \{v \in V(m) \mid \tilde{e}\,v \in \mathcal{L}(m)\},$$

then

$$U(m) = \mathbf{F}\overline{u}, \qquad W(m) = \mathbf{F}(q)u + \mathcal{L}(m).$$

(2) *If $\mathcal{L}$ is a free $\mathbf{A}_0$-submodule of $V(m)$ such that $\tilde{e}\mathcal{L} \subset \mathcal{L}$, $\tilde{f}\mathcal{L} \subset \mathcal{L}$, and $\mathcal{L}_{(m)} = \mathbf{A}_0 u$, then $\mathcal{L} = \mathcal{L}(m)$.*

Proof. (1) Clearly, $\mathbf{F}\overline{u} \subset U(m)$. For the other inclusion, note that any $v \in \mathcal{L}(m)$ can be written as

$$v = a_0 u + a_1 fu + \cdots + a_m f^{(m)}u \quad \text{with } a_k \in \mathbf{A}_0 \ (k = 0, 1, \ldots, m).$$

If $\tilde{e}\,\overline{v} = 0$, then we have

$$\tilde{e}\,v = a_1 u + a_2 fu + \cdots + a_m f^{(m-1)}u \in q\mathcal{L}(m),$$

which implies all $a_k \in \mathbf{J}_0$ for $k = 1, 2, \ldots, m$. Hence $\overline{v} = \overline{a_0}\,\overline{u} \in \mathbf{F}\overline{u}$.

Similarly, it is clear that $\mathbf{F}(q)u + \mathcal{L}(m) \subset W(m)$. For any $v \in W(m)$, write

$$v = a_0 u + a_1 fu + \cdots + a_m f^{(m)}u \quad \text{with } a_k \in \mathbf{F}(q) \ (k = 0, 1, \ldots, m).$$

Since $\tilde{e}v = a_1 u + a_2 fu + \cdots + a_m f^{(m-1)}u \in \mathcal{L}(m)$, all $a_k \in \mathbf{A}_0$ for $k = 1, 2, \ldots, m$. Hence $v \in \mathbf{F}(q)u + \mathcal{L}(m)$.

(2) Since $V(m)$ has a weight space decomposition, using $\tilde{e}\mathcal{L} \subset \mathcal{L}$, we may show that $\mathcal{L}$ has a weight space decomposition. From this, we have $\mathcal{L}(m)_{(m)} = \mathbf{A}_0 u = \mathcal{L}_{(m)}$, which implies $\mathcal{L}(m) \subset \mathcal{L}$. For the other inclusion, suppose $r < m$ and let $v \in \mathcal{L}_{(r)}$. By induction, we have $\tilde{e}v \in \mathcal{L}_{(r+2)} = \mathcal{L}(m)_{(r+2)}$. It follows from (1) that $v \in \mathcal{L}(m)_{(r)}$. $\square$

We will now prove some special cases of Theorem 4.3.2, which in turn will be used to prove the general statement in Theorem 4.3.2.

Lemma 4.3.4. *Let $V(m)$ be the $(m+1)$-dimensional irreducible $U_q(\mathfrak{sl}_2)$-module with highest weight m and highest weight vector u. Then, for any crystal basis $(\mathcal{L}, \mathcal{B})$ of $V(m)$, there exists an isomorphism of crystal bases*

$$\Psi : (\mathcal{L}, \mathcal{B}) \xrightarrow{\sim} (\mathcal{L}(m), \mathcal{B}(m)).$$

Proof. Since $V(m)_{(m)} = \mathbf{F}(q)u$, we may assume that $\mathcal{L}_{(m)} = \mathbf{A}_0 u = \mathcal{L}(m)_{(m)}$ and $\mathcal{B}_{(m)} = \{\bar{u}\} = \mathcal{B}(m)_{(m)}$. Thus it suffices to show that $\mathcal{L}(m) = \mathcal{L}$, $\mathcal{B}(m) = \mathcal{B}$. By Lemma 4.3.3, we have $\mathcal{L}(m) = \mathcal{L}$. By the definition of crystal bases, it is clear that $\mathcal{B}(m) \subset \mathcal{B}$. Moreover, we have

$$\#\mathcal{B}(m) = \dim_{\mathbf{F}(q)} V(m) = \#\mathcal{B} = m+1.$$

Hence we conclude that $\mathcal{B}(m) = \mathcal{B}$. □

Lemma 4.3.4 implies that Lemma 4.3.3 would still hold even if we replaced $(\mathcal{L}(m), \mathcal{B}(m))$ with any crystal basis $(\mathcal{L}, \mathcal{B})$ of $V(m)$. More generally, we have:

Lemma 4.3.5. *Let $M \cong V(m)^{\oplus \nu}$ be a finite dimensional $U_q(\mathfrak{sl}_2)$-module and let $(\mathcal{L}, \mathcal{B}) \cong (\mathcal{L}(m)^{\oplus \nu}, \mathcal{B}(m)^{\oplus \nu})$ be the crystal basis of M given in Theorem 4.3.1.*

(1) *If*

$$U(m) = \{\overline{v} \in \mathcal{L}/q\mathcal{L} \mid \tilde{e}\,\overline{v} = 0\},$$
$$W(m) = \{v \in M \mid \tilde{e}\, v \in \mathcal{L}\},$$

then

$$U(m) = \mathcal{L}_{(m)}/q\mathcal{L}_{(m)}, \qquad W(m) = M_{(m)} + \mathcal{L}.$$

(2) *If $\mathcal{L}'$ is a free $\mathbf{A}_0$-submodule of M such that $\tilde{e}\mathcal{L}' \subset \mathcal{L}'$, $\tilde{f}\mathcal{L}' \subset \mathcal{L}'$ and $\mathcal{L}'_{(m)} = \mathcal{L}_{(m)}$, then $\mathcal{L}' = \mathcal{L}$.*

Proof. The proof will be left to the readers as an exercise (Exercise 4.5). □

Lemma 4.3.6. *Let $M \cong V(m)^{\oplus \nu}$ be a finite dimensional $U_q(\mathfrak{sl}_2)$-module. Then, for any crystal basis $(\mathcal{L}, \mathcal{B})$ of M, there exists an isomorphism of crystal bases*

$$\Psi : (\mathcal{L}, \mathcal{B}) \xrightarrow{\sim} \left(\mathcal{L}(m)^{\oplus \nu}, \mathcal{B}(m)^{\oplus \nu}\right).$$

Proof. Recall that M has the weight space decomposition

$$M = \bigoplus_{r=-m}^{m} M_{(r)} = \bigoplus_{k=0}^{m} M_{(m-2k)},$$

where $M_{(m-2k)} = \{v \in M \mid Kv = q^{m-2k}v\}$, and that

$$\dim_{\mathbf{F}(q)} M_{(r)} = \operatorname{rank}_{\mathbf{A}_0} \mathcal{L}_{(r)} = \#\mathcal{B}_{(r)} = \nu \quad (-m \le r \le m).$$

Let $\mathcal{B}_{(m)} = \{b_1, \ldots, b_\nu\}$ and choose $v_j \in \mathcal{L}_{(m)}$ such that $\overline{v_j} = b_j$ ($j = 1, \ldots, \nu$). By Nakayama's Lemma, the vectors $v_1, \ldots, v_\nu$ form an $\mathbf{A}_0$-basis of $\mathcal{L}_{(m)}$.

Let $M_j = U_q(\mathfrak{sl}_2)v_j \cong V(m)$ ($j = 1, 2, \ldots, \nu$), $\mathcal{L}_j = \bigoplus_{k=0}^{m} \mathbf{A}_0 f^{(k)} v_j$ and

$$\mathcal{B}_j = \{\overline{f^{(k)} v_j} \mid k = 0, 1, \ldots, m\}.$$

Then $M = M_1 \oplus \cdots \oplus M_\nu$ and for each $j = 1, \ldots, \nu$, $(\mathcal{L}_j, \mathcal{B}_j)$ is a crystal basis of M_j which is isomorphic to $(\mathcal{L}(m), \mathcal{B}(m))$. Hence it suffices to show that

$$\mathcal{L} = \mathcal{L}_1 \oplus \cdots \oplus \mathcal{L}_\nu \quad \text{and} \quad \mathcal{B} = \mathcal{B}_1 \sqcup \cdots \sqcup \mathcal{B}_\nu.$$

Clearly,

$$\mathcal{L}_1 \oplus \cdots \oplus \mathcal{L}_\nu \subset \mathcal{L} \quad \text{and} \quad \mathcal{B}_1 \sqcup \cdots \sqcup \mathcal{B}_\nu \subset \mathcal{B}.$$

Hence Lemma 4.3.5 implies $\mathcal{L}_1 \oplus \cdots \oplus \mathcal{L}_\nu = \mathcal{L}$. Moreover,

$$\#(\mathcal{B}_1 \sqcup \cdots \sqcup \mathcal{B}_\nu)_{(r)} = \nu = \dim_{\mathbf{F}(q)} M_{(r)} = \#\mathcal{B}_{(r)}$$

for all $r = -m, -m+2, \ldots, m-2, m$. It follows that

$$\mathcal{B}_1 \sqcup \cdots \sqcup \mathcal{B}_\nu = \mathcal{B}.$$

This completes the proof. □

Proof of Theorem 4.3.2. Let $M \cong \bigoplus_{m \geq 0} V(m)^{\oplus \nu(m)}$ and let $(\mathcal{L}, \mathcal{B})$ be a crystal basis of M. Choose a maximal weight m_0 of M and set $V = U_q(\mathfrak{sl}_2) M_{(m_0)}$. Let $\{v_1, \ldots, v_\nu\}$ be an $\mathbf{A}_0$-basis of $\mathcal{L}_{(m_0)}$, and let $V_j = U_q(\mathfrak{sl}_2) v_j \cong V(m_0)$. Then we have

$$V = V_1 \oplus \cdots \oplus V_\nu \cong V(m_0)^{\oplus \nu},$$

where

$$\nu = \nu(m_0) = \dim_{\mathbf{F}(q)} M_{(m_0)} = \operatorname{rank}_{\mathbf{A}_0} \mathcal{L}_{(m_0)} = \#\mathcal{B}_{(m_0)}.$$

Let $(\mathcal{L}_j, \mathcal{B}_j) \cong (\mathcal{L}(m_0), \mathcal{B}(m_0))$ be the crystal basis of V_j defined in Example 4.2.6 using the highest weight vector v_j ($j = 1, \ldots, \nu$). Then by Theorem 4.2.10, $(\mathcal{L}_1 \oplus \cdots \oplus \mathcal{L}_\nu, \mathcal{B}_1 \sqcup \cdots \sqcup \mathcal{B}_\nu)$ is a crystal basis of V.

Let W be the complementary submodule of V in M and set

$$\mathcal{L}_V = \mathcal{L} \cap V, \qquad \mathcal{B}_V = \mathcal{B} \cap (\mathcal{L}_V / q\mathcal{L}_V),$$
$$\mathcal{L}_W = \mathcal{L} \cap W, \qquad \mathcal{B}_W = \mathcal{B} \cap (\mathcal{L}_W / q\mathcal{L}_W).$$

We will show that $\mathcal{L} = \mathcal{L}_V \oplus \mathcal{L}_W$ and $\mathcal{B} = \mathcal{B}_V \sqcup \mathcal{B}_W$. Then, Theorem 4.2.10 would imply that $(\mathcal{L}_V, \mathcal{B}_V)$ is a crystal basis of V and $(\mathcal{L}_W, \mathcal{B}_W)$ is a crystal basis of W. We have seen in Lemma 4.3.6 that there exists an isomorphism of crystal bases $(\mathcal{L}_V, \mathcal{B}_V) \cong (\mathcal{L}(m_0)^{\oplus \nu}, \mathcal{B}(m_0)^{\oplus \nu})$. Now the uniqueness of crystal basis will follow by repeating the above procedure on W.

We will prove our assertion by induction. First, note that $(\mathcal{L}_V)_{(m_0)} = \mathcal{L}_{(m_0)}$ and $(\mathcal{L}_W)_{(m_0)} = \{0\}$. Hence $\mathcal{L}_{(m_0)} = (\mathcal{L}_V \oplus \mathcal{L}_W)_{(m_0)}$. Moreover, since $\tilde{e}\mathcal{L}_V \subset \mathcal{L}_V$ and $\tilde{f}\mathcal{L}_V \subset \mathcal{L}_V$, Lemma 4.3.3 yields

$$\mathcal{L}_V = \mathcal{L}_1 \oplus \cdots \oplus \mathcal{L}_\nu \cong \mathcal{L}(m_0)^{\oplus \nu}.$$

Suppose $r < m_0$ and let $u \in \mathcal{L}_{(r)}$. Write $u = v + w$ with $v \in V_{(r)}$, $w \in W_{(r)}$. We would like to show that $v, w \in \mathcal{L}$. Since $V = V_1 \oplus \cdots \oplus V_\nu$, every $v \in V$ can be written uniquely as $v = x_1 + \cdots + x_\nu$ with $x_j \in V_j$ $(j = 1, 2, \ldots, \nu)$. By induction, we have

$$\tilde{e}u = \tilde{e}v + \tilde{e}w \in \mathcal{L}_{(r+2)} = (\mathcal{L}_V)_{(r+2)} \oplus (\mathcal{L}_W)_{(r+2)}.$$

Since the $U_q(\mathfrak{sl}_2)$-submodules are closed under the Kashiwara operators $\tilde{e}$ and $\tilde{f}$, we have $\tilde{e}v \in V_{(r+2)}$, $\tilde{e}w \in W_{(r+2)}$, which implies $\tilde{e}v \in (\mathcal{L}_V)_{(r+2)}$, $\tilde{e}w \in (\mathcal{L}_W)_{(r+2)}$. It follows that

$$\tilde{e}v = \tilde{e}x_1 + \cdots + \tilde{e}x_\nu \in (\mathcal{L}_V)_{(r+2)} = (\mathcal{L}_1)_{(r+2)} \oplus \cdots \oplus (\mathcal{L}_\nu)_{(r+2)},$$

which implies $\tilde{e}x_j \in (\mathcal{L}_j)_{(r+2)}$ $(j = 1, 2, \ldots, \nu)$. Hence by Lemma 4.3.5, we get

$$v = x_1 + \cdots + x_\nu \in (\mathcal{L}_1)_{(r)} \oplus \cdots \oplus (\mathcal{L}_\nu)_{(r)} = (\mathcal{L}_V)_{(r)}.$$

Therefore $w = u - v \in \mathcal{L} \cap W = \mathcal{L}_W$, as desired.

It remains to show that $\mathcal{B} = \mathcal{B}_V \sqcup \mathcal{B}_W$. Let $b \in \mathcal{B}$ and $\mathrm{wt}(b) = r$. If $r = m_0$, then since $\mathcal{B}_{(m_0)} = (\mathcal{B}_V)_{(m_0)}$ and $(\mathcal{B}_W)_{(m_0)} = \emptyset$, we are done.

Suppose $r < m_0$. If $b' = \tilde{e}b \neq 0$, then $b' \in \mathcal{B}_{(r+2)}$. By induction, $b' \in \mathcal{B}_V$ or $b' \in \mathcal{B}_W$. Hence $b = \tilde{f}b' \in \mathcal{B}_V$ or $b = \tilde{f}b' \in \mathcal{B}_W$.

If $b' = \tilde{e}b = 0$, then write $b = b_1 + b_2$ with $b_1 \in \mathcal{L}_V/q\mathcal{L}_V$, $b_2 \in \mathcal{L}_W/q\mathcal{L}_W$. Since $\tilde{e}b = 0$, we have $\tilde{e}b_1 = \tilde{e}b_2 = 0$. Since $\mathcal{L}_V = \mathcal{L}_1 \oplus \cdots \oplus \mathcal{L}_\nu$, Lemma 4.3.5 implies $b_1 = 0$. Hence $b = b_2 \in \mathcal{B} \cap (\mathcal{L}_W/q\mathcal{L}_W) = \mathcal{B}_W$, as desired. □

4.4. Tensor product rule

In this section, we will develop one of the nicest combinatorial features of crystal bases—the ***tensor product rule***—which is described in the following theorem.

Theorem 4.4.1. *Let M_j be a $U_q(\mathfrak{g})$-module in the category $\mathcal{O}^q_{\mathrm{int}}$ and let $(\mathcal{L}_j, \mathcal{B}_j)$ be a crystal basis of M_j $(j = 1, 2)$. Set $\mathcal{L} = \mathcal{L}_1 \otimes_{\mathbf{A}_0} \mathcal{L}_2$ and $\mathcal{B} = \mathcal{B}_1 \times \mathcal{B}_2$.*

Then $(\mathcal{L}, \mathcal{B})$ *is a crystal basis of* $M_1 \otimes_{\mathbf{F}(q)} M_2$*, where the action of Kashiwara operators* $\tilde{e}_i$ *and* $\tilde{f}_i$ *on* $\mathcal{B}$ $(i \in I)$ *is given by*

$$(4.11)\qquad \begin{aligned} \tilde{e}_i(b_1 \otimes b_2) &= \begin{cases} \tilde{e}_i b_1 \otimes b_2 & \text{if } \varphi_i(b_1) \geq \varepsilon_i(b_2), \\ b_1 \otimes \tilde{e}_i b_2 & \text{if } \varphi_i(b_1) < \varepsilon_i(b_2), \end{cases} \\ \tilde{f}_i(b_1 \otimes b_2) &= \begin{cases} \tilde{f}_i b_1 \otimes b_2 & \text{if } \varphi_i(b_1) > \varepsilon_i(b_2), \\ b_1 \otimes \tilde{f}_i b_2 & \text{if } \varphi_i(b_1) \leq \varepsilon_i(b_2). \end{cases} \end{aligned}$$

Therefore, we have

$$\begin{aligned} \mathrm{wt}(b_1 \otimes b_2) &= \mathrm{wt}(b_1) + \mathrm{wt}(b_2), \\ \varepsilon_i(b_1 \otimes b_2) &= \max(\varepsilon_i(b_1), \varepsilon_i(b_2) - \langle h_i, \mathrm{wt}(b_1) \rangle), \\ \varphi_i(b_1 \otimes b_2) &= \max(\varphi_i(b_2), \varphi_i(b_1) + \langle h_i, \mathrm{wt}(b_2) \rangle). \end{aligned}$$

Here, we write $b_1 \otimes b_2$ *for* $(b_1, b_2) \in \mathcal{B}_1 \times \mathcal{B}_2$ *and we understand* $b_1 \otimes 0 = 0 \otimes b_2 = 0$.

The crystal graph $\mathcal{B}_1 \times \mathcal{B}_2$ of $M_1 \otimes M_2$ with the Kashiwara operators defined by (4.11) will be denoted by $\mathcal{B}_1 \otimes \mathcal{B}_2$. The tensor product rule can be interpreted as follows: given a vector in $\mathcal{B}_1 \otimes \mathcal{B}_2$, reading the column first and the row next, we draw the arrow to the right as far as it can go and then go down.

Before giving a proof of Theorem 4.4.1, we illustrate how one can apply the tensor product rule (4.11).

Example 4.4.2.

(1) Let $V(2)$ be the three-dimensional irreducible $U_q(\mathfrak{sl}_2)$-module and let $(\mathcal{L}(2), \mathcal{B}(2))$ be the crystal basis of $V(2)$. Then the crystal graph $\mathcal{B}(2) \otimes \mathcal{B}(2)$ is given as follows.

$$\begin{array}{ccccc} \overline{u} \otimes \overline{u} & \longrightarrow & \overline{fu} \otimes \overline{u} & \longrightarrow & \overline{f^{(2)}u} \otimes \overline{u} \\ & & & & \downarrow \\ \overline{u} \otimes \overline{fu} & \longrightarrow & \overline{fu} \otimes \overline{fu} & & \overline{f^{(2)}u} \otimes \overline{fu} \\ & & \downarrow & & \downarrow \\ \overline{u} \otimes \overline{f^{(2)}u} & & \overline{fu} \otimes \overline{f^{(2)}u} & & \overline{f^{(2)}u} \otimes \overline{f^{(2)}u} \end{array}$$

We read the column first and then read the row; for example, the vector in the second row and in the third column is understood as $\overline{f^{(2)}u} \otimes \overline{fu}$.

Note that the decomposition of the crystal graph $\mathcal{B}(2) \otimes \mathcal{B}(2)$ into a disjoint union of its connected components coincides with the decomposition of the $U_q(\mathfrak{sl}_2)$-module $V(2) \otimes V(2)$ into a direct sum of its irreducible components.

$$V(2) \otimes V(2) \cong V(4) \oplus V(2) \oplus V(0).$$

(2) Let V be the three-dimensional vector representation of the quantum group $U_q(\mathfrak{sl}_3)$ and let $(\mathcal{L}, \mathcal{B})$ be its crystal basis defined in Example 4.2.7. Then the crystal graph $\mathcal{B} \otimes \mathcal{B}$ is given as follows.

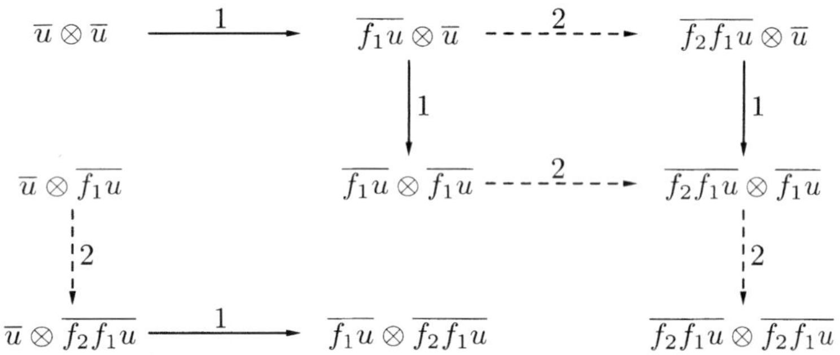

Again, note that the decomposition of the crystal graph $\mathcal{B} \otimes \mathcal{B}$ into a disjoint union of its connected components coincides with the decomposition of the $U_q(\mathfrak{sl}_3)$-module $V \otimes V$ into a direct sum of irreducible components.

$$V \otimes V \cong S^2(V) \oplus \Lambda^2(V) \cong V(2\epsilon_1) \oplus V(\epsilon_1 + \epsilon_2),$$

where $V(\lambda)$ denotes the irreducible highest weight $U_q(\mathfrak{sl}_3)$-module with highest weight λ.

We first prove the $U_q(\mathfrak{sl}_2)$-version of Theorem 4.4.1.

Theorem 4.4.3. *Let $V(a)$ (respectively, $V(b)$) be the $(a+1)$-dimensional (respectively, $(b+1)$-dimensional) irreducible $U_q(\mathfrak{sl}_2)$-module with highest weight a (respectively, b) and highest weight vector u (respectively, v). Let $(\mathcal{L}(a), \mathcal{B}(a))$ (respectively, $(\mathcal{L}(b), \mathcal{B}(b))$) be the crystal basis of $V(a)$ (respectively, $V(b)$) given in Example 4.2.6. Set $\mathcal{L} = \mathcal{L}(a) \otimes_{\mathbf{A}_0} \mathcal{L}(b)$ and $\mathcal{B} = \mathcal{B}(a) \times \mathcal{B}(b) = \mathcal{B}(a) \otimes \mathcal{B}(b)$. Then $(\mathcal{L}, \mathcal{B})$ is a crystal basis of $V(a) \otimes_{\mathbf{F}(q)} V(b)$, where the action of Kashiwara operators $\tilde{e}$ and $\tilde{f}$ on $\mathcal{B}(a) \otimes \mathcal{B}(b)$ is given by*

$$\tag{4.12} \begin{aligned} \tilde{e}(\overline{f^{(k)}u} \otimes \overline{f^{(l)}v}) &= \begin{cases} \overline{f^{(k-1)}u} \otimes \overline{f^{(l)}v} & \text{if } k+l \leq a, \\ \overline{f^{(k)}u} \otimes \overline{f^{(l-1)}v} & \text{if } k+l > a, \end{cases} \\ \tilde{f}(\overline{f^{(k)}u} \otimes \overline{f^{(l)}v}) &= \begin{cases} \overline{f^{(k+1)}u} \otimes \overline{f^{(l)}v} & \text{if } k+l < a, \\ \overline{f^{(k)}u} \otimes \overline{f^{(l+1)}v} & \text{if } k+l \geq a. \end{cases} \end{aligned}$$

Therefore, we have

$$\begin{aligned}
\mathrm{wt}(\overline{f^{(k)}u} \otimes \overline{f^{(l)}v}) &= a + b - 2k - 2l, \\
\varepsilon(\overline{f^{(k)}u} \otimes \overline{f^{(l)}v}) &= k + \max(k + l - a, 0), \\
\varphi(\overline{f^{(k)}u} \otimes \overline{f^{(l)}v}) &= b - l + \max(a - k - l, 0).
\end{aligned}$$

Proof. We will prove the theorem by induction on $b > 0$. Suppose $b = 1$. The maximal vectors in $V(a) \otimes V(1)$ are

$$E_{a+1} = u \otimes v, \qquad E_{a-1} = u \otimes fv - q^a \frac{q^2 - 1}{q^{2a} - 1} fu \otimes v.$$

Let W_1 and W_2 be the $U_q(\mathfrak{sl}_2)$-submodules of $V(a) \otimes V(1)$ generated by E_{a+1} and E_{a-1}, respectively. Then

$$V(a) \otimes V(1) = W_1 \oplus W_2 \cong V(a+1) \oplus V(a-1),$$

and W_1 (respectively, W_2) has a basis $\{E_{a+1}^{(r)} \,|\, 0 \le r \le a+1\}$ (respectively, $\{E_{a-1}^{(s)} \,|\, 0 \le s \le a-1\}$), where

$$\begin{aligned}
E_{a+1}^{(r)} &= f^{(r)}(E_{a+1}) \\
&= f^{(r)}u \otimes v + q^{a+1-r} f^{(r-1)}u \otimes fv, \\
E_{a-1}^{(s)} &= f^{(s)}(E_{a-1}) \qquad (4.13)\\
&= \frac{q^{2(a-s)} - 1}{q^{2a} - 1} f^{(s)}u \otimes fv - q^{a-s} \frac{q^{2(s+1)} - 1}{q^{2a} - 1} f^{(s+1)}u \otimes v
\end{aligned}$$

for $0 \le r \le a+1, \;\; 0 \le s \le a-1$.

Set $\mathcal{L}_1 = \bigoplus_{r=0}^{a+1} \mathbf{A}_0 E_{a+1}^{(r)}$ and $\mathcal{L}_2 = \bigoplus_{s=0}^{a-1} \mathbf{A}_0 E_{a-1}^{(s)}$. One can verify that every vector of the form $f^{(k)}u \otimes f^{(l)}v$ can be expressed as an $\mathbf{A}_0$-linear combination of $E_{a+1}^{(r)}$ and $E_{a-1}^{(s)}$ ($0 \le r \le a+1$, $0 \le s \le a-1$) (Exercise 4.6). Hence

$$\mathcal{L}_1 \oplus \mathcal{L}_2 = \mathcal{L}(a) \otimes_{\mathbf{A}_0} \mathcal{L}(1) = \mathcal{L}$$

and

$$\mathcal{L}_1/q\mathcal{L}_1 = \mathcal{L}_1/\left(q\mathcal{L} \cap W_1\right), \qquad \mathcal{L}_2/q\mathcal{L}_2 = \mathcal{L}_2/\left(q\mathcal{L} \cap W_2\right).$$

Let

$$\begin{aligned}
\mathcal{B}_1 &= \{\overline{E_{a+1}^{(r)}} \,|\, 0 \le r \le a+1\} \subset \mathcal{L}_1/q\mathcal{L}_1 = \mathcal{L}_1/\left(q\mathcal{L} \cap W_1\right), \\
\mathcal{B}_2 &= \{\overline{E_{a-1}^{(s)}} \,|\, 0 \le s \le a-1\} \subset \mathcal{L}_2/q\mathcal{L}_2 = \mathcal{L}_2/\left(q\mathcal{L} \cap W_2\right).
\end{aligned}$$

Since

$$\begin{aligned}
\overline{E_{a+1}^{(r)}} &= \begin{cases} \overline{f^{(r)}u} \otimes \overline{v} & (0 \le r \le a), \\ \overline{f^{(a)}u} \otimes \overline{fv} & (r = a+1), \end{cases} \\
\overline{E_{a-1}^{(s)}} &= \overline{f^{(s)}u} \otimes \overline{fv} \qquad (0 \le s \le a-1),
\end{aligned}$$

we have

$$\mathcal{B}_1 \sqcup \mathcal{B}_2 = \{\overline{f^{(k)}u} \otimes \overline{v}, \overline{f^{(k)}u} \otimes \overline{fv} \,|\, 0 \le k \le a\} = \mathcal{B}(a) \otimes \mathcal{B}(1).$$

Moreover, $(\mathcal{L}_1, \mathcal{B}_1) \cong (\mathcal{L}(a+1), \mathcal{B}(a+1))$ is a crystal basis for W_1. And $(\mathcal{L}_2, \mathcal{B}_2) \cong (\mathcal{L}(a-1), \mathcal{B}(a-1))$ is a crystal basis for W_2. Therefore, by Theorem 4.2.10, $(\mathcal{L}, \mathcal{B})$ is a crystal basis of $V(a) \otimes V(1)$.

It remains to verify that the tensor product rule (4.12) holds in this case. Indeed, if $s = 0$, $r < a$, then

$$\tilde{f}(\overline{f^{(r)}u} \otimes \overline{v}) = \tilde{f}\left(\overline{E^{(r)}_{a+1}}\right) = \overline{E^{(r+1)}_{a+1}} = \overline{f^{(r+1)}u} \otimes \overline{v};$$

if $s = 0$, $r = a$, then

$$\tilde{f}(\overline{f^{(a)}u} \otimes \overline{v}) = \tilde{f}\left(\overline{E^{(a)}_{a+1}}\right) = \overline{E^{(a+1)}_{a+1}} = \overline{f^{(a)}u} \otimes \overline{fv};$$

if $s = 1$, $r < a-1$, then

$$\tilde{f}(\overline{f^{(r)}u} \otimes \overline{fv}) = \tilde{f}\left(\overline{E^{(r)}_{a-1}}\right) = \overline{E^{(r+1)}_{a-1}} = \overline{f^{(r+1)}u} \otimes \overline{fv};$$

if $s = 1$, $r = a-1$, then

$$\tilde{f}(\overline{f^{(a-1)}u} \otimes \overline{fv}) = \tilde{f}\left(\overline{E^{(a-1)}_{a-1}}\right) = 0 = \overline{f^{(a-1)}u} \otimes \overline{f^{(2)}v};$$

if $s = 1$, $r = a$, then

$$\tilde{f}(\overline{f^{(a)}u} \otimes \overline{fv}) = \tilde{f}\left(\overline{E^{(a+1)}_{a+1}}\right) = 0 = \overline{f^{(a)}u} \otimes \overline{f^{(2)}v},$$

which proves (4.12) for $\tilde{f}$.

Similarly, we can verify the tensor product rule for $\tilde{e}$.

By Theorem 4.2.10, it can be easily seen that we actually proved the following: if $(\mathcal{L}, \mathcal{B})$ is a crystal basis of a finite dimensional $U_q(\mathfrak{sl}_2)$-module M, then $(\mathcal{L} \otimes_{\mathbf{A}_0} \mathcal{L}(1), \mathcal{B} \otimes \mathcal{B}(1))$ is a crystal basis of $M \otimes V(1)$ and the tensor product rule holds for the crystal graph $\mathcal{B} \otimes \mathcal{B}(1)$.

Now assume that $b > 1$ and that our theorem is true for $V(a) \otimes V(b-1)$. By the induction hypothesis, $(\mathcal{L}(a) \otimes_{\mathbf{A}_0} \mathcal{L}(b-1), \mathcal{B}(a) \otimes \mathcal{B}(b-1))$ is a crystal basis of $V(a) \otimes V(b-1)$ and the tensor product rule holds for $\mathcal{B}(a) \otimes \mathcal{B}(b-1)$. Hence by the preceding paragraph,

$$(\mathcal{L}(a) \otimes_{\mathbf{A}_0} \mathcal{L}(b-1) \otimes_{\mathbf{A}_0} \mathcal{L}(1), \mathcal{B}(a) \otimes \mathcal{B}(b-1) \otimes B(1))$$

is a crystal basis of $V(a) \otimes V(b-1) \otimes V(1)$ and the tensor product rule holds for the crystal graph $\mathcal{B}(a) \otimes \mathcal{B}(b-1) \otimes \mathcal{B}(1)$. Also, we have seen that

$(\mathcal{L}(b-1)\otimes_{\mathbf{A}_0}\mathcal{L}(1), \mathcal{B}(b-1)\otimes\mathcal{B}(1))$ is a crystal basis of $V(b-1)\otimes V(1)$ and that

$$\mathcal{L}(b-1)\otimes_{\mathbf{A}_0}\mathcal{L}(1)\cong\mathcal{L}(b)\oplus\mathcal{L}(b-2),$$
$$\mathcal{B}(b-1)\otimes\mathcal{B}(1)\cong\mathcal{B}(b)\sqcup\mathcal{B}(b-2).$$

Moreover observe that

$$V(a)\otimes V(b-1)\otimes V(1)\cong(V(a)\otimes V(b))\oplus(V(a)\otimes V(b-2)),$$
$$\mathcal{L}(a)\otimes\mathcal{L}(b-1)\otimes\mathcal{L}(1)\cong(\mathcal{L}(a)\otimes\mathcal{L}(b))\oplus(\mathcal{L}(a)\otimes\mathcal{L}(b-2)),$$
$$\mathcal{B}(a)\otimes\mathcal{B}(b-1)\otimes\mathcal{B}(1)\cong(\mathcal{B}(a)\otimes\mathcal{B}(b))\sqcup(\mathcal{B}(a)\otimes\mathcal{B}(b-2)).$$

Therefore, by Theorem 4.2.10, $(\mathcal{L}(a)\otimes\mathcal{L}(b), \mathcal{B}(a)\otimes\mathcal{B}(b))$ is a crystal basis of $V(a)\otimes V(b)$ and $(\mathcal{L}(a)\otimes\mathcal{L}(b-2), \mathcal{B}(a)\otimes\mathcal{B}(b-2))$ is a crystal basis of $V(a)\otimes V(b-2)$.

To complete the proof, it remains to prove that the tensor product rule (4.12) holds for $b>1$ on $\mathcal{B}(a)\otimes\mathcal{B}(b)$. Consider the embedding of crystal graphs

$$\begin{aligned}\Psi:\mathcal{B}(a)\otimes\mathcal{B}(b)&\hookrightarrow\mathcal{B}(a)\otimes\mathcal{B}(b-1)\otimes\mathcal{B}(1)\\&\xrightarrow{\sim}(\mathcal{B}(a)\otimes\mathcal{B}(b))\sqcup(\mathcal{B}(a)\otimes\mathcal{B}(b-2))\end{aligned}$$

defined by

$$\begin{aligned}&\overline{f^{(k)}u}\otimes\overline{f^{(l)}v}\longmapsto\overline{f^{(k)}u}\otimes\overline{f^{(l)}w}\otimes\overline{z}\quad(0\le l\le b-1),\\&\overline{f^{(k)}u}\otimes\overline{f^{(b)}v}\longmapsto\overline{f^{(k)}u}\otimes\overline{f^{(b-1)}w}\otimes\overline{fz},\end{aligned}$$

where u, v, w, z are highest weight vectors of $U_q(\mathfrak{sl}_2)$-modules $V(a)$, $V(b)$, $V(b-1)$, and $V(1)$, respectively. By the induction hypothesis, the tensor product rule holds for $\mathcal{B}(a)\otimes\mathcal{B}(b-1)$ and $(\mathcal{B}(a)\otimes\mathcal{B}(b-1))\otimes\mathcal{B}(1)$. Since Ψ is an embedding of crystal graphs, it suffices to verify that the tensor product rule holds for the image of $\mathcal{B}(a)\otimes\mathcal{B}(b)$ in $\mathcal{B}(a)\otimes\mathcal{B}(b-1)\otimes\mathcal{B}(1)$.

If $0\le l<b-1$, then since

$$\varphi(\overline{f^{(k)}u}\otimes\overline{f^{(l)}w})=(b-1-l)+\max(a-k-l,0)>0=\varepsilon(\overline{z}),$$

we have

$$\begin{aligned}\tilde{f}(\overline{f^{(k)}u}\otimes\overline{f^{(l)}v})&\overset{\Psi}{\longmapsto}\tilde{f}(\overline{f^{(k)}u}\otimes\overline{f^{(l)}w}\otimes\overline{z})=\tilde{f}(\overline{f^{(k)}u}\otimes\overline{f^{(l)}w})\otimes\overline{z}\\&=\begin{cases}\overline{f^{(k+1)}u}\otimes\overline{f^{(l)}w}\otimes\overline{z}\overset{\Psi^{-1}}{\longmapsto}\overline{f^{(k+1)}u}\otimes\overline{f^{(l)}v}&\text{if } k+l<a,\\\overline{f^{(k)}u}\otimes\overline{f^{(l+1)}w}\otimes\overline{z}\overset{\Psi^{-1}}{\longmapsto}\overline{f^{(k)}u}\otimes\overline{f^{(l+1)}v}&\text{if } k+l\ge a.\end{cases}\end{aligned}$$

If $l = b-1$, then

$$\tilde{f}(\overline{f^{(k)}u} \otimes \overline{f^{(b-1)}v}) \overset{\Psi}{\longmapsto} \tilde{f}(\overline{f^{(k)}u} \otimes \overline{f^{(b-1)}w} \otimes \overline{z}) \; .$$

$$= \begin{cases} \tilde{f}(\overline{f^{(k)}u} \otimes \overline{f^{(b-1)}w} \otimes \overline{z}) & = \overline{f^{(k+1)}u} \otimes \overline{f^{(b-1)}w} \otimes \overline{z} \\ & \overset{\Psi^{-1}}{\longmapsto} \overline{f^{(k+1)}u} \otimes \overline{f^{(b-1)}v} \quad \text{if } k+l < a, \\ \overline{f^{(k)}u} \otimes \overline{f^{(b-1)}w} \otimes \overline{fz} & \overset{\Psi^{-1}}{\longmapsto} \overline{f^{(k)}u} \otimes \overline{f^{(b)}v} \quad \text{if } k+l \geq a. \end{cases}$$

If $l = b$, $k+b < a$, then since

$$\varphi(\overline{f^{(k)}u} \otimes \overline{f^{(b-1)}w}) = a-k-b+l > 1 = \varepsilon(\overline{fz}),$$

we have

$$\begin{aligned} \tilde{f}(\overline{f^{(k)}u} \otimes \overline{f^{(b)}v}) &\overset{\Psi}{\longmapsto} \tilde{f}(\overline{f^{(k)}u} \otimes \overline{f^{(b-1)}w} \otimes \overline{fz}) \\ &= \tilde{f}(\overline{f^{(k)}u} \otimes \overline{f^{(b-1)}w}) \otimes \overline{fz} \\ &= \overline{f^{(k+1)}u} \otimes \overline{f^{(b-1)}w} \otimes \overline{fz} \\ &\overset{\Psi^{-1}}{\longmapsto} \overline{f^{(k+1)}u} \otimes \overline{f^{(b)}v}, \end{aligned}$$

and if $l = b$, $k+b \geq a$, then since

$$\varphi(\overline{f^{(k)}u} \otimes \overline{f^{(b-1)}w}) \leq a-k-b+l \leq 1 = \varepsilon(\overline{fz}),$$

we have

$$\begin{aligned} \tilde{f}(\overline{f^{(k)}u} \otimes \overline{f^{(b)}v}) &\overset{\Psi}{\longmapsto} \tilde{f}(\overline{f^{(k)}u} \otimes \overline{f^{(b-1)}w} \otimes \overline{fz}) \\ &= \overline{f^{(k)}u} \otimes \overline{f^{(b-1)}w} \otimes \overline{f^{(2)}z} = 0, \end{aligned}$$

as desired.

Similarly, the tensor product rule for $\tilde{e}$ can be verified. □

Proof of Theorem 4.4.1: Tensor Product Rule.
It is easy to verify that

(1) $\mathbf{F}(q) \otimes_{\mathbf{A}_0} (\mathcal{L}_1 \otimes_{\mathbf{A}_0} \mathcal{L}_2) \cong M_1 \otimes_{\mathbf{F}(q)} M_2$,

(2) $\mathcal{L}_1 \otimes_{\mathbf{A}_0} \mathcal{L}_2 = \bigoplus_{\tau \in P} \left(\bigoplus_{\lambda+\mu=\tau} (\mathcal{L}_1)_\lambda \otimes_{\mathbf{A}_0} (\mathcal{L}_2)_\mu \right)$,

(3) $\mathcal{B}_1 \times \mathcal{B}_2$ is an $\mathbf{F}$-basis of $\mathcal{L}_1/q\mathcal{L}_1 \otimes_{\mathbf{F}} \mathcal{L}_2/q\mathcal{L}_2$, which is isomorphic to $(\mathcal{L}_1 \otimes_{\mathbf{A}_0} \mathcal{L}_2)/q(\mathcal{L}_1 \otimes_{\mathbf{A}_0} \mathcal{L}_2)$,

(4) $\mathcal{B}_1 \times \mathcal{B}_2 = \bigsqcup_{\tau \in P} \left(\bigsqcup_{\lambda+\mu=\tau} (\mathcal{B}_1)_\lambda \times (\mathcal{B}_2)_\mu \right)$.

Comparing these statements with the definition of crystal bases, we see that it remains to show

(5) $\tilde{e}_i \mathcal{L} \subset \mathcal{L}$, $\tilde{f}_i \mathcal{L} \subset \mathcal{L}$,

(6) $\tilde{e}_i \mathcal{B} \subset \mathcal{B} \cup \{0\}$, $\tilde{f}_i \mathcal{B} \subset \mathcal{B} \cup \{0\}$,

(7) the tensor product rule (4.11) holds.

We first focus on proving (5). Fix $i \in I$. It suffices to show $\tilde{e}_i(u \otimes v) \in \mathcal{L}$ and $\tilde{f}_i(u \otimes v) \in \mathcal{L}$ for any $u \in \mathcal{L}_1$ and $v \in \mathcal{L}_2$ of homogeneous weights. Using Proposition 4.2.11, we may further assume that $u = f_i^{(k)} u_0$ and $v = f_i^{(l)} v_0$ for some $u_0, v_0 \in \mathcal{L} \cap \ker e_i$. Consider the $U_q(\mathfrak{g}_{(i)})$-submodule of M_1 (respectively, M_2) generated by u_0 (respectively, v_0) which is isomorphic to $V(a)$ (respectively, $V(b)$) for some $a, b \in \mathbf{Z}_{\geq 0}$. Note that $u = f_i^{(k)} u_0 \in \mathcal{L}(a)$ and $v = f_i^{(l)} v_0 \in \mathcal{L}(b)$. By Theorem 4.4.3, we have

$$\tilde{e}_i(u \otimes v) \in \mathcal{L}(a) \otimes_{\mathbf{A}_0} \mathcal{L}(b) \subset \mathcal{L}_1 \otimes_{\mathbf{A}_0} \mathcal{L}_2 = \mathcal{L},$$
$$\tilde{f}_i(u \otimes v) \in \mathcal{L}(a) \otimes_{\mathbf{A}_0} \mathcal{L}(b) \subset \mathcal{L}_1 \otimes_{\mathbf{A}_0} \mathcal{L}_2 = \mathcal{L},$$

which proves our claim.

Similarly, one can prove (6). The assertion (7) follows immediately from (4.12) in Theorem 4.4.3. □

The tensor product rule yields the following corollary, which is very useful in combinatorial representation theory.

Corollary 4.4.4.

(1) *Let M_j be a $U_q(\mathfrak{g})$-module in the category $\mathcal{O}^q_{\text{int}}$ and let $(\mathcal{L}_j, \mathcal{B}_j)$ be a crystal basis of M_j for $j = 1, 2$. Then $b_1 \otimes b_2 \in \mathcal{B}_1 \otimes \mathcal{B}_2$ is a maximal vector (i.e., $\tilde{e}_i(b_1 \otimes b_2) = 0$ for all $i \in I$) if and only if $\tilde{e}_i b_1 = 0$ and $\langle h_i, \text{wt}(b_1) \rangle \geq \varepsilon_i(b_2)$ for all $i \in I$.*

(2) *Let M_j be a $U_q(\mathfrak{g})$-module in the category $\mathcal{O}^q_{\text{int}}$ and let $(\mathcal{L}_j, \mathcal{B}_j)$ be a crystal basis of M_j for $j = 1, \dots, N$. Then the vector $b_1 \otimes \cdots \otimes b_N \in \mathcal{B}_1 \otimes \cdots \otimes \mathcal{B}_N$ is a maximal vector if and only if $b_1 \otimes \cdots \otimes b_k$ is a maximal vector for all $k = 1, \dots, N$.*

Proof. We leave it as an exercise to the readers (Exercise 4.7). □

The tensor product rule gives a very convenient combinatorial description of the action of Kashiwara operators on the multifold tensor product of crystal graphs. Let M_j be a $U_q(\mathfrak{g})$-module in the category $\mathcal{O}^q_{\text{int}}$ with a crystal basis $(\mathcal{L}_j, \mathcal{B}_j)$ for $j = 1, \dots, N$.

Fix $i \in I$ and consider a vector $b = b_1 \otimes \cdots \otimes b_N \in \mathcal{B}_1 \otimes \cdots \otimes \mathcal{B}_N$. To each $b_j \in \mathcal{B}_j$ ($j = 1, \dots, N$), we assign a sequence of $-$'s and $+$'s with as many $-$'s as $\varepsilon_i(b_j)$ followed by as many $+$'s as $\varphi_i(b_j)$:

$$\begin{aligned} b &= b_1 \otimes b_2 \otimes \cdots \otimes b_N \\ &\longmapsto (\underbrace{-, \dots, -}_{\varepsilon_i(b_1)}, \underbrace{+, \dots, +}_{\varphi_i(b_1)}, \dots\dots, \underbrace{-, \dots, -}_{\varepsilon_i(b_N)}, \underbrace{+, \dots, +}_{\varphi_i(b_N)}). \end{aligned}$$

In this sequence, we cancel out all $(+,-)$-pairs to obtain a sequence of $-$'s followed by $+$'s:

$$i\text{-sgn}(b) := (-,-,\ldots,-,+,+,\ldots,+). \tag{4.14}$$

The sequence i-sgn(b) is called the ***i-signature*** of b.

Now the tensor product rule tells us that $\tilde{e}_i$ acts on b_j corresponding to the right-most $-$ in i-sgn(b) and $\tilde{f}_i$ acts on b_k corresponding to the left-most $+$ in i-sgn(b):

$$\begin{aligned} \tilde{e}_i b &= b_1 \otimes \cdots \otimes \tilde{e}_i b_j \otimes \cdots \otimes b_N, \\ \tilde{f}_i b &= b_1 \otimes \cdots \otimes \tilde{f}_i b_k \otimes \cdots \otimes b_N. \end{aligned} \tag{4.15}$$

Example 4.4.5. Consider the four-dimensional irreducible $U_q(\mathfrak{sl}_2)$-module $V(3)$ with the crystal basis $(\mathcal{L}(3), \mathcal{B}(3))$. We take the tensor product $\mathcal{B}(3) \otimes \mathcal{B}(3) \otimes \mathcal{B}(3) \otimes \mathcal{B}(3)$ of the crystal basis. Let $b = \overline{fu} \otimes \overline{f^{(3)}u} \otimes \overline{f^{(2)}u} \otimes \overline{fu}$, where u denotes the highest weight vector of $V(3)$. Then we get a sequence of $-$'s and $+$'s:

$$b = \overline{fu} \otimes \overline{f^{(3)}u} \otimes \overline{f^{(2)}u} \otimes \overline{fu} \longmapsto (-++,---,--+,-++).$$

By canceling out all $(+,-)$-pairs, we obtain the signature of b:

$$\text{sgn}(b) = (-,-,--,++).$$

It follows that

$$\begin{aligned} \tilde{e}b &= \overline{fu} \otimes \overline{f^{(3)}u} \otimes \overline{fu} \otimes \overline{fu}, \\ \tilde{f}b &= \overline{fu} \otimes \overline{f^{(3)}u} \otimes \overline{f^{(2)}u} \otimes \overline{f^{(2)}u}. \end{aligned}$$

4.5. Crystals

Recall that crystal bases $(\mathcal{L}, \mathcal{B})$ of $U_q(\mathfrak{g})$-modules in the category $\mathcal{O}^q_{\text{int}}$ are characterized by the maps

(i) $\text{wt} : \mathcal{B} \longrightarrow P$, $b \in \mathcal{B}_\lambda \longmapsto \text{wt}(b) = \lambda$,

(ii) the Kashiwara operators $\tilde{e}_i, \tilde{f}_i : \mathcal{B} \longrightarrow \mathcal{B} \cup \{0\}$,

(iii) $\varepsilon_i, \varphi_i : \mathcal{B} \longrightarrow \mathbf{Z}$ defined by

$$\varepsilon_i(b) = \max\{k \geq 0 \mid \tilde{e}_i^k b \in \mathcal{B}\}, \qquad \varphi_i(b) = \max\{k \geq 0 \mid \tilde{f}_i^k b \in \mathcal{B}\}$$

satisfying the properties

$$
\begin{aligned}
&\varphi_i(b) = \varepsilon_i(b) + \langle h_i, \mathrm{wt}(b)\rangle, \\
&\tilde{e}_i \mathcal{B}_\lambda \subset \mathcal{B}_{\lambda+\alpha_i} \cup \{0\}, \qquad \tilde{f}_i \mathcal{B}_\lambda \subset \mathcal{B}_{\lambda-\alpha_i} \cup \{0\}, \\
&\varepsilon_i(\tilde{e}_i b) = \varepsilon_i(b) - 1, \qquad \varphi_i(\tilde{e}_i b) = \varphi_i(b) + 1 \quad \text{if } \tilde{e}_i b \in \mathcal{B}, \\
&\varepsilon_i(\tilde{f}_i b) = \varepsilon_i(b) + 1, \qquad \varphi_i(\tilde{f}_i b) = \varphi_i(b) - 1 \quad \text{if } \tilde{f}_i b \in \mathcal{B}, \\
&\tilde{f}_i b = b' \quad \text{if and only if} \quad b = \tilde{e}_i b' \quad \text{for all } i \in I \text{ and } b, b' \in \mathcal{B}.
\end{aligned}
\tag{4.16}
$$

We will now define the abstract notion of *crystals* associated with a Cartan datum.

Definition 4.5.1. Let I be a finite index set and let $A = (a_{ij})_{i,j\in I}$ be a generalized Cartan matrix with the Cartan datum $(A, \Pi, \Pi^\vee, P, P^\vee)$. A ***crystal*** associated with the Cartan datum $(A, \Pi, \Pi^\vee, P, P^\vee)$ is a set $\mathcal{B}$ together with the maps $\mathrm{wt} : \mathcal{B} \to P$, $\tilde{e}_i, \tilde{f}_i : \mathcal{B} \to \mathcal{B} \cup \{0\}$, and $\varepsilon_i, \varphi_i : \mathcal{B} \to \mathbf{Z} \cup \{-\infty\}$ $(i \in I)$ satisfying the following properties:

(1) $\varphi_i(b) = \varepsilon_i(b) + \langle h_i, \mathrm{wt}(b)\rangle$ for all $i \in I$,

(2) $\mathrm{wt}(\tilde{e}_i b) = \mathrm{wt}\, b + \alpha_i$ if $\tilde{e}_i b \in \mathcal{B}$,

(3) $\mathrm{wt}(\tilde{f}_i b) = \mathrm{wt}\, b - \alpha_i$ if $\tilde{f}_i b \in \mathcal{B}$,

(4) $\varepsilon_i(\tilde{e}_i b) = \varepsilon_i(b) - 1$, $\varphi_i(\tilde{e}_i b) = \varphi_i(b) + 1$ if $\tilde{e}_i b \in \mathcal{B}$,

(5) $\varepsilon_i(\tilde{f}_i b) = \varepsilon_i(b) + 1$, $\varphi_i(\tilde{f}_i b) = \varphi_i(b) - 1$ if $\tilde{f}_i b \in \mathcal{B}$,

(6) $\tilde{f}_i b = b'$ if and only if $b = \tilde{e}_i b'$ for $b, b' \in \mathcal{B}$, $i \in I$.

(7) if $\varphi_i(b) = -\infty$ for $b \in \mathcal{B}$, then $\tilde{e}_i b = \tilde{f}_i b = 0$.

In this case, we often say that $\mathcal{B}$ is a $U_q(\mathfrak{g})$-***crystal***, where $U_q(\mathfrak{g})$ denotes the quantum group associated with the Cartan datum $(A, \Pi, \Pi^\vee, P, P^\vee)$. We set $\mathcal{B}_\lambda = \{b \in \mathcal{B} \mid \mathrm{wt}(b) = \lambda\}$ so that $\mathcal{B} = \bigsqcup_{\lambda\in P} \mathcal{B}_\lambda$. A $U_q(\mathfrak{g})$-crystal $\mathcal{B}$ is said to be ***semiregular*** if, for any $b \in \mathcal{B}$ and $i \in I$, we have

$$\varepsilon_i(b) = \max\{k \geq 0 \mid \tilde{e}_i^k b \in \mathcal{B}\}, \qquad \varphi_i(b) = \max\{k \geq 0 \mid \tilde{f}_i^k b \in \mathcal{B}\}.$$

Example 4.5.2.

(1) The crystal graph $\mathcal{B}$ of a $U_q(\mathfrak{g})$-module in the category $\mathcal{O}^q_{\mathrm{int}}$ is a $U_q(\mathfrak{g})$-crystal. In particular, the crystal graph $\mathcal{B}(\lambda)$ of the irreducible highest weight $U_q(\mathfrak{g})$-module $V(\lambda)$ with $\lambda \in P^+$ is a $U_q(\mathfrak{g})$ crystal.

(2) For $\lambda \in P$, let $T_\lambda = \{t_\lambda\}$ and for all $i \in I$, define

$$\mathrm{wt}(t_\lambda) = \lambda, \qquad \tilde{e}_i t_\lambda = \tilde{f}_i t_\lambda = 0, \qquad \varepsilon_i(t_\lambda) = \varphi_i(t_\lambda) = -\infty.$$

Then T_λ is a $U_q(\mathfrak{g})$-crystal.

(3) For $i \in I$, let $\mathcal{B}^{(i)} = \{b_i(n) \mid n \in \mathbf{Z}\}$ and define

$$\mathrm{wt}(b_i(n)) = n\alpha_i,$$

$$\tilde{e}_j(b_i(n)) = \begin{cases} b_i(n+1) & \text{if } j = i, \\ 0 & \text{if } j \neq i, \end{cases}$$

$$\tilde{f}_j(b_i(n)) = \begin{cases} b_i(n-1) & \text{if } j = i, \\ 0 & \text{if } j \neq i, \end{cases}$$

$$\varepsilon_j(b_i(n)) = \begin{cases} -n & \text{if } j = i, \\ -\infty & \text{if } j \neq i, \end{cases}$$

$$\varphi_j(b_i(n)) = \begin{cases} n & \text{if } j = i, \\ -\infty & \text{if } j \neq i. \end{cases}$$

Then $\mathcal{B}^{(i)}$ is a $U_q(\mathfrak{g})$-crystal.

(4) Let $M(0)$ be the Verma module over the quantum group $U_q(\mathfrak{sl}_2)$ with highest weight 0 and highest weight vector u. It has a basis consisting of the vectors of the form $f^{(k)}u$ for $k = 0, 1, 2, \ldots$. Let $\mathcal{B}(\infty) = \{f^{(k)}u \mid k = 0, 1, 2, \ldots\}$ and define

$$\mathrm{wt}(f^{(k)}u) = -2k,$$

$$\tilde{e}(f^{(k)}u) = f^{(k-1)}u, \qquad \tilde{f}(f^{(k)}u) = f^{(k+1)}u,$$

$$\varepsilon(f^{(k)}u) = k, \qquad \varphi(f^{(k)}u) = -k.$$

Then $\mathcal{B}(\infty)$ becomes a $U_q(\mathfrak{sl}_2)$-crystal with crystal graph given below.

$$u \longrightarrow fu \longrightarrow f^{(2)}u \longrightarrow f^{(3)}u \longrightarrow \cdots\cdots$$

Motivated by the tensor product rule for Kashiwara operators on crystal graphs, we define the *tensor product of crystals* as follows.

Definition 4.5.3. The ***tensor product*** $\mathcal{B}_1 \otimes \mathcal{B}_2$ of crystals $\mathcal{B}_1$ and $\mathcal{B}_2$ is defined to be the set $\mathcal{B}_1 \times \mathcal{B}_2$ whose crystal structure is defined by

(1) $\mathrm{wt}(b_1 \otimes b_2) = \mathrm{wt}(b_1) + \mathrm{wt}(b_2)$,

(2) $\varepsilon_i(b_1 \otimes b_2) = \max(\varepsilon_i(b_1), \varepsilon_i(b_2) - \langle h_i, \mathrm{wt}(b_1)\rangle)$,

(3) $\varphi_i(b_1 \otimes b_2) = \max(\varphi_i(b_2), \varphi_i(b_1) + \langle h_i, \mathrm{wt}(b_2)\rangle)$,

(4) $\tilde{e}_i(b_1 \otimes b_2) = \begin{cases} \tilde{e}_i b_1 \otimes b_2 & \text{if } \varphi_i(b_1) \geq \varepsilon_i(b_2), \\ b_1 \otimes \tilde{e}_i b_2 & \text{if } \varphi_i(b_1) < \varepsilon_i(b_2), \end{cases}$

(5) $\tilde{f}_i(b_1 \otimes b_2) = \begin{cases} \tilde{f}_i b_1 \otimes b_2 & \text{if } \varphi_i(b_1) > \varepsilon_i(b_2), \\ b_1 \otimes \tilde{f}_i b_2 & \text{if } \varphi_i(b_1) \leq \varepsilon_i(b_2). \end{cases}$

Here, we write $b_1 \otimes b_2$ for $(b_1, b_2) \in \mathcal{B}_1 \times \mathcal{B}_2$, and we understand $b_1 \otimes 0 = 0 \otimes b_2 = 0$.

Example 4.5.4.

(1) Let $\mathcal{B}_j$ be the crystal graph of a $U_q(\mathfrak{g})$-module M_j in the category $\mathcal{O}^q_{\text{int}}$ $(j = 1, 2)$. Then the $U_q(\mathfrak{g})$-crystal structure on $\mathcal{B}_1 \otimes \mathcal{B}_2$ is given by Theorem 4.4.1.

(2) For any $U_q(\mathfrak{g})$-crystal $\mathcal{B}$, the crystal structure on $\mathcal{B} \otimes T_\lambda$ is given by

$$\begin{aligned} &\operatorname{wt}(b \otimes t_\lambda) = \operatorname{wt}(b) + \lambda, \\ &\tilde{e}_i(b \otimes t_\lambda) = \tilde{e}_i b \otimes t_\lambda, && \tilde{f}_i(b \otimes t_\lambda) = \tilde{f}_i b \otimes t_\lambda, \\ &\varepsilon_i(b \otimes t_\lambda) = \varepsilon_i(b), && \varphi_i(b \otimes t_\lambda) = \varphi_i(b) + \langle h_i, \lambda \rangle. \end{aligned}$$

Similarly, the crystal structure on $T_\lambda \otimes \mathcal{B}$ is given by

$$\begin{aligned} &\operatorname{wt}(t_\lambda \otimes b) = \lambda + \operatorname{wt}(b), \\ &\tilde{e}_i(t_\lambda \otimes b) = t_\lambda \otimes \tilde{e}_i b, && \tilde{f}_i(t_\lambda \otimes b) = t_\lambda \otimes \tilde{f}_i b, \\ &\varepsilon_i(t_\lambda \otimes b) = \varepsilon_i(b) - \langle h_i, \lambda \rangle, && \varphi_i(t_\lambda \otimes b) = \varphi_i(b). \end{aligned}$$

(3) Let $B(\infty)$ be the crystal graph of the Verma module $M(0)$ over $U_q(\mathfrak{sl}_2)$ with highest weight 0 and highest weight vector u. Then the crystal structure on $\mathcal{B}(\infty) \otimes \mathcal{B}(\infty)$ is given by

$$\begin{aligned} \operatorname{wt}(f^{(k)}u \otimes f^{(l)}u) &= -2(k + l), \\ \tilde{e}(f^{(k)}u \otimes f^{(l)}u) &= f^{(k)}u \otimes f^{(l-1)}u, \\ \tilde{f}(f^{(k)}u \otimes f^{(l)}u) &= f^{(k)}u \otimes f^{(l+1)}u, \\ \varepsilon(f^{(k)}u \otimes f^{(l)}u) &= l + 2k, \\ \varphi(f^{(k)}u \otimes f^{(l)}u) &= -l. \end{aligned}$$

Definition 4.5.5. Let $\mathcal{B}_1$, $\mathcal{B}_2$ be crystals associated with the Cartan datum $(A, \Pi, \Pi^\vee, P, P^\vee)$. A ***crystal morphism*** (or ***morphism of crystals***) $\Psi : \mathcal{B}_1 \to \mathcal{B}_2$ is a map $\Psi : \mathcal{B}_1 \cup \{0\} \to \mathcal{B}_2 \cup \{0\}$ such that

(1) $\Psi(0) = 0$,

(2) if $b \in \mathcal{B}_1$ and $\Psi(b) \in \mathcal{B}_2$, then $\operatorname{wt}(\Psi(b)) = \operatorname{wt}(b)$, $\varepsilon_i(\Psi(b)) = \varepsilon_i(b)$, and $\varphi_i(\Psi(b)) = \varphi_i(b)$ for all $i \in I$,

(3) if $b, b' \in \mathcal{B}_1$, $\Psi(b), \Psi(b') \in \mathcal{B}_2$ and $\tilde{f}_i b = b'$, then $\tilde{f}_i \Psi(b) = \Psi(b')$ and $\Psi(b) = \tilde{e}_i \Psi(b')$ for all $i \in I$.

Definition 4.5.6.

(1) A crystal morphism is called ***strict*** if it commutes with all $\tilde{e}_i$ and $\tilde{f}_i$ $(i \in I)$.

(2) A crystal morphism $\Psi : \mathcal{B}_1 \to \mathcal{B}_2$ is called an ***embedding*** if Ψ induces an injective map from $\mathcal{B}_1 \cup \{0\}$ to $\mathcal{B}_2 \cup \{0\}$.

(3) A crystal morphism $\Psi : \mathcal{B}_1 \to \mathcal{B}_2$ is called an ***isomorphism*** if it is a bijection from $\mathcal{B}_1 \cup \{0\}$ to $\mathcal{B}_2 \cup \{0\}$.

Example 4.5.7. Let $\mathcal{B}(m) = \{f^{(k)}u \,|\, 0 \leq k \leq m\}$ be the crystal graph of the $(m+1)$-dimensional irreducible $U_q(\mathfrak{sl}_2)$-module with highest weight m and highest weight vector u and let $\mathcal{B}(\infty) = \{f^{(k)}v \,|\, k \geq 0\}$ be the crystal graph of the Verma module over $U_q(\mathfrak{sl}_2)$ with highest weight 0 and highest weight vector v. Then the map $\Psi : \mathcal{B}(m) \longrightarrow \mathcal{B}(\infty) \otimes T_m$ defined by $f^{(k)}u \longmapsto f^{(k)}v \otimes t_m \quad (0 \leq k \leq m)$ is an embedding of $U_q(\mathfrak{sl}_2)$-crystals, but this crystal morphism is *not* strict.

We close this section with an important property of strict crystal morphisms.

Proposition 4.5.8. *Let $\lambda \in P^+$ be a dominant integral weight and $\mathcal{B}(\lambda)$ the crystal graph of the irreducible highest weight $U_q(\mathfrak{g})$-module $V(\lambda)$ with highest weight λ. Suppose that $\mathcal{B}$ is a connected $U_q(\mathfrak{g})$-crystal. If there exists a strict crystal morphism $\Psi : \mathcal{B}(\lambda) \longrightarrow \mathcal{B}$ such that $\Psi(\mathcal{B}(\lambda)) \subset \mathcal{B}$, then Ψ is a crystal isomorphism.*

Proof. The proof is left to the readers as an exercise (Exercise 4.9). □

Exercises

4.1. Show that $(\mathcal{L}(m), \mathcal{B}(m))$, as defined by (4.5), forms a crystal basis of the $(m+1)$-dimensional $U_q(\mathfrak{sl}_2)$-module $V(m)$.

4.2. Verify that $(\mathcal{L}, \mathcal{B})$ given in Example 4.2.7 forms a crystal basis for the vector representation V of $U_q(\mathfrak{sl}_n)$.

4.3. Prove Proposition 4.2.8.

4.4. Complete the proof of Theorem 4.2.10.

4.5. Prove Lemma 4.3.5.

4.6. Let $u \in V(a)$ and $v \in V(b)$ be the highest weight vectors of irreducible $U_q(\mathfrak{sl}_2)$-modules. For each $l = a + b - 2s$ with $0 \leq s \leq \min(a, b)$, let E_l be the maximal vector of weight l in $V(a) \otimes V(b)$, unique up to a scalar multiple.

(a) Verify that

$$E_l = \sum_{\alpha=0}^{s} (-1)^\alpha q^{\alpha(a-s+1)} \left(\prod_{\beta=1}^{\alpha} \frac{q^{2(b-s+\beta)} - 1}{q^{2(a-\beta+1)} - 1} \right) f^{(\alpha)} v \otimes f^{(s-\alpha)} w.$$

(b) Show that $E_l^{(r)} = f^{(r)}(E_l) \in \mathcal{L}(a) \otimes \mathcal{L}(b)$ for all $l = a + b - 2s$ $(0 \leq s \leq \min(a, b))$ and $0 \leq r \leq l$.

(c) Show that every vector of the form $f^{(k)} u \otimes f^{(l)} v$ can be expressed as an $\mathbf{A}_0$-linear combination of the vectors $E_l^{(r)} = f^{(r)}(E_l)$, where $l = a + b - 2s$ $(0 \leq s \leq \min(a, b))$ and $0 \leq r \leq l$.

4.7. Prove Corollary 4.4.4, which characterizes the maximal vectors in the tensor product of crystal graphs.

4.8. Verify all the statements given in Example 4.5.4.

4.9. Prove Proposition 4.5.8 concerning strict crystal morphisms.

Chapter 5

Existence and Uniqueness of Crystal Bases

In this chapter, we will prove the existence and uniqueness of crystal bases for quantum groups. We will first give a sketch of the proof for the existence theorem, and, assuming that the existence theorem holds, we will prove the uniqueness of crystal bases. The full proof of the existence theorem will be given at the end of this chapter using Kashiwara's *grand-loop argument*.

5.1. Existence of crystal bases

We will now discuss the existence of crystal bases for $U_q(\mathfrak{g})$-modules in the category $\mathcal{O}^q_{\text{int}}$. The proof of the existence theorem relies on Kashiwara's ***grand-loop argument*** [**39**], which is rather lengthy and technical. Thus, in this section, we will only give a sketch of the proof. The whole grand-loop argument will be given in Section 5.3.

Let $\lambda \in P^+$ be a dominant integral weight and let $V(\lambda)$ be the irreducible highest weight $U_q(\mathfrak{g})$-module with highest weight λ and highest weight vector v_λ. Let $\mathcal{L}(\lambda)$ be the free $\mathbf{A}_0$-submodule of $V(\lambda)$ spanned by the vectors of the form $\tilde{f}_{i_1} \cdots \tilde{f}_{i_r} v_\lambda$ $(r \geq 0,\ i_k \in I)$, and set

$$\mathcal{B}(\lambda) = \{\tilde{f}_{i_1} \cdots \tilde{f}_{i_r} v_\lambda + q\mathcal{L}(\lambda) \in \mathcal{L}(\lambda)/q\mathcal{L}(\lambda) \mid r \geq 0, i_k \in I\} \setminus \{0\}.$$

Theorem 5.1.1. *The pair* $(\mathcal{L}(\lambda), \mathcal{B}(\lambda))$ *is a crystal basis of* $V(\lambda)$.

Sketch of the proof. It is easy to see that $\mathcal{L}(\lambda)$ has the weight space decomposition $\mathcal{L}(\lambda) = \bigoplus_{\mu \leq \lambda} \mathcal{L}(\lambda)_\mu$, where $\mathcal{L}(\lambda)_\mu$ is the free $\mathbf{A}_0$-submodule of $\mathcal{L}(\lambda)$ spanned by the vectors of the form $\tilde{f}_{i_1}\cdots\tilde{f}_{i_r}v_\lambda$ such that $\lambda - (\alpha_{i_1}+\cdots+\alpha_{i_r}) = \mu$. Similarly, $\mathcal{B}(\lambda)$ has the weight set decomposition $\mathcal{B}(\lambda) = \bigsqcup_{\mu\leq\lambda}\mathcal{B}(\lambda)_\mu$, where

$$\mathcal{B}(\lambda)_\mu = \{\tilde{f}_{i_1}\cdots\tilde{f}_{i_r}v_\lambda + q\mathcal{L}(\lambda) \in \mathcal{B}(\lambda) \mid \lambda - (\alpha_{i_1}+\cdots+\alpha_{i_r}) = \mu\}.$$

Next, we will show by induction that the $\mathbf{F}(q)$-span of $\mathcal{L}(\lambda)$ is $V(\lambda)$. Since $\mathcal{L}(\lambda)_\lambda = \mathbf{A}_0 v_\lambda$, we have $\mathbf{F}(q)\otimes_{\mathbf{A}_0}\mathcal{L}(\lambda)_\lambda = \mathbf{F}(q)\otimes_{\mathbf{A}_0}\mathbf{A}_0 v_\lambda \cong \mathbf{F}(q)v_\lambda = V(\lambda)_\lambda$. Suppose $\mu < \lambda$ and assume that $\mathbf{F}(q)\otimes_{\mathbf{A}_0}\mathcal{L}(\lambda)_\tau \cong V(\lambda)_\tau$ for all $\tau > \mu$. By definition, $\mathcal{L}(\lambda)_\mu = \sum_{i\in I}\tilde{f}_i\mathcal{L}(\lambda)_{\mu+\alpha_i}$. It follows that

$$\begin{aligned}\mathbf{F}(q)\otimes_{\mathbf{A}_0}\mathcal{L}(\lambda)_\mu &= \mathbf{F}(q)\otimes_{\mathbf{A}_0}\left(\sum_{i\in I}\tilde{f}_i\mathcal{L}(\lambda)_{\mu+\alpha_i}\right)\\ &= \sum_{i\in I}\tilde{f}_i\left(\mathbf{F}(q)\otimes_{\mathbf{A}_0}\mathcal{L}(\lambda)_{\mu+\alpha_i}\right)\\ &\cong \sum_{i\in I}\tilde{f}_i V(\lambda)_{\mu+\alpha_i} = \sum_{i\in I} f_i V(\lambda)_{\mu+\alpha_i}\\ &= V(\lambda)_\mu.\end{aligned}$$

By definition, we have $\tilde{f}_i\mathcal{L}(\lambda)\subset\mathcal{L}(\lambda)$ and $\tilde{f}_i\mathcal{B}(\lambda)\subset\mathcal{B}(\lambda)\sqcup\{0\}$ for all $i\in I$. Hence in order to prove that $(\mathcal{L}(\lambda),\mathcal{B}(\lambda))$ is a crystal basis of $V(\lambda)$, it remains to prove the following statements.

(1) $\tilde{e}_i\mathcal{L}(\lambda)\subset\mathcal{L}(\lambda)$ for all $i\in I$.

(2) $\tilde{e}_i\mathcal{B}(\lambda)\subset\mathcal{B}(\lambda)\sqcup\{0\}$ for all $i\in I$.

(3) For any $b, b'\in\mathcal{B}(\lambda)$ and $i\in I$, $\tilde{f}_i b = b'$ if and only if $b = \tilde{e}_i b'$.

(4) $\mathcal{B}(\lambda)$ is an $\mathbf{F}$-basis of $\mathcal{L}(\lambda)/q\mathcal{L}(\lambda)$.

In Section 5.3, we will prove these statements using an interlocking induction argument on weights, called the *grand-loop argument*, discovered by Kashiwara ([**39**]). □

Therefore, by Theorem 4.2.10, we have the *existence theorem* for the crystal bases of $U_q(\mathfrak{g})$-modules in the category $\mathcal{O}^q_{\text{int}}$.

Theorem 5.1.2. *Let M be a $U_q(\mathfrak{g})$-module in the category $\mathcal{O}^q_{\text{int}}$ which is isomorphic to $\bigoplus_{\lambda\in P^+}V(\lambda)^{\oplus\nu(\lambda)}$. Then there exists a crystal basis $(\mathcal{L},\mathcal{B})$ of M such that*

$$(\mathcal{L},\mathcal{B}) \xrightarrow{\sim} \left(\bigoplus_{\lambda\in P^+}\mathcal{L}(\lambda)^{\oplus\nu(\lambda)},\ \bigsqcup_{\lambda\in P^+}\mathcal{B}(\lambda)^{\oplus\nu(\lambda)}\right),$$

where $(\mathcal{L}(\lambda),\mathcal{B}(\lambda))$ denotes the crystal basis of $V(\lambda)$ defined in Theorem 5.1.1.

Example 5.1.3. Let $V(\lambda)$ be the irreducible highest weight module over $U_q(\mathfrak{sl}_3)$ with highest weight λ and highest weight vector v_λ, where $\lambda \in P^+$ is the linear functional defined by $\lambda(h_1) = \lambda(h_2) = 1$. By the definition of the Kashiwara operators, we can easily see that

$$\begin{aligned} &\tilde{f}_1 v_\lambda = f_1 v_\lambda, && \tilde{f}_2 v_\lambda = f_2 v_\lambda, \\ &\tilde{f}_2(f_1 v_\lambda) = f_2 f_1 v_\lambda, && \tilde{f}_2^2(f_1 v_\lambda) = f_2^{(2)} f_1 v_\lambda, \\ &\tilde{f}_1(f_2 v_\lambda) = f_1 f_2 v_\lambda, && \tilde{f}_1^2(f_2 v_\lambda) = f_1^{(2)} f_2 v_\lambda, \\ &\tilde{f}_1(f_2^{(2)} f_1 v_\lambda) = f_1 f_2^{(2)} f_1 v_\lambda, && \tilde{f}_2(f_1^{(2)} f_2 v_\lambda) = f_2 f_1^{(2)} f_2 v_\lambda. \end{aligned}$$

The Serre relations can be written as

$$f_1^{(2)} f_2 - f_1 f_2 f_1 + f_2 f_1^{(2)} = 0, \qquad f_2^{(2)} f_1 - f_2 f_1 f_2 + f_1 f_2^{(2)} = 0.$$

Hence we get

$$\begin{aligned} f_2 f_1^{(2)} f_2 v_\lambda &= (f_1 f_2 f_1 - f_1^{(2)} f_2) f_2 v_\lambda = f_1 f_2 f_1 f_2 v_\lambda \\ &= f_1 (f_2^{(2)} f_1 + f_1 f_2^{(2)}) v_\lambda = f_1 f_2^{(2)} f_1 v_\lambda. \end{aligned}$$

Therefore we get a *part* of the crystal graph.

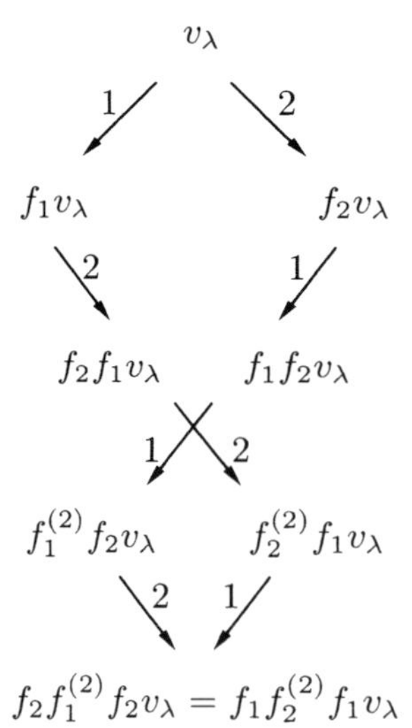

We can show that this is actually *all* of the crystal graph by proving that

$$\tilde{f}_1(f_2 f_1 v_\lambda) \equiv 0, \qquad \tilde{f}_2(f_1 f_2 v_\lambda) \equiv 0 \mod q\mathcal{L}(\lambda).$$

For example, write

$$f_2 f_1 v_\lambda = \left(f_2 f_1 v_\lambda - \frac{q}{1+q^2} f_1 f_2 v_\lambda \right) + f_1 \left(\frac{q}{1+q^2} f_2 v_\lambda \right).$$

Then

$$\begin{aligned}
\tilde{f}_1(f_2f_1v_\lambda) &= f_1f_2f_1v_\lambda - \frac{q}{1+q^2}f_1^2f_2v_\lambda + f_1^{(2)}\left(\frac{q}{1+q^2}f_2v_\lambda\right)\\
&= f_1f_2f_1v_\lambda - f_1^{(2)}f_2v_\lambda + \frac{q}{1+q^2}f_1^{(2)}f_2v_\lambda\\
&= f_2f_1^{(2)}v_\lambda + \frac{q}{1+q^2}f_1^{(2)}f_2v_\lambda\\
&= \frac{q}{1+q^2}f_1^{(2)}f_2v_\lambda\\
&\equiv 0 \mod q\mathcal{L}(\lambda).
\end{aligned}$$

Similarly, we can show that $\tilde{f}_2(f_1f_2v_\lambda) \equiv 0 \mod q\mathcal{L}(\lambda)$.

We now aim to provide a *recognition theorem* for $\mathcal{L}(\lambda)$ with $\lambda \in P^+$. Observe that the map

$$q^h \longmapsto q^h, \qquad e_i \longmapsto q_iK_i^{-1}f_i, \qquad f_i \longmapsto q_i^{-1}K_ie_i \tag{5.1}$$

defines an antiautomorphism of $U_q(\mathfrak{g})$. Thus there exists a unique symmetric bilinear form $(\ ,\ )$ on $V(\lambda)$ $(\lambda \in P^+)$ satisfying

$$\begin{aligned}
(q^hu, v) &= (u, q^hv),\\
(e_iu, v) &= (u, q_iK_i^{-1}f_iv),\\
(f_iu, v) &= (u, q_i^{-1}K_ie_iv),\\
(v_\lambda, v_\lambda) &= 1
\end{aligned} \tag{5.2}$$

for $i \in I$, $h \in P^\vee$, $u, v \in V(\lambda)$ (Exercise 5.1). Note that $(V(\lambda)_\mu, V(\lambda)_\tau) = 0$ unless $\mu = \tau$.

Lemma 5.1.4. *The symmetric bilinear form $(\ ,\ )$ on $V(\lambda)$ defined by (5.2) is nondegenerate.*

Proof. For $\alpha = \sum k_i\alpha_i \in Q_+$, the height of α will be denoted by $|\alpha| = \sum k_i$. Using induction on $|\alpha|$, we will show that the symmetric bilinear form $(\ ,\ )$ is nondegenerate on $V(\lambda)_{\lambda-\alpha}$ for all $\alpha \in Q_+$.

If $|\alpha| = 0$, then our assertion is trivial. So suppose that we have $|\alpha| > 0$ and $(u, V(\lambda)_{\lambda-\alpha}) = 0$ for $u \in V(\lambda)_{\lambda-\alpha}$. Since $V(\lambda)_{\lambda-\alpha} = \sum f_iV(\lambda)_{\lambda-\alpha+\alpha_i}$, we have

$$(u, f_iV(\lambda)_{\lambda-\alpha+\alpha_i}) = 0 \quad \text{for all } i \in I.$$

It follows that

$$0 = (q_i^{-1}K_ie_iu, V(\lambda)_{\lambda-\alpha+\alpha_i}) = q_i^{\langle h_i, \lambda-\alpha\rangle+1}(e_iu, V(\lambda)_{\lambda-\alpha+\alpha_i}).$$

By the induction hypothesis, $e_iu = 0$ for all $i \in I$, which implies $u = 0$. □

Lemma 5.1.5. *Fix a dominant integral weight* $\lambda \in P^+$ *and let* $(\mathcal{L}(\lambda), \mathcal{B}(\lambda))$ *be the crystal basis of* $V(\lambda)$ *defined by Theorem* 5.1.1. *Then we have*

$$(\mathcal{L}(\lambda), \mathcal{L}(\lambda)) \subset \mathbf{A}_0.$$

Proof. We will prove

$$(\mathcal{L}(\lambda)_{\lambda-\alpha}, \mathcal{L}(\lambda)_{\lambda-\alpha}) \subset \mathbf{A}_0$$

for all $\alpha \in Q_+$ by induction on $|\alpha|$. If $|\alpha| = 0$, then our assertion is trivial. If $|\alpha| > 0$, since $\mathcal{L}(\lambda)_{\lambda-\alpha} = \sum \tilde{f}_i \mathcal{L}(\lambda)_{\lambda-\alpha+\alpha_i}$, it suffices to prove

$$(\tilde{f}_i u, v) \equiv (u, \tilde{e}_i v) \mod q\mathbf{A}_0 \tag{5.3}$$

for $u \in \mathcal{L}(\lambda)_{\lambda-\alpha+\alpha_i}$, $v \in \mathcal{L}(\lambda)_{\lambda-\alpha}$.

By Proposition 4.2.11, we may assume that $u = f_i^{(m)} u_0$, $v = f_i^{(n)} v_0$, where $e_i u_0 = e_i v_0 = 0$, $\langle h_i, \lambda-\alpha+(m+1)\alpha_i\rangle \geq m$, and $\langle h_i, \lambda-\alpha+n\alpha_i\rangle \geq n$. Observe that

$$\begin{aligned}
(\tilde{f}_i u, v) &= (f_i^{(m+1)} u_0, f_i^{(n)} v_0) \\
&= \frac{1}{[n]_{q_i}!}\left((q_i^{-1}K_i e_i)^n f_i^{(m+1)} u_0, v_0\right) \\
&= \frac{1}{[n]_{q_i}!}(q_i^{-n^2} K_i^n e_i^n f_i^{(m+1)} u_0, v_0) \\
&= q_i^{-n^2}(K_i^n e_i^{(n)} f_i^{(m+1)} u_0, v_0) \\
&= \delta_{m+1,n} q_i^{-n^2} \begin{bmatrix} \langle h_i, \lambda-\alpha+(m+1)\alpha_i\rangle \\ n \end{bmatrix}_{q_i} (K_i^n u_0, v_0) \\
&= \begin{cases} q_i^{n^2+n\langle h_i,\lambda-\alpha\rangle} \begin{bmatrix} 2n+\langle h_i, \lambda-\alpha\rangle \\ n \end{bmatrix}_{q_i} (u_0, v_0) & \text{if } n = m+1, \\ 0 & \text{if } n \neq m+1. \end{cases}
\end{aligned}$$

We know that $u_0, v_0 \in \mathcal{L}(\lambda)$, so by the induction hypothesis, $(u_0, v_0) \in \mathbf{A}_0$. Also, by direct calculation (Exercise 5.2), one can verify that

$$q_i^{n^2+n\langle h_i,\lambda-\alpha\rangle} \begin{bmatrix} 2n+\langle h_i, \lambda-\alpha\rangle \\ n \end{bmatrix}_{q_i} \in 1 + q\mathbf{A}_0.$$

It follows that

$$(f_i^{(m+1)} u_0, f_i^{(n)} v_0) \equiv \delta_{m+1,n}(u_0, v_0) \mod q\mathbf{A}_0.$$

Similarly, we can show that

$$(u, \tilde{e}_i v) = (f_i^{(m)} u_0, f_i^{(n-1)} v_0) \equiv \delta_{m,n-1}(u_0, v_0) \mod q\mathbf{A}_0.$$

Therefore, we obtain

$$(\tilde{f}_i u, v) \equiv (u, \tilde{e}_i v) \mod q\mathbf{A}_0,$$

as desired. □

Let $(\ ,\ )_0$ be the $\mathbf{F}$-valued symmetric bilinear form on $\mathcal{L}(\lambda)/q\mathcal{L}(\lambda)$ given by taking the crystal limit of $(\ ,\ )$ on $\mathcal{L}(\lambda)$. Then we have:

Lemma 5.1.6. *The crystal $\mathcal{B}(\lambda)$ forms an orthonormal basis of $\mathcal{L}(\lambda)/q\mathcal{L}(\lambda)$ with respect to $(\ ,\ )_0$. In particular, if $\mathbf{F} = \mathbf{Q}$, then $(\ ,\ )_0$ is positive definite on $\mathcal{L}(\lambda)/q\mathcal{L}(\lambda)$.*

Proof. We will prove $(b, b')_0 = \delta_{b,b'}$ for all $b, b' \in \mathcal{B}(\lambda)_{\lambda-\alpha}$ $(\alpha \in Q_+)$ by induction on $|\alpha|$.

If $|\alpha| = 0$, our assertion is trivial. If $|\alpha| > 0$, choose $i \in I$ such that $\tilde{e}_i b \neq 0$. Then, by Theorem 5.1.1 and (5.3), we obtain

$$(b, b')_0 = (\tilde{f}_i \tilde{e}_i b, b')_0 = (\tilde{e}_i b, \tilde{e}_i b')_0 = \delta_{\tilde{e}_i b, \tilde{e}_i b'} = \delta_{b,b'}.$$

This completes the proof. □

Now the crystal lattice $\mathcal{L}(\lambda)$ can be characterized as follows.

Proposition 5.1.7. *For $\lambda \in P^+$, we have*

$$\mathcal{L}(\lambda) = \{u \in V(\lambda) \mid (u, \mathcal{L}(\lambda)) \subset \mathbf{A}_0\}. \tag{5.4}$$

Moreover, if $\mathbf{F} = \mathbf{Q}$, then we have

$$\mathcal{L}(\lambda) = \{u \in V(\lambda) \mid (u, u) \in \mathbf{A}_0\}. \tag{5.5}$$

Proof. Let

$$\begin{aligned} \mathcal{L}_1 &= \{u \in V(\lambda) \mid (u, \mathcal{L}(\lambda)) \subset \mathbf{A}_0\}, \\ \mathcal{L}_2 &= \{u \in V(\lambda) \mid (u, u) \in \mathbf{A}_0\}. \end{aligned}$$

It is clear that $\mathcal{L}(\lambda) \subset \mathcal{L}_1$ and $\mathcal{L}(\lambda) \subset \mathcal{L}_2$. So we need to show $\mathcal{L}_1 \subset \mathcal{L}(\lambda)$ and $\mathcal{L}_2 \subset \mathcal{L}(\lambda)$.

We first show that $\mathcal{L}_1 \subset \mathcal{L}(\lambda)$. Let $u \in V(\lambda)$ be such that $(u, \mathcal{L}(\lambda)) \subset \mathbf{A}_0$ and take the smallest $n \geq 0$ such that $q^n u \in \mathcal{L}(\lambda)$. If $n > 0$, then

$$(q^n u, \mathcal{L}(\lambda) \equiv 0 \mod q\mathbf{A}_0.$$

Since $(\ ,\ )_0$ is nondegenerate, $q^n u \equiv 0 \mod q\mathcal{L}(\lambda)$; i.e., $q^{n-1}u \in \mathcal{L}(\lambda)$, which is a contradiction to the minimality of n. Therefore $n = 0$ and $u \in \mathcal{L}(\lambda)$.

If $\mathbf{F} = \mathbf{Q}$, the remaining inclusion $\mathcal{L}_2 \subset \mathcal{L}(\lambda)$ can be shown in a similar manner using the positive definiteness of $(\ ,\ )_0$ (Exercise 5.3). □

5.2. Uniqueness of crystal bases

We will now prove the uniqueness of the crystal bases for $U_q(\mathfrak{g})$-modules in the category $\mathcal{O}^q_{\text{int}}$. Assuming the existence theorem, the uniqueness theorem can be proved in a very similar way to the proof of the uniqueness theorem for crystal bases of $U_q(\mathfrak{sl}_2)$-modules. The proof uses the content of the existence theorem. This aspect is different from the usual existence and uniqueness theorems, which may be proved independently.

Theorem 5.2.1. *Let M be a $U_q(\mathfrak{g})$-module in the category $\mathcal{O}^q_{\text{int}}$ with a crystal basis $(\mathcal{L}, \mathcal{B})$. If $M \cong \bigoplus_{\lambda \in P^+} V(\lambda)^{\oplus \nu(\lambda)}$ $(\nu(\lambda) \in \mathbf{Z}_{\geq 0})$, then there exists an isomorphism of crystal bases*

$$\Psi : (\mathcal{L}, \mathcal{B}) \xrightarrow{\sim} \left(\bigoplus_{\lambda \in P^+} \mathcal{L}(\lambda)^{\oplus \nu(\lambda)}, \bigsqcup_{\lambda \in P^+} \mathcal{B}(\lambda)^{\oplus \nu(\lambda)} \right).$$

The proof of Theorem 5.2.1 is very similar to that of Theorem 4.3.2. We first prove:

Lemma 5.2.2. *Let $V(\lambda)$ be the irreducible highest weight $U_q(\mathfrak{g})$-module with highest weight $\lambda \in P^+$ and highest weight vector v_λ. Let $(\mathcal{L}(\lambda), \mathcal{B}(\lambda))$ be the crystal basis of $V(\lambda)$ defined in Theorem 5.1.1.*

(1) *If*

$$U(\lambda) = \{u \in \mathcal{L}(\lambda)/q\mathcal{L}(\lambda) \mid \tilde{e}_i u = 0 \ \text{ for all } i \in I\},$$
$$W(\lambda) = \{v \in V(\lambda) \mid \tilde{e}_i v \in \mathcal{L}(\lambda) \ \text{ for all } i \in I\},$$

then we have

$$U(\lambda) = \mathbf{F}\overline{v_\lambda}, \qquad W(\lambda) = \mathbf{F}(q) v_\lambda + \mathcal{L}(\lambda).$$

(2) *If $\mathcal{L}$ is a free $\mathbf{A}_0$-submodule of $V(\lambda)$ with a weight space decomposition $\mathcal{L} = \bigoplus_{\mu \leq \lambda} \mathcal{L}_\mu$ such that $\mathcal{L}_\lambda = \mathbf{A}_0 v_\lambda$ and that $\tilde{e}_i \mathcal{L} \subset \mathcal{L}$, $\tilde{f}_i \mathcal{L} \subset \mathcal{L}$ for all $i \in I$, then $\mathcal{L} = \mathcal{L}(\lambda)$.*

Proof. (1) For the first equality, it suffices to show that if $u \in \mathcal{L}(\lambda)_\mu / q\mathcal{L}(\lambda)_\mu$ with $\mu < \lambda$ satisfies $\tilde{e}_i u{=}0$ for all $i \in I$, then $u = 0$. Write

$$u = \sum_{b \in \mathcal{B}(\lambda)_\mu} c_b b \quad \text{with } c_b \in \mathbf{F}.$$

Then we have

$$0 = \tilde{e}_i u = \sum_{\tilde{e}_i b \neq 0} c_b \tilde{e}_i b,$$

which implies $c_b = 0$ whenever $\tilde{e}_i b \neq 0$. But for any $b \in \mathcal{B}(\lambda)_\mu$ with $\mu < \lambda$, there exists an $i \in I$ such that $\tilde{e}_i b \neq 0$. Therefore $c_b = 0$ for all $b \in \mathcal{B}(\lambda)_\mu$ and hence $u = 0$.

For the second equality, we will show that if $v \in V(\lambda)_\mu$ with $\mu < \lambda$ and $\tilde{e}_i v \in \mathcal{L}(\lambda)$ for all $i \in I$, then $v \in \mathcal{L}(\lambda)$. Let $\{v_1, \dots, v_\nu\}$ be an $\mathbf{A}_0$-basis of $\mathcal{L}(\lambda)_\mu$ and write $v = c_1 v_1 + \cdots + c_\nu v_\nu$ with $c_j \in \mathbf{F}(q)$ (which is possible because $V(\lambda)_\mu = \mathbf{F}(q) \otimes_{\mathbf{A}_0} \mathcal{L}(\lambda)_\mu$). Choose the smallest nonnegative integer N such that $q^N v \in \mathcal{L}(\lambda)_\mu$. Note that $\tilde{e}_i(q^N v) = q^N \tilde{e}_i v \in q^N \mathcal{L}(\lambda)$ for all $i \in I$. If $N > 0$, then $\tilde{e}_i(\overline{q^N v}) = 0$ for all $i \in I$, which implies that $\overline{q^N v} = 0$ by the first equality. It follows that $q^N v \in q\mathcal{L}(\lambda)$ and hence $q^{N-1} v \in \mathcal{L}(\lambda)$, which is a contradiction to the minimality of N. Therefore $N = 0$ and $v \in \mathcal{L}(\lambda)_\mu$.

(2) Since $\mathcal{L}_\lambda = \mathbf{A}_0 v_\lambda = \mathcal{L}(\lambda)_\lambda$, it is clear that $\mathcal{L}(\lambda) \subset \mathcal{L}$. For the other inclusion, suppose $\mu < \lambda$ and let $v \in \mathcal{L}_\mu$. By induction, we have $\tilde{e}_i v \in \mathcal{L}_{\mu+\alpha_i} = \mathcal{L}(\lambda)_{\mu+\alpha_i}$ for all $i \in I$. It follows from (1) that $v \in \mathcal{L}(\lambda)_\mu$. □

We will now prove some special cases of Theorem 5.2.1, which in turn will be used to prove the general statement in Theorem 5.2.1.

Lemma 5.2.3. *Let $V(\lambda)$ be the irreducible highest weight $U_q(\mathfrak{g})$-module with highest weight $\lambda \in P^+$ and highest weight vector v_λ. Then, for any crystal basis $(\mathcal{L}, \mathcal{B})$ of $V(\lambda)$, there exists an isomorphism of crystal bases $(\mathcal{L}, \mathcal{B}) \xrightarrow{\sim} (\mathcal{L}(\lambda), \mathcal{B}(\lambda))$, where $(\mathcal{L}(\lambda), \mathcal{B}(\lambda))$ is the crystal basis of $V(\lambda)$ defined in Theorem* 5.1.1.

Proof. Since $V(\lambda)_\lambda = \mathbf{F}(q) v_\lambda$, we may assume that $\mathcal{L}_\lambda = \mathbf{A}_0 v_\lambda = \mathcal{L}(\lambda)_\lambda$. By Lemma 5.2.2, $\mathcal{L} = \mathcal{L}(\lambda)$. By definition of $\mathcal{B}(\lambda)$, it is clear that $\mathcal{B}(\lambda) \subset \mathcal{B}$. Moreover, we have

$$\#\mathcal{B}(\lambda)_\mu = \dim_{\mathbf{F}(q)} V(\lambda)_\mu = \#\mathcal{B}_\mu$$

for all $\mu \le \lambda$. Therefore, we must have $\mathcal{B}(\lambda)_\mu = \mathcal{B}_\mu$ for all $\mu \le \lambda$, which proves our assertion. □

By Lemma 5.2.3, we see that Lemma 5.2.2 holds for any crystal basis $(\mathcal{L}, \mathcal{B})$ of $V(\lambda)$. More generally, we have:

Lemma 5.2.4. *Suppose $M \cong V(\lambda)^{\oplus \nu}$ for some $\lambda \in P^+$ and let $(\mathcal{L}, \mathcal{B}) \cong (\mathcal{L}(\lambda)^{\oplus \nu}, \mathcal{B}(\lambda)^{\oplus \nu})$ be the crystal basis of M given by Theorem* 5.1.2.

(1) *If*

$$U(\lambda) = \{\overline{v} \in \mathcal{L}/q\mathcal{L} \mid \tilde{e}_i \overline{v} = 0 \quad \text{for all } i \in I\},$$
$$W(\lambda) = \{v \in M \mid \tilde{e}_i v \in \mathcal{L} \quad \text{for all } i \in I\},$$

then

$$U(\lambda) = \mathcal{L}_\lambda / q\mathcal{L}_\lambda, \qquad W(\lambda) = M_\lambda + \mathcal{L}.$$

(2) *If $\mathcal{L}'$ is a free $\mathbf{A}_0$-submodule of M with weight space decomposition $\mathcal{L}' = \bigoplus_{\mu \le \lambda} \mathcal{L}'_\mu$ such that $\mathcal{L}'_\lambda = \mathcal{L}_\lambda$ and that $\tilde{e}_i \mathcal{L}' \subset \mathcal{L}'$, $\tilde{f}_i \mathcal{L}' \subset \mathcal{L}'$ for all $i \in I$, then $\mathcal{L}' = \mathcal{L}$.*

Proof. The proof will be left to the readers as an exercise (Exercise 5.4). □

Lemma 5.2.5. *Suppose $M \cong V(\lambda)^{\oplus \nu}$ for some $\lambda \in P^+$. Then, for any crystal basis $(\mathcal{L}, \mathcal{B})$ of M, there exists an isomorphism of crystal bases*

$$\Psi : (\mathcal{L}, \mathcal{B}) \xrightarrow{\sim} (\mathcal{L}(\lambda)^{\oplus \nu}, \mathcal{B}(\lambda)^{\oplus \nu}) .$$

Proof. Recall that M has the weight space decomposition $M = \bigoplus_{\mu \le \lambda} M_\mu$ and that $\dim_{\mathbf{F}(q)} M_\lambda = \operatorname{rank}_{\mathbf{A}_0} \mathcal{L}_\lambda = \#\mathcal{B}_\lambda = \nu$. Let $\mathcal{B}_\lambda = \{b_1, \dots, b_\nu\}$ and choose $v_j \in \mathcal{L}_\lambda$ such that $\overline{v_j} = b_j$ $(j = 1, \dots, \nu)$. By Nakayama's Lemma, the vectors $v_1, \dots, v_\nu$ form an $\mathbf{A}_0$-basis of $\mathcal{L}_\lambda$.

Since $e_i v_j = 0$ for all $i \in I$, each v_j generates a $U_q(\mathfrak{g})$-submodule M_j which is isomorphic to $V(\lambda)$. Let $\mathcal{L}_j$ be the free $\mathbf{A}_0$-submodule of M_j spanned by the vectors of the form $\tilde{f}_{i_1} \cdots \tilde{f}_{i_r} v_j$ $(i_k \in I, r \ge 0)$, and let

$$\mathcal{B}_j = \{\tilde{f}_{i_1} \cdots \tilde{f}_{i_r} v_j + q\mathcal{L}_j \mid i_k \in I, r \ge 0\} \setminus \{0\}.$$

Then $M = M_1 \oplus \cdots \oplus M_\nu$ and for each $j = 1, \dots, \nu$, $(\mathcal{L}_j, \mathcal{B}_j)$ is a crystal basis of M_j which is isomorphic to $(\mathcal{L}(\lambda), \mathcal{B}(\lambda))$. Thus to obtain the desired isomorphism of crystal bases

$$\Psi : (\mathcal{L}, \mathcal{B}) \xrightarrow{\sim} (\mathcal{L}(\lambda)^{\oplus \nu}, \mathcal{B}(\lambda)^{\oplus \nu}) ,$$

it suffices to show that

$$\mathcal{L} = \mathcal{L}_1 \oplus \cdots \oplus \mathcal{L}_\nu, \qquad \mathcal{B} = \mathcal{B}_1 \sqcup \cdots \sqcup \mathcal{B}_\nu.$$

Clearly,

$$\mathcal{L}_1 \oplus \cdots \oplus \mathcal{L}_\nu \subset \mathcal{L} \quad \text{and} \quad \mathcal{B}_1 \sqcup \cdots \sqcup \mathcal{B}_\nu \subset \mathcal{B}.$$

By Lemma 5.2.2, we have $\mathcal{L}_1 \oplus \cdots \oplus \mathcal{L}_\nu = \mathcal{L}$. Moreover, note that

$$\begin{aligned} \#(\mathcal{B}_1 \sqcup \cdots \sqcup \mathcal{B}_\nu)_\mu &= \nu\ (\#\mathcal{B}(\lambda)_\mu) = \nu\ (\dim_{\mathbf{F}(q)} V(\lambda)_\mu) \\ &= \dim_{\mathbf{F}(q)} M_\mu = \#\mathcal{B}_\mu \end{aligned}$$

for all $\mu \le \lambda$. Hence we must have

$$(\mathcal{B}_1 \sqcup \cdots \sqcup \mathcal{B}_\nu)_\mu = \mathcal{B}_\mu$$

for all $\mu \le \lambda$, which proves our claim. □

Proof of Theorem 5.2.1. Let M be a $U_q(\mathfrak{g})$-module in the category $\mathcal{O}^q_{\text{int}}$ and let $(\mathcal{L}, \mathcal{B})$ be a crystal basis of M. Let $\lambda \in P^+$ be a maximal weight of M; i.e., $M_{\lambda + \alpha_i} = \{0\}$ for all $i \in I$. Note that if $\lambda \in P$ is a maximal weight, then it must be an element of P^+ (Exercise 5.5). Set $V = U_q(\mathfrak{g}) M_\lambda$. Let $\{v_1, \dots, v_\nu\}$ be an $\mathbf{A}_0$-basis of $\mathcal{L}_\lambda$, and let $V_j = U_q(\mathfrak{g}) v_j \cong V(\lambda)$. Then we have

$$V = V_1 \oplus \cdots \oplus V_\nu \cong V(\lambda)^{\oplus \nu},$$

where

$$\nu = \dim_{\mathbf{F}(q)} M_\lambda = \operatorname{rank}_{\mathbf{A}_0} \mathcal{L}_\lambda = \#\mathcal{B}_\lambda.$$

Let $(\mathcal{L}_j, \mathcal{B}_j) \cong (\mathcal{L}(\lambda), \mathcal{B}(\lambda))$ be the crystal basis of V_j defined in Theorem 5.1.1 using the maximal vector v_j ($j = 1, \dots, \nu$). Then by Theorem 4.2.10 (1), $(\mathcal{L}_1 \oplus \cdots \oplus \mathcal{L}_\nu, \mathcal{B}_1 \sqcup \cdots \sqcup \mathcal{B}_\nu)$ is a crystal basis of $V = V_1 \oplus \cdots \oplus V_\nu$.

Let W be the complementary submodule of V in M and set

$$\mathcal{L}_V = \mathcal{L} \cap V, \qquad \mathcal{B}_V = \mathcal{B} \cap (\mathcal{L}_V / q\mathcal{L}_V),$$
$$\mathcal{L}_W = \mathcal{L} \cap W, \qquad \mathcal{B}_W = \mathcal{B} \cap (\mathcal{L}_W / q\mathcal{L}_W).$$

We will show that $\mathcal{L} = \mathcal{L}_V \oplus \mathcal{L}_W$ and $\mathcal{B} = \mathcal{B}_V \sqcup \mathcal{B}_W$. Then by Theorem 4.2.10 (2), $(\mathcal{L}_V, \mathcal{B}_V)$ would be a crystal basis of V and $(\mathcal{L}_W, \mathcal{B}_W)$ would be a crystal basis of W. By Lemma 5.2.5, we have an isomorphism of crystal bases $(\mathcal{L}_V, \mathcal{B}_V) \cong (\mathcal{L}(\lambda)^{\oplus\nu}, \mathcal{B}(\lambda)^{\oplus\nu})$. Then by an induction argument, which we explain later, applying this *single step* repeatedly will give the uniqueness of crystal bases.

First, note that if $\mu \not\leq \lambda$, then $(\mathcal{L}_V)_\mu = \{0\}$ and $(\mathcal{L}_W)_\mu = \mathcal{L}_\mu$. Hence we may assume that $\mu \leq \lambda$. We will prove our assertion by induction. If $\mu = \lambda$, we get $(\mathcal{L}_V)_\lambda = \mathcal{L}_\lambda$ and $(\mathcal{L}_W)_\lambda = \{0\}$, which implies $\mathcal{L}_\lambda = (\mathcal{L}_V \oplus \mathcal{L}_W)_\lambda$. Moreover, since $\tilde{e}_i \mathcal{L}_V \subset \mathcal{L}_V$ and $\tilde{f}_i \mathcal{L}_V \subset \mathcal{L}_V$ for all $i \in I$, Lemma 5.2.2 yields

$$\mathcal{L}_V = \mathcal{L}_1 \oplus \cdots \oplus \mathcal{L}_\nu \cong \mathcal{L}(\lambda)^{\oplus\nu}.$$

Suppose $\mu < \lambda$ and assume that $\mathcal{L}_\tau = (\mathcal{L}_V \oplus \mathcal{L}_W)_\tau$ for all $\mu < \tau \leq \lambda$. Let $u \in \mathcal{L}_\mu$ and write $u = v + w$ with $v \in V_\mu$, $w \in W_\mu$. We would like to show that $v, w \in \mathcal{L}$. Since $V = V_1 \oplus \cdots \oplus V_\nu$, v can be written uniquely as $v = x_1 + \cdots + x_\nu$ with $x_j \in V_j$ ($j = 1, 2, \dots, \nu$). By induction, we have

$$\tilde{e}_i u = \tilde{e}_i v + \tilde{e}_i w \in \mathcal{L}_{\mu+\alpha_i} = (\mathcal{L}_V)_{\mu+\alpha_i} \oplus (\mathcal{L}_W)_{\mu+\alpha_i}$$

for all $i \in I$. Since the $U_q(\mathfrak{g})$-submodules are closed under the Kashiwara operators $\tilde{e}_i$ and $\tilde{f}_i$, we have $\tilde{e}_i v \in (\mathcal{L}_V)_{\mu+\alpha_i}$ and $\tilde{e}_i w \in (\mathcal{L}_W)_{\mu+\alpha_i}$ ($i \in I$). It follows that

$$\tilde{e}_i v = \tilde{e}_i x_1 + \cdots + \tilde{e}_i x_\nu \in (\mathcal{L}_V)_{\mu+\alpha_i} = (\mathcal{L}_1)_{\mu+\alpha_i} \oplus \cdots \oplus (\mathcal{L}_\nu)_{\mu+\alpha_i},$$

which implies $\tilde{e}_i x_j \in (\mathcal{L}_j)_{\mu+\alpha_i}$ for all $i \in I$, $j = 1, 2, \dots, \nu$. Hence by Lemma 5.2.2, we get

$$v = x_1 + \cdots + x_\nu \in (\mathcal{L}_1)_\mu \oplus \cdots \oplus (\mathcal{L}_\nu)_\mu = (\mathcal{L}_V)_\mu.$$

Therefore $w = u - v \in \mathcal{L} \cap W = \mathcal{L}_W$ as desired.

We should now show $\mathcal{B} = \mathcal{B}_V \sqcup \mathcal{B}_W$. Let $b \in \mathcal{B}$ and $\mathrm{wt}(b) = \mu$. If $\mu \not\leq \lambda$, then $(\mathcal{B}_V)_\mu = \emptyset$ and $(\mathcal{B}_W)_\mu = \mathcal{B}_\mu$. Hence we may assume that $\mu \leq \lambda$. If $\mu = \lambda$, then since $\mathcal{B}_\lambda = (\mathcal{B}_V)_\lambda$ and $(\mathcal{B}_W)_\lambda = \emptyset$, we are done.

Suppose $\mu < \lambda$. If $b' = \tilde{e}_i b \neq 0$ for some $i \in I$, then $b' \in \mathcal{B}_{\mu+\alpha_i}$, and by induction $b' \in \mathcal{B}_V$ or $b' \in \mathcal{B}_W$. Hence $b = \tilde{f}_i b' \in \mathcal{B}_V$ or $b = \tilde{f}_i b' \in \mathcal{B}_W$.

If $b' = \tilde{e}_i b = 0$ for all $i \in I$, then write $b = b_1 + b_2$ with $b_1 \in \mathcal{L}_V/q\mathcal{L}_V$, $b_2 \in \mathcal{L}_W/q\mathcal{L}_W$. Since $\tilde{e}_i b = 0$, we have $\tilde{e}_i b_1 = \tilde{e}_i b_2 = 0$ for all $i \in I$. Since $\mathcal{L}_V = \mathcal{L}_1 \oplus \cdots \oplus \mathcal{L}_\nu$, Lemma 5.2.2 implies $b_1 = 0$. Hence, $b = b_2 \in \mathcal{B} \cap (\mathcal{L}_W/q\mathcal{L}_W) = \mathcal{B}_W$ as desired.

We now present the induction argument we mentioned earlier. Let λ be a *singular weight* of M in the sense that $\lambda + \alpha \notin \mathrm{wt}(M)$ for any $\alpha \in Q_+ \setminus \{0\}$. Note that every singular weight is a maximal weight. Hence, by the definition of $\mathcal{O}^q_{\mathrm{int}}$, there are only finitely many singular weights for any $U_q(\mathfrak{g})$-module M belonging to $\mathcal{O}^q_{\mathrm{int}}$.

Recall from Theorem 3.5.4 on complete reducibility, that we may write

$$M = \bigoplus_{j \in J} V(\lambda_j)^{\oplus m_j},$$

where $\lambda_j \in P^+$ and J is an (at most countable) index set. If J is finite, the induction is trivial. If J is infinite, we rename the indices as follows. We first choose all the singular weights of M and name them $\mu_0, \mu_1, \ldots, \mu_r$ in some order denoting their outer multiplicities by $n_0, n_1, \ldots, n_r$. The $U_q(\mathfrak{g})$-module resulting from M after taking away the irreducible components corresponding to these weights (counted with outer multiplicities) again belongs to $\mathcal{O}^q_{\mathrm{int}}$. Thus we may continue this process to obtain two sequences $\{\mu_j\}_{j=0}^{\infty}$ and $\{n_j\}_{j=0}^{\infty}$. By the definition of $\mathcal{O}^q_{\mathrm{int}}$, one can see that these sequences exhaust all of λ_j and m_j $(j \in J)$ so that

$$M = \bigoplus_{j=0}^{\infty} V(\mu_j)^{\oplus n_j}.$$

Now we may apply our *single step argument* inductively to complete the proof. □

5.3. Kashiwara's grand-loop argument

In this section, we will give a detailed proof of the existence theorem for crystal bases (Theorem 5.1.1) using the ***grand-loop argument*** discovered by Kashiwara. The original grand-loop argument consists of 14 interlocking inductive statements, including those concerning the crystal bases for $U_q^-(\mathfrak{g})$. In this book, since we focus on the crystal bases for $U_q(\mathfrak{g})$-modules in the category $\mathcal{O}^q_{\mathrm{int}}$, we will slightly modify the original argument and prove seven inductive statements only (cf. [**22**]). Still, the spirit of the argument is the same as the original one: our argument depends heavily on the fundamental properties of crystal bases for $U_q^-(\mathfrak{g})$.

For dominant integral weights $\lambda, \mu \in P^+$, we define the $U_q(\mathfrak{g})$-module homomorphisms

$$\Phi_{\lambda,\mu} : V(\lambda+\mu) \longrightarrow V(\lambda) \otimes V(\mu),$$
$$\Psi_{\lambda,\mu} : V(\lambda) \otimes V(\mu) \longrightarrow V(\lambda+\mu)$$

by

$$\Phi_{\lambda,\mu}(v_{\lambda+\mu}) = v_\lambda \otimes v_\mu, \quad \Psi_{\lambda,\mu}(v_\lambda \otimes v_\mu) = v_{\lambda+\mu}. \tag{5.6}$$

It is clear that

$$\Psi_{\lambda,\mu} \circ \Phi_{\lambda,\mu} = \mathrm{id}_{V(\lambda+\mu)}\,. \tag{5.7}$$

Let $(\ ,\ )$ denote the symmetric bilinear form on irreducible highest weight $U_q(\mathfrak{g})$-modules with dominant integral weights defined in Section 5.1 and define a symmetric bilinear form on $V(\lambda) \otimes V(\mu)$ by

$$(u_1 \otimes u_2, v_1 \otimes v_2) = (u_1, v_1)(u_2, v_2)$$

for $u_1, v_1 \in V(\lambda)$, $u_2, v_2 \in V(\mu)$. Then the form on $V(\lambda) \otimes V(\mu)$ satisfies the first three conditions in (5.2). Moreover, we have

$$(\Phi_{\lambda,\mu}(u), w) = (u, \Psi_{\lambda,\mu}(w)) \tag{5.8}$$

for all $u \in V(\lambda+\mu)$, $w \in V(\lambda) \otimes V(\mu)$ (Exercise 5.6).

For a nonnegative integer $r \in \mathbf{Z}_{\geq 0}$, let

$$Q_+(r) = \{\alpha \in Q_+ \mid |\alpha| \leq r\}. \tag{5.9}$$

Given $r \in \mathbf{Z}_{\geq 0}$, we will denote $\lambda \gg 0$ if $\lambda - \alpha \in P^+$ for all $\alpha \in Q_+(r)$ and $\lambda - \Lambda_i \in P^+$ for all $i \in I$. Note that for any $\lambda, \mu \gg 0$, there is an $\mathbf{F}(q)$-linear isomorphism $\sum_{\tau \geq \lambda-\alpha} V(\lambda)_\tau \xrightarrow{\sim} \sum_{\tau \geq \mu-\alpha} V(\mu)_\tau$ that commutes with e_i and f_i, and hence with $\tilde{e}_i$ and $\tilde{f}_i$ $(i \in I)$ (whenever the actions are defined).

For $\lambda, \mu \in P^+$ and $\alpha \in Q_+(r)$, using an interlocking induction argument, we will show that the following seven statements are true, which would complete the proof of the existence theorem for crystal bases.

- $\mathbf{A(r)}$: $\tilde{e}_i \mathcal{L}(\lambda)_{\lambda-\alpha} \subset \mathcal{L}(\lambda)$ for all $i \in I$;
- $\mathbf{B(r)}$: $\tilde{e}_i \mathcal{B}(\lambda)_{\lambda-\alpha} \subset \mathcal{B}(\lambda) \sqcup \{0\}$ for all $i \in I$;
- $\mathbf{C(r)}$: $\tilde{f}_i b = b'$ if and only if $b = \tilde{e}_i b'$ for $i \in I$, $b \in \mathcal{B}(\lambda)_{\lambda-\alpha+\alpha_i}$, and $b' \in \mathcal{B}(\lambda)_{\lambda-\alpha}$;
- $\mathbf{D(r)}$: $\mathcal{B}(\lambda)_{\lambda-\alpha}$ is an $\mathbf{F}$-basis of $\mathcal{L}(\lambda)_{\lambda-\alpha}/q\mathcal{L}(\lambda)_{\lambda-\alpha}$;
- $\mathbf{E(r)}$: $\Phi_{\lambda,\mu}(\mathcal{L}(\lambda+\mu)_{\lambda+\mu-\alpha}) \subset \mathcal{L}(\lambda) \otimes \mathcal{L}(\mu)$;
- $\mathbf{F(r)}$: $\Psi_{\lambda,\mu}((\mathcal{L}(\lambda) \otimes \mathcal{L}(\mu))_{\lambda+\mu-\alpha}) \subset \mathcal{L}(\lambda+\mu)$;
- $\mathbf{G(r)}$: $\Psi_{\lambda,\mu}((\mathcal{B}(\lambda) \otimes \mathcal{B}(\mu))_{\lambda+\mu-\alpha}) \subset \mathcal{B}(\lambda+\mu) \cup \{0\}$.

Note that the above statements are true if $r = 0$ or $r = 1$ (Exercise 5.7). Thus we assume that $r \geq 2$ and the statements $\mathbf{A(s)}, \mathbf{B(s)}, \ldots, \mathbf{G(s)}$ are true for $s < r$.

Lemma 5.3.1. *For $\alpha \in Q_+(r-1)$, let $u = \sum_{k=0}^{N} f_i^{(k)} u_k \in V(\lambda)_{\lambda-\alpha}$ be a weight vector, where $u_k \in \ker e_i \cap V(\lambda)_{\lambda-\alpha+k\alpha_i}$ and $\langle h_i, \lambda-\alpha+k\alpha_i\rangle \geq k \geq 0$.*

(1) *If $u \in \mathcal{L}(\lambda)$, then $u_k \in \mathcal{L}(\lambda)$ for all $k \geq 0$.*

(2) *If $u + q\mathcal{L}(\lambda) \in \mathcal{B}(\lambda)$, then there exists a nonnegative integer k_0 such that*

$$u \equiv f_i^{(k_0)} u_{k_0} \mod q\mathcal{L}(\lambda) \quad \textit{and} \quad u_k \in q\mathcal{L}(\lambda) \quad \textit{for all } k \neq k_0.$$

Proof. (1) We will use induction on $N \geq 0$. If $N = 0$, then $u = u_0 \in \mathcal{L}(\lambda)$. If $N > 0$, by $\mathbf{A(r-1)}$, we have $\tilde{e}_i u = \sum_{k=1}^{N} f_i^{(k-1)} u_k \in \mathcal{L}(\lambda)$. By the induction hypothesis, $u_k \in \mathcal{L}(\lambda)$ for $k = 1, \ldots, N$. Hence, $u_0 = u - \sum_{k=1}^{N} f_i^{(k)} u_k \in \mathcal{L}(\lambda)$, which proves our claim.

(2) If $N = 0$, our assertion is trivial. If $N > 0$, by $\mathbf{B(r-1)}$, we have $\tilde{e}_i u + q\mathcal{L}(\lambda) \in \mathcal{B}(\lambda) \cup \{0\}$. If $\tilde{e}_i u \in q\mathcal{L}(\lambda)$, then (1) implies $u_k \in q\mathcal{L}(\lambda)$ for all $k \geq 1$. Hence $k_0 = 0$ and $u \equiv u_0 \mod q\mathcal{L}(\lambda)$. If $\tilde{e}_i u \notin q\mathcal{L}(\lambda)$, then $\tilde{e}_i u + q\mathcal{L}(\lambda) \in \mathcal{B}(\lambda)$, and by the induction hypothesis, there exists $k_0 \geq 1$ such that

$$\tilde{e}_i u \equiv f_i^{(k_0-1)} u_{k_0} \mod q\mathcal{L}(\lambda) \quad \text{and} \quad u_k \in q\mathcal{L}(\lambda) \quad \text{for all } k \neq k_0.$$

Hence, by $\mathbf{C(r-1)}$, we obtain

$$u \equiv \tilde{f}_i \tilde{e}_i u \equiv \tilde{f}_i f_i^{(k_0-1)} u_{k_0} \equiv f_i^{(k_0)} u_{k_0} \mod q\mathcal{L}(\lambda),$$

and $u_0 = u - \sum_{k=1}^{N} f_i^{(k)} u_k \in q\mathcal{L}(\lambda)$. □

Lemma 5.3.2. *Suppose $\alpha, \beta \in Q_+(r-1)$ and $b \in \mathcal{B}(\lambda)_{\lambda-\alpha}$, $b' \in \mathcal{B}(\mu)_{\mu-\beta}$.*

(1) *For each $i \in I$, we have*

$$\tilde{e}_i(\mathcal{L}(\lambda)_{\lambda-\alpha} \otimes \mathcal{L}(\mu)_{\mu-\beta}) \subset \mathcal{L}(\lambda) \otimes \mathcal{L}(\mu),$$
$$\tilde{f}_i(\mathcal{L}(\lambda)_{\lambda-\alpha} \otimes \mathcal{L}(\mu)_{\mu-\beta}) \subset \mathcal{L}(\lambda) \otimes \mathcal{L}(\mu).$$

(2) *For each $i \in I$, the tensor product rule holds:*

$$\tilde{e}_i(b \otimes b') = \begin{cases} \tilde{e}_i b \otimes b' & \textit{if } \varphi_i(b) \geq \varepsilon_i(b'), \\ b \otimes \tilde{e}_i b' & \textit{if } \varphi_i(b) < \varepsilon_i(b'), \end{cases}$$

$$\tilde{f}_i(b \otimes b') = \begin{cases} \tilde{f}_i b \otimes b' & \textit{if } \varphi_i(b) > \varepsilon_i(b'), \\ b \otimes \tilde{f}_i b' & \textit{if } \varphi_i(b) \leq \varepsilon_i(b') \end{cases}$$

in $(\mathcal{L}(\lambda) \otimes \mathcal{L}(\mu))/q(\mathcal{L}(\lambda) \otimes \mathcal{L}(\mu))$.

(3) *For each $i \in I$, we have*

$$\tilde{e}_i(\mathcal{B}(\lambda)_{\lambda-\alpha} \otimes \mathcal{B}(\mu)_{\mu-\beta}) \subset (\mathcal{B}(\lambda) \otimes \mathcal{B}(\mu)) \cup \{0\},$$
$$\tilde{f}_i(\mathcal{B}(\lambda)_{\lambda-\alpha} \otimes \mathcal{B}(\mu)_{\mu-\beta}) \subset (\mathcal{B}(\lambda) \otimes \mathcal{B}(\mu)) \cup \{0\}.$$

(4) *If $\tilde{e}_i(b \otimes b') \neq 0$, then $b \otimes b' = \tilde{f}_i \tilde{e}_i(b \otimes b')$.*

(5) *If $\tilde{e}_i(b \otimes b') = 0$ for all $i \in I$, then $b = v_\lambda$.*

(6) *For each $i \in I$, we have $\tilde{f}_i(b \otimes v_\mu) = \tilde{f}_i b \otimes v_\mu$ or $\tilde{f}_i b = 0$.*

(7) *For any sequence of indices $i_1, \dots, i_r \in I$, we have*

$$\tilde{f}_{i_1} \cdots \tilde{f}_{i_r}(v_\lambda \otimes v_\mu) \equiv (\tilde{f}_{i_1} \cdots \tilde{f}_{i_r} v_\lambda) \otimes v_\mu \mod q\mathcal{L}(\lambda) \otimes \mathcal{L}(\mu)$$

or $\tilde{f}_{i_1} \cdots \tilde{f}_{i_r} v_\lambda = 0$.

Proof. (1) By Lemma 5.3.1 (1), it suffices to prove the following statement: if $u \in \mathcal{L}(\lambda)_{\lambda-\alpha+k\alpha_i}$, $v \in \mathcal{L}(\mu)_{\mu-\beta+l\alpha_i}$ with $e_i u = e_i v = 0$, $\langle h_i, \lambda - \alpha + k\alpha_i \rangle \geq k \geq 0$, $\langle h_i, \mu - \beta + l\alpha_i \rangle \geq l \geq 0$, then

$$\tilde{e}_i(f_i^{(k)} u \otimes f_i^{(l)} v) \in \mathcal{L}(\lambda) \otimes \mathcal{L}(\mu),$$
$$\tilde{f}_i(f_i^{(k)} u \otimes f_i^{(l)} v) \in \mathcal{L}(\lambda) \otimes \mathcal{L}(\mu).$$

Let $\mathcal{L}$ be the free $\mathbf{A}_0$-submodule spanned by the vectors $f_i^{(s)} u \otimes f_i^{(t)} v$ $(s, t \geq 0)$. Then by the tensor product rule (Theorem 4.4.3), we have $\tilde{e}_i \mathcal{L} \subset \mathcal{L}$, $\tilde{f}_i \mathcal{L} \subset \mathcal{L}$. Hence our assertion follows immediately from the fact that $\mathcal{L} \subset \mathcal{L}(\lambda) \otimes \mathcal{L}(\mu)$.

(2) We may assume that $b = f_i^{(k)} u + q\mathcal{L}(\lambda)$ and $b' = f_i^{(l)} v + q\mathcal{L}(\mu)$, where $e_i u = e_i v = 0$. Let $\mathcal{L}$ be the free $\mathbf{A}_0$-submodule spanned by the vectors $f_i^{(s)} u \otimes f_i^{(t)} v$ $(s, t \geq 0)$ and set $a = \langle h_i, \lambda - \alpha + k\alpha_i \rangle$. Then by the tensor product rule (Theorem 4.4.3), we have

$$\tilde{e}_i(b \otimes b') \equiv \begin{cases} \tilde{e}_i b \otimes b' & \mod q_i\mathcal{L} \quad \text{if } k + l \leq a, \\ b \otimes \tilde{e}_i b' & \mod q_i\mathcal{L} \quad \text{if } k + l > a, \end{cases}$$

$$\tilde{f}_i(b \otimes b') \equiv \begin{cases} \tilde{f}_i b \otimes b' & \mod q_i\mathcal{L} \quad \text{if } k + l < a, \\ b \otimes \tilde{f}_i b' & \mod q_i\mathcal{L} \quad \text{if } k + l \geq a. \end{cases}$$

Since $\mathcal{L} \subset \mathcal{L}(\lambda) \otimes \mathcal{L}(\mu)$, our assertion follows immediately.

(3), (4), (5) These are immediate consequences of (2). Note that, by Lemma 5.3.1 (2), if $b \in \mathcal{B}(\lambda)$ satisfies $\tilde{e}_i b = 0$ for all $i \in I$, then we must have $b = v_\lambda + q\mathcal{L}(\lambda)$.

(6) Since $\varepsilon_i(v_\mu) = 0$, we have $\tilde{f}_i(b \otimes v_\mu) = \tilde{f}_i b \otimes v_\mu$ unless $k = a$, which is equivalent to $\tilde{f}_i b = 0$.

(7) This is an immediate consequence of (6). □

We will now prove the statement $\mathbf{E(r)}$.

Proposition 5.3.3 ($\mathbf{E(r)}$). *For all $\alpha \in Q_+(r)$, we have*

$$\Phi_{\lambda,\mu}(\mathcal{L}(\lambda+\mu)_{\lambda+\mu-\alpha}) \subset \mathcal{L}(\lambda) \otimes \mathcal{L}(\mu).$$

Proof. Recall that

$$\mathcal{L}(\lambda+\mu)_{\lambda+\mu-\alpha} = \sum \tilde{f}_i \mathcal{L}(\lambda+\mu)_{\lambda+\mu-\alpha+\alpha_i}.$$

By $\mathbf{E(r-1)}$, we have

$$\Phi_{\lambda,\mu}(\mathcal{L}(\lambda+\mu)_{\lambda+\mu-\alpha+\alpha_i}) \subset (\mathcal{L}(\lambda) \otimes \mathcal{L}(\mu))_{\lambda+\mu-\alpha+\alpha_i}.$$

Hence our assertion follows from Lemma 5.3.2 (1). □

The following technical lemma will play a crucial role in the grand-loop argument.

Lemma 5.3.4. *For any sequence of indices $i_1, i_2, \ldots, i_r \in I$, set $\lambda = \Lambda_{i_{r-1}}$. Then we have*

$$\tilde{f}_{i_1} \cdots \tilde{f}_{i_r}(v_\lambda \otimes v_\mu) \equiv b \otimes b' \mod q\mathcal{L}(\lambda) \otimes \mathcal{L}(\mu)$$

for some $b \in \mathcal{B}(\lambda)_{\lambda-\alpha}$, $b' \in \mathcal{B}(\mu)_{\mu-\beta} \cup \{0\}$ and $\alpha, \beta \in Q_+(r-1) \setminus \{0\}$.

Proof. If $i_{r-1} \neq i_r$, then since $f_{i_r} v_\lambda = 0$, we get

$$\begin{aligned}\tilde{f}_{i_r}(v_\lambda \otimes v_\mu) &= f_{i_r}(v_\lambda \otimes v_\mu) = (f_{i_r} \otimes 1 + K_{i_r} \otimes f_{i_r})(v_\lambda \otimes v_\mu) \\ &= v_\lambda \otimes \tilde{f}_{i_r} v_\mu.\end{aligned}$$

Since $\tilde{e}_{i_{r-1}} \tilde{f}_{i_r} v_\mu = 0$ and $\tilde{f}_{i_{r-1}} v_\lambda = f_{i_{r-1}} v_\lambda \neq 0$, we have

$$\begin{aligned}\tilde{f}_{i_{r-1}} \tilde{f}_{i_r}(v_\lambda \otimes v_\mu) &= (f_{i_{r-1}} \otimes 1 + K_{i_{r-1}} \otimes f_{i_{r-1}})(v_\lambda \otimes f_{i_r} v_\mu) \\ &= f_{i_{r-1}} v_\lambda \otimes \tilde{f}_{i_r} v_\mu + q_{i_{r-1}} v_\lambda \otimes f_{i_{r-1}} f_{i_r} v_\mu \\ &\equiv \tilde{f}_{i_{r-1}} v_\lambda \otimes \tilde{f}_{i_r} v_\mu \mod q\mathcal{L}(\lambda) \otimes \mathcal{L}(\mu).\end{aligned}$$

If $i_{r-1} = i_r$, then $f_{i_r}^{(2)} v_\lambda = 0$, and hence we obtain

$$\begin{aligned}&\tilde{f}_{i_r}^2(v_\lambda \otimes v_\mu) \\ &= f_{i_r}^{(2)}(v_\lambda \otimes v_\mu) \\ &= \frac{1}{[2]_{q_{i_r}}} \left(f_{i_r}^2 v_\lambda \otimes v_\mu + (q_{i_r} + q_{i_r}^{-1}) f_{i_r} v_\lambda \otimes f_{i_r} v_\mu + q_{i_r}^2 v_\lambda \otimes f_{i_r}^2 v_\mu\right) \\ &\equiv \tilde{f}_{i_r} v_\lambda \otimes \tilde{f}_{i_r} v_\mu \mod q\mathcal{L}(\lambda) \otimes \mathcal{L}(\mu).\end{aligned}$$

Hence Lemma 5.3.2 yields our assertion. □

Proposition 5.3.5. *For all $\alpha \in Q_+(r)$, we have*

$$(\mathcal{L}(\lambda) \otimes \mathcal{L}(\mu))_{\lambda+\mu-\alpha} = \sum \tilde{f}_i((\mathcal{L}(\lambda) \otimes \mathcal{L}(\mu))_{\lambda+\mu-\alpha+\alpha_i}) + v_\lambda \otimes \mathcal{L}(\mu)_{\mu-\alpha}.$$

Proof. Let

$$\mathcal{L} = (\mathcal{L}(\lambda) \otimes \mathcal{L}(\mu))_{\lambda+\mu-\alpha}$$

and

$$\mathcal{L}' = \sum \tilde{f}_i((\mathcal{L}(\lambda) \otimes \mathcal{L}(\mu))_{\lambda+\mu-\alpha+\alpha_i}) + v_\lambda \otimes \mathcal{L}(\mu)_{\mu-\alpha}.$$

We already know that $\mathcal{L}' \subset \mathcal{L}$ (Lemma 5.3.2).

For the other inclusion, recall that, by Lemma 5.3.2 (5), for any $\beta \in Q_+(r-1)$ and $b \otimes b' \in \mathcal{B}(\lambda)_{\lambda-\beta} \otimes \mathcal{B}(\mu)_{\mu-\alpha+\beta}$, there exists an index $i \in I$ such that $\tilde{e}_i(b \otimes b') \neq 0$. By Lemma 5.3.2 (4), we have

$$b \otimes b' \equiv \tilde{f}_i \tilde{e}_i(b \otimes b') \mod q\mathcal{L}(\lambda) \otimes \mathcal{L}(\mu).$$

Hence $\mathcal{L}(\lambda)_{\lambda-\beta} \otimes \mathcal{L}(\mu)_{\mu-\alpha+\beta} \subset \mathcal{L}' + q\mathcal{L}$, and we obtain

$$\begin{aligned} \mathcal{L} &= \mathcal{L}(\lambda)_{\lambda-\alpha} \otimes v_\mu + \sum_{\beta \in Q_+(r-1)\setminus\{0\}} \mathcal{L}(\lambda)_{\lambda-\beta} \otimes \mathcal{L}(\mu)_{\mu-\alpha+\beta} \\ &\quad + v_\lambda \otimes \mathcal{L}(\mu)_{\mu-\beta} \\ &\subset \mathcal{L}(\lambda)_{\lambda-\alpha} \otimes v_\mu + \mathcal{L}' + q\mathcal{L}. \end{aligned}$$

Note that for any $\tilde{f}_{i_1} \cdots \tilde{f}_{i_r} v_\lambda \in \mathcal{B}(\lambda)_{\lambda-\alpha}$, Lemma 5.3.2 (6) gives

$$\tilde{f}_{i_1} \cdots \tilde{f}_{i_r} v_\lambda \otimes v_\mu \equiv \tilde{f}_{i_1}((\tilde{f}_{i_2} \cdots \tilde{f}_{i_r} v_\lambda) \otimes v_\mu) \mod q\mathcal{L}(\lambda) \otimes \mathcal{L}(\mu).$$

It follows that $\mathcal{L}(\lambda)_{\lambda-\alpha} \otimes v_\mu \subset \mathcal{L}' + q\mathcal{L}$, and hence $\mathcal{L} = \mathcal{L}' + q\mathcal{L}$. Therefore, by Nakayama's Lemma, we conclude that $\mathcal{L} = \mathcal{L}'$. □

For $\lambda, \mu \in P^+$, define a linear transformation $S_{\lambda,\mu} : V(\lambda) \otimes V(\mu) \longrightarrow V(\lambda)$ by

$$\begin{aligned} &S_{\lambda,\mu}(u \otimes v_\mu) = u \quad \text{for all } u \in V(\lambda), \\ &S_{\lambda,\mu}\left(V(\lambda) \otimes \sum f_i V(\mu)\right) = 0. \end{aligned} \tag{5.10}$$

It is straightforward to verify that $S_{\lambda,\mu}$ is a $U_q^-(\mathfrak{g})$-modules homomorphism (Exercise 5.8).

Lemma 5.3.6.

(1) *For $\lambda, \mu \in P^+$, we have*

$$S_{\lambda,\mu}(\mathcal{L}(\lambda) \otimes \mathcal{L}(\mu)) = \mathcal{L}(\lambda).$$

(2) *For all $\alpha \in Q_+(r-1)$ and $w \in (\mathcal{L}(\lambda) \otimes \mathcal{L}(\mu))_{\lambda+\mu-\alpha}$, we have*

$$S_{\lambda,\mu}\tilde{f}_i(w) \equiv \tilde{f}_i S_{\lambda,\mu}(w) \mod q\mathcal{L}(\lambda).$$

Proof. (1) Since $\mathcal{L}(\mu)_\mu = \mathbf{A}_0 v_\mu$, our assertion follows immediately from the definition of $S_{\lambda,\mu}$.

(2) We may assume that $w = f_i^{(k)}u \otimes f_i^{(l)}v$ with $u \in \mathcal{L}(\lambda)$, $v \in \mathcal{L}(\mu)$, $e_i u = e_i v = 0$. Let $\mathcal{L}$ be the free $\mathbf{A}_0$-submodule spanned by the vectors $f_i^{(s)}u \otimes f_i^{(t)}v$ $(s, t \geq 0)$. Then the tensor product rule (Theorem 4.4.3) yields

$$\tilde{f}_i w \equiv f_i^{(k+1)}u \otimes f_i^{(l)}v \quad \text{or} \quad \tilde{f}_i w \equiv f_i^{(k)}u \otimes f_i^{(l+1)}v \mod q\mathcal{L}.$$

Observe that $S_{\lambda,\mu}(\tilde{f}_i w) \equiv \tilde{f}_i S_{\lambda,\mu}(w) \equiv 0 \bmod q\mathcal{L}$ unless $l = 0$ and $v \in \mathcal{L}(\mu)_\mu$. Thus we may assume that $v = v_\mu$. By Lemma 5.3.2, we have

$$\tilde{f}_i(f_i^{(k)}u \otimes v_\mu) \equiv \begin{cases} f_i^{(k+1)}u \otimes v_\mu \mod q\mathcal{L} & \text{if } f_i^{(k+1)}u \neq 0, \\ f_i^{(k)}u \otimes f_i v_\mu \mod q\mathcal{L} & \text{if } f_i^{(k+1)}u = 0. \end{cases}$$

It follows that

$$S_{\lambda,\mu}(\tilde{f}_i w) \equiv f_i^{(k+1)}u \equiv \tilde{f}_i f_i^{(k)}u \equiv \tilde{f}_i S_{\lambda,\mu}(w) \mod q\mathcal{L}(\lambda) \otimes \mathcal{L}(\mu).$$

□

We will now prove the statement $\mathbf{A}(\mathbf{r})$. Given a dominant integral weight $\lambda \in P^+$, let T be a finite subset of P^+ *containing* λ *and all the fundamental weights.* Note that, for each $\mu \in T$, there is a nonnegative integer $n_\mu \geq 0$ such that $\tilde{e}_i \mathcal{L}(\mu)_{\mu-\alpha} \subset q^{-n_\mu}\mathcal{L}(\mu)$.

Moreover, since

$$\sum_{\tau \geq \mu-\alpha} \mathcal{L}(\mu)_\tau \xrightarrow{\sim} \sum_{\tau \geq \mu'-\alpha} \mathcal{L}(\mu')_\tau \quad \text{for all } \mu, \mu' \gg 0,$$

there is a nonnegative integer $n \geq 0$ such that $\tilde{e}_i \mathcal{L}(\mu)_{\mu-\alpha} \subset q^{-n}\mathcal{L}(\mu)$ for all $\mu \gg 0$. (Here, "$\xrightarrow{\sim}$" denotes the linear isomorphism that commutes with e_i and f_i whenever the actions are defined.) Therefore, given $\alpha \in Q_+(r)$, there is a nonnegative integer $N \geq 0$ such that

$$\tilde{e}_i \mathcal{L}(\mu)_{\mu-\alpha} \subset q^{-N}\mathcal{L}(\mu) \quad \text{for all } \mu \in T, \ \mu \gg 0. \tag{5.11}$$

Lemma 5.3.7. *Given $\alpha \in Q_+(r)$, suppose $N \geq 0$ is a nonnegative integer satisfying* (5.11). *Then for all $\lambda \in T$ and $\mu \gg 0$, we have*

$$\tilde{e}_i(\mathcal{L}(\lambda) \otimes \mathcal{L}(\mu))_{\lambda+\mu-\alpha} \subset q^{-N}\mathcal{L}(\lambda) \otimes \mathcal{L}(\mu).$$

Proof. Let $u \in \mathcal{L}(\lambda)_{\lambda-\beta}$ and $v \in \mathcal{L}(\mu)_{\mu-\gamma}$ with $\alpha = \beta + \gamma$. If $\beta \neq 0$ or $\gamma \neq 0$, then we have already proved our assertion in Lemma 5.3.2 (1).

If $\beta = 0$, then $\gamma = \alpha$ and we may assume $u = v_\lambda$. Write $v = \sum_{l \geq 0} f_i^{(l)} v_l$ with $e_i v_l = 0$, $\langle h_i, \mu - \alpha + l\alpha_i \rangle \geq l \geq 0$. By (5.11), we have

$$\tilde{e}_i v = \sum_{l \geq 1} f_i^{(l)} v_l \in q^{-N} \mathcal{L}(\mu).$$

It follows that $v_l \in q^{-N} \mathcal{L}(\mu)$ for all $l \geq 1$ (Exercise 5.9).

Let $\mathcal{L}$ be the free $\mathbf{A}_0$-submodule spanned by the vectors $f_i^{(s)} v_\lambda \otimes f_i^{(t)} v_l$ $(s, t \geq 0)$. By the tensor product rule (Theorem 4.4.3), we have $\tilde{e}_i \mathcal{L} \subset \mathcal{L}$ and $\tilde{f}_i \mathcal{L} \subset \mathcal{L}$. Note that $\mathcal{L} \subset q^{-N} \mathcal{L}(\lambda) \otimes \mathcal{L}(\mu)$. It follows that

$$\begin{aligned} \tilde{e}_i(v_\lambda \otimes v) &= \sum_{l \geq 0} \tilde{e}_i(v_\lambda \otimes f_i^{(l)} v_l) \\ &= \sum_{l \geq 1} \tilde{e}_i(v_\lambda \otimes f_i^{(l)} v_l) \in \mathcal{L} \subset q^{-N} \mathcal{L}(\lambda) \otimes \mathcal{L}(\mu). \end{aligned}$$

If $\beta = \alpha$ and $\gamma = 0$, a similar argument yields $\tilde{e}_i(u \otimes v_\mu) \in q^{-N} \mathcal{L}(\lambda) \otimes \mathcal{L}(\mu)$ (Exercise 5.10). $\square$

Lemma 5.3.8. *Given $\alpha \in Q_+(r)$, suppose $N \geq 0$ is a nonnegative integer satisfying* (5.11). *Then we have*

(1) $\tilde{e}_i \mathcal{L}(\mu)_{\mu - \alpha} \subset q^{-N+1} \mathcal{L}(\mu)$ *for all* $\mu \gg 0$,
(2) $\tilde{e}_i \mathcal{L}(\mu)_{\mu - \alpha} \subset q^{-N+1} \mathcal{L}(\mu)$ *for all* $\mu \in T$.

Proof. (1) Let $u = \tilde{f}_{i_1} \cdots \tilde{f}_{i_r} v_\mu \in \mathcal{L}(\mu)_{\mu - \alpha}$ and set $\mu_0 = \Lambda_{i_{r-1}} \in T$, $\tau = \mu - \mu_0 \gg 0$. By Lemma 5.3.4, we have

$$w = \tilde{f}_{i_1} \cdots \tilde{f}_{i_r}(v_{\mu_0} \otimes v_\tau) \equiv v \otimes v' \mod q\mathcal{L}(\mu_0) \otimes \mathcal{L}(\tau)$$

for some $v \in \mathcal{L}(\mu_0)_{\mu_0 - \beta}$, $v' \in \mathcal{L}(\tau)_{\tau - \gamma}$, $\beta, \gamma \in Q_+(r-1) \setminus \{0\}$ with $\alpha = \beta + \gamma$. By Lemma 5.3.2, we have $\tilde{e}_i(v \otimes v') \in \mathcal{L}(\mu_0) \otimes \mathcal{L}(\tau)$. Hence Lemma 5.3.7 gives

$$\begin{aligned} \tilde{e}_i w &\in \mathcal{L}(\mu_0) \otimes \mathcal{L}(\tau) + q\tilde{e}_i(\mathcal{L}(\mu_0) \otimes \mathcal{L}(\tau))_{\mu_0 + \tau - \alpha} \\ &\subset \mathcal{L}(\mu_0) \otimes \mathcal{L}(\tau) + q^{-N+1} \mathcal{L}(\mu_0) \otimes \mathcal{L}(\tau) = q^{-N+1} \mathcal{L}(\mu_0) \otimes \mathcal{L}(\tau). \end{aligned}$$

Note that we actually have

$$\tilde{e}_i w = \tilde{e}_i \tilde{f}_{i_1} \cdots \tilde{f}_{i_r}(v_{\mu_0} \otimes v_\tau) \in q^{-N+1}(\mathcal{L}(\mu_0) \otimes \mathcal{L}(\tau))_{\mu_0 + \tau - \alpha + \alpha_i}.$$

Hence, applying $\Psi_{\mu_0, \tau} : V(\mu_0) \otimes V(\tau) \longrightarrow V(\mu_0 + \tau) = V(\mu)$, the statement $\mathbf{F(r-1)}$ yields

$$\tilde{e}_i u = \tilde{e}_i \tilde{f}_{i_1} \cdots \tilde{f}_{i_r} v_\mu \in q^{-N+1} \mathcal{L}(\mu).$$

(2) Let $u = \tilde{f}_{i_1} \cdots \tilde{f}_{i_r} v_\mu \in \mathcal{L}(\mu)_{\mu - \alpha}$. If $u \in q\mathcal{L}(\lambda)$, then our assertion follows immediately from (5.11). If $u \notin q\mathcal{L}(\mu)$, then for any $\tau \in P^+$,

Lemma 5.3.2 (7) gives

$$\tilde{f}_{i_1} \cdots \tilde{f}_{i_r}(v_\mu \otimes v_\tau) \equiv u \otimes v_\tau \mod q\mathcal{L}(\mu) \otimes \mathcal{L}(\tau). \tag{5.12}$$

If $\tau \gg 0$, then (1) implies

$$\tilde{e}_i \tilde{f}_{i_1} \cdots \tilde{f}_{i_r} v_{\mu+\tau} \in q^{-N+1}\mathcal{L}(\mu+\tau).$$

Applying $\Phi_{\mu,\tau}$, $\mathbf{E}(\mathbf{r}-\mathbf{1})$ gives

$$\tilde{e}_i \tilde{f}_{i_1} \cdots \tilde{f}_{i_r}(v_\mu \otimes v_\tau) \in q^{-N+1}\mathcal{L}(\mu) \otimes \mathcal{L}(\tau).$$

Hence (5.12) and Lemma 5.3.7 yield

$$\begin{aligned}\tilde{e}_i(u \otimes v_\tau) &\in q^{-N+1}\mathcal{L}(\mu) \otimes \mathcal{L}(\tau) + q\tilde{e}_i(\mathcal{L}(\mu) \otimes \mathcal{L}(\tau)) \\ &\subset q^{-N+1}\mathcal{L}(\mu) \otimes \mathcal{L}(\tau).\end{aligned}$$

Write $u = \sum_{k \geq 0} f_i^{(k)} u_k$ with $e_i u_k = 0$. Then $\tilde{e}_i u = \sum_{k \geq 1} f_i^{(k-1)} u_k \in q^{-N}\mathcal{L}(\mu)$, which implies that $u_k \in q^{-N}\mathcal{L}(\mu)$ for all $k \geq 1$ (Exercise 5.9).

Let $\mathcal{L}$ be the free $\mathbf{A}_0$-submodule spanned by the vectors $f_i^{(s)} u_k \otimes f_i^{(t)} v_\tau$ $(s, t \geq 0,\ k \geq 1)$. Then $\mathcal{L} \subset q^{-N}\mathcal{L}(\mu) \otimes \mathcal{L}(\tau)$ and by the tensor product rule (Theorem 4.4.3), we have

$$\begin{aligned}\tilde{e}_i(u \otimes v_\tau) &= \sum_{k \geq 1} \tilde{e}_i(f_i^{(k)} u_k \otimes v_\tau) \\ &\equiv \sum_{k \geq 1} f_i^{(k-1)} u_k \otimes v_\tau \equiv \tilde{e}_i u \otimes v_\tau \mod q\mathcal{L}.\end{aligned}$$

Hence we get

$$\tilde{e}_i u \otimes v_\tau \equiv \tilde{e}_i(u \otimes v_\tau) \mod q^{-N+1}\mathcal{L}(\mu) \otimes \mathcal{L}(\tau).$$

Since $\tilde{e}_i u \otimes v_\tau \in q^{-N+1}(\mathcal{L}(\mu) \otimes \mathcal{L}(\tau))_{\mu+\tau-\alpha+\alpha_i}$, by Lemma 5.3.6, we obtain $\tilde{e}_i u \in q^{-N+1}\mathcal{L}(\mu)$, as desired. □

Therefore, by applying Lemma 5.3.8 repeatedly, we conclude that $N = 0$ and the statement $\mathbf{A}(\mathbf{r})$ follows immediately.

Proposition 5.3.9 ($\mathbf{A}(\mathbf{r})$). *For all $\lambda \in P^+$ and $\alpha \in Q_+^{(r)}$, we have*

$$\tilde{e}_i \mathcal{L}(\lambda)_{\lambda-\alpha} \subset \mathcal{L}(\lambda).$$

As an immediate corollary, we can prove:

Corollary 5.3.10. *Let $\lambda, \mu \in P^+$ and $\alpha, \beta \in Q_+(r)$.*

(1) *If $u = \sum_{k \geq 0} f_i^{(k)} u_k \in \mathcal{L}(\lambda)_{\lambda-\alpha}$ with $e_i u_k = 0$, $\langle h_i, \lambda - \alpha + k\alpha_i \rangle \geq k \geq 0$, then $u_k \in \mathcal{L}(\lambda)$ for all $k \geq 0$.*

(2) *For each* $i \in I$, *we have*

$$\tilde{e}_i(\mathcal{L}(\lambda)_{\lambda-\alpha} \otimes \mathcal{L}(\mu)_{\mu-\beta}) \subset \mathcal{L}(\lambda) \otimes \mathcal{L}(\mu),$$
$$\tilde{f}_i(\mathcal{L}(\lambda)_{\lambda-\alpha} \otimes \mathcal{L}(\mu)_{\mu-\beta}) \subset \mathcal{L}(\lambda) \otimes \mathcal{L}(\mu).$$

Proof. The proof is the same as Lemmas 5.3.1 and 5.3.2 (Exercise 5.11). □

To prove the statement $\mathbf{B}(\mathbf{r})$, we first prove:

Lemma 5.3.11. *Let* $\lambda, \mu \in P^+$ *and* $\alpha \in Q_+(r)$.

(1) *For all* $u \in \mathcal{L}(\lambda)_{\lambda-\alpha}$, *we have*

$$\tilde{e}_i(u \otimes v_\mu) \equiv \tilde{e}_i u \otimes v_\mu \mod q\mathcal{L}(\lambda) \otimes \mathcal{L}(\mu).$$

(2) *If* $\lambda \gg 0$, *then* $\tilde{e}_i \mathcal{B}(\lambda)_{\lambda-\alpha} \subset \mathcal{B}(\lambda) \cup \{0\}$.

Proof. (1) By Corollary 5.3.10 (1), we may assume $u = f_i^{(k)} w$ with $e_i w = 0$, $w \in \mathcal{L}(\lambda)_{\lambda-\alpha+k\alpha_i}$. Let $\mathcal{L}$ be the free $\mathbf{A}_0$-submodule spanned by the vectors $f_i^{(s)} w \otimes f_i^{(t)} v_\mu$ $(s, t \geq 0)$. Then the tensor product rule (Theorem 4.4.3) gives

$$\tilde{e}_i(f_i^{(k)} w \otimes v_\mu) \equiv f_i^{(k-1)} w \otimes v_\mu \mod q\mathcal{L}.$$

Since $\mathcal{L} \subset \mathcal{L}(\lambda) \otimes \mathcal{L}(\mu)$, our assertion follows immediately.

(2) Let $u = \tilde{f}_{i_1} \cdots \tilde{f}_{i_r} v_\lambda$ with $\lambda \gg 0$. Set $\lambda_0 = \Lambda_{i_{r-1}}$ and take $\mu = \lambda - \lambda_0 \gg 0$. By Lemma 5.3.4, we have

$$\tilde{f}_{i_1} \cdots \tilde{f}_{i_r}(v_{\lambda_0} \otimes v_\mu) \equiv v \otimes v' \mod q\mathcal{L}(\lambda_0) \otimes \mathcal{L}(\mu)$$

for some $v \in \mathcal{L}(\lambda_0)_{\lambda_0-\beta}$, $v' \in \mathcal{L}(\mu)_{\mu-\gamma}$, $\beta, \gamma \in Q_+(r-1) \setminus \{0\}$, $\alpha = \beta + \gamma$ such that $v + q\mathcal{L}(\lambda_0) \in \mathcal{B}(\lambda_0)$, $v' + q\mathcal{L}(\mu) \in \mathcal{B}(\mu) \cup \{0\}$. Hence the tensor product rule (Theorem 4.4.3) gives

$$\begin{aligned}\tilde{e}_i \tilde{f}_{i_1} \cdots \tilde{f}_{i_r}(v_{\lambda_0} \otimes v_\mu) &\equiv \tilde{e}_i(v \otimes v') \\ &\equiv \tilde{e}_i v \otimes v' \text{ or } v \otimes \tilde{e}_i v' \mod q\mathcal{L}(\lambda_0) \otimes \mathcal{L}(\mu).\end{aligned}$$

By $\mathbf{B}(\mathbf{r-1})$, we have

$$\tilde{e}_i \tilde{f}_{i_1} \cdots \tilde{f}_{i_r}(v_{\lambda_0} \otimes v_\mu) + q\mathcal{L}(\lambda_0) \otimes \mathcal{L}(\mu) \in (\mathcal{B}(\lambda_0) \otimes \mathcal{B}(\mu)) \cup \{0\}.$$

Hence, by applying $\Psi_{\lambda_0,\mu}$, the statement $\mathbf{G}(\mathbf{r-1})$ yields

$$\tilde{e}_i(u + q\mathcal{L}(\lambda)) = \tilde{e}_i \tilde{f}_{i_1} \cdots \tilde{f}_{i_r} v_\lambda + q\mathcal{L}(\lambda) \in \mathcal{B}(\lambda) \cup \{0\}.$$

□

We now prove the statement $\mathbf{B}(\mathbf{r})$.

Proposition 5.3.12 ($\mathbf{B}(\mathbf{r})$). *For all* $\lambda \in P^+$ *and* $\alpha \in Q_+(r)$, *we have*

$$\tilde{e}_i \mathcal{B}(\lambda)_{\lambda-\alpha} \subset \mathcal{B}(\lambda) \sqcup \{0\}.$$

Proof. Let $b = \tilde{f}_{i_1} \cdots \tilde{f}_{i_r} v_\lambda \in \mathcal{B}(\lambda)_{\lambda-\alpha}$ and take $\mu \gg 0$. Then Lemma 5.3.2 (7) yields

$$\tilde{f}_{i_1} \cdots \tilde{f}_{i_r}(v_\lambda \otimes v_\mu) \equiv b \otimes v_\mu \mod q\mathcal{L}(\lambda) \otimes \mathcal{L}(\mu).$$

Combining Corollary 5.3.10 (2) and Lemma 5.3.11 (1), we have

$$\begin{aligned}\tilde{e}_i \tilde{f}_{i_1} \cdots \tilde{f}_{i_r}(v_\lambda \otimes v_\mu) &\equiv \tilde{e}_i(b \otimes v_\mu) \\ &\equiv \tilde{e}_i b \otimes v_\mu \mod q\mathcal{L}(\lambda) \otimes \mathcal{L}(\mu).\end{aligned} \tag{5.13}$$

On the other hand, since $\lambda + \mu \gg 0$, Lemma 5.3.11 (2) gives

$$\tilde{e}_i \tilde{f}_{i_1} \cdots \tilde{f}_{i_r} v_{\lambda+\mu} + q\mathcal{L}(\lambda+\mu) \in \mathcal{B}(\lambda+\mu) \cup \{0\}.$$

If $\tilde{e}_i \tilde{f}_{i_1} \cdots \tilde{f}_{i_r} v_{\lambda+\mu} \in q\mathcal{L}(\lambda+\mu)$, then by the statement $\mathbf{E}(\mathbf{r})$, we get

$$0 \equiv \tilde{e}_i \tilde{f}_{i_1} \cdots \tilde{f}_{i_r}(v_\lambda \otimes v_\mu) \equiv \tilde{e}_i b \otimes v_\mu \mod q\mathcal{L}(\lambda) \otimes \mathcal{L}(\mu),$$

which implies $\tilde{e}_i b = 0$.

If $\tilde{e}_i \tilde{f}_{i_1} \cdots \tilde{f}_{i_r} v_{\lambda+\mu} \notin q\mathcal{L}(\lambda+\mu)$, we can write

$$\tilde{e}_i \tilde{f}_{i_1} \cdots \tilde{f}_{i_r} v_{\lambda+\mu} + q\mathcal{L}(\lambda+\mu) = \tilde{f}_{j_2} \cdots \tilde{f}_{j_r} v_{\lambda+\mu} + q\mathcal{L}(\lambda+\mu)$$

for some $j_2, \ldots, j_r \in I$. Combining (5.13) and the statement $\mathbf{E}(\mathbf{r})$ yields

$$\tilde{e}_i b \otimes v_\mu \equiv \tilde{e}_i \tilde{f}_{i_1} \cdots \tilde{f}_{i_r}(v_\lambda \otimes v_\mu) \equiv \tilde{f}_{j_2} \cdots \tilde{f}_{j_r}(v_\lambda \otimes v_\mu) \mod q\mathcal{L}(\lambda) \otimes \mathcal{L}(\mu).$$

Hence, by applying the map $S_{\lambda,\mu}$, we obtain

$$\tilde{e}_i b \equiv \tilde{f}_{j_2} \cdots \tilde{f}_{j_r} v_\lambda \mod q\mathcal{L}(\lambda)$$

as desired. □

To prove the remaining inductive statements, we need to make use of some fundamental properties of the symmetric bilinear form $(\ ,\ )$ defined in Section 5.1. Let $\lambda \in P^+$ be a dominant integral weight, and let $(\ ,\ )$ be the symmetric bilinear form on $V(\lambda)$ satisfying (5.2). By the same argument as in Lemma 5.1.5, one can prove

$$(\tilde{f}_i u, v) \equiv (u, \tilde{e}_i v) \mod q\mathbf{A}_0$$

for all $u \in \mathcal{L}(\lambda)_{\lambda-\alpha+\alpha_i}$, $v \in \mathcal{L}(\lambda)_{\lambda-\alpha}$, $\alpha \in Q_+(r)$, which implies

$$(\mathcal{L}(\lambda), \mathcal{L}(\lambda)) \subset \mathbf{A}_0.$$

Set

$$\begin{aligned}\mathcal{L}^\vee &= \{u \in V(\lambda) \mid (u, \mathcal{L}(\lambda)) \subset \mathbf{A}_0\}, \\ \mathcal{L}^{\vee\vee} &= \{u \in V(\lambda) \mid (u, \mathcal{L}(\lambda)^\vee) \subset \mathbf{A}_0\}.\end{aligned}$$

It is easy to see that $\mathcal{L}(\lambda) \subset \mathcal{L}(\lambda)^{\vee\vee} \subset \mathcal{L}(\lambda)^\vee$ (Exercise 5.12). Actually, we have:

Lemma 5.3.13.

$$\mathcal{L}(\lambda)^{\vee\vee} = \mathcal{L}(\lambda) \quad \textit{for all } \lambda \in P^+.$$

Proof. We have seen in the proof of Theorem 5.1.1 that $\mathcal{L}(\lambda)_{\lambda-\alpha}$ is a free $\mathbf{A}_0$-submodule of $V(\lambda)_{\lambda-\alpha}$ such that $\mathbf{F}(q) \otimes \mathcal{L}(\lambda)_{\lambda-\alpha} \cong V(\lambda)_{\lambda-\alpha}$. Let us take $\{u_1, \ldots, u_n\}$ to be an $\mathbf{A}_0$-basis of $\mathcal{L}(\lambda)_{\lambda-\alpha}$ which is also an $\mathbf{F}(q)$-basis of $V(\lambda)_{\lambda-\alpha}$. Since $(\ ,\ )$ is nondegenerate on $V(\lambda)_{\lambda-\alpha}$, there is an $\mathbf{F}(q)$-basis $\{v_1, \ldots, v_n\}$ of $V(\lambda)_{\lambda-\alpha}$ which is dual to $\{u_1, \ldots, u_n\}$. Clearly, $v_1, \ldots, v_n \in \mathcal{L}(\lambda)^\vee$. Moreover, if $v = \sum c_i v_i \in \mathcal{L}(\lambda)^\vee$ ($c_i \in \mathbf{F}(q)$), then we have $(u_i, v) = c_i \in \mathbf{A}_0$, which implies $\{v_1, \ldots, v_n\}$ is an $\mathbf{A}_0$-basis of $\mathcal{L}(\lambda)^\vee$.

Similarly, one can prove $\{u_1, \ldots, u_n\}$ is an $\mathbf{A}_0$-basis of $\mathcal{L}(\lambda)^{\vee\vee}_{\lambda-\alpha}$. Therefore, we conclude $\mathcal{L}(\lambda)^{\vee\vee}_{\lambda-\alpha} = \mathcal{L}(\lambda)_{\lambda-\alpha}$, which proves our claim. $\square$

Remark 5.3.14. In Section 5.1, we have seen that $\mathcal{L}(\lambda) = \mathcal{L}(\lambda)^\vee$. But this can be proved only *after* the existence theorem is proved.

We are now ready to prove the special case of the statement $\mathbf{F}(\mathbf{r})$, which in turn will be used to prove the general case of $\mathbf{F}(\mathbf{r})$.

Lemma 5.3.15. *For all $\lambda \in P^+$, $\mu \gg 0$ and $\alpha \in Q_+(r)$, we have*

$$\Psi_{\lambda,\mu}((\mathcal{L}(\lambda) \otimes \mathcal{L}(\mu))_{\lambda+\mu-\alpha}) = \mathcal{L}(\lambda+\mu)_{\lambda+\mu-\alpha}.$$

Proof. By the statement $\mathbf{E}(\mathbf{r})$, it is obvious that

$$\begin{aligned}\mathcal{L}(\lambda+\mu)_{\lambda+\mu-\alpha} &= \Psi_{\lambda,\mu} \circ \Phi_{\lambda,\mu}(\mathcal{L}(\lambda+\mu)_{\lambda+\mu-\alpha}) \\ &\subset \Psi_{\lambda,\mu}((\mathcal{L}(\lambda) \otimes \mathcal{L}(\mu))_{\lambda+\mu-\alpha}).\end{aligned}$$

For the other inclusion, recall that (Proposition 5.3.5)

$$\begin{aligned}(\mathcal{L}(\lambda) \otimes \mathcal{L}(\mu))_{\lambda+\mu-\alpha} =& \sum \tilde{f}_i((\mathcal{L}(\lambda) \otimes \mathcal{L}(\mu))_{\lambda+\mu-\alpha+\alpha_i}) \\ &+ v_\lambda \otimes \mathcal{L}(\mu)_{\mu-\alpha}.\end{aligned}$$

By the induction hypothesis $\mathbf{F}(\mathbf{r-1})$, we have

$$\begin{aligned}\Psi_{\lambda,\mu}&\Big(\sum \tilde{f}_i(\mathcal{L}(\lambda) \otimes \mathcal{L}(\mu))_{\lambda+\mu-\alpha+\alpha_i}\Big) \\ &= \sum \tilde{f}_i \Psi_{\lambda,\mu}((\mathcal{L}(\lambda) \otimes \mathcal{L}(\mu))_{\lambda+\mu-\alpha+\alpha_i}) \\ &\subset \sum \tilde{f}_i \mathcal{L}(\lambda+\mu)_{\lambda+\mu-\alpha+\alpha_i} \\ &= \mathcal{L}(\lambda+\mu)_{\lambda+\mu-\alpha}.\end{aligned}$$

Thus it remains to show $\Psi_{\lambda,\mu}(v_\lambda \otimes \mathcal{L}(\mu)_{\mu-\alpha}) \subset \mathcal{L}(\lambda+\mu)_{\lambda+\mu-\alpha}$.

Let $u \in \mathcal{L}(\lambda+\mu)^\vee_{\lambda+\mu-\alpha}$ and write $u = P_\alpha v_{\lambda+\mu}$, where $P_\alpha \in U_q^-(\mathfrak{g})_{-\alpha}$ is a polynomial in f_i's ($i \in I$). Observe that

$$\Delta(P_\alpha) = P_\alpha \otimes 1 + (\text{intermediate terms}) + K_\alpha \otimes P_\alpha,$$

which yields

$$\begin{aligned}(\Phi_{\lambda,\mu}(u), v_\lambda \otimes \mathcal{L}(\mu)_{\mu-\alpha}) &= (\Delta(P_\alpha)(v_\lambda \otimes v_\mu), v_\lambda \otimes \mathcal{L}(\mu)_{\mu-\alpha}) \\ &= (P_\alpha v_\lambda \otimes v_\mu + \cdots + K_\alpha v_\lambda \otimes P_\alpha v_\mu, v_\lambda \otimes \mathcal{L}(\mu)_{\mu-\alpha}) \\ &= q^{\langle h_\alpha, \lambda\rangle}(P_\alpha v_\mu, \mathcal{L}(\mu)_{\mu-\alpha}).\end{aligned}$$

Since $\mu \gg 0$, we have $P_\alpha v_{\lambda+\mu} \in \mathcal{L}(\lambda+\mu)^\vee_{\lambda+\mu-\alpha}$ if and only if $P_\alpha v_\mu \in \mathcal{L}(\mu)^\vee_{\mu-\alpha}$ (Exercise 5.13). It follows that

$$\begin{aligned}(u, \Psi_{\lambda,\mu}(v_\lambda \otimes \mathcal{L}(\mu)_{\mu-\alpha})) &= (\Phi_{\lambda,\mu}(u), v_\lambda \otimes \mathcal{L}(\mu)_{\mu-\alpha}) \\ &= q^{\langle h_\alpha, \lambda\rangle}(P_\alpha v_\mu, \mathcal{L}(\mu)_{\mu-\alpha}) \subset \mathbf{A}_0.\end{aligned}$$

Therefore, by Lemma 5.3.13, we have

$$\Psi_{\lambda,\mu}(v_\lambda \otimes \mathcal{L}(\mu)_{\mu-\alpha}) \subset \mathcal{L}(\lambda+\mu)^{\vee\vee}_{\lambda+\mu-\alpha} = \mathcal{L}(\lambda+\mu)_{\lambda+\mu-\alpha},$$

which completes the proof. □

In the next two propositions, we will prove the statements $\mathbf{C(r)}$ and $\mathbf{D(r)}$.

Proposition 5.3.16 ($\mathbf{C(r)}$).

(1) *Suppose $\lambda \gg 0$ and $\alpha \in Q_+(r)$. Then for all $i \in I$, $b \in \mathcal{B}(\lambda)_{\lambda-\alpha+\alpha_i}$ and $b' \in \mathcal{B}(\lambda)_{\lambda-\alpha}$, we have $\tilde{f}_i b = b'$ if and only if $b = \tilde{e}_i b'$.*

(2) *Suppose $\lambda \in P^+$ and $\alpha \in Q_+(r)$. Then for all $i \in I$, $b \in \mathcal{B}(\lambda)_{\lambda-\alpha+\alpha_i}$ and $b' \in \mathcal{B}(\lambda)_{\lambda-\alpha}$, we have $\tilde{f}_i b = b'$ if and only if $b = \tilde{e}_i b'$.*

Proof. (1) ($\Longrightarrow$) Let $\tilde{f}_i b = b' \in \mathcal{B}(\lambda)_{\lambda-\alpha}$. Since $b \in \mathcal{B}(\lambda)_{\lambda-\alpha+\alpha_i}$, by Lemma 5.3.1 (2), there exists $k_0 \geq 0$ such that $b \equiv f_i^{(k_0)} b_0$ with $\tilde{e}_i b_0 = 0$. Hence $b' \equiv f_i^{(k_0+1)} b_0$ and $\tilde{e}_i b' \equiv f_i^{(k_0)} b_0 \equiv b$.

($\Longleftarrow$) Let $b' = \tilde{f}_{i_1} \cdots \tilde{f}_{i_r} v_\lambda \in \mathcal{B}(\lambda)_{\lambda-\alpha}$ and suppose $\tilde{e}_i b' = b \neq 0$. Set $\lambda_0 = \Lambda_{i_{r-1}}$ and $\mu = \lambda - \lambda_0 \gg 0$. Then Lemma 5.3.4 yields

$$\tilde{f}_{i_1} \cdots \tilde{f}_{i_r}(v_{\lambda_0} \otimes v_\mu) \equiv v \otimes v' \mod q\mathcal{L}(\lambda_0) \otimes \mathcal{L}(\mu)$$

for some $v \in \mathcal{L}(\lambda_0)_{\lambda_0-\beta}$, $v' \in \mathcal{L}(\mu)_{\mu-\gamma}$, $\beta, \gamma \in Q_+(r-1) \setminus \{0\}$, $\alpha = \beta + \gamma$ such that $b_1 = v + q\mathcal{L}(\lambda_0) \in \mathcal{B}(\lambda_0)$, $b_2 = v' + q\mathcal{L}(\mu) \in \mathcal{B}(\mu) \cup \{0\}$. By Corollary 5.3.10 (2), we have

$$\tilde{e}_i \tilde{f}_{i_1} \cdots \tilde{f}_{i_r}(v_{\lambda_0} \otimes v_\mu) \equiv \tilde{e}_i(v \otimes v') \mod q\mathcal{L}(\lambda_0) \otimes \mathcal{L}(\mu),$$

and by $\mathbf{G(r-1)}$, we obtain

$$\tilde{e}_i b' = \tilde{e}_i \tilde{f}_{i_1} \cdots \tilde{f}_{i_r} v_\lambda \equiv \tilde{e}_i \Psi_{\lambda_0,\mu}(v \otimes v') \mod q\mathcal{L}(\lambda).$$

Hence $\tilde{e}_i(v \otimes v') \notin q\mathcal{L}(\lambda_0) \otimes \mathcal{L}(\mu)$ and Lemma 5.3.2 (4) gives

$$\begin{aligned}\tilde{f}_{i_1} \cdots \tilde{f}_{i_r}(v_{\lambda_0} \otimes v_\mu) &\equiv v \otimes v' \equiv \tilde{f}_i \tilde{e}_i (v \otimes v') \\ &\equiv \tilde{f}_i \tilde{e}_i \tilde{f}_{i_1} \cdots \tilde{f}_{i_r}(v_{\lambda_0} \otimes v_\mu) \mod q\mathcal{L}(\lambda_0) \otimes \mathcal{L}(\mu).\end{aligned}$$

Therefore, by Lemma 5.3.15, we obtain

$$\tilde{f}_{i_1} \cdots \tilde{f}_{i_r} v_\lambda \equiv \tilde{f}_i \tilde{e}_i \tilde{f}_{i_1} \cdots \tilde{f}_{i_r} v_\lambda \mod q\mathcal{L}(\lambda),$$

which implies $b' = \tilde{f}_i b$.

(2) ($\Longrightarrow$) The proof is the same as (1).

($\Longleftarrow$) Let $b' = \tilde{f}_{i_1} \cdots \tilde{f}_{i_r} v_\lambda \in \mathcal{B}(\lambda)_{\lambda-\alpha}$ and suppose $\tilde{e}_i b' = b \neq 0$. Write $b = \tilde{e}_i b' = \tilde{f}_{j_2} \cdots \tilde{f}_{j_r} v_\lambda \in \mathcal{B}(\lambda)_{\lambda-\alpha+\alpha_i}$ for some $j_2, \ldots, j_r \in I$.

Take $\mu \gg 0$. By Lemma 5.3.2 (7), we have

$$\tilde{f}_{i_1} \cdots \tilde{f}_{i_r}(v_\lambda \otimes v_\mu) \equiv (\tilde{f}_{i_1} \cdots \tilde{f}_{i_r} v_\lambda) \otimes v_\mu \equiv b' \otimes v_\mu \mod q\mathcal{L}(\lambda) \otimes \mathcal{L}(\mu),$$

and by Lemma 5.3.11 (1), we get

$$\begin{aligned}\tilde{e}_i \tilde{f}_{i_1} \cdots \tilde{f}_{i_r}(v_\lambda \otimes v_\mu) &\equiv \tilde{e}_i(b' \otimes v_\mu) \equiv \tilde{e}_i b' \otimes v_\mu \\ &\equiv \tilde{f}_{j_2} \cdots \tilde{f}_{j_r} v_\lambda \otimes v_\mu \mod q\mathcal{L}(\lambda) \otimes \mathcal{L}(\mu).\end{aligned}$$

Hence, by the statement $\mathbf{F(r-1)}$, we obtain

$$\tilde{e}_i \tilde{f}_{i_1} \cdots \tilde{f}_{i_r} v_{\lambda+\mu} \equiv \tilde{f}_{j_2} \cdots \tilde{f}_{j_r} v_{\lambda+\mu} \mod q\mathcal{L}(\lambda+\mu).$$

Note that $\tilde{f}_{j_2} \cdots \tilde{f}_{j_r} v_{\lambda+\mu} \notin q\mathcal{L}(\lambda+\mu)$; otherwise, by applying $\Phi_{\lambda,\mu}$, we would get $\tilde{f}_{j_2} \cdots \tilde{f}_{j_r}(v_\lambda \otimes v_\mu) \in q\mathcal{L}(\lambda) \otimes \mathcal{L}(\mu)$, and applying $S_{\lambda,\mu}$ would yield $\tilde{f}_{j_2} \cdots \tilde{f}_{j_r} v_{\lambda+\mu} \in q\mathcal{L}(\lambda)$, which is a contradiction.

Since $\lambda + \mu \gg 0$, by (1), we have

$$\tilde{f}_{i_1} \cdots \tilde{f}_{i_r} v_{\lambda+\mu} \equiv \tilde{f}_i \tilde{e}_i \tilde{f}_{i_1} \cdots \tilde{f}_{i_r} v_{\lambda+\mu} \equiv \tilde{f}_i \tilde{f}_{j_2} \cdots \tilde{f}_{j_r} v_{\lambda+\mu} \mod q\mathcal{L}(\lambda+\mu).$$

By applying $\Phi_{\lambda,\mu}$, we get

$$\begin{aligned}\tilde{f}_{i_1} \cdots \tilde{f}_{i_r}(v_\lambda \otimes v_\mu) &\equiv \tilde{f}_i \tilde{f}_{j_2} \cdots \tilde{f}_{j_r}(v_\lambda \otimes v_\mu) \\ &\equiv \tilde{f}_i(\tilde{f}_{j_2} \cdots \tilde{f}_{j_r} v_\lambda \otimes v_\mu) \\ &\equiv \tilde{f}_i(\tilde{e}_i b' \otimes v_\mu) \mod q\mathcal{L}(\lambda) \otimes \mathcal{L}(\mu).\end{aligned}$$

Therefore, by applying $S_{\lambda,\mu}$, we obtain

$$b' = \tilde{f}_{i_1} \cdots \tilde{f}_{i_r} v_\lambda \equiv \tilde{f}_i \tilde{e}_i b' = \tilde{f}_i b \mod q\mathcal{L}(\lambda).$$

□

Proposition 5.3.17 (D(r)). *For all $\lambda \in P^+$ and $\alpha \in Q_+(r)$, $\mathcal{B}(\lambda)_{\lambda-\alpha}$ is an $\mathbf{F}$-basis of $\mathcal{L}(\lambda)_{\lambda-\alpha}/q\mathcal{L}(\lambda)_{\lambda-\alpha}$.*

Proof. Suppose that we have an $\mathbf{F}$-linear dependence relation

$$\sum_{b \in \mathcal{B}(\lambda)_{\lambda-\alpha}} a_b b = 0 \quad \text{with } a_b \in \mathbf{F}.$$

Since $\tilde{e}_i \mathcal{B}(\lambda)_{\lambda-\alpha} \subset \mathcal{B}(\lambda) \sqcup \{0\}$ for all $i \in I$, we get

$$0 = \tilde{e}_i \left(\sum_{b \in \mathcal{B}(\lambda)_{\lambda-\alpha}} a_b b \right) = \sum_{\tilde{e}_i b \neq 0} a_b \tilde{e}_i b.$$

By the induction hypothesis $\mathbf{D(r-1)}$, the set $\{\tilde{e}_i b \,|\, \tilde{e}_i b \neq 0\}$ is linearly independent over $\mathbf{F}$. Therefore, $a_b = 0$ whenever $\tilde{e}_i b \neq 0$. But, for each $b \in \mathcal{B}(\lambda)_{\lambda-\alpha}$, there exists an $i \in I$ such that $\tilde{e}_i b \neq 0$. Hence $a_b = 0$ for all $b \in \mathcal{B}(\lambda)_{\lambda-\alpha}$, which completes the proof. □

Lemma 5.3.18. *Let $\lambda \in P^+$ and $\alpha \in Q_+(r)$.*

(1) *If $u \in \mathcal{L}(\lambda)_{\lambda-\alpha}/q\mathcal{L}(\lambda)_{\lambda-\alpha}$ and $\tilde{e}_i u = 0$ for all $i \in I$, then $u = 0$.*
(2) *If $u \in V(\lambda)_{\lambda-\alpha}$ and $\tilde{e}_i u \in \mathcal{L}(\lambda)$ for all $i \in I$, then $u \in \mathcal{L}(\lambda)$.*

Proof. (1) Write $u = \sum_{b \in \mathcal{B}(\lambda)_{\lambda-\alpha}} a_b b$ with $a_b \in \mathbf{F}$. Then $\sum_{\tilde{e}_i b \neq 0} a_b(\tilde{e}_i b) = 0$ for all $i \in I$. By the same argument as in Proposition 5.3.17, all $a_b = 0$, and hence $u = 0$.

(2) Choose the smallest $N \geq 0$ such that $q^N u \in \mathcal{L}(\lambda)$. If $N > 0$, then $\tilde{e}_i(q^N u) = q^N \tilde{e}_i u \in q\mathcal{L}(\lambda)$ for all $i \in I$. Hence, by (1), we would have $q^N u \in q\mathcal{L}(\lambda)$; i.e., $q^{N-1} u \in \mathcal{L}(\lambda)$, which is a contradiction to the minimality of N. Therefore, $N = 0$ and $u \in \mathcal{L}(\lambda)$. □

Proposition 5.3.19 (F(r)). *For all $\lambda, \mu \in P^+$ and $\alpha \in Q_+(r)$, we have*

$$\Psi_{\lambda,\mu}((\mathcal{L}(\lambda) \otimes \mathcal{L}(\mu))_{\lambda+\mu-\alpha}) \subset \mathcal{L}(\lambda+\mu).$$

Proof. By Corollary 5.3.10 (2), we have

$$\tilde{e}_i(\mathcal{L}(\lambda) \otimes \mathcal{L}(\mu))_{\lambda+\mu-\alpha} \subset (\mathcal{L}(\lambda) \otimes \mathcal{L}(\mu))_{\lambda+\mu-\alpha+\alpha_i}$$

for all $i \in I$. Then, by applying $\Psi_{\lambda,\mu}$, the statement $\mathbf{F(r-1)}$ yields

$$\tilde{e}_i \Psi(\mathcal{L}(\lambda) \otimes \mathcal{L}(\mu))_{\lambda+\mu-\alpha} \subset \Psi(\mathcal{L}(\lambda) \otimes \mathcal{L}(\mu))_{\lambda+\mu-\alpha+\alpha_i} \subset \mathcal{L}(\lambda+\mu)$$

for all $i \in I$. Hence our assertion follows from Lemma 5.3.18 (2). □

So far, we have proved all the statements $\mathbf{A(r)}, \mathbf{B(r)}, \dots, \mathbf{F(r)}$. Then using these results, one can prove that Lemma 5.3.2 holds for all $\alpha \in Q_+(r)$ (Exercise 5.14). In particular, we have:

Lemma 5.3.20.

(1) *For all* $i \in I$, $\lambda, \mu \in P^+$ *and* $\alpha \in Q_+(r)$, *we have*
$$\tilde{e}_i(\mathcal{B}(\lambda) \otimes \mathcal{B}(\mu))_{\lambda+\mu-\alpha} \subset (\mathcal{B}(\lambda) \otimes \mathcal{B}(\mu)) \sqcup \{0\}.$$

(2) *If* $b \otimes b' \in (\mathcal{B}(\lambda) \otimes \mathcal{B}(\mu))_{\lambda+\mu-\alpha}$ *and* $\tilde{e}_i(b \otimes b') \neq 0$, *then*
$$b \otimes b' = \tilde{f}_i \tilde{e}_i(b \otimes b').$$

Finally, we complete the grand-loop argument by proving the statement $\mathbf{G(r)}$.

Proposition 5.3.21 ($\mathbf{G(r)}$). *For all* $i \in I$, $\lambda, \mu \in P^+$ *and* $\alpha \in Q_+(r)$, *we have*
$$\Psi_{\lambda,\mu}((\mathcal{B}(\lambda) \otimes \mathcal{B}(\mu))_{\lambda+\mu-\alpha}) \subset \mathcal{B}(\lambda+\mu) \sqcup \{0\}.$$

Proof. Let $b \otimes b' \in (\mathcal{B}(\lambda) \otimes \mathcal{B}(\mu))_{\lambda+\mu-\alpha}$. If $\tilde{e}_i(b \otimes b') = 0$ for all $i \in I$, then $\tilde{e}_i \Psi_{\lambda,\mu}(b \otimes b') = 0$ for all $i \in I$, and Lemma 5.3.18 (1) implies $\Psi_{\lambda,\mu}(b \otimes b') = 0$.

If $\tilde{e}_i(b \otimes b') \neq 0$ for some $i \in I$, then by Lemma 5.3.20 and $\mathbf{G(r-1)}$, we have
$$\begin{aligned}\Psi_{\lambda,\mu}(b \otimes b') &= \Psi_{\lambda,\mu}(\tilde{f}_i \tilde{e}_i(b \otimes b')) = \tilde{f}_i \Psi_{\lambda,\mu}(\tilde{e}_i(b \otimes b')) \\ &\in \tilde{f}_i(\mathcal{B}(\lambda+\mu) \sqcup \{0\}) \subset \mathcal{B}(\lambda+\mu) \sqcup \{0\},\end{aligned}$$
which completes our interlocking induction argument. □

Exercises

5.1. Show that there exists a unique symmetric bilinear form $(\ ,\ )$ on $V(\lambda)$ ($\lambda \in P^+$) satisfying the conditions in (5.2).

5.2. Deduce that
$$q^{\frac{1}{2}n(n+1)-\frac{1}{2}k(k+1)-\frac{1}{2}(n-k)(n-k+1)} \begin{bmatrix} n \\ k \end{bmatrix}_q \in 1 + q\mathbf{A}_0.$$

5.3. Complete the proof of Proposition 5.1.7.

5.4. Prove Lemma 5.2.4.

5.5. Show that any maximal weight of a module belonging to the category $\mathcal{O}^q_{\mathrm{int}}$ must be a dominant integral weight.

5.6. Let $\Phi_{\lambda,\mu} : V(\lambda+\mu) \longrightarrow V(\lambda) \otimes V(\mu)$ and $\Psi : V(\lambda) \otimes V(\mu) \longrightarrow V(\lambda+\mu)$ be the $U_q(\mathfrak{g})$-module homomorphisms defined by (5.6). Show that we have
$$(\Phi_{\lambda,\mu}(u), w) = (u, \Psi_{\lambda,\mu}(w))$$

for all $u \in V(\lambda+\mu)$, $w \in V(\lambda) \otimes V(\mu)$.

5.7. Verify that the statements $\mathbf{A(r)}$, $\mathbf{B(r)}$, ..., $\mathbf{G(r)}$ are true for $r = 0$ and $r = 1$.

5.8. Show that the linear map $S_{\lambda,\mu} : V(\lambda) \otimes V(\mu) \longrightarrow V(\lambda)$ defined by (5.10) is a $U_q^-(\mathfrak{g})$-module homomorphism.

5.9. Let $v \in \mathcal{L}(\mu)_{\mu-\alpha}$ and write $v = \sum_{l\geq 0} f_i^{(l)} v_l$ with each v_l satisfying $e_i v_l = 0$ and $\langle h_i, \mu - \alpha + l\alpha_i \rangle \geq l \geq 0$. If $\tilde{e}_i v \in q^{-N}\mathcal{L}(\mu)$, show that $v_l \in q^{-N}\mathcal{L}(\mu)$ for all $l \geq 1$.

5.10. Complete the proof of Lemma 5.3.7.

5.11. Prove Corollary 5.3.10.

5.12. Show that $\mathcal{L}(\lambda) \subset \mathcal{L}(\lambda)^{\vee\vee} \subset \mathcal{L}(\lambda)^{\vee}$.

5.13. (a) For all $\lambda \gg 0$, $\mu \in P^+$, and $\alpha \in Q_+(r)$, show that

$$\Psi_{\lambda,\mu}((\mathcal{L}(\lambda) \otimes \mathcal{L}(\mu))_{\lambda+\mu-\alpha}) = \mathcal{L}(\lambda+\mu)_{\lambda+\mu-\alpha}.$$

(b) Use (a) to prove that $P_\alpha v_{\lambda+\mu} \in \mathcal{L}(\lambda+\mu)^{\vee}_{\lambda+\mu-\alpha}$ if and only if $P_\alpha v_\mu \in \mathcal{L}(\mu)^{\vee}_{\mu-\alpha}$ for all $\mu \gg 0$, $\alpha \in Q_+(r)$.

5.14. Assume that the statements $\mathbf{A(r)}$, $\mathbf{B(r)}$, ..., $\mathbf{F(r)}$ are true. Prove that Lemma 5.3.2 holds for all $\alpha \in Q_+(r)$.

Chapter 6

Global Bases

The purpose of this chapter is to *globalize* the theory of crystal bases. Given a dominant integral weight $\lambda \in P^+$, we have seen that there exists a unique crystal basis $\mathcal{B}(\lambda)$ for the irreducible highest weight $U_q(\mathfrak{g})$-module $V(\lambda)$, which can be viewed as a local basis at $q = 0$. By a similar argument, we can construct a local basis $\mathcal{B}(\lambda)^-$ for $V(\lambda)$ at $q = \infty$. In this chapter, we will take $\mathbf{F} = \mathbf{Q}$ and show that there exists a unique $\mathbf{Q}(q)$-basis $\mathcal{G}(\lambda)$ for $V(\lambda)$ which is *balanced* in the sense that it gives both local bases $\mathcal{B}(\lambda)$ and $\mathcal{B}(\lambda)^-$. We will call $\mathcal{G}(\lambda)$ the *global basis* for $V(\lambda)$.

6.1. Balanced triple

Recall that $\mathbf{A}_0$ is the subring of $\mathbf{Q}(q)$ consisting of rational functions in q that are regular at $q = 0$. Let $\mathbf{A}_\infty$ be the subring of $\mathbf{Q}(q)$ consisting of rational functions in q that are regular at $q = \infty$. Hence $\mathbf{A}_0/q\mathbf{A}_0 \cong \mathbf{Q}$ and $\mathbf{A}_\infty/q^{-1}\mathbf{A}_\infty \cong \mathbf{Q}$ by the evaluation at $q = 0$ and $q = \infty$, respectively. We also denote by $\mathbf{A} = \mathbf{Q}[q, q^{-1}]$, the ring of Laurent polynomials in q.

Let R be a ring and let $S \subset R$ be a subring. Given an R-module M and an S-module $N \subset M$, we say N is a ***free S-lattice*** of M if N is a free S-module and the canonical map

$$R \otimes_S N \longrightarrow M$$

sending $r \otimes v \longmapsto r \cdot v$ is an isomorphism of R-modules.

Throughout this chapter, we will fix a finite dimensional $\mathbf{Q}(q)$-vector space V, and denote by $\mathcal{L}_0$, $\mathcal{L}_\infty$, and $V^{\mathbf{A}}$, respectively, a free $\mathbf{A}_0$-lattice, $\mathbf{A}_\infty$-lattice, and $\mathbf{A}$-lattice of V.

We first prove:

Lemma 6.1.1. *The canonical maps*

$$\mathbf{A}_0 \otimes_{\mathbf{Q}[q]} (V^{\mathbf{A}} \cap \mathcal{L}_0) \longrightarrow \mathcal{L}_0,$$
$$\mathbf{A}_\infty \otimes_{\mathbf{Q}[q^{-1}]} (V^{\mathbf{A}} \cap \mathcal{L}_\infty) \longrightarrow \mathcal{L}_\infty$$

are isomorphisms.

Proof. We will prove the first one only. The second assertion can be proved in a similar manner (Exercise 6.1).

Since $\mathcal{L}_0 \subset V$ and $V \cong \mathbf{Q}(q) \otimes_{\mathbf{A}} V^{\mathbf{A}}$ under a canonical map, given any $u \in \mathcal{L}_0$, there exists a nonzero $f(q) \in \mathbf{Q}[q]$ for which $f(q)u \in V^{\mathbf{A}}$. Since $V^{\mathbf{A}}$ is an $\mathbf{A}$-module, by dividing by a suitable power of q, we may assume $f(0) \neq 0$, which implies $\frac{1}{f(q)} \in \mathbf{A}_0$. Thus we obtain an element

$$\frac{1}{f(q)} \otimes f(q)u \in \mathbf{A}_0 \otimes_{\mathbf{Q}[q]} (V^{\mathbf{A}} \cap \mathcal{L}_0)$$

which is mapped to u under the canonical map. That is, the first canonical map is surjective. Moreover it is injective because it may be written as the composition of maps

$$\mathbf{A}_0 \otimes_{\mathbf{Q}[q]} (V^{\mathbf{A}} \cap \mathcal{L}_0) \hookrightarrow \mathbf{A}_0 \otimes_{\mathbf{Q}[q]} \mathcal{L}_0 \cong \mathcal{L}_0,$$

where the last isomorphism is a canonical one. □

Definition 6.1.2. Let V be a $\mathbf{Q}(q)$-vector space and let $\mathcal{L}_0$, $\mathcal{L}_\infty$, and $V^{\mathbf{A}}$, respectively, be a free $\mathbf{A}_0$-lattice, $\mathbf{A}_\infty$-lattice, and $\mathbf{A}$-lattice of V. Set $E = V^{\mathbf{A}} \cap \mathcal{L}_0 \cap \mathcal{L}_\infty$. Note that E is a $\mathbf{Q}$-vector space. Then $(V^{\mathbf{A}}, \mathcal{L}_0, \mathcal{L}_\infty)$ is called a ***balanced triple*** for V if the following conditions are satisfied:

(1) E is a free $\mathbf{Q}$-lattice of the $\mathbf{A}_0$-module $\mathcal{L}_0$,

(2) E is a free $\mathbf{Q}$-lattice of the $\mathbf{A}_\infty$-module $\mathcal{L}_\infty$,

(3) E is a free $\mathbf{Q}$-lattice of the $\mathbf{A}$-module $V^{\mathbf{A}}$.

In other words, $(V^{\mathbf{A}}, \mathcal{L}_0, \mathcal{L}_\infty)$ is a balanced triple if we have

$$\mathbf{A}_0 \otimes_{\mathbf{Q}} E \cong \mathcal{L}_0, \qquad \mathbf{A}_\infty \otimes_{\mathbf{Q}} E \cong \mathcal{L}_\infty, \qquad \mathbf{A} \otimes_{\mathbf{Q}} E \cong V^{\mathbf{A}}, \tag{6.1}$$

where each of these isomorphisms is a canonical one.

Remark 6.1.3. Observe that if $(V^{\mathbf{A}}, \mathcal{L}_0, \mathcal{L}_\infty)$ is a balanced triple, then we have

$$\mathbf{Q}(q) \otimes_{\mathbf{Q}} E \cong \mathbf{Q}(q) \otimes_{\mathbf{A}} \mathbf{A} \otimes_{\mathbf{Q}} E \cong \mathbf{Q}(q) \otimes_{\mathbf{A}} V^{\mathbf{A}} \cong V. \tag{6.2}$$

Theorem 6.1.4. *Let V, $\mathcal{L}_0$, $\mathcal{L}_\infty$, and $V^{\mathbf{A}}$ be given as before and set $E = V^{\mathbf{A}} \cap \mathcal{L}_0 \cap \mathcal{L}_\infty$. Then the following statements are equivalent.*

(1) *$(V^{\mathbf{A}}, \mathcal{L}_0, \mathcal{L}_\infty)$ is a balanced triple.*

(2) *The canonical map* $E \longrightarrow \mathcal{L}_0/q\mathcal{L}_0$ *is an isomorphism.*

(3) *The canonical map* $E \longrightarrow \mathcal{L}_\infty/q^{-1}\mathcal{L}_\infty$ *is an isomorphism.*

Proof. (1) $\Rightarrow$ (2) : By (6.1), we have

$$\begin{aligned} E &\cong \mathbf{Q} \otimes_{\mathbf{Q}} E \cong \mathbf{Q} \otimes_{\mathbf{A}_0} \mathbf{A}_0 \otimes_{\mathbf{Q}} E \\ &\cong \mathbf{Q} \otimes_{\mathbf{A}_0} \mathcal{L}_0 \cong \mathbf{A}_0/q\mathbf{A}_0 \otimes_{\mathbf{A}_0} \mathcal{L}_0 \xrightarrow{\sim} \mathcal{L}_0/q\mathcal{L}_0. \end{aligned}$$

The proof of (1) $\Rightarrow$ (3) is similar (Exercise 6.2).

(2) $\Rightarrow$ (1) : Suppose $E \xrightarrow{\sim} \mathcal{L}_0/q\mathcal{L}_0$ under the canonical map. We claim that

$$\left(\bigoplus_{k=0}^{n} \mathbf{Q}q^k\right) \otimes_{\mathbf{Q}} E \cong V^{\mathbf{A}} \cap \mathcal{L}_0 \cap q^n\mathcal{L}_\infty$$

under the canonical map. We will prove our claim by induction on n. If $n = 0$, our assertion is obvious. If $n > 0$, consider the following diagram.

$$\begin{array}{ccccccccc} 0 & \longrightarrow & (\bigoplus_{k=1}^{n} \mathbf{Q}q^k) \otimes_{\mathbf{Q}} E & \longrightarrow & (\bigoplus_{k=0}^{n} \mathbf{Q}q^k) \otimes_{\mathbf{Q}} E & \longrightarrow & E & \longrightarrow & 0 \\ & & \Big\downarrow \alpha & & \Big\downarrow \beta & & \Big\downarrow \wr & & \\ 0 & \longrightarrow & V^{\mathbf{A}} \cap q\mathcal{L}_0 \cap q^n\mathcal{L}_\infty & \longrightarrow & V^{\mathbf{A}} \cap \mathcal{L}_0 \cap q^n\mathcal{L}_\infty & \longrightarrow & \mathcal{L}_0/q\mathcal{L}_0 & & \end{array}$$

It is easy to verify that the diagram is commutative and that the rows are exact (Exercise 6.3). Note that α is an isomorphism by induction hypothesis. Hence, by diagram chasing, β is an isomorphism, which is our claim. Therefore, for any $a \leq b$, we have

$$\left(\bigoplus_{k=0}^{b-a} \mathbf{Q}q^k\right) \otimes_{\mathbf{Q}} E \cong V^{\mathbf{A}} \cap \mathcal{L}_0 \cap q^{b-a}\mathcal{L}_\infty,$$

which yields

$$\left(\bigoplus_{a\leq k\leq b} \mathbf{Q}q^k\right) \otimes_{\mathbf{Q}} E \cong V^{\mathbf{A}} \cap q^a\mathcal{L}_0 \cap q^b\mathcal{L}_\infty \tag{6.3}$$

under the canonical map. By taking $a = 0$, $b \to \infty$ (respectively, $a \to -\infty$, $b = 0$ and $a \to -\infty$, $b \to \infty$), we can show

$$\begin{aligned} &\mathbf{Q}[q] \otimes_{\mathbf{Q}} E \cong V^{\mathbf{A}} \cap \mathcal{L}_0, \\ &\mathbf{Q}[q^{-1}] \otimes_{\mathbf{Q}} E \cong V^{\mathbf{A}} \cap \mathcal{L}_\infty, \\ &\mathbf{Q}[q, q^{-1}] \otimes_{\mathbf{Q}} E \cong V^{\mathbf{A}} \end{aligned} \tag{6.4}$$

with all isomorphisms the canonical ones. To be precise, let us just consider the first one as an example. The injectivity is straightforward from (6.3).

The surjectivity also follows from the same equation and the simple fact $\bigcup_{n=0}^{\infty} q^n \mathcal{L}_\infty = V$. Therefore, Lemma 6.1.1 yields

$$\mathbf{A}_0 \otimes_{\mathbf{Q}} E \cong \mathbf{A}_0 \otimes_{\mathbf{Q}[q]} \mathbf{Q}[q] \otimes_{\mathbf{Q}} E \cong \mathbf{A}_0 \otimes_{\mathbf{Q}[q]} (V^{\mathbf{A}} \cap \mathcal{L}_0) \cong \mathcal{L}_0$$

and

$$\mathbf{A}_\infty \otimes_{\mathbf{Q}} E \cong \mathbf{A}_\infty \otimes_{\mathbf{Q}[q^{-1}]} \mathbf{Q}[q^{-1}] \otimes_{\mathbf{Q}} E \cong \mathbf{A}_\infty \otimes_{\mathbf{Q}[q^{-1}]} (V^{\mathbf{A}} \cap \mathcal{L}_\infty) \cong \mathcal{L}_\infty.$$

The proof of (3) $\Rightarrow$ (1) is similar (Exercise 6.2). □

Suppose that $(V^{\mathbf{A}}, \mathcal{L}_0, \mathcal{L}_\infty)$ is a balanced triple for an $\mathbf{Q}(q)$-vector space V, and let

$$G : \mathcal{L}_0/q\mathcal{L}_0 \longrightarrow E = V^{\mathbf{A}} \cap \mathcal{L}_0 \cap \mathcal{L}_\infty$$

be the inverse of the canonical isomorphism $E \xrightarrow{\sim} \mathcal{L}_0/q\mathcal{L}_0$.

Proposition 6.1.5. *If $\mathcal{B}$ is a $\mathbf{Q}$-basis of $\mathcal{L}_0/q\mathcal{L}_0$, then $\mathcal{G}(\mathcal{B}) = \{G(b) \mid b \in \mathcal{B}\}$ is an $\mathbf{A}$-basis of $V^{\mathbf{A}}$. Hence it is a $\mathbf{Q}(q)$-basis of V.*

Proof. Since $\mathcal{L}_0/q\mathcal{L}_0 = \bigoplus_{b\in\mathcal{B}} \mathbf{Q}b$, we have $E = \bigoplus_{b\in\mathcal{B}} \mathbf{Q}G(b)$. It follows that

$$V^{\mathbf{A}} \cong \mathbf{A} \otimes_{\mathbf{Q}} E = \bigoplus_{b\in\mathcal{B}} \mathbf{Q}[q, q^{-1}]G(b),$$

$$V \cong \mathbf{Q}(q) \otimes_{\mathbf{A}} V^{\mathbf{A}} = \bigoplus_{b\in\mathcal{B}} \mathbf{Q}(q)G(b).$$

This completes the proof. □

Definition 6.1.6. Suppose that $(V^{\mathbf{A}}, \mathcal{L}_0, \mathcal{L}_\infty)$ is a balanced triple for a $\mathbf{Q}(q)$-vector space V.

(1) A $\mathbf{Q}$-basis $\mathcal{B}$ of $\mathcal{L}_0/q\mathcal{L}_0$ is called a ***local basis*** of V at $q = 0$.

(2) In this case, $\mathcal{G}(\mathcal{B})$ is called the ***global basis*** of V corresponding to the local basis $\mathcal{B}$.

Remark 6.1.7. Observe that the image $\overline{\mathcal{B}}$ of $\mathcal{G}(\mathcal{B})$ under the canonical isomorphism $E \xrightarrow{\sim} \mathcal{L}_\infty/q^{-1}\mathcal{L}_\infty$ is a $\mathbf{Q}$-basis of $\mathcal{L}_\infty/q^{-1}\mathcal{L}_\infty$ which can be viewed as a local basis of V at $q = \infty$. (Exercise 6.4).

6.2. Global basis for $V(\lambda)$

In this section, we will construct the *global basis* $\mathcal{G}(\lambda)$ for the irreducible highest weight $U_q(\mathfrak{g})$-module $V(\lambda)$ with highest weight $\lambda \in P^+$ corresponding to the crystal basis $\mathcal{B}(\lambda)$.

Let us use the notation

$$\begin{Bmatrix} x \\ t \end{Bmatrix}_{q_i} = \frac{1}{[t]_{q_i}!}\prod_{k=1}^{t}[x; 1-t]_{q_i}.$$

We then denote by $U_{\mathbf{A}}(\mathfrak{g})$ (respectively, $U_{\mathbf{A}}^{+}(\mathfrak{g})$ and $U_{\mathbf{A}}^{-}(\mathfrak{g})$) the $\mathbf{A}$-subalgebra of $U_q(\mathfrak{g})$ generated by $e_i^{(n)}$, $f_i^{(n)}$, q^h, $\begin{Bmatrix} K_i q^{-n} \\ m \end{Bmatrix}_{q_i}$ (respectively, $e_i^{(n)}$ and $f_i^{(n)}$) for $i \in I$, $m, n \in \mathbf{Z}_{\geq 0}$, and $h \in P^{\vee}$. We also denote by $U_{\mathbf{A}}^{0}(\mathfrak{g})$ the $\mathbf{A}$-subalgebra of $U_q(\mathfrak{g})$ generated by q^h and $\begin{Bmatrix} K_i q^{-n} \\ m \end{Bmatrix}_{q_i}$. Then we have the triangular decomposition (cf. Exercise 3.6):

$$U_{\mathbf{A}}(\mathfrak{g}) \cong U_{\mathbf{A}}^{-}(\mathfrak{g}) \otimes_{\mathbf{A}} U_{\mathbf{A}}^{0}(\mathfrak{g}) \otimes_{\mathbf{A}} U_{\mathbf{A}}^{+}(\mathfrak{g}),$$

as $\mathbf{A}$-modules.

Consider the $\mathbf{Q}$-algebra automorphism $^{-} : U_q(\mathfrak{g}) \longrightarrow U_q(\mathfrak{g})$ defined by

$$(6.5)\quad q \mapsto q^{-1}, \qquad q^h \mapsto q^{-h}, \qquad e_i \mapsto e_i, \qquad f_i \mapsto f_i \quad (h \in P^{\vee},\ i \in I).$$

Then we get a $\mathbf{Q}$-linear automorphism $^{-} : V(\lambda) \longrightarrow V(\lambda)$ defined by

$$(6.6)\qquad P v_\lambda \longmapsto \overline{P} v_\lambda \quad \text{for } P \in U_q^{-}(\mathfrak{g}),$$

where v_λ denotes the highest weight vector of $V(\lambda)$.

Let $(\mathcal{L}(\lambda), \mathcal{B}(\lambda))$ be the crystal basis of $V(\lambda)$ given in Theorem 5.1.1 and let $\mathcal{L}(\lambda)^{-} = \overline{\mathcal{L}(\lambda)}$. Thus $\mathcal{L}(\lambda)$ (respectively, $\mathcal{L}(\lambda)^{-}$) is a free $\mathbf{A}_0$-lattice (respectively, $\mathbf{A}_\infty$-lattice) of $V(\lambda)$.

Set $V(\lambda)^{\mathbf{A}} = U_{\mathbf{A}}(\mathfrak{g}) v_\lambda = U_{\mathbf{A}}^{-}(\mathfrak{g}) v_\lambda$. Note that $\overline{V(\lambda)^{\mathbf{A}}} = V(\lambda)^{\mathbf{A}}$ (Exercise 6.5). Now we can state the main theorem of this chapter.

Theorem 6.2.1. *$(V(\lambda)^{\mathbf{A}}, \mathcal{L}(\lambda), \mathcal{L}(\lambda)^{-})$ is a balanced triple for $V(\lambda)$.*

The proof of Theorem 6.2.1 will be given in Section 6.5. Here, we discuss some of the consequences of this theorem. Since $(V(\lambda)^{\mathbf{A}}, \mathcal{L}(\lambda), \mathcal{L}(\lambda)^{-})$ is a balanced triple for $V(\lambda)$, the canonical map

$$E = V(\lambda)^{\mathbf{A}} \cap \mathcal{L}(\lambda) \cap \mathcal{L}(\lambda)^{-} \longrightarrow \mathcal{L}(\lambda)/q\mathcal{L}(\lambda)$$

is an isomorphism.

As in Section 6.1, let

$$G : \mathcal{L}(\lambda)/q\mathcal{L}(\lambda) \longrightarrow E$$

be the inverse of this canonical isomorphism, and set

$$(6.7)\qquad \mathcal{G}(\lambda) = \{G(b) \mid b \in \mathcal{B}(\lambda)\}.$$

Then $\mathcal{G}(\lambda)$ is the *global basis* of $V(\lambda)$ corresponding to the crystal basis $\mathcal{B}(\lambda)$. By definition, we have

$$(6.8)\qquad G(b) \equiv b \mod q\mathcal{L}(\lambda) \quad \text{for all } b \in \mathcal{B}(\lambda).$$

Also, by (6.4) in the proof of Theorem 6.1.4, we get

$$\begin{aligned} V(\lambda)^{\mathbf{A}} \cap \mathcal{L}(\lambda) &= \bigoplus_{b\in\mathcal{B}(\lambda)} \mathbf{Q}[q]G(b), \\ V(\lambda)^{\mathbf{A}} &= \bigoplus_{b\in\mathcal{B}(\lambda)} \mathbf{A}G(b). \end{aligned} \tag{6.9}$$

For each $b \in \mathcal{B}(\lambda)$, set

$$Q(b) = \frac{G(b) - \overline{G(b)}}{q - q^{-1}}.$$

Clearly, $G(b), \overline{G(b)} \in E = V(\lambda)^{\mathbf{A}} \cap \mathcal{L}(\lambda) \cap \mathcal{L}(\lambda)^{-}$. Write

$$G(b) = \sum_j c_j(q) P_j v_\lambda,$$

where $c_j(q) \in \mathbf{A}$ and P_j is a monomial in $f_i^{(n)}$ $(i \in I,\, n \geq 1)$. Then we have

$$\overline{G(b)} = \sum_j c_j(q^{-1}) P_j v_\lambda.$$

It follows that

$$Q(b) = \frac{G(b) - \overline{G(b)}}{q - q^{-1}} = \sum_j \frac{c_j(q) - c_j(q^{-1})}{q - q^{-1}} P_j v_\lambda \in V(\lambda)^{\mathbf{A}}.$$

Moreover, since

$$\frac{1}{q - q^{-1}} = \frac{q}{q^2 - 1} \in q\mathbf{A}_0 \cap \mathbf{A}_\infty,$$

we have

$$Q(b) = \frac{q}{q^2 - 1}(G(b) - \overline{G(b)}) \in q\mathcal{L}(\lambda) \cap \mathcal{L}(\lambda)^{-}.$$

Thus $Q(b)$ is mapped to 0 under the canonical isomorphism

$$E \xrightarrow{\sim} \mathcal{L}(\lambda)/q\mathcal{L}(\lambda),$$

which implies $Q(b) = 0$. Therefore we get

$$\overline{G(b)} = G(b) \quad \text{for all } b \in \mathcal{B}(\lambda). \tag{6.10}$$

To summarize, we obtain:

Theorem 6.2.2. *Let $V(\lambda)$ be the irreducible highest weight $U_q(\mathfrak{g})$-module with highest weight $\lambda \in P^+$ and let $(\mathcal{L}(\lambda), \mathcal{B}(\lambda))$ be the crystal basis of $V(\lambda)$. Then there exists a unique $\mathbf{A}$-basis $\mathcal{G}(\lambda) = \{G(b) \mid b \in \mathcal{B}(\lambda)\}$ of $V(\lambda)^{\mathbf{A}}$ parameterized by $\mathcal{B}(\lambda)$ satisfying*

(1) $G(b) \equiv b \mod q\mathcal{L}(\lambda)$,

(2) $\overline{G(b)} = G(b)$ *for all* $b \in \mathcal{B}(\lambda)$.

Proof. We only have to show the uniqueness of $\mathcal{G}(\lambda)$. Suppose $\mathcal{G}_0(\lambda) = \{G_0(b) \mid b \in \mathcal{B}(\lambda)\}$ is an $\mathbf{A}$-basis of $V(\lambda)^{\mathbf{A}}$ such that

$$\begin{aligned} G_0(b) &\equiv b \mod q\mathcal{L}(\lambda), \\ \overline{G_0(b)} &= G_0(b) \quad \text{for all } b \in \mathcal{B}(\lambda). \end{aligned}$$

Then $G_0(b) \in E = V(\lambda)^{\mathbf{A}} \cap \mathcal{L}(\lambda) \cap \mathcal{L}(\lambda)^{-}$. Since the inverse of the canonical isomorphism $E \xrightarrow{\sim} \mathcal{L}(\lambda)/q\mathcal{L}(\lambda)$ is unique, we must have $G(b) = G_0(b)$ for all $b \in \mathcal{B}(\lambda)$. □

Example 6.2.3.

(1) Let $V(2) = \mathbf{Q}(q)u \oplus \mathbf{Q}(q)fu \oplus \mathbf{Q}(q)f^{(2)}u$ be the three-dimensional irreducible $U_q(\mathfrak{sl}_2)$-module with crystal basis $(\mathcal{L}(2), \mathcal{B}(2))$, and consider the tensor product $V(2) \otimes V(2)$. Then $u \otimes u$ would generate the five-dimensional irreducible submodule $V \cong V(4)$ inside $V(2) \otimes V(2)$. Recall that the crystal graph $\mathcal{B}$ of V appears as the connected component of $\mathcal{B}(2) \otimes \mathcal{B}(2)$ containing $u \otimes u$.

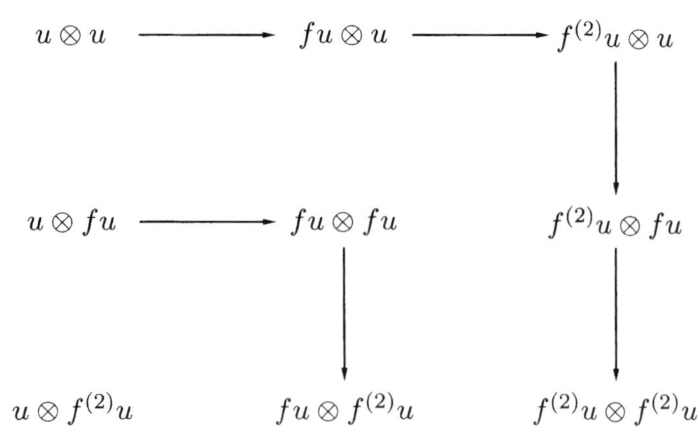

By definition of the automorphism $- : V \longrightarrow V$, we have

$$\begin{aligned} &\overline{u \otimes u} = u \otimes u, \\ &\overline{f(u \otimes u)} = f(u \otimes u), \\ &\qquad \vdots \\ &\overline{f^{(4)}(u \otimes u)} = f^{(4)}(u \otimes u). \end{aligned}$$

Observe that

$$\begin{aligned}
f(u\otimes u) &= fu\otimes u + q^2 u\otimes fu \\
&\equiv fu\otimes u \mod q\mathcal{L}(2)\otimes\mathcal{L}(2), \\
f^{(2)}(u\otimes u) &= f^{(2)}u\otimes u + qfu\otimes fu + q^4 u\otimes f^{(2)}u \\
&\equiv f^{(2)}u\otimes u \mod q\mathcal{L}(2)\otimes\mathcal{L}(2), \\
f^{(3)}(u\otimes u) &= f^{(2)}u\otimes fu + q^2 fu\otimes f^{(2)}u \\
&\equiv f^{(2)}u\otimes fu \mod q\mathcal{L}(2)\otimes\mathcal{L}(2), \\
f^{(4)}(u\otimes u) &= f^{(2)}u\otimes f^{(2)}u \\
&\equiv f^{(2)}u\otimes f^{(2)}u \mod q\mathcal{L}(2)\otimes\mathcal{L}(2).
\end{aligned}$$

Hence the global basis $\mathcal{G}(4)$ of V is given by

$$\begin{aligned}
&G(u\otimes u) = u\otimes u, \\
&G(fu\otimes u) = fu\otimes u + q^2 u\otimes fu, \\
&G(f^{(2)}u\otimes u) = f^{(2)}u\otimes u + qfu\otimes fu + q^4 u\otimes f^{(2)}u, \\
&G(f^{(2)}u\otimes fu) = f^{(2)}u\otimes fu + q^2 fu\otimes f^{(2)}u, \\
&G(f^{(2)}u\otimes f^{(2)}u) = f^{(2)}u\otimes f^{(2)}u.
\end{aligned}$$

(2) In this example the vector $u\otimes fu$ is not a maximal vector in $V(2)\otimes V(2)$. The desired maximal vector is

$$u\otimes fu - q^2 fu\otimes u \equiv u\otimes fu \mod q\mathcal{L}(2)\otimes\mathcal{L}(2),$$

which generates the three-dimensional irreducible submodule $W\cong V(2)$ inside $V(2)\otimes V(2)$.

Observe that

$$\begin{aligned}
f(u\otimes fu - q^2 fu\otimes u) &= (1-q^2)fu\otimes fu - q(1+q^2)f^{(2)}u\otimes u \\
&\quad + q(1+q^2)u\otimes f^{(2)}u \\
&\equiv fu\otimes fu \mod q\mathcal{L}(2)\otimes\mathcal{L}(2), \\
f^{(2)}(u\otimes fu - q^2 fu\otimes u) &= fu\otimes f^{(2)}u - q^2 f^{(2)}u\otimes fu \\
&\equiv fu\otimes f^{(2)}u \mod q\mathcal{L}(2)\otimes\mathcal{L}(2).
\end{aligned}$$

Therefore, the global basis $\mathcal{G}(2)$ of W is given by

$$\begin{aligned}
G(u\otimes fu) &= u\otimes fu - q^2 fu\otimes u, \\
G(fu\otimes fu) &= (1-q^2)fu\otimes fu - q(1+q^2)f^{(2)}u\otimes u \\
&\quad + q(1+q^2)u\otimes f^{(2)}u, \\
G(fu\otimes f^{(2)}u) &= fu\otimes f^{(2)}u - q^2 f^{(2)}u\otimes fu.
\end{aligned}$$

Example 6.2.4. Let $V = \mathbf{Q}(q)u \oplus \mathbf{Q}(q)f_1u \oplus \mathbf{Q}(q)f_2f_1u$ be the three-dimensional vector representation of $U_q(\mathfrak{sl}_3)$ with crystal basis $(\mathcal{L}, \mathcal{B})$ and consider the tensor product $V \otimes V$. Then $u \otimes u$ would generate the irreducible $U_q(\mathfrak{sl}_3)$-module $W \cong V(2\varepsilon_1)$ inside $V \otimes V$. As we have seen in Example 4.4.2, the crystal graph of W appears as the connected component of $\mathcal{B} \otimes \mathcal{B}$ containing $u \otimes u$. Observe that

$$\begin{aligned}
f_1(u \otimes u) &= f_1u \otimes u + qu \otimes f_1u \\
&\equiv f_1u \otimes u \mod q\mathcal{L} \otimes \mathcal{L}, \\
f_1^{(2)}(u \otimes u) &= f_1u \otimes f_1u \\
&\equiv f_1u \otimes f_1u \mod q\mathcal{L} \otimes \mathcal{L}, \\
f_2f_1(u \otimes u) &= f_2f_1u \otimes u + qu \otimes f_2f_1u \\
&\equiv f_2f_1u \otimes u \mod q\mathcal{L} \otimes \mathcal{L}, \\
f_2f_1^{(2)}(u \otimes u) &= f_2f_1u \otimes f_1u + qf_1u \otimes f_2f_1u \\
&\equiv f_2f_1u \otimes f_1u \mod q\mathcal{L} \otimes \mathcal{L}, \\
f_2^{(2)}f_1^{(2)}(u \otimes u) &= f_2f_1u \otimes f_2f_1u \\
&\equiv f_2f_1u \otimes f_2f_1u \mod q\mathcal{L} \otimes \mathcal{L}.
\end{aligned}$$

Hence the global basis $\mathcal{G}$ of W is given by

$$\begin{aligned}
&G(u \otimes u) = u \otimes u, \\
&G(f_1u \otimes u) = f_1u \otimes u + qu \otimes f_1u, \\
&G(f_1u \otimes f_1u) = f_1u \otimes f_1u, \\
&G(f_2f_1u \otimes u) = f_2f_1u \otimes u + qu \otimes f_2f_1u, \\
&G(f_2f_1u \otimes f_1u) = f_2f_1u \otimes f_1u + qf_1u \otimes f_2f_1u, \\
&G(f_2f_1u \otimes f_2f_1u) = f_2f_1u \otimes f_2f_1u.
\end{aligned}$$

6.3. Polarization on $U_q^-(\mathfrak{g})$

Fix $i \in I$. We first define the modified root operators e_i' and e_i'' on $U_q^-(\mathfrak{g})$.

Lemma 6.3.1. *For any $P \in U_q^-(\mathfrak{g})$, there exist unique $Q, R \in U_q^-(\mathfrak{g})$ such that*

$$[e_i, P] = \frac{K_iQ - K_i^{-1}R}{q_i - q_i^{-1}}. \tag{6.11}$$

Proof. The uniqueness of Q and R follows from the triangular decomposition

$$U_q(\mathfrak{g}) \cong U_q^-(\mathfrak{g}) \otimes U_q^0(\mathfrak{g}) \otimes U_q^+(\mathfrak{g}).$$

If $P = 1$, our assertion is trivial with $Q = R = 0$.

Suppose that our lemma is true for P. Then for any $j \in I$, we have

$$\begin{aligned}[e_i, f_jP] &= [e_i, f_j]P + f_j[e_i, P] \\ &= \delta_{ij}\frac{K_i - K_i^{-1}}{q_i - q_i^{-1}}P + \frac{f_j(K_iQ - K_i^{-1}R)}{q_i - q_i^{-1}} \\ &= \frac{K_i(\delta_{ij}P + q_i^{a_{ij}}f_jQ) - K_i^{-1}(\delta_{ij}P + q_i^{-a_{ij}}f_jR)}{q_i - q_i^{-1}},\end{aligned}$$

which proves our lemma by induction. □

We define the endomorphisms e_i', $e_i'' : U_q^-(\mathfrak{g}) \to U_q^-(\mathfrak{g})$ by

$$e_i'(P) = R, \qquad e_i''(P) = Q. \tag{6.12}$$

Observe that the proof of Lemma 6.3.1 implies

$$e_i'f_j = q_i^{-a_{ij}}f_je_i' + \delta_{ij}, \qquad e_i''f_j = q_i^{a_{ij}}f_je_i'' + \delta_{ij}. \tag{6.13}$$

Moreover, for any $i, j, k \in I$, we have

$$\begin{aligned}e_i'e_j''f_k &= e_i'(q_j^{a_{jk}}f_ke_j'' + \delta_{jk}) \\ &= q_k^{a_{kj}}(q_i^{-a_{ik}}f_ke_i' + \delta_{ki})e_j'' + \delta_{jk}e_i' \\ &= q_k^{-a_{ki}+a_{kj}}f_ke_i'e_j'' + \delta_{ki}q_i^{a_{ij}}e_j'' + \delta_{jk}e_i'.\end{aligned}$$

Similarly, we obtain

$$e_j''e_i'f_k = q_i^{-a_{ki}+a_{kj}}f_ke_j''e_i' + \delta_{jk}q_i^{-a_{ij}}e_i' + \delta_{ik}e_j''.$$

Set $S = e_i'e_j'' - q_i^{a_{ij}}e_j''e_i'$ $(i, j \in I)$. Then

$$Sf_k = q_k^{-a_{ki}+a_{kj}}f_kS \quad \text{for all } k \in I.$$

Since $S \cdot 1 = 0$, we obtain $S = 0$. It follows that

$$e_i'e_j'' = q_i^{a_{ij}}e_j''e_i' \quad \text{for all } i, j \in I. \tag{6.14}$$

One can prove that there exists a unique symmetric bilinear form $(\ ,\ )$ on $U_q^-(\mathfrak{g})$ satisfying

$$\begin{aligned}&(1, 1) = 1, \\ &(f_iP, Q) = (P, e_i'Q)\end{aligned} \tag{6.15}$$

for all $P, Q \in U_q^-(\mathfrak{g})$ and $i \in I$ (Exercise 6.6).

Lemma 6.3.2. *For all $P, Q \in U_q^-(\mathfrak{g})$ and $i \in I$, we have*

$$(Pf_i, Q) = (P, K_ie_i''QK_i^{-1}). \tag{6.16}$$

Proof. It is easy to verify that our assertion holds when $P = 1$. Assume that P satisfies (6.16) for all $Q \in U_q^-(\mathfrak{g})$. We will show that f_jP also satisfies (6.16) for all $j \in I$.

Note that (6.14) implies

$$K_i e_i'' e_j' K_i^{-1} = e_j' K_i e_i'' K_i^{-1} \quad \text{for all } \ i, j \in I.$$

Hence we have

$$\begin{aligned}(f_jPf_i, Q) &= (Pf_i, e_j'Q) = (P, K_ie_i''e_j'QK_i^{-1}) \\ &= (P, (K_ie_i''e_j'K_i^{-1})(K_iQK_i^{-1})) = (P, e_j'K_ie_i''K_i^{-1}K_iQK_i^{-1}) \\ &= (P, e_j'K_ie_i''QK_i^{-1}) = (f_jP, K_ie_i''QK_i^{-1}),\end{aligned}$$

as desired. □

Let $\star : U_q(\mathfrak{g}) \to U_q(\mathfrak{g})$ denote the $\mathbf{Q}(q)$-linear antiautomorphism of $U_q(\mathfrak{g})$ given by

$$(6.17) \qquad e_i^\star = e_i, \qquad f_i^\star = f_i, \qquad (q^h)^\star = q^{-h}.$$

Recall that

$$[e_i, P] = \frac{K_i(e_i''P) - K_i^{-1}(e_i'P)}{q_i - q_i^{-1}}.$$

By taking $\star$, we obtain

$$\begin{aligned}[P^\star, e_i] &= \frac{(e_i''P)^\star K_i^{-1} - (e_i'P)^\star K_i}{q_i - q_i^{-1}} \\ &= \frac{K_i^{-1}(K_i(e_i''P)^\star K_i^{-1}) - K_i(K_i^{-1}(e_i'P)^\star K_i)}{q_i - q_i^{-1}}.\end{aligned}$$

Hence we obtain

$$(6.18) \qquad e_i'(P^\star) = K_i(e_i''P)^\star K_i^{-1}, \qquad e_i''(P^\star) = K_i^{-1}(e_iP)^\star K_i.$$

Lemma 6.3.3. *For any $P, Q \in U_q^-(\mathfrak{g})$, we have*

$$(6.19) \qquad (P^\star, Q^\star) = (P, Q).$$

Proof. If $P = 1$, our assertion is clear. Assume that P satisfies (6.19) for all $Q \in U_q^-(\mathfrak{g})$. Then by (6.16) and (6.18), we have

$$\begin{aligned}((Pf_i)^\star, Q^\star) &= (f_iP^\star, Q^\star) = (P^\star, e_i'Q^\star) \\ &= (P^\star, K_i(e_i''Q)^\star K_i^{-1}) = (Pf_i, Q),\end{aligned}$$

which proves our claim by induction. □

Recall that we write $\lambda \gg 0$ if $\lambda - \alpha \in P^+$ for all $\alpha \in Q_+(r)$ and $\lambda - \Lambda_i \in P^+$ for all $i \in I$ (see Section 5.3).

Lemma 6.3.4. *Let $P, Q \in U_q^-(\mathfrak{g})_{-\beta}$ for some $\beta \in Q_+(r)$ and assume that $\lambda \gg 0$. Then we have*

$$(Pv_\lambda, Qv_\lambda) \equiv (P, Q) \mod q\mathbf{A}_0, \tag{6.20}$$

where v_λ denotes the highest weight vector in $V(\lambda)$.

Proof. If $P = 1$, our assertion is clear. Suppose that P satisfies (6.20) for all $R \in U_q^-(\mathfrak{g})_{-\gamma}$ with $|\gamma| < |\beta|$. Then we have

$$\begin{aligned}
(f_i Pv_\lambda, Qv_\lambda) &= (Pv_\lambda, q_i^{-1} K_i e_i Qv_\lambda) \\
&= \left(Pv_\lambda, q_i^{-1} K_i \Big(Qe_i + \frac{K_i(e_i''Q) - K_i^{-1}(e_i'Q)}{q_i - q_i^{-1}}\Big) v_\lambda\right) \\
&= \left(Pv_\lambda, \frac{K_i^2(e_i''Q)v_\lambda - (e_i'Q)v_\lambda}{q_i^2 - 1}\right) \\
&= \left(Pv_\lambda, \frac{q_i^{2\langle h_i, \lambda - \beta + \alpha_i\rangle}(e_i''Q)v_\lambda - (e_i'Q)v_\lambda}{q_i^2 - 1}\right) \\
&= \frac{q_i^{2\langle h_i, \lambda - \beta + \alpha_i\rangle}}{q_i^2 - 1}(Pv_\lambda, (e_i''Q)v_\lambda) + \frac{1}{1 - q_i^2}(Pv_\lambda, (e_i'Q)v_\lambda) \\
&\equiv (Pv_\lambda, (e_i'Q)v_\lambda) \equiv (P, e_i'Q) = (f_i P, Q) \mod q\mathbf{A}_0,
\end{aligned}$$

which proves our claim by induction. □

Corollary 6.3.5. *If $\lambda \gg 0$ and $Pv_\lambda \in \mathcal{L}(\lambda)$, then $P^\star v_\lambda \in \mathcal{L}(\lambda)$.*

Proof. Since $Pv_\lambda \in \mathcal{L}(\lambda)$, we have $(Pv_\lambda, Pv_\lambda) \in \mathbf{A}_0$ (see Proposition 5.1.7). By Lemmas 6.3.3 and 6.3.4, we obtain

$$(P^\star v_\lambda, P^\star v_\lambda) \equiv (P^\star, P^\star) = (P, P) \equiv (Pv_\lambda, Pv_\lambda) \mod q\mathbf{A}_0.$$

Hence $(P^\star v_\lambda, P^\star v_\lambda) \in \mathbf{A}_0$ and, by Proposition 5.1.7, we get $P^\star v_\lambda \in \mathcal{L}(\lambda)$. □

In the rest of this section, we will present a couple of combinatorial lemmas that will be needed in the proof of Theorem 6.2.1.

Lemma 6.3.6 ([**2, 39**]).

(1) *For $a_1, a_2, b_1, b_2 \geq 0$, we have*

$$\begin{aligned}
&\sum_{k \geq 0} \frac{[a_1 + a_2 + b_1 + b_2 + k]_q!}{[k]_q![a_1 - k]_q![a_2 - k]_q![b_1 + k]_q![b_2 + k]_q!} \\
&\quad = \frac{[a_1 + a_2 + b_1 + b_2]_q![a_1 + a_2 + b_1]_q![a_1 + a_2 + b_2]_q!}{[a_1]_q![a_2]_q![a_1 + b_1]_q![a_2 + b_1]_q![a_1 + b_2]_q![a_2 + b_2]_q!}.
\end{aligned}$$

(2) *For $m \geq 0$ and $n \geq 1$, we have*

$$\sum_{k=0}^{m}(-1)^k \begin{bmatrix} k+n-1 \\ k \end{bmatrix}_q \begin{bmatrix} k+m+n \\ m \end{bmatrix}_q \begin{bmatrix} m+n \\ k+n \end{bmatrix}_q = 1.$$

Proof. The identity in (1) is well known (see, for example, [**2**, p. 37]). The proof of (2), which uses (1), is left to the readers as an exercise (Exercise 6.7). □

Using (6.13), we may observe that $e_i' U_{\mathbf{A}}^-(\mathfrak{g}) \subset U_{\mathbf{A}}^-(\mathfrak{g})$ (Exercise 6.8). By an argument similar to the one in Lemma 4.1.1, every $u \in U_q^-(\mathfrak{g})$ can be written uniquely in the form

$$u = \sum_{k \geq 0} f_i^{(k)} u_k,$$

where $e_i' u_k = 0$ for all $k \geq 0$. Moreover, if $u \in U_{\mathbf{A}}^-(\mathfrak{g})$, one can show that each u_k belongs to $U_{\mathbf{A}}^-(\mathfrak{g})$ (Exercise 6.9).

For $n \geq 1$ and $i \in I$, define

$$(f_i^n U_q^-(\mathfrak{g}))^{\mathbf{A}} = f_i^n U_q^-(\mathfrak{g}) \cap U_{\mathbf{A}}^-(\mathfrak{g}).$$

By the above observation, it is easy to see that

$$(f_i^n U_q^-(\mathfrak{g}))^{\mathbf{A}} = \sum_{k \geq n} f_i^{(k)} U_{\mathbf{A}}^-(\mathfrak{g}).$$

For $\lambda \in P^+$, we define

$$(f_i^n V(\lambda))^{\mathbf{A}} = (f_i^n U_q^-(\mathfrak{g}))^{\mathbf{A}} v_\lambda = \sum_{k \geq n} f_i^{(k)} V(\lambda)^{\mathbf{A}}.$$

Then we have:

Lemma 6.3.7. *Fix $i \in I$ and $\mu \in P$.*

(1) *For $u \in V(\lambda)_\mu$ with $n = -\mu(h_i) \geq 1$, we have*

$$u = \sum_{k \geq n} (-1)^{k-n} \begin{bmatrix} k-1 \\ k-n \end{bmatrix}_{q_i} f_i^{(k)} e_i^{(k)} u.$$

(2) *If $n = -\mu(h_i) \geq 0$, then we have*

$$V(\lambda)_\mu^{\mathbf{A}} = \sum_{k \geq n} f_i^{(k)} V(\lambda)_{\mu + k\alpha_i}^{\mathbf{A}} = (f_i^n V(\lambda))_\mu^{\mathbf{A}}.$$

Proof. (1) We may assume that $u = f_i^{(m)}v$ with $v \in \ker e_i \cap V(\lambda)_{\mu+m\alpha_i}$ with $m \geq n$. Then by Exercise 3.6 and Lemma 6.3.6(2), we have

$$\begin{aligned}
&\sum_{k\geq n}(-1)^{k-n}\begin{bmatrix} k-1 \\ k-n \end{bmatrix}_{q_i} f_i^{(k)}e_i^{(k)}u \\
&= \sum_{k\geq n}(-1)^{k-n}\begin{bmatrix} k-1 \\ k-n \end{bmatrix}_{q_i} f_i^{(k)}e_i^{(k)}f_i^{(m)}v \\
&= \sum_{k=n}^{m}(-1)^{k-n}\begin{bmatrix} k-1 \\ k-n \end{bmatrix}_{q_i} f_i^{(k)}\begin{bmatrix} (k-m)+(2m-n) \\ k \end{bmatrix}_{q_i} f_i^{(m-k)}v \\
&= \left(\sum_{k=n}^{m}(-1)^{k-n}\begin{bmatrix} k-1 \\ k-n \end{bmatrix}_{q_i}\begin{bmatrix} k+m-n \\ k \end{bmatrix}_{q_i}\begin{bmatrix} m \\ k \end{bmatrix}_{q_i}\right) f_i^{(m)}v \\
&= f_i^{(m)}v = u.
\end{aligned}$$

(2) This follows immediately from (1). □

6.4. Triviality of vector bundles over $\mathbf{P}^1$

Let V be a finite dimensional $\mathbf{Q}(q)$-vector space and let $\mathcal{L}_0$ (respectively, $\mathcal{L}_\infty$) be a free $\mathbf{A}_0$-lattice (respectively, $\mathbf{A}_\infty$-lattice) of V as in Section 6.1.

Lemma 6.4.1. *Let M be an $\mathbf{A}$-submodule of V and let $E = M \cap \mathcal{L}_0 \cap \mathcal{L}_\infty$. Suppose that the canonical map*

$$E \longrightarrow (M \cap \mathcal{L}_0)/(M \cap q\mathcal{L}_0)$$

is an isomorphism. Then we have canonical isomorphisms

(1) $\mathbf{Q}[q] \otimes_{\mathbf{Q}} E \cong M \cap \mathcal{L}_0$,

(2) $\mathbf{Q}[q^{-1}] \otimes_{\mathbf{Q}} E \cong M \cap \mathcal{L}_\infty$,

(3) $\mathbf{A} \otimes_{\mathbf{Q}} E \cong M$,

(4) $E \xrightarrow{\sim} (M \cap \mathcal{L}_\infty)/(M \cap q^{-1}\mathcal{L}_\infty)$.

Proof. Since our argument is similar to that of Theorem 6.1.4, we only give a sketch of the proof.

We first prove the canonical isomorphism

$$\left(\bigoplus_{k=0}^{n} \mathbf{Q}q^k\right) \otimes_{\mathbf{Q}} E \cong M \cap \mathcal{L}_0 \cap q^n\mathcal{L}_\infty$$

by induction on n. If $n = 0$, our assertion is obvious. If $n > 0$, we show that the following diagram is commutative with exact rows and all maps canonical (Exercise 6.10).

$$\begin{array}{ccccccccc}
0 & \longrightarrow & (\bigoplus_{k=1}^{n} \mathbf{Q}q^k) \otimes_{\mathbf{Q}} E & \longrightarrow & (\bigoplus_{k=0}^{n} \mathbf{Q}q^k) \otimes_{\mathbf{Q}} E & \longrightarrow & E & \longrightarrow & 0 \\
 & & \downarrow \alpha & & \downarrow \beta & & \downarrow \wr & & \\
0 & \longrightarrow & M \cap q\mathcal{L}_0 \cap q^n \mathcal{L}_\infty & \longrightarrow & M \cap \mathcal{L}_0 \cap q^n \mathcal{L}_\infty & \longrightarrow & (M \cap \mathcal{L}_0)/(M \cap q\mathcal{L}_0) & &
\end{array}$$

Since α is an isomorphism by the induction hypothesis, by diagram chasing one can show that β is also an isomorphism. Hence, as in the proof of Theorem 6.1.4, for any $a \leq b$, we have

$$\left(\bigoplus_{a \leq k \leq b} \mathbf{Q}q^k \right) \otimes_{\mathbf{Q}} E \cong M \cap q^a \mathcal{L}_0 \cap q^b \mathcal{L}_\infty,$$

which yields

$$\begin{aligned}
\mathbf{Q}[q] \otimes_{\mathbf{Q}} E &\cong M \cap \mathcal{L}_0, \\
\mathbf{Q}[q^{-1}] \otimes_{\mathbf{Q}} E &\cong M \cap \mathcal{L}_\infty, \\
\mathbf{A} \otimes_{\mathbf{Q}} E &\cong M,
\end{aligned}$$

under canonical maps. Therefore, we obtain

$$\begin{aligned}
E &\cong \mathbf{Q} \otimes_{\mathbf{Q}} E \\
&\cong (\mathbf{Q}[q^{-1}] \otimes_{\mathbf{Q}} E)/(q^{-1}\mathbf{Q}[q^{-1}] \otimes_{\mathbf{Q}} E) \\
&\xrightarrow{\sim} (M \cap \mathcal{L}_\infty)/(M \cap q^{-1}\mathcal{L}_\infty),
\end{aligned}$$

which completes the proof. □

Lemma 6.4.2. *Let M be an $\mathbf{A}$-submodule of V and let $E = M \cap \mathcal{L}_0 \cap \mathcal{L}_\infty$. Suppose that the canonical map*

$$E \longrightarrow (M \cap \mathcal{L}_0)/(M \cap q\mathcal{L}_0)$$

is an isomorphism. Then we have the canonical isomorphisms

(1) $\mathbf{A}_0 \otimes_{\mathbf{Q}} E \cong (\mathbf{Q}(q) \otimes_{\mathbf{A}} M) \cap \mathcal{L}_0$,

(2) $\mathbf{A}_\infty \otimes_{\mathbf{Q}} E \cong (\mathbf{Q}(q) \otimes_{\mathbf{A}} M) \cap \mathcal{L}_\infty$,

(3) $E \xrightarrow{\sim} (\mathbf{Q}(q) \otimes_{\mathbf{A}} M) \cap \mathcal{L}_0/(\mathbf{Q}(q) \otimes_{\mathbf{A}} M) \cap q\mathcal{L}_0$,

(4) $E \xrightarrow{\sim} ((\mathbf{Q}(q) \otimes_{\mathbf{A}} M) \cap \mathcal{L}_\infty)/((\mathbf{Q}(q) \otimes_{\mathbf{A}} M) \cap q^{-1}\mathcal{L}_\infty)$.

Proof. Let $S = \{f(q) \in \mathbf{Q}[q] \mid f(0) \neq 0\}$. We first observe that $\mathbf{Q}(q) \otimes_{\mathbf{A}} M \cong S^{-1}M$. To see this, let $\frac{f(q)}{g(q)} \otimes u \in \mathbf{Q}(q) \otimes_{\mathbf{A}} M$ with $f(q), g(q) \in \mathbf{Q}[q]$, $u \in M$. Then we may write

$$\frac{f(q)}{g(q)} \otimes u = q^{-N} \frac{f(q)}{h(q)} \otimes u = \frac{1}{h(q)} \otimes q^{-N} f(q) u$$

with $h(0) \neq 0$ for some $N \in \mathbf{Z}$. Hence the map

$$\frac{f(q)}{g(q)} \otimes u \longmapsto \frac{1}{h(q)} q^{-N} f(q) u$$

defines an isomorphism $\mathbf{Q}(q) \otimes_{\mathbf{A}} M \cong S^{-1}M$.

Now, it is easy to verify that $(S^{-1}M) \cap \mathcal{L}_0 = S^{-1}(M \cap \mathcal{L}_0)$ (Exercise 6.11). Therefore, by Lemma 6.4.1, we have

$$\begin{aligned}\mathbf{A}_0 \otimes_{\mathbf{Q}} E &= S^{-1}(\mathbf{Q}[q] \otimes_{\mathbf{Q}} E) \cong S^{-1}(M \cap \mathcal{L}_0)\\ &= S^{-1}M \cap \mathcal{L}_0 \cong (\mathbf{Q}(q) \otimes_{\mathbf{A}} M) \cap \mathcal{L}_0,\end{aligned}$$

and

$$\begin{aligned}E = \mathbf{Q} \otimes_{\mathbf{Q}} E &\cong (\mathbf{A}_0 \otimes_{\mathbf{Q}} E)/(q\mathbf{A}_0 \otimes_{\mathbf{Q}} E)\\ &\xrightarrow{\sim} \frac{(\mathbf{Q}(q) \otimes_{\mathbf{A}} M) \cap \mathcal{L}_0}{(\mathbf{Q}(q) \otimes_{\mathbf{A}} M) \cap q\mathcal{L}_0},\end{aligned}$$

which proves (1) and (3). One can prove (2) and (4) by a similar argument. □

Lemma 6.4.3. *Let M be an $\mathbf{A}$-submodule of V and let E be a finite dimensional $\mathbf{Q}$-vector space. Suppose that there is an (injective) $\mathbf{Q}$-linear map $\varphi : E \to M \cap \mathcal{L}_0 \cap \mathcal{L}_\infty$ satisfying the following conditions:*

(1) *the canonical maps*

$$E \xrightarrow{\varphi} M \cap \mathcal{L}_0 \cap \mathcal{L}_\infty \longrightarrow \mathcal{L}_0/q\mathcal{L}_0,$$
$$E \xrightarrow{\varphi} M \cap \mathcal{L}_0 \cap \mathcal{L}_\infty \longrightarrow \mathcal{L}_\infty/q\mathcal{L}_\infty$$

are injective;

(2) $M = \mathbf{A}\varphi(E) \cong \mathbf{A} \otimes_{\mathbf{Q}} E$.

Then we have the canonical isomorphisms

(1) $E \overset{\varphi}{\cong} M \cap \mathcal{L}_0 \cap \mathcal{L}_\infty \xrightarrow{\sim} (M \cap \mathcal{L}_0)/(M \cap q\mathcal{L}_0)$,

(2) $E \overset{\varphi}{\cong} M \cap \mathcal{L}_0 \cap \mathcal{L}_\infty \xrightarrow{\sim} (M \cap \mathcal{L}_\infty)/(M \cap q^{-1}\mathcal{L}_\infty)$.

Proof. We first claim that

$$(\mathbf{Q}(q) \otimes_{\mathbf{Q}} E) \cap \mathcal{L}_0 = \mathbf{A}_0 \otimes_{\mathbf{Q}} E.$$

By our assumption on φ, we have a canonical embedding

$$\begin{aligned}\mathbf{A}_0 \otimes_{\mathbf{Q}} E &\hookrightarrow \mathbf{A}_0 \otimes_{\mathbf{Q}} \mathcal{L}_0/q\mathcal{L}_0 \cong \mathbf{A}_0 \otimes_{\mathbf{Q}} \mathbf{Q} \otimes_{\mathbf{A}_0} \mathcal{L}_0\\ &= \mathbf{A}_0 \otimes_{\mathbf{A}_0} \mathcal{L}_0 = \mathcal{L}_0.\end{aligned}$$

Since the second map (which is an isomorphism) is the left inverse of the canonical embedding, we get

$$\mathbf{A}_0 \otimes_{\mathbf{Q}} E \subset (\mathbf{Q}(q) \otimes_{\mathbf{Q}} E) \cap \mathcal{L}_0.$$

Conversely, let $v = \sum_i g_i \otimes v_i$ be a nonzero element in $(\mathbf{Q}(q) \otimes_{\mathbf{Q}} E) \cap \mathcal{L}_0$, where $g_i \in \mathbf{Q}(q)$ and $\{v_i \mid i = 1, \ldots, \dim_{\mathbf{Q}} E\}$ is a $\mathbf{Q}$-basis of E. Then there is a nonnegative integer N such that

$$q^N v = \sum_i g_i' \otimes v_i \in (\mathbf{A}_0 \otimes_{\mathbf{Q}} E) \cap q^N \mathcal{L}_0,$$

with all $g_i' \in \mathbf{A}_0$. Since the canonical map $E \longrightarrow \mathcal{L}_0/q\mathcal{L}_0$ is injective, the induced map $\mathbf{Q}(q) \otimes_{\mathbf{Q}} E \to \mathbf{Q}(q) \otimes_{\mathbf{Q}} (\mathcal{L}_0/q\mathcal{L}_0) = V$ is also injective. If $N > 0$, then $q^N v \in (\mathbf{A}_0 \otimes_{\mathbf{Q}} E) \cap q^N \mathcal{L}_0$ and is mapped to zero under this map. It follows that $q^N v = 0 = \sum_i g_i' \otimes v_i$. Since $\{v_i \mid i = 1, \ldots, \dim_{\mathbf{Q}} E\}$ is a $\mathbf{Q}$-basis of E, we must have all $g_i' = 0$, a contradiction. Hence $N = 0$ and $v = \sum_i g_i' \otimes v_i \in \mathbf{A}_0 \otimes_{\mathbf{Q}} E$.

Next observe that

$$\begin{aligned}
\mathbf{Q}[q] \otimes_{\mathbf{Q}} E &= (\mathbf{A} \otimes_{\mathbf{Q}} E) \cap (\mathbf{A}_0 \otimes_{\mathbf{Q}} E) \\
&\cong (\mathbf{A} \otimes_{\mathbf{Q}} E) \cap (\mathbf{Q}(q) \otimes_{\mathbf{Q}} E) \cap \mathcal{L}_0 \\
&= (\mathbf{A} \otimes_{\mathbf{Q}} E) \cap \mathcal{L}_0 \\
&\cong M \cap \mathcal{L}_0.
\end{aligned}$$

Similarly we have

$$\mathbf{Q}[q^{-1}] \otimes_{\mathbf{Q}} E \cong M \cap \mathcal{L}_\infty.$$

Therefore we may obtain the following isomorphisms:

$$\begin{aligned}
E &= \mathbf{Q} \otimes_{\mathbf{Q}} E = (\mathbf{Q}[q] \cap \mathbf{Q}[q^{-1}]) \otimes_{\mathbf{Q}} E \\
&= (\mathbf{Q}[q] \otimes_{\mathbf{Q}} E) \cap (\mathbf{Q}[q^{-1}] \otimes_{\mathbf{Q}} E) \\
&\cong (M \cap \mathcal{L}_0) \cap (M \cap \mathcal{L}_\infty) \\
&= M \cap \mathcal{L}_0 \cap \mathcal{L}_\infty, \\
E &= \mathbf{Q} \otimes_{\mathbf{Q}} E = \mathbf{Q}[q] \otimes_{\mathbf{Q}} E / q\mathbf{Q}[q] \otimes_{\mathbf{Q}} E \\
&\xrightarrow{\sim} (M \cap \mathcal{L}_0)/(M \cap q\mathcal{L}_0), \\
E &= F \otimes E = \mathbf{Q}[q^{-1}] \otimes_{\mathbf{Q}} E / q^{-1}\mathbf{Q}[q^{-1}] \otimes_{\mathbf{Q}} E \\
&\xrightarrow{\sim} (M \cap \mathcal{L}_\infty)/(M \cap q^{-1}\mathcal{L}_\infty).
\end{aligned}$$

This completes the proof. □

Lemma 6.4.4. *Let $N \subset M$ be $\mathbf{A}$-submodules of V and let F be a $\mathbf{Q}$-vector space. Suppose that the canonical map $N \cap \mathcal{L}_0 \cap \mathcal{L}_\infty \longrightarrow (N \cap \mathcal{L}_0)/(N \cap q\mathcal{L}_0)$ is an isomorphism and that there is an (injective) $\mathbf{Q}$-linear map*

$$\varphi : F \longrightarrow M \cap (\mathcal{L}_0 + N) \cap (\mathcal{L}_\infty + N)$$

satisfying the following conditions:

(1) *the canonical maps*

$$\varphi_0 : F \longrightarrow M \cap (\mathcal{L}_0 + N) \cap (\mathcal{L}_\infty + N) \longrightarrow (\mathcal{L}_0 + N)/(q\mathcal{L}_0 + N),$$
$$\varphi_\infty : F \longrightarrow M \cap (\mathcal{L}_0 + N) \cap (\mathcal{L}_\infty + N) \longrightarrow (\mathcal{L}_\infty + N)/(q^{-1}\mathcal{L}_\infty + N)$$

are injective;

(2) $M = \mathbf{A}\varphi(F) + N$.

Then we have:

(1) *the canonical map* $M \cap \mathcal{L}_0 \cap \mathcal{L}_\infty \longrightarrow (M \cap \mathcal{L}_0)/(M \cap q\mathcal{L}_0)$ *is an isomorphism;*

(2) *the sequence*

$$0 \to (N \cap \mathcal{L}_0)/(N \cap q\mathcal{L}_0) \to (M \cap \mathcal{L}_0)/(M \cap q\mathcal{L}_0) \xrightarrow{g} (\mathcal{L}_0 + N)/(q\mathcal{L}_0 + N)$$

is exact;

(3) $\varphi_0(F) = g((M \cap \mathcal{L}_0)/(M \cap q\mathcal{L}_0))$.

Proof. Since $N \cap \mathcal{L}_0 \cap \mathcal{L}_\infty \xrightarrow{\sim} (N \cap \mathcal{L}_0)/(N \cap q\mathcal{L}_0)$, Lemma 6.4.1 gives

$$\begin{aligned} N &\cong \mathbf{A} \otimes_{\mathbf{Q}} (N \cap \mathcal{L}_0 \cap \mathcal{L}_\infty) \\ &= \mathbf{Q}[q] \otimes_{\mathbf{Q}} (N \cap \mathcal{L}_0 \cap \mathcal{L}_\infty) + \mathbf{Q}[q^{-1}] \otimes_{\mathbf{Q}} (N \cap \mathcal{L}_0 \cap \mathcal{L}_\infty) \\ &\cong (N \cap \mathcal{L}_0) + (N \cap \mathcal{L}_\infty), \end{aligned}$$

where the first and the last canonical isomorphisms are inverses to each other. Hence, we obtain

$$N = (N \cap \mathcal{L}_0) + (N \cap \mathcal{L}_\infty).$$

We first claim that

$$M \cap (\mathcal{L}_0 + N) \cap (\mathcal{L}_\infty + N) = N + (M \cap \mathcal{L}_0 \cap \mathcal{L}_\infty).$$

It is clear that

$$M \cap (\mathcal{L}_0 + N) \cap (\mathcal{L}_\infty + N) \supset N + (M \cap \mathcal{L}_0 \cap \mathcal{L}_\infty).$$

To prove the other inclusion, let

$$x \in M \cap (\mathcal{L}_0 + N) \cap (\mathcal{L}_\infty + N)$$

and write

$$x = u + v = y + z \quad \text{with } u \in \mathcal{L}_0,\ y \in \mathcal{L}_\infty,\ v, z \in N \subset M.$$

Then

$$u = x - v = y + z - v \in (M \cap \mathcal{L}_0) \cap (\mathcal{L}_\infty + N),$$

and hence

$$x = u + v \in N + M \cap \mathcal{L}_0 \cap (\mathcal{L}_\infty + N).$$

It follows that

$$M \cap (\mathcal{L}_0 + N) \cap (\mathcal{L}_\infty + N) \subset N + M \cap \mathcal{L}_0 \cap (\mathcal{L}_\infty + N).$$

Since $N = N \cap \mathcal{L}_0 + N \cap \mathcal{L}_\infty$, we have $\mathcal{L}_\infty + N = \mathcal{L}_\infty + N \cap \mathcal{L}_0$ and hence

$$N + M \cap \mathcal{L}_0 \cap (\mathcal{L}_\infty + N) \subset N + M \cap \mathcal{L}_0 \cap (\mathcal{L}_\infty + N \cap \mathcal{L}_0).$$

If $x \in N + M \cap \mathcal{L}_0 \cap (\mathcal{L}_\infty + N \cap \mathcal{L}_0)$, we may write

$$x = y + z_1 + z_2, \quad \text{where } y \in N,\ z_1 \in N \cap \mathcal{L}_0,\ z_2 \in M \cap \mathcal{L}_0 \cap \mathcal{L}_\infty.$$

Hence

$$x = y + z_1 + z_2 \in N + M \cap \mathcal{L}_0 \cap \mathcal{L}_\infty,$$

which proves our claim.

By use of this claim, we have a $\mathbf{Q}$-linear map

$$\varphi : F \to M \cap (\mathcal{L}_0 + N) \cap (\mathcal{L}_\infty + N) = N + (M \cap \mathcal{L}_0 \cap \mathcal{L}_\infty)$$

such that $M = \mathbf{A}\varphi(F) + N$. Note that the elements in $\varphi(F) \cap N$ are not needed to generate M. Hence we may assume that $\varphi(F) \subset M \cap \mathcal{L}_0 \cap \mathcal{L}_\infty$. Now it is straightforward to verify that the following diagram is commutative with exact rows (Exercise 6.12).

$$\begin{array}{ccccccccc}
0 & \longrightarrow & N \cap \mathcal{L}_0 \cap \mathcal{L}_\infty & \longrightarrow & F \oplus (N \cap \mathcal{L}_0 \cap \mathcal{L}_\infty) & \longrightarrow & F & \longrightarrow & 0 \\
 & & \Big\downarrow \mathrm{id} & & \Big\downarrow \alpha & & \Big\downarrow \varphi_0 & & \\
0 & \longrightarrow & N \cap \mathcal{L}_0 \cap \mathcal{L}_\infty & \xrightarrow{\ \pi\ } & \mathcal{L}_0/q\mathcal{L}_0 & \xrightarrow{\ g\ } & (\mathcal{L}_0 + N)/(q\mathcal{L}_0 + N) & &
\end{array}$$

Since

$$\begin{aligned} \operatorname{Im} \pi = \pi(N \cap \mathcal{L}_0 \cap \mathcal{L}_\infty) &= (N \cap \mathcal{L}_0)/(N \cap q\mathcal{L}_0) \\ &\subset (M \cap \mathcal{L}_0)/(M \cap q\mathcal{L}_0), \end{aligned}$$

we get a sequence

$$0 \to N \cap \mathcal{L}_0 \cap \mathcal{L}_\infty \xrightarrow{\pi} (M \cap \mathcal{L}_0)/(M \cap q\mathcal{L}_0) \xrightarrow{g} (\mathcal{L}_0 + N)/(q\mathcal{L}_0 + N).$$

Note that

$$\begin{aligned} \ker g &= \{u + M \cap q\mathcal{L}_0 \mid u \in M \cap \mathcal{L}_0, u + q\mathcal{L}_0 + N = 0\} \\ &= \{u + M \cap q\mathcal{L}_0 \mid u \in (q\mathcal{L}_0 + N) \cap (M \cap \mathcal{L}_0)\} \\ &= \frac{(N \cap \mathcal{L}_0) + (M \cap q\mathcal{L}_0)}{M \cap q\mathcal{L}_0} \\ &\cong (N \cap \mathcal{L}_0)/(N \cap q\mathcal{L}_0) = \operatorname{Im} \pi, \end{aligned}$$

which proves (2).

Let $E = F \oplus (N \cap \mathcal{L}_0 \cap \mathcal{L}_\infty)$. Since φ_0 is injective, $\alpha : E \to \mathcal{L}_0/q\mathcal{L}_0$ is injective. Similarly, $\alpha_\infty : E \to \mathcal{L}_\infty/q^{-1}\mathcal{L}_\infty$ is injective. Moreover, since $N \cap \mathcal{L}_0 \cap \mathcal{L}_\infty \xrightarrow{\sim} (N \cap \mathcal{L}_0)/(N \cap q\mathcal{L}_0)$, Lemma 6.4.1 yields

$$\begin{aligned}\mathbf{A} \otimes_{\mathbf{Q}} E &= (\mathbf{A} \otimes_{\mathbf{Q}} F) \oplus (\mathbf{A} \otimes_{\mathbf{Q}} (N \cap \mathcal{L}_0 \cap \mathcal{L}_\infty)) \\ &\cong \mathbf{A}\varphi(F) + N = M.\end{aligned}$$

Therefore, by Lemma 6.4.3, we obtain

$$E = F \oplus (N \cap \mathcal{L}_0 \cap \mathcal{L}_\infty) \cong M \cap \mathcal{L}_0 \cap \mathcal{L}_\infty \xrightarrow{\sim} (M \cap \mathcal{L}_0)/(M \cap q\mathcal{L}_0). \tag{6.21}$$

Thus (1) is proved.

To prove (3), note that the map

$$\alpha : E = F \oplus (N \cap \mathcal{L}_0 \cap \mathcal{L}_\infty) \longrightarrow \mathcal{L}_0/q\mathcal{L}_0$$

is given by

$$a + u \longmapsto \varphi(a) + u + q\mathcal{L}_0 \quad \text{for } a \in F,\ u \in N \cap \mathcal{L}_0 \cap \mathcal{L}_\infty.$$

That is, α is the isomorphism given in (6.21). Thus,

$$\alpha(E) = (M \cap \mathcal{L}_0)/(M \cap q\mathcal{L}_0),$$

which implies

$$g\left(\frac{(M \cap \mathcal{L}_0)}{(M \cap q\mathcal{L}_0)}\right) = g \circ \alpha(E) = \varphi_0(F).$$

This completes the proof. □

Remark 6.4.5. The lemmas we have proved in this section have geometric interpretations in the language of vector bundles over $\mathbf{P}^1$.

Let

$$X = \mathbf{P}^1 = \operatorname{Spec} \mathbf{Q}[q] \cup \operatorname{Spec} \mathbf{Q}[q^{-1}]$$

and define a torsion free coherent $\mathcal{O}_X$-module $\mathcal{F}$ by

$$\Gamma(\operatorname{Spec} \mathbf{Q}[q], \mathcal{F}) = M \cap \mathcal{L}_0, \qquad \Gamma(\operatorname{Spec} \mathbf{Q}[q^{-1}], \mathcal{F}) = M \cap \mathcal{L}_\infty.$$

Then $\mathcal{F}$ is well defined, since

$$\begin{aligned}M &\cong \mathbf{Q}[q, q^{-1}] \otimes_{\mathbf{Q}[q]} (M \cap \mathcal{L}_0) \\ &\cong \mathbf{Q}[q, q^{-1}] \otimes_{\mathbf{Q}[q^{-1}]} (M \cap \mathcal{L}_\infty).\end{aligned}$$

Note that $\Gamma(X, \mathcal{F}) = E$ and

$$\mathcal{F}_0 = \mathcal{O}_{X,0} \otimes_{\mathbf{Q}[q]} \Gamma(\operatorname{Spec} \mathbf{Q}[q], \mathcal{F}) \cong \mathbf{A}_0 \otimes_{\mathbf{Q}[q]} (M \cap \mathcal{L}_0).$$

Let $\iota_0 : \operatorname{Spec} \mathbf{Q} \longrightarrow X$ be the section given by $q = 0$. Then

$$\begin{aligned}\Gamma(\operatorname{Spec} \mathbf{Q}, \iota_0^* \mathcal{F}) &= \mathbf{Q} \otimes_{\mathbf{A}_0} \mathbf{A}_0 \otimes_{\mathbf{Q}[q]} (M \cap \mathcal{L}_0) \\ &\cong (M \cap \mathcal{L}_0)/(M \cap q\mathcal{L}_0).\end{aligned}$$

Therefore Lemma 6.4.1 can be interpreted as follows:
if $\Gamma(X, \mathcal{F}) \cong \Gamma(\operatorname{Spec} \mathbf{Q}, \iota_0^* \mathcal{F})$, then we must have $\mathcal{F} \cong \Gamma(X, \mathcal{F}) \otimes_{\mathbf{Q}} \mathcal{O}_X$.

When M is an $\mathbf{A}$-lattice of V, it is easy to show that $(M, \mathcal{L}_0, \mathcal{L}_\infty)$ is a balanced triple if and only if $\mathcal{F} \cong \Gamma(X, \mathcal{F}) \otimes_{\mathbf{Q}} \mathcal{O}_X$.

6.5. Existence of global bases

In this section we will prove the existence theorem for global bases (Theorem 6.2.1). Actually, most of this section will be devoted to proving the following inductive statements, from which the existence theorem follows.

Theorem 6.5.1.
$\mathbf{A}(\mathbf{r})$: *For all* $\lambda \in P^+$ *and* $\alpha \in Q_+(r)$, *the canonical map*

$$V(\lambda)^{\mathbf{A}}_{\lambda-\alpha} \cap \mathcal{L}(\lambda) \cap \mathcal{L}(\lambda)^- \longrightarrow (V(\lambda)^{\mathbf{A}}_{\lambda-\alpha} \cap \mathcal{L}(\lambda))/(V(\lambda)^{\mathbf{A}}_{\lambda-\alpha} \cap q\mathcal{L}(\lambda))$$

is an isomorphism.
$\mathbf{B}(\mathbf{r})$: *For* $\alpha \in Q_+(r)$, *let*

$$G : (V(\lambda)^{\mathbf{A}}_{\lambda-\alpha} \cap \mathcal{L}(\lambda))/(V(\lambda)^{\mathbf{A}}_{\lambda-\alpha} \cap q\mathcal{L}(\lambda)) \longrightarrow V(\lambda)^{\mathbf{A}}_{\lambda-\alpha} \cap \mathcal{L}(\lambda) \cap \mathcal{L}(\lambda)^-$$

be the inverse of the canonical isomorphism given in $\mathbf{A}(\mathbf{r})$. *Then for any* $b \in \mathcal{B}(\lambda)_{\lambda-\alpha} \cap \tilde{f}_i^n \mathcal{B}(\lambda)$ *with* $n \geq 0$, *we have* $G(b) \in f_i^n V(\lambda)$.

Proof of Theorem 6.2.1. Assume that the statement $\mathbf{A}(\mathbf{r})$ is true. Then by Lemma 6.4.2 (3), we have

$$V(\lambda)^{\mathbf{A}}_{\lambda-\alpha} \cap \mathcal{L}(\lambda) \cap \mathcal{L}(\lambda)^- \xrightarrow{\sim} \frac{(\mathbf{Q}(q) \otimes_{\mathbf{A}} V(\lambda)^{\mathbf{A}}_{\lambda-\alpha}) \cap \mathcal{L}(\lambda)}{(\mathbf{Q}(q) \otimes_{\mathbf{A}} V(\lambda)^{\mathbf{A}}_{\lambda-\alpha}) \cap q\mathcal{L}(\lambda)}. \tag{6.22}$$

Since $\mathbf{Q}(q) \otimes_{\mathbf{A}} V(\lambda)^{\mathbf{A}}_{\lambda-\alpha} \cong V(\lambda)_{\lambda-\alpha}$, we obtain

$$V(\lambda)^{\mathbf{A}}_{\lambda-\alpha} \cap \mathcal{L}(\lambda) \cap \mathcal{L}(\lambda)^- \xrightarrow{\sim} \mathcal{L}(\lambda)_{\lambda-\alpha}/q\mathcal{L}(\lambda)_{\lambda-\alpha}.$$

Hence, by Theorem 6.1.4, we conclude that $(V(\lambda)^{\mathbf{A}}_{\lambda-\alpha}, \mathcal{L}(\lambda)_{\lambda-\alpha}, \mathcal{L}(\lambda)^-_{\lambda-\alpha})$ is a balanced triple, as stated in Theorem 6.2.1. $\square$

For the rest of this section, we will focus on proving the statements $\mathbf{A}(\mathbf{r})$ and $\mathbf{B}(\mathbf{r})$. Fix a dominant integral weight $\lambda \in P^+$. If $r = 0$, our assertions are obvious. Suppose $r > 0$ and assume that the statements $\mathbf{A}(\mathbf{r}-\mathbf{1})$ and $\mathbf{B}(\mathbf{r}-\mathbf{1})$ are true. Then $(V(\lambda)^{\mathbf{A}}_{\lambda-\alpha}, \mathcal{L}(\lambda)_{\lambda-\alpha}, \mathcal{L}(\lambda)^-_{\lambda-\alpha})$ is a balanced triple for all $\alpha \in Q_+(r-1)$. Hence the same argument preceding Theorem 6.2.2 gives:

Lemma 6.5.2. *For all* $\alpha \in Q_+(r-1)$, *we have*

(1) $V(\lambda)^{\mathbf{A}}_{\lambda-\alpha} \cap \mathcal{L}(\lambda) = \bigoplus_{b \in \mathcal{B}(\lambda)_{\lambda-\alpha}} \mathbf{Q}[q] G(b)$,

(2) $V(\lambda)^{\mathbf{A}}_{\lambda-\alpha} = \bigoplus_{b \in \mathcal{B}(\lambda)_{\lambda-\alpha}} \mathbf{A} G(b)$,

(3) $\overline{G(b)} = G(b)$ *for all* $b \in \mathcal{B}(\lambda)_{\lambda-\alpha}$.

Proof. We leave the proof of this lemma to the readers as an exercise (Exercise 6.13). □

To prove $\mathbf{A}(\mathbf{r})$ and $\mathbf{B}(\mathbf{r})$, we would like to define a map

$$G: \frac{V(\lambda)^{\mathbf{A}}_{\lambda-\alpha} \cap \mathcal{L}(\lambda)}{V(\lambda)^{\mathbf{A}}_{\lambda-\alpha} \cap q\mathcal{L}(\lambda)} \longrightarrow V(\lambda)^{\mathbf{A}}_{\lambda-\alpha} \cap \mathcal{L}(\lambda) \cap \mathcal{L}(\lambda)^{-} \quad (\alpha \in Q_+(r)),$$

which will turn out to be the inverse of the canonical map

$$V(\lambda)^{\mathbf{A}}_{\lambda-\alpha} \cap \mathcal{L}(\lambda) \cap \mathcal{L}(\lambda)^{-} \longrightarrow \frac{V(\lambda)^{\mathbf{A}}_{\lambda-\alpha} \cap \mathcal{L}(\lambda)}{V(\lambda)^{\mathbf{A}}_{\lambda-\alpha} \cap q\mathcal{L}(\lambda)}.$$

For this purpose, the following proposition will play a crucial role.

Proposition 6.5.3. *For all* $\alpha \in Q_+(r)$, $n \geq 1$ *and* $i \in I$, *we have the following canonical isomorphisms*:

$$\begin{aligned}(f_i^n V(\lambda))^{\mathbf{A}}_{\lambda-\alpha} \cap \mathcal{L}(\lambda) \cap \mathcal{L}(\lambda)^{-} &\xrightarrow{\sim} \frac{(f_i^n V(\lambda))^{\mathbf{A}}_{\lambda-\alpha} \cap \mathcal{L}(\lambda)}{(f_i^n V(\lambda))^{\mathbf{A}}_{\lambda-\alpha} \cap q\mathcal{L}(\lambda)} \\ &\cong \bigoplus_{b \in \mathcal{B}(\lambda)_{\lambda-\alpha} \cap \tilde{f}_i^n \mathcal{B}(\lambda)} \mathbf{Q}b.\end{aligned}$$

Proof. We will prove our assertion by the descending induction on n. Note that if $n > r$, then $(f_i^n V(\lambda))^{\mathbf{A}}_{\lambda-\alpha} = 0$ and $\mathcal{B}(\lambda)_{\lambda-\alpha} \cap \tilde{f}_i^n \mathcal{B}(\lambda) \neq \emptyset$. Hence our assertion is trivial.

Assume that $n \leq r$ and our assertion is true for $n+1$:

$$\begin{aligned}(f_i^{n+1} V(\lambda))^{\mathbf{A}}_{\lambda-\alpha} \cap \mathcal{L}(\lambda) \cap \mathcal{L}(\lambda)^{-} &\xrightarrow{\sim} \frac{(f_i^{n+1} V(\lambda))^{\mathbf{A}}_{\lambda-\alpha} \cap \mathcal{L}(\lambda)}{(f_i^{n+1} V(\lambda))^{\mathbf{A}}_{\lambda-\alpha} \cap q\mathcal{L}(\lambda)} \\ &\cong \bigoplus_{b \in \mathcal{B}(\lambda)_{\lambda-\alpha} \cap \tilde{f}_i^{n+1} \mathcal{B}(\lambda)} \mathbf{Q}b.\end{aligned} \tag{6.23}$$

Recall that $\langle h_i, \lambda-\alpha \rangle = p' - p$, where p (respectively, p') is the maximal non-negative integer such that $\lambda - \alpha + p\alpha_i$ (respectively, $\lambda - \alpha - p'\alpha_i$) is a weight of $V(\lambda)$. Suppose $n + \langle h_i, \lambda - \alpha \rangle < 0$. Then $p - p' = -\langle h_i, \lambda - \alpha \rangle > n \geq 1$. We claim that

$$\mathcal{B}(\lambda)_{\lambda-\alpha} \cap \tilde{f}_i^n \mathcal{B}(\lambda) = \mathcal{B}(\lambda)_{\lambda-\alpha} \cap \tilde{f}_i^{-\langle h_i, \lambda-\alpha \rangle} \mathcal{B}(\lambda), \tag{6.24}$$

$$(f_i^n V(\lambda))^{\mathbf{A}}_{\lambda-\alpha} = (f_i^{-\langle h_i, \lambda-\alpha \rangle} V(\lambda))^{\mathbf{A}}_{\lambda-\alpha}. \tag{6.25}$$

It is easy to see that the inclusion $\supset$ holds in both (6.24) and (6.25). For the other inclusion, let $b = \tilde{f}_i^n b_1 \in \mathcal{B}(\lambda)_{\lambda-\alpha} \cap \tilde{f}_i^n \mathcal{B}(\lambda)$ for some $b_1 \in \mathcal{B}(\lambda)_{\lambda-\alpha+n\alpha_i}$. Then

$$\begin{aligned}\tilde{e}_i^p b &= \tilde{e}_i^{p-n} b_1 = \tilde{e}_i^{p+\langle h_i,\lambda-\alpha\rangle} \tilde{e}_i^{-\langle h_i,\lambda-\alpha\rangle-n} b_1 \\ &= \tilde{e}_i^{p'} \tilde{e}_i^{-\langle h_i,\lambda-\alpha\rangle-n} b_1,\end{aligned}$$

which implies

$$b' := \tilde{e}_i^{-\langle h_i,\lambda-\alpha\rangle-n} b_1 = \tilde{e}_i^{p-p'} b \in \mathcal{B}(\lambda).$$

Hence

$$b = \tilde{f}_i^{p'-p} b' = \tilde{f}_i^{-\langle h_i,\lambda-\alpha\rangle} b' \in \mathcal{B}(\lambda)_{\lambda-\alpha} \cap \tilde{f}_i^{-\langle h_i,\lambda-\alpha\rangle} \mathcal{B}(\lambda),$$

which proves (6.24).

Moreover, by Lemma 6.3.7 (2), we have

$$(f_i^{-\langle h_i,\lambda-\alpha\rangle} V(\lambda))^{\mathbf{A}}_{\lambda-\alpha} = \sum_{k \geq -\langle h_i,\lambda-\alpha\rangle} f_i^{(k)} V(\lambda)^{\mathbf{A}}_{\lambda-\alpha-k\alpha_i} = V(\lambda)^{\mathbf{A}}_{\lambda-\alpha}.$$

Since

$$(f_i^n V(\lambda))^{\mathbf{A}}_{\lambda-\alpha} \subset V(\lambda)^{\mathbf{A}}_{\lambda-\alpha},$$

(6.25) follows immediately. Thus we may assume that $n + \langle h_i, \lambda - \alpha\rangle \geq 0$. By definition, we have

$$\begin{aligned}(f_i^n V(\lambda))^{\mathbf{A}}_{\lambda-\alpha} &= \sum_{k \geq n} f_i^{(k)} V(\lambda)^{\mathbf{A}}_{\lambda-\alpha+k\alpha_i} \\ &= f_i^{(n)} V(\lambda)^{\mathbf{A}}_{\lambda-\alpha+n\alpha_i} + \sum_{k \geq n+1} f_i^{(k)} V(\lambda)^{\mathbf{A}}_{\lambda-\alpha+k\alpha_i} \\ &= f_i^{(n)} V(\lambda)^{\mathbf{A}}_{\lambda-\alpha+n\alpha_i} + (f_i^{n+1} V(\lambda))^{\mathbf{A}}_{\lambda-\alpha}.\end{aligned}$$

Since $n \geq 1$, Lemma 6.5.2 (2) gives

$$V(\lambda)^{\mathbf{A}}_{\lambda-\alpha+n\alpha_i} = \bigoplus_{b \in \mathcal{B}(\lambda)_{\lambda-\alpha+n\alpha_i}} \mathbf{A} G(b).$$

If $b \in \mathcal{B}(\lambda)_{\lambda-\alpha+n\alpha_i}$ and $\tilde{e}_i b \neq 0$, then $b \in \tilde{f}_i \mathcal{B}(\lambda) \cap \mathcal{B}(\lambda)_{\lambda-\alpha+n\alpha_i}$. Hence by $\mathbf{B(r-1)}$, $G(b) \in (f_i V(\lambda))^{\mathbf{A}}$, which yields $f_i^{(n)} G(b) \in (f_i^{n+1} V(\lambda))^{\mathbf{A}}_{\lambda-\alpha}$ (Exercise 6.14).

Let $\mathcal{B}_0 = \{b \in \mathcal{B}(\lambda)_{\lambda-\alpha+n\alpha_i} \mid \tilde{e}_i b = 0\}$. Then the above discussion implies

$$(f_i^n V(\lambda))^{\mathbf{A}}_{\lambda-\alpha} = \bigoplus_{b \in \mathcal{B}_0} \mathbf{A} f_i^{(n)} G(b) + (f_i^{n+1} V(\lambda))^{\mathbf{A}}_{\lambda-\alpha}. \tag{6.26}$$

We will apply Lemma 6.4.4 with $V = V(\lambda)_{\lambda-\alpha}$, $M = (f_i^n V(\lambda))^{\mathbf{A}}_{\lambda-\alpha}$, $N = (f_i^{n+1} V(\lambda))^{\mathbf{A}}_{\lambda-\alpha}$, $\mathcal{L}_0 = \mathcal{L}(\lambda)_{\lambda-\alpha}$, $\mathcal{L}_\infty = \mathcal{L}(\lambda)^-_{\lambda-\alpha}$ and $F = \bigoplus_{b \in \mathcal{B}_0} \mathbf{Q} f_i^{(n)} G(b)$. Let us verify that these data satisfy the conditions of Lemma 6.4.4.

By the induction hypothesis (6.23), we have

$$N \cap \mathcal{L}_0 \cap \mathcal{L}_\infty \xrightarrow{\sim} (N \cap \mathcal{L}_0)/(N \cap q\mathcal{L}_0) \cong \bigoplus_{b \in \mathcal{B}(\lambda)_{\lambda-\alpha} \cap \tilde{f}_i^{n+1}\mathcal{B}(\lambda)} \mathbf{Q}b.$$

Let

$$\varphi : F \to M \cap (\mathcal{L}_0 + N) \cap (\mathcal{L}_\infty + N) = N + M \cap \mathcal{L}_0 \cap \mathcal{L}_\infty$$

be the $\mathbf{Q}$-linear map defined by

$$f_i^{(n)}G(b) \longmapsto f_i^{(n)}G(b) \in M \cap \mathcal{L}_0 \cap \mathcal{L}_\infty \quad \text{for } b \in \mathcal{B}_0.$$

To see if this map is well defined, note that

$$\begin{aligned} f_i^{(n)}G(b) = \tilde{f}_i^n G(b) &\in f_i^{(n)}V(\lambda)^{\mathbf{A}}_{\lambda-\alpha+n\alpha_i} \cap \mathcal{L}(\lambda) \\ &\subset (f_i^n V(\lambda))^{\mathbf{A}}_{\lambda-\alpha} \cap \mathcal{L}(\lambda). \end{aligned}$$

Since $b \in \mathcal{B}_0 \subset \mathcal{B}(\lambda)_{\lambda-\alpha+n\alpha_i}$, by the induction hypothesis, we have $\overline{G(b)} = G(b)$, which implies

$$\overline{f_i^{(n)}G(b)} = f_i^{(n)}G(b) \in \mathcal{L}(\lambda) \cap \mathcal{L}(\lambda)^-.$$

Hence φ is well defined.

By (6.26), we have

$$M = \mathbf{A} \otimes_{\mathbf{Q}} F + N = \mathbf{A}\varphi(F) + N.$$

To apply Lemma 6.4.4, we still need to verify that the canonical maps

$$F \longrightarrow N + M \cap \mathcal{L}_0 \cap \mathcal{L}_\infty \longrightarrow (\mathcal{L}_0 + N)/(q\mathcal{L}_0 + N)$$

and

$$F \longrightarrow N + M \cap \mathcal{L}_0 \cap \mathcal{L}_\infty \longrightarrow (\mathcal{L}_0 + N)/(q^{-1}\mathcal{L}_\infty + N)$$

are injective. Let $H = (\mathcal{L}_0 + N)/(q\mathcal{L}_0 + N)$. Since $G(b) \equiv b \mod q\mathcal{L}_0$ for $b \in \mathcal{B}_0$, the canonical map $F \to H$ is given by $f_i^{(n)}G(b) \mapsto \tilde{f}_i^n b$. Note that

$$\begin{aligned} H = (\mathcal{L}_0 + N)/(q\mathcal{L}_0 + N) &\cong \frac{(\mathcal{L}_0 + N)/N}{(q\mathcal{L}_0 + N)/N} \\ &\cong \frac{\mathcal{L}_0/(\mathcal{L}_0 \cap N)}{q\mathcal{L}_0/(q\mathcal{L}_0 \cap N)} \cong \frac{\mathcal{L}_0/q\mathcal{L}_0}{(\mathcal{L}_0 \cap N)/(q\mathcal{L}_0 \cap N)}. \end{aligned}$$

Since

$$N \cap \mathcal{L}_0 \cap \mathcal{L}_\infty \xrightarrow{\sim} (N \cap \mathcal{L}_0)/(N \cap q\mathcal{L}_0) \cong \bigoplus_{b \in \mathcal{B}(\lambda)_{\lambda-\alpha} \cap \tilde{f}_i^{n+1}\mathcal{B}(\lambda)} \mathbf{Q}b,$$

we have

$$H = \bigoplus_{b \in (\mathcal{B}(\lambda)_{\lambda-\alpha} \cap \tilde{f}_i^n \mathcal{B}(\lambda)) \setminus \tilde{f}_i^{n+1}\mathcal{B}(\lambda)} \mathbf{Q}b.$$

Observe that

$$(\mathcal{B}(\lambda)_{\lambda-\alpha} \cap \tilde{f}_i^n \mathcal{B}(\lambda)) \setminus \tilde{f}_i^{n+1}\mathcal{B}(\lambda) = \tilde{f}_i^n \mathcal{B}_0.$$

Hence the map

$$F = \bigoplus_{b\in\mathcal{B}_0} \mathbf{Q} f_i^{(n)} G(b) \longrightarrow H = \bigoplus_{b\in(\mathcal{B}(\lambda)_{\lambda-\alpha}\cap\tilde{f}_i^n\mathcal{B}(\lambda))\setminus\tilde{f}_i^{n+1}\mathcal{B}(\lambda)} \mathbf{Q}b$$

given by $f_i^{(n)}G(b) \longmapsto \tilde{f}_i^n b$ is injective. Actually, it is an isomorphism.

Similarly, the map $F \to \overline{H} = (\mathcal{L}_\infty + N)/(q^{-1}\mathcal{L}_\infty + N)$ is injective. Therefore, by Lemma 6.4.4 (and its proof), we have

$$\begin{array}{ccccc}
E = F\oplus(N\cap\mathcal{L}_0\cap\mathcal{L}_\infty) & \xrightarrow{\sim} & M\cap\mathcal{L}_0\cap\mathcal{L}_\infty & \xrightarrow{\sim} & \frac{M\cap\mathcal{L}_0}{M\cap q\mathcal{L}_0} \\
\downarrow\wr & & & & \uparrow\wr \\
F\oplus \frac{N\cap\mathcal{L}_0}{N\cap q\mathcal{L}_0} & \xrightarrow{\sim} & H\oplus\frac{N\cap\mathcal{L}_0}{N\cap q\mathcal{L}_0} & = & \bigoplus_{b\in\mathcal{B}(\lambda)_{\lambda-\alpha}\cap\tilde{f}_i^n\mathcal{B}(\lambda)}\mathbf{Q}b
\end{array},$$

which yields

$$\begin{aligned}(f_i^n V(\lambda))^{\mathbf{A}}_{\lambda-\alpha}\cap\mathcal{L}(\lambda)\cap\mathcal{L}(\lambda)^- &\xrightarrow{\sim} \frac{(f_i^n V(\lambda))^{\mathbf{A}}_{\lambda-\alpha}\cap\mathcal{L}(\lambda)}{(f_i^n V(\lambda))^{\mathbf{A}}_{\lambda-\alpha}\cap q\mathcal{L}(\lambda)} \\ &\cong \bigoplus_{b\in\mathcal{B}(\lambda)_{\lambda-\alpha}\cap\tilde{f}_i^n\mathcal{B}(\lambda)} \mathbf{Q}b.\end{aligned}$$

This completes the proof. □

For $\alpha\in Q_+(r)$ and $i\in I$, let

$$G_i : \bigoplus_{b\in\mathcal{B}(\lambda)_{\lambda-\alpha}\cap\tilde{f}_i\mathcal{B}(\lambda)} \mathbf{Q}b \longrightarrow (f_i V(\lambda))^{\mathbf{A}}_{\lambda-\alpha}\cap\mathcal{L}(\lambda)\cap\mathcal{L}(\lambda)^-$$

be the inverse of the isomorphism

$$(f_i V(\lambda))^{\mathbf{A}}_{\lambda-\alpha}\cap\mathcal{L}(\lambda)\cap\mathcal{L}(\lambda)^- \longrightarrow \bigoplus_{b\in\mathcal{B}(\lambda)_{\lambda-\alpha}\cap\tilde{f}_i\mathcal{B}(\lambda)} \mathbf{Q}b$$

given in Proposition 6.5.3. Then by Lemma 6.4.1 we have

$$(f_i^n V(\lambda))^{\mathbf{A}}_{\lambda-\alpha} = \bigoplus_{b\in\mathcal{B}(\lambda)_{\lambda-\alpha}\cap\tilde{f}_i^n\mathcal{B}(\lambda)} \mathbf{A}G_i(b) \quad \text{for all } n\geq 1.$$

We would like to show that $G_i(b) = G_j(b)$ for all $b\in\mathcal{B}(\lambda)_{\lambda-\alpha}\cap\tilde{f}_i\mathcal{B}(\lambda)\cap\tilde{f}_j\mathcal{B}(\lambda)$ and $i,j\in I$.

Lemma 6.5.4. *Given $r>0$, let $\mu\gg 0$ be a dominant integral weight satisfying $\mu(h_i)\geq\lambda(h_i)$ for all $i\in I$, and consider the $U_q^-(\mathfrak{g})$-module homomorphism $\pi=\pi_{\mu,\lambda} : V(\mu)\longrightarrow V(\lambda)$ given by $v_\mu\longmapsto v_\lambda$. Then, for all $\alpha\in Q_+(r)$, we have*

(1) $\pi(\tilde{f}_i(Pv_\mu))\equiv\tilde{f}_i(Pv_\lambda) \mod q\mathcal{L}(\lambda)$ *for all $P\in U_q^-(\mathfrak{g})$,*

(2) $\pi(\mathcal{L}(\mu)_{\mu-\alpha}) = \mathcal{L}(\lambda)_{\lambda-\alpha}$.

Proof. (1) Choose $\tau \gg 0$ such that $\lambda + \tau \gg 0$ and that there exists a $U_q^-(\mathfrak{g})$-module homomorphism

$$\Psi : V(\mu) \longrightarrow V(\lambda+\tau)$$

given by $v_\mu \longmapsto v_{\lambda+\tau}$. (We may take $\tau = \mu - \lambda$.) Since $\mu \gg 0$ and $\lambda+\tau \gg 0$, Ψ defines a (weight preserving) $\mathbf{Q}(q)$-linear isomorphism

$$\sum_{\nu \geq \mu-\alpha} V(\mu)_\nu \xrightarrow{\sim} \sum_{\nu \geq \lambda+\tau-\alpha} V(\lambda+\tau)_\nu$$

that commutes with e_i and f_i ($i \in I$). Hence it commutes with $\tilde{e}_i$ and $\tilde{f}_i$, which implies

$$\Psi(\mathcal{L}(\mu)_{\mu-\alpha}) = \mathcal{L}(\lambda+\tau)_{\mu-\alpha},$$
$$\Psi(\tilde{f}_i(Pv_\mu)) = \tilde{f}_i\Psi(Pv_\mu) = \tilde{f}_i(Pv_{\lambda+\tau})$$

for all $P \in U_q^-(\mathfrak{g})$.

Consider the composition of the maps

$$V(\mu) \xrightarrow{\Psi} V(\lambda+\tau) \xrightarrow{\Phi_{\lambda,\tau}} V(\lambda) \otimes V(\tau) \xrightarrow{S_{\lambda,\tau}} V(\lambda),$$

where $\Phi_{\lambda,\tau}$ and $S_{\lambda,\tau}$ are the maps given in Section 5.3. Note that this map is given by

$$\begin{aligned} Pv_\mu \longmapsto Pv_{\lambda+\tau} &\longmapsto P(v_\lambda \otimes v_\tau) \\ &= Pv_\lambda \otimes v_\tau + \sum P'v_\lambda \otimes P''v_\tau \longmapsto Pv_\lambda. \end{aligned}$$

Hence $S_{\lambda,\tau} \circ \Phi_{\lambda,\tau} \circ \Psi = \pi_{\mu,\lambda}$. Therefore, by Lemma 5.3.6, we have

$$\begin{aligned} \pi(\tilde{f}_i(Pv_\mu)) &= S_{\lambda,\tau} \circ \Phi_{\lambda,\tau} \circ \Psi(\tilde{f}_i(Pv_\mu)) \\ &= S_{\lambda,\tau} \circ \Phi_{\lambda,\tau}(\tilde{f}_i(Pv_{\lambda+\tau})) \\ &= S_{\lambda,\tau}(\tilde{f}_i P(v_\lambda \otimes v_\tau)) \\ &\equiv \tilde{f}_i S_{\lambda,\tau}(P(v_\lambda \otimes v_\tau)) \mod q\mathcal{L}(\lambda) \\ &\equiv \tilde{f}_i(Pv_\lambda) \mod q\mathcal{L}(\lambda), \end{aligned}$$

which proves (1).

(2) By (1), we have

$$\tilde{f}_{i_1} \cdots \tilde{f}_{i_r} v_\lambda \equiv \pi(\tilde{f}_{i_1} \cdots \tilde{f}_{i_r} v_\mu) \mod q\mathcal{L}(\lambda),$$

which implies

$$\mathcal{L}(\lambda)_{\lambda-\alpha} \subset \pi(\mathcal{L}(\mu)_{\mu-\alpha}) + q\mathcal{L}(\lambda)_{\lambda-\alpha}.$$

On the other hand, by Proposition 5.3.3 and Lemma 5.3.6, we have

$$\begin{aligned}\pi(\mathcal{L}(\mu)_{\mu-\alpha}) &= S_{\lambda,\tau} \circ \Phi_{\lambda,\tau} \circ \Psi(\mathcal{L}(\mu)_{\mu-\alpha}) \\ &= S_{\lambda,\tau} \circ \Phi_{\lambda,\tau}(\mathcal{L}(\lambda+\tau)_{\lambda+\tau-\alpha}) \\ &\subset S_{\lambda,\tau}((\mathcal{L}(\lambda) \otimes \mathcal{L}(\tau))_{\lambda+\tau-\alpha}) \\ &= \mathcal{L}(\lambda)_{\lambda-\alpha}.\end{aligned}$$

Hence, by Nakayama's Lemma, we obtain

$$\mathcal{L}(\lambda)_{\lambda-\alpha} = \pi(\mathcal{L}(\mu)_{\mu-\alpha}).$$

This completes the proof. □

Corollary 6.5.5. *Given $r > 0$, let $\mu \gg 0$ be a dominant integral weight satisfying $\mu(h_i) \geq \lambda(h_i)$ for all $i \in I$, and let $\overline{\pi} : \mathcal{L}(\mu)/q\mathcal{L}(\mu) \to \mathcal{L}(\lambda)/q\mathcal{L}(\lambda)$ be the $\mathbf{Q}$-linear map induced by π. Then, for all $\alpha \in Q_+(r)$, we have*

(1) $\overline{\pi} \circ \tilde{f}_i(b) = \tilde{f}_i \circ \overline{\pi}(b)$ *for all* $b \in \mathcal{L}(\mu)_{\mu-\alpha}/q\mathcal{L}(\mu)_{\mu-\alpha}$,

(2) $\overline{\pi}(\mathcal{B}(\mu)_{\mu-\alpha}) \setminus \{0\} = \mathcal{B}(\lambda)_{\lambda-\alpha}$.

Proof. Our assertions follow immediately from Lemma 6.5.4 (Exercise 6.15). □

Lemma 6.5.6. *For all $\alpha \in Q_+(r)$ and $b \in \mathcal{B}(\lambda)_{\lambda-\alpha} \cap \tilde{f}_i\mathcal{B}(\lambda) \cap \tilde{f}_j\mathcal{B}(\lambda)$ with $i, j \in I$, we have $G_i(b) = G_j(b)$.*

Proof. Let $b = \tilde{f}_i \cdots \tilde{f}_k v_\lambda \in \mathcal{B}(\lambda)_{\lambda-\alpha} \cap \tilde{f}_i\mathcal{B}(\lambda) \cap \tilde{f}_j\mathcal{B}(\lambda)$. Take $\mu \gg 0$ such that $\mu(h_i) \geq \lambda(h_i)$ for all $i \in I$ and let $b' = \tilde{f}_i \cdots \tilde{f}_k v_\mu$. Then

$$G_i(b') = Pv_\mu \in (f_iV(\mu))^{\mathbf{A}}_{\mu-\alpha} \cap \mathcal{L}(\mu) \cap \mathcal{L}(\mu)^-,$$

where

$$P \in (f_iU_q^-(\mathfrak{g}))^{\mathbf{A}}_{-\alpha} = f_iU_q^-(\mathfrak{g})_{-\alpha+\alpha_i} \cap U_{\mathbf{A}}^-(\mathfrak{g}).$$

Since $G_i(b') \equiv b' \mod q\mathcal{L}(\mu)$, by applying π, and $\overline{\pi}$, we get

$$Pv_\lambda \equiv b \mod q\mathcal{L}(\lambda).$$

Hence, by the definition of G_i, we have

$$G_i(b) = Pv_\lambda \in (f_iV(\lambda))^{\mathbf{A}}_{\lambda-\alpha} \cap \mathcal{L}(\lambda) \cap \mathcal{L}(\lambda)^-.$$

Choose $\nu \in P^+$ such that $\langle h_k, \nu\rangle = 0$ and $\langle h_l, \nu\rangle = \langle h_l, \mu\rangle \gg 0$ for all $l \neq k$. Let $b'' = \tilde{f}_i \cdots \tilde{f}_k v_\nu$. Then, since $\mu(h_i) \geq \nu(h_i)$ for all $i \in I$, the same argument as above gives

$$G_i(b'') = Pv_\nu \in (f_iV(\nu))^{\mathbf{A}}_{\nu-\alpha} \cap \mathcal{L}(\nu) \cap \mathcal{L}(\nu)^-.$$

On the other hand, by the choice of ν, we observe $b'' = 0$; i.e., $G_i(b'') \equiv 0 \mod q\mathcal{L}(\nu)$. Since

$$(f_iV(\nu))^{\mathbf{A}}_{\nu-\alpha} \cap \mathcal{L}(\nu) \cap \mathcal{L}(\nu)^- \xrightarrow{\sim} \frac{(f_iV(\nu))^{\mathbf{A}}_{\nu-\alpha} \cap \mathcal{L}(\nu)}{(f_iV(\nu))^{\mathbf{A}}_{\nu-\alpha} \cap q\mathcal{L}(\nu)},$$

we conclude that $G_i(b'') = Pv_\nu = 0$.

Note that $V(\nu) \cong U_q^-(\mathfrak{g})/U_q^-(\mathfrak{g})f_k$ as a $U_q^-(\mathfrak{g})$-module. It follows that $P \in U_q^-(\mathfrak{g})f_k \cap f_iU_q^-(\mathfrak{g}) \cap U_{\mathbf{A}}^-(\mathfrak{g})$. By applying the antiautomorphism $\star$ given in (6.17), we have

$$P^\star \in f_kU_q^-(\mathfrak{g}) \cap U_{\mathbf{A}}^-(\mathfrak{g}) = (f_kU_q^-(\mathfrak{g}))^{\mathbf{A}},$$

which implies $P^\star v_\mu \in (f_kV(\mu))^{\mathbf{A}}_{\mu-\alpha}$. Moreover, by Corollary 6.3.5, we have

$$G_i(b)^\star = P^\star v_\mu \in f_kV(\mu)^{\mathbf{A}}_{\mu-\alpha} \cap \mathcal{L}(\mu) \cap \mathcal{L}(\mu)^-.$$

Similarly, if $G_j(b') = Qv_\mu$ for some $Q \in U_{\mathbf{A}}^-(\mathfrak{g})$, then $G_j(b) = Qv_\lambda$ and $G_j(b'') = Qv_\nu = 0$. Hence, we have

$$G_j(b)^\star = Q^\star v_\mu \in (f_kV(\mu))^{\mathbf{A}}_{\mu-\alpha} \cap \mathcal{L}(\mu) \cap \mathcal{L}(\mu)^-.$$

Since

$$Pv_\mu \equiv Qv_\mu \equiv b' \mod q\mathcal{L}(\mu),$$

we have $(P-Q)v_\mu \equiv 0 \mod q\mathcal{L}(\mu)$, implying $(P^\star - Q^\star)v_\mu \equiv 0 \mod q\mathcal{L}(\mu)$. Then Proposition 6.5.3 gives $(P^\star - Q^\star)v_\mu = 0$. Since $\mu \gg 0$, we conclude that $P^\star = Q^\star$, which yields $P = Q$. Therefore we obtain $Pv_\lambda = Qv_\lambda$, as desired. □

We shall now complete the proof of Theorem 6.5.1. Let $b \in \mathcal{B}(\lambda)_{\lambda-\alpha}$ with $\alpha \in Q_+(r)$. Note that $\tilde{e}_ib \neq 0$ for some $i \in I$; i.e., $b \in \mathcal{B}(\lambda)_{\lambda-\alpha} \cap \tilde{f}_i\mathcal{B}(\lambda)$ for some $i \in I$. Hence, by Lemma 6.5.6, we may define a map

$$G : \mathcal{L}(\lambda)_{\lambda-\alpha}/q\mathcal{L}(\lambda)_{\lambda-\alpha} \longrightarrow V(\lambda)^{\mathbf{A}}_{\lambda-\alpha} \cap \mathcal{L}(\lambda) \cap \mathcal{L}(\lambda)^-$$

by $G(b) = G_i(b)$ for $b \in \mathcal{B}(\lambda)_{\lambda-\alpha} \cap \tilde{f}_i\mathcal{B}(\lambda)$.

By definition, we have

$$G(b) \equiv b \mod q\mathcal{L}(\lambda), \tag{6.27}$$

and

$$(f_i^nV(\lambda))^{\mathbf{A}}_{\lambda-\alpha} = \bigoplus_{b \in \mathcal{B}(\lambda)_{\lambda-\alpha} \cap \tilde{f}_i^n\mathcal{B}(\lambda)} \mathbf{A}G(b) \quad \text{for } n \geq 1. \tag{6.28}$$

Let $M = V(\lambda)^{\mathbf{A}}_{\lambda-\alpha}$, $E = \bigoplus_{b \in \mathcal{B}(\lambda)_{\lambda-\alpha}} \mathbf{Q}b = \mathcal{L}(\lambda)_{\lambda-\alpha}/q\mathcal{L}(\lambda)_{\lambda-\alpha}$, and consider the map $G : E \longrightarrow V(\lambda)^{\mathbf{A}}_{\lambda-\alpha} \cap \mathcal{L}(\lambda) \cap \mathcal{L}(\lambda)^-$ given above. Since

$V(\lambda)^{\mathbf{A}}_{\lambda-\alpha} = \sum_{i\in I}(f_i V(\lambda))^{\mathbf{A}}_{\lambda-\alpha}$, we obtain

$$V(\lambda)^{\mathbf{A}}_{\lambda-\alpha} = \sum_{b\in\mathcal{B}(\lambda)_{\lambda-\alpha}} \mathbf{A}G(b).$$

Moreover, by (6.27), the map

$$E = \bigoplus_{b\in\mathcal{B}(\lambda)_{\lambda-\alpha}} \mathbf{Q}b \longrightarrow V(\lambda)^{\mathbf{A}}_{\lambda-\alpha}\mathcal{L}(\lambda)\cap\mathcal{L}(\lambda)^{-}$$
$$\longrightarrow \mathcal{L}(\lambda)_{\lambda-\alpha}/q\mathcal{L}(\lambda)_{\lambda-\alpha}$$

given by $b \longmapsto G(b) \longmapsto G(b) + q\mathcal{L}(\lambda)_{\lambda-\alpha}$ is injective. Hence the map

$$E = \bigoplus_{b\in\mathcal{B}(\lambda)_{\lambda-\alpha}} \mathbf{Q}b \longrightarrow V(\lambda)^{\mathbf{A}}_{\lambda-\alpha}\cap\mathcal{L}(\lambda)\cap\mathcal{L}(\lambda)^{-}$$
$$\longrightarrow \mathcal{L}(\lambda)^{-}_{\lambda-\alpha}/q^{-1}\mathcal{L}(\lambda)^{-}_{\lambda-\alpha}$$

given by $b \longmapsto \overline{G(b)} = G(b) \longmapsto G(b) + q^{-1}\mathcal{L}(\lambda)^{-}_{\lambda-\alpha}$ is also injective. Therefore, by Lemma 6.4.3, we conclude that the canonical map

$$V(\lambda)^{\mathbf{A}}_{\lambda-\alpha}\cap\mathcal{L}(\lambda)\cap\mathcal{L}(\lambda)^{-} \longrightarrow \frac{V(\lambda)^{\mathbf{A}}_{\lambda-\alpha}\cap\mathcal{L}(\lambda)}{V(\lambda)^{\mathbf{A}}_{\lambda-\alpha}\cap q\mathcal{L}(\lambda)}$$

is an isomorphism, which is the statement $\mathbf{A}(\mathbf{r})$.

Finally, the statement $\mathbf{B}(\mathbf{r})$ follows from (6.28): for any $b \in \mathcal{B}(\lambda)_{\lambda-\alpha}\cap \tilde{f}_i^n\mathcal{B}(\lambda)$, we have $G(b) \in f_i^n V(\lambda)$.

This completes the proof of Theorem 6.5.1.

Exercises

6.1. Show that there is a canonical isomorphism

$$\mathbf{A}_\infty \otimes_{\mathbf{Q}[q^{-1}]} (V^{\mathbf{A}}\cap\mathcal{L}_\infty) \xrightarrow{\sim} \mathcal{L}_\infty.$$

6.2. Show that $(V^{\mathbf{A}}, \mathcal{L}_0, \mathcal{L}_\infty)$ is a balanced triple for a finite dimensional $\mathbf{Q}(q)$-vector space V if and only if the canonical map

$$E = V^{\mathbf{A}}\cap\mathcal{L}_0\cap\mathcal{L}_\infty \longrightarrow \mathcal{L}_\infty/q^{-1}\mathcal{L}_\infty$$

is an isomorphism.

6.3. Verify that the diagram in the proof of Theorem 6.1.4 is commutative with exact rows.

6.4. Suppose that $(V^{\mathbf{A}}, \mathcal{L}_0, \mathcal{L}_\infty)$ is a balanced triple for a finite dimensional $\mathbf{Q}(q)$-vector space V. Show that if $\mathcal{G}(\mathcal{B})$ is the global basis of V associated with a local basis $\mathcal{B}$ at $q = 0$, then the image $\overline{\mathcal{B}}$ of $\mathcal{G}(\mathcal{B})$ under the canonical isomorphism

$$V^{\mathbf{A}} \cap \mathcal{L}_0 \cap \mathcal{L}_\infty \xrightarrow{\sim} \mathcal{L}_\infty / q^{-1} \mathcal{L}_\infty$$

is a local basis at $q = \infty$.

6.5. Verify that $\overline{V(\lambda)^{\mathbf{A}}} = V(\lambda)^{\mathbf{A}}$.

6.6. Show that there is a unique symmetric bilinear form $(\ ,\ )$ on $U_q^-(\mathfrak{g})$ satisfying

$$(1, 1) = 1,$$
$$(f_i P, Q) = (P, e_i' Q)$$

for all $P, Q \in U_q^-(\mathfrak{g})$ and $i \in I$.

6.7. Prove Lemma 6.3.6.

6.8. Show that $e_i' U_{\mathbf{A}}^-(\mathfrak{g}) \subset U_{\mathbf{A}}^-(\mathfrak{g})$ for all $i \in I$.

6.9. Show that if $u = \sum f_i^{(k)} u_k \in U_{\mathbf{A}}^-(\mathfrak{g})$ with $e_i' u_k = 0$ for all k, then all u_k belong to $U_{\mathbf{A}}^-(\mathfrak{g})$.

6.10. Verify that the diagram in the proof of Lemma 6.4.1 is commutative with exact rows.

6.11. Let M be an $\mathbf{A}$-submodule of V and let $S = \{f \in \mathbf{Q}[q] \mid f(0) \neq 0\}$. Verify that $S^{-1}M \cap \mathcal{L}_0 = S^{-1}(M \cap \mathcal{L}_0)$.

6.12. Verify that the commutative diagram in the proof of Lemma 6.4.4 is commutative with exact rows.

6.13. Prove Lemma 6.5.2.

6.14. Show that if $G(b) \in (f_i V(\lambda))^{\mathbf{A}}$, then $f_i^{(n)} G(b) \in (f_i^{n+1} V(\lambda))^{\mathbf{A}}$.

6.15. Prove Corollary 6.5.5.

Chapter 7

Young Tableaux and Crystals

The crystal basis theory has a great many applications to combinatorial representation theory. Over the past 100 years, it has been discovered that there is a close connection between the representation theory of general linear Lie groups and the combinatorics of Young diagrams and Young tableaux (see, for example, [**46, 52, 55, 57–59**]). In this chapter, we investigate this connection in the language of crystal basis theory for $U_q(\mathfrak{gl}_n)$. The crystal graph of a finite dimensional irreducible $U_q(\mathfrak{gl}_n)$-module will be realized as the set of semistandard Young tableaux of a given shape, and using the tensor product rule for Kashiwara operators, we will give a combinatorial rule for decomposing the tensor product of finite dimensional $U_q(\mathfrak{gl}_n)$-modules into a direct sum of irreducible components. This chapter (and the next one) should be regarded as the crystal basis theoretic interpretation of the classical works on representation theory and combinatorics. For example, the combinatorial algorithm for tensor product decomposition coincides with the classical *Littlewood-Richardson rule.*

7.1. The quantum group $U_q(\mathfrak{gl}_n)$

Recall that the *general linear Lie algebra* $\mathfrak{gl}_n(\mathbf{C})$ is the Lie algebra of $n \times n$ matrices over $\mathbf{C}$ with the bracket defined by

$$[X, Y] = XY - YX$$

for $X, Y \in \mathfrak{gl}_n(\mathbf{C})$. Let E_{ij} denote the elementary matrix having 1 at the (i, j)-entry and 0 elsewhere. Let $I = \{1, 2, \ldots, n-1\}$ be an index set and let

$$e_i = E_{i,i+1}, \qquad f_i = E_{i+1,i}, \qquad h_i = E_{ii} - E_{i+1,i+1} \quad \text{for } i \in I.$$

The Lie algebra $\mathfrak{gl}_n(\mathbf{C})$ is generated by the elements e_i, f_i $(i \in I)$, E_{jj} $(j = 1, 2, \ldots, n)$, and the *special linear Lie algebra* $\mathfrak{sl}_n(\mathbf{C})$ is the subalgebra of $\mathfrak{gl}_n(\mathbf{C})$ generated by the elements e_i, f_i, h_i $(i \in I)$.

The *maximal toral subalgebra* for $\mathfrak{gl}_n(\mathbf{C})$ is given by $\mathfrak{h} = \mathbf{C}E_{1,1} \oplus \cdots \oplus \mathbf{C}E_{n,n}$. For each $i = 1, 2, \ldots, n$, define a linear map $\epsilon_i : \mathfrak{h} \to \mathbf{C}$ by

$$\epsilon_i(h) = \lambda_i, \quad \text{where } h = \operatorname{diag}(\lambda_j \mid j = 1, \ldots, n) \in \mathfrak{h}. \tag{7.1}$$

As we have seen in Section 1.4, for $h = \operatorname{diag}(\lambda_j \mid j = 1, \ldots, n) \in \mathfrak{h}$ and $i, j = 1, \ldots, n$, we have

$$[h, E_{ij}] = (\lambda_i - \lambda_j)E_{ij} = (\epsilon_i - \epsilon_j)(h)E_{ij}.$$

Hence the general linear Lie algebra $\mathfrak{gl}_n(\mathbf{C})$ has the *triangular decomposition*

$$\begin{aligned}\mathfrak{gl}_n(\mathbf{C}) &= \left(\bigoplus_{\alpha \in \Phi_-} \mathfrak{g}_\alpha\right) \oplus \mathfrak{h} \oplus \left(\bigoplus_{\alpha \in \Phi_+} \mathfrak{g}_\alpha\right) \\ &= \left(\bigoplus_{i>j} \mathbf{C}E_{ij}\right) \oplus \left(\bigoplus_{i=1}^{n} \mathbf{C}E_{ii}\right) \oplus \left(\bigoplus_{i<j} \mathbf{C}E_{ij}\right),\end{aligned}$$

where $\Phi_+ = \{\epsilon_i - \epsilon_j \mid i < j\}$ (respectively, $\Phi_- = \{\epsilon_i - \epsilon_j \mid i > j\}$) denotes the set of all *positive roots* (respectively, *negative roots*) of $\mathfrak{gl}_n(\mathbf{C})$.

The *simple roots* of $\mathfrak{gl}_n(\mathbf{C})$ are given by

$$\alpha_i = \epsilon_i - \epsilon_{i+1} \quad (i = 1, 2, \ldots, n-1), \tag{7.2}$$

and the free abelian group $Q = \mathbf{Z}\alpha_1 \oplus \cdots \oplus \mathbf{Z}\alpha_{n-1}$ is called the *root lattice* of $\mathfrak{gl}_n(\mathbf{C})$. We will also use the notation $Q_+ = \sum_{i=1}^{n-1} \mathbf{Z}_{\geq 0}\alpha_i$ and $Q_- = -Q_+$. The *Cartan matrix* $A = (a_{ij})_{i,j \in I} = (\alpha_j(h_i))_{i,j \in I}$ and the *Dynkin diagram* are given by

$$A = (a_{ij})_{i,j=1,\ldots,n-1} = \begin{pmatrix} 2 & -1 & 0 & 0 & \cdots & 0 \\ -1 & 2 & -1 & 0 & \cdots & 0 \\ 0 & -1 & 2 & -1 & \cdots & 0 \\ \vdots & \vdots & & \ddots & & \vdots \\ 0 & & \cdots & & 2 & -1 \\ 0 & & \cdots & & -1 & 2 \end{pmatrix},$$

The general linear Lie algebra $\mathfrak{gl}_n(\mathbf{C})$ is isomorphic to the Lie algebra generated by the elements e_i, f_i $(i \in I)$ and $\mathfrak{h}$ with the defining relations

$$
(7.3) \qquad \begin{aligned}
&[h, h'] = 0 \quad \text{for } h, h' \in \mathfrak{h}, \\
&[e_i, f_j] = \delta_{i,j} h_i \quad \text{for } i, j \in I, \\
&[h, e_i] = \alpha_i(h) e_i, \qquad [h, f_i] = -\alpha_i(h) f_i \quad \text{for } i, j \in I, \\
&[e_i, [e_i, e_j]] = [f_i, [f_i, f_j]] = 0 \quad \text{for } |i - j| = 1, \\
&[e_i, e_j] = [f_i, f_j] = 0 \quad \text{for } |i - j| > 1.
\end{aligned}
$$

The $\mathbf{Z}$-lattice $P = \mathbf{Z}\epsilon_1 \oplus \cdots \oplus \mathbf{Z}\epsilon_n$ lying inside $\mathfrak{h}^*$ is called the *weight lattice* of $\mathfrak{gl}_n(\mathbf{C})$. We call $P^\vee = \mathbf{Z}E_{11} \oplus \cdots \oplus \mathbf{Z}E_{nn}$ the *dual weight lattice* of $\mathfrak{gl}_n(\mathbf{C})$. We now define the corresponding quantum group $U_q(\mathfrak{gl}_n)$ as follows.

Definition 7.1.1. The ***quantum general linear algebra*** $U_q(\mathfrak{gl}_n)$ is the associative algebra with 1 over $\mathbf{C}(q)$ generated by the elements e_i, f_i $(i = 1, 2, \ldots, n-1)$ and q^h $(h \in P^\vee)$ with the defining relations

$$
(7.4) \qquad \begin{aligned}
&q^0 = 1, \qquad q^h q^{h'} = q^{h+h'} \quad (0, h, h' \in P^\vee), \\
&q^h e_i q^{-h} = q^{\alpha_i(h)} e_i, \qquad q^h f_i q^{-h} = q^{-\alpha_i(h)} f_i, \\
&e_i f_j - f_j e_i = \delta_{i,j} \frac{K_i - K_i^{-1}}{q - q^{-1}}, \quad \text{where } K_i = q^{h_i}, \\
&e_i^2 e_j - (q + q^{-1}) e_i e_j e_i + e_j e_i^2 = 0 \quad \text{for } |i - j| = 1, \\
&f_i^2 f_j - (q + q^{-1}) f_i f_j f_i + f_j f_i^2 = 0 \quad \text{for } |i - j| = 1, \\
&e_i e_j - e_j e_i = f_i f_j - f_j f_i = 0 \quad \text{for } |i - j| > 1.
\end{aligned}
$$

The ***quantum special linear algebra*** $U_q(\mathfrak{sl}_n)$ is the subalgebra of $U_q(\mathfrak{gl}_n)$ generated by e_i, f_i, $K_i^{\pm 1}$ $(i = 1, 2, \ldots, n-1)$.

The algebra $U_q(\mathfrak{gl}_n)$ has a Hopf algebra structure with the comultiplication Δ, counit ε, and antipode S defined by

$$
(7.5) \qquad \begin{aligned}
&\Delta(q^h) = q^h \otimes q^h, \\
&\Delta(e_i) = e_i \otimes K_i^{-1} + 1 \otimes e_i, \qquad \Delta(f_i) = f_i \otimes 1 + K_i \otimes f_i, \\
&\varepsilon(q^h) = 1, \qquad \varepsilon(e_i) = \varepsilon(f_i) = 0, \\
&S(q^h) = q^{-h}, \qquad S(e_i) = -e_i K_i, \qquad S(f_i) = -K_i^{-1} f_i
\end{aligned}
$$

for $h \in P^\vee$ and $i = 1, 2, \ldots, n-1$, and it has the *triangular decomposition*

$$
U_q(\mathfrak{gl}_n) \cong U_q^-(\mathfrak{gl}_n) \otimes U_q^0(\mathfrak{gl}_n) \otimes U_q^+(\mathfrak{gl}_n),
$$

where $U_q^+(\mathfrak{gl}_n)$ (respectively, $U_q^-(\mathfrak{gl}_n)$) is the subalgebra of $U_q(\mathfrak{gl}_n)$ generated by e_i (respectively, f_i) for $i \in I$ and $U_q^0(\mathfrak{gl}_n)$ is the subalgebra of $U_q(\mathfrak{gl}_n)$ generated by q^h $(h \in P^\vee)$.

7.2. The category $\mathcal{O}_{\rm int}^{\geq 0}$

Let $\mathbf{V} = \mathbf{C}(q)v_1 \oplus \cdots \oplus \mathbf{C}(q)v_n$ be an n-dimensional vector space over $\mathbf{C}(q)$ with basis $v_1, \ldots, v_n$ and define the $U_q(\mathfrak{gl}_n)$-action on $\mathbf{V}$ by

$$
\begin{aligned}
&q^h v_j = q^{\epsilon_j(h)} v_j \quad \text{for } h \in P^\vee,\ j = 1, \ldots, n, \\
&e_i v_j = \begin{cases} v_i & \text{if } j = i+1, \\ 0 & \text{otherwise,} \end{cases} \\
&f_i v_j = \begin{cases} v_{i+1} & \text{if } j = i, \\ 0 & \text{otherwise.} \end{cases}
\end{aligned} \tag{7.6}
$$

Then $\mathbf{V}$ becomes a $U_q(\mathfrak{gl}_n)$-module, called the ***vector representation*** of $U_q(\mathfrak{gl}_n)$. Let $\mathbf{L} = \bigoplus_{j=1}^n \mathbf{A}_0 v_j$ and $\mathbf{B} = \{\boxed{j} = v_j + q\mathbf{L} \mid j = 1, 2, \ldots, n\}$. Then $(\mathbf{L}, \mathbf{B})$ is a crystal basis of $\mathbf{V}$ with crystal graph (Exercise 7.1)

$$
\mathbf{B}\ :\quad \boxed{1} \xrightarrow{\ 1\ } \boxed{2} \xrightarrow{\ 2\ } \cdots \xrightarrow{\ n-2\ } \boxed{n-1} \xrightarrow{\ n-1\ } \boxed{n}\ .
$$

The vector representation $\mathbf{V}$ is an example of highest weight $U_q(\mathfrak{gl}_n)$-modules in the category $\mathcal{O}_{\rm int}^{\geq 0}$ defined below. The crystal graph $\mathbf{B}$ will play a fundamental role in constructing concrete realizations of crystal graphs for finite dimensional irreducible $U_q(\mathfrak{gl}_n)$-modules in the category $\mathcal{O}_{\rm int}^{\geq 0}$.

The ***weight modules*** and ***highest weight modules*** are defined as in Section 3.2. For $\lambda = \lambda_1\epsilon_1 + \cdots + \lambda_n\epsilon_n \in P$ ($\lambda_j \in \mathbf{Z}$), the ***Verma module*** of highest weight λ will be denoted by $M(\lambda)$ and $V(\lambda)$ will denote the ***irreducible highest weight module*** of highest weight λ.

Set

$$
P_{\geq 0} = \{\lambda = \lambda_1\epsilon_1 + \cdots + \lambda_n\epsilon_n \in P \mid \lambda_j \geq 0 \quad \text{for all } j = 1, \ldots, n\}. \tag{7.7}
$$

Definition 7.2.1. The ***category*** $\mathcal{O}_{\rm int}^{\geq 0}$ consists of finite dimensional $U_q(\mathfrak{gl}_n)$-modules M with a weight space decomposition $M = \bigoplus_{\lambda \in P} M_\lambda$ such that $\mathrm{wt}(M) \subset P_{\geq 0}$. The morphisms in the category $\mathcal{O}_{\rm int}^{\geq 0}$ are the usual $U_q(\mathfrak{gl}_n)$-module homomorphisms.

Example 7.2.2. The vector representation $\mathbf{V}$ is a highest weight $U_q(\mathfrak{gl}_n)$-module with highest weight ϵ_1 and highest weight vector v_1. Since the weights of $\mathbf{V}$ are ϵ_j ($j = 1, \ldots, n$), it belongs to the category $\mathcal{O}_{\rm int}^{\geq 0}$.

We define the set $P_{\geq 0}^+$ of ***dominant integral weights*** to be

$$
P_{\geq 0}^+ = \{\lambda = \lambda_1\epsilon_1 + \cdots + \lambda_n\epsilon_n \in P_{\geq 0} \mid \lambda_1 \geq \lambda_2 \geq \cdots \geq \lambda_n \geq 0\}. \tag{7.8}
$$

Let $\lambda = \lambda_1\epsilon_1 + \cdots + \lambda_n\epsilon_n \in P_{\geq 0}^+$ be a dominant integral weight and let $V(\lambda)$ be the irreducible highest weight module with highest weight λ. We first show that $V(\lambda)$ belongs to the category $\mathcal{O}_{\rm int}^{\geq 0}$.

Since $\lambda(h_i) = \lambda_i - \lambda_{i+1} \in \mathbf{Z}_{\geq 0}$ for all $i = 1, 2, \ldots, n-1$, viewed as a $U_q(\mathfrak{sl}_n)$-module, $V(\lambda)$ is finite dimensional. Let $\mu = \mu_1\epsilon_1 + \cdots + \mu_n\epsilon_n \in P$ be a weight of $V(\lambda)$. Since the simple reflection r_i acts as $(i, i+1)$ on P $(i = 1, 2, \ldots, n-1)$, we may assume that $\mu_1 \geq \mu_2 \geq \cdots \geq \mu_n$. Write $\mu = \lambda - (k_1\alpha_1 + \cdots + k_{n-1}\alpha_{n-1})$ for some $k_i \in \mathbf{Z}_{\geq 0}$. Then we have

$$\begin{aligned}
\mu_1 &= \lambda_1 - k_1, \\
\mu_2 &= \lambda_2 + k_1 - k_2, \\
&\vdots \\
\mu_{n-1} &= \lambda_{n-1} + k_{n-2} - k_{n-1}, \\
\mu_n &= \lambda_n + k_{n-1}.
\end{aligned}$$

Clearly $\mu_n \geq 0$, which implies all $\mu_i \geq 0$. Hence $\mu \in P_{\geq 0}$ and therefore $V(\lambda)$ belongs to the category $\mathcal{O}_{\text{int}}^{\geq 0}$.

Conversely, let V be a finite dimensional irreducible $U_q(\mathfrak{gl}_n)$-module in the category $\mathcal{O}_{\text{int}}^{\geq 0}$. Then there is a maximal weight of V, i.e., there is a weight $\lambda = \lambda_1\epsilon_1 + \cdots + \lambda_n\epsilon_n$ of V such that $\lambda + \alpha_i$ is not a weight of V for all $i = 1, 2, \ldots, n-1$. Let v_λ be a weight vector of weight λ. Then $e_i \cdot v_\lambda = 0$ for all $i = 1, 2, \ldots, n-1$, and hence v_λ generates a nontrivial highest weight submodule $W = U_q(\mathfrak{gl}_n) \cdot v_\lambda \cong V(\lambda)$ which must be the same as V. Since V is finite dimensional, viewed as a $U_q(\mathfrak{sl}_n)$-module, we must have $\lambda(h_i) = \lambda_i - \lambda_{i+1} \in \mathbf{Z}_{\geq 0}$ for all $i = 1, 2, \ldots, n-1$, which implies $\lambda \in P_{\geq 0}^+$. Therefore, we obtain:

Theorem 7.2.3. *For each $\lambda = \lambda_1\epsilon_1 + \cdots + \lambda_n\epsilon_n \in P_{\geq 0}^+$, there exists a unique irreducible $U_q(\mathfrak{gl}_n)$-module $V(\lambda)$ in the category $\mathcal{O}_{\text{int}}^{\geq 0}$. Conversely, every finite dimensional irreducible $U_q(\mathfrak{gl}_n)$-module in the category $\mathcal{O}_{\text{int}}^{\geq 0}$ has the form $V(\lambda)$ for some dominant integral weight $\lambda \in P_{\geq 0}^+$.*

Remark 7.2.4. By Theorem 7.2.3, the irreducible $U_q(\mathfrak{gl}_n)$-modules in the category $\mathcal{O}_{\text{int}}^{\geq 0}$ can be parameterized by the partitions $\lambda = (\lambda_1 \geq \lambda_2 \geq \cdots \geq \lambda_n \geq 0)$ with at most n parts, which in turn corresponds to Young diagrams with at most n rows. Our goal in this chapter is to construct a realization of the crystal graph $\mathcal{B}(\lambda)$ of $V(\lambda)$ in terms of *semistandard Young tableaux* of shape λ.

Remark 7.2.5. Let $\lambda = \lambda_1\epsilon_1 + \cdots + \lambda_n\epsilon_n$ and $\mu = \mu_1\epsilon_1 + \cdots + \mu_n\epsilon_n$ be two elements in $P_{\geq 0}$ such that $\lambda_1 + \cdots + \lambda_n = \mu_1 + \cdots + \mu_n$. We define a partial ordering $\leq$ by setting $\lambda \geq \mu$ if and only if $\lambda_1 + \cdots + \lambda_k \geq \mu_1 + \cdots + \mu_k$ for all $k = 1, 2, \ldots, n$. Then it is straightforward to verify that $\lambda \geq \mu$ if and only if $\lambda - \mu \in Q_+ = \sum \mathbf{Z}_{\geq 0}\alpha_i$ (Exercise 7.3).

Remark 7.2.6. Define the ***fundamental weights*** by

$$\omega_j = \epsilon_1 + \cdots + \epsilon_j \quad (j = 1, \ldots, n).$$

Then any dominant integral weight $\lambda = \lambda_1\epsilon_1 + \cdots + \lambda_n\epsilon_n \in P^+_{\geq 0}$ can be expressed in terms of fundamental weights:

$$\lambda = (\lambda_1 - \lambda_2)\omega_1 + \cdots + (\lambda_{n-1} - \lambda_n)\omega_{n-1} + \lambda_n\omega_n.$$

Conversely, any dominant integral weight $\lambda = a_1\omega_1 + \cdots + a_n\omega_n$ with $a_i \in \mathbf{Z}_{\geq 0}$ can be expressed as

$$\lambda = (a_1 + \cdots + a_n)\epsilon_1 + (a_2 + \cdots + a_n)\epsilon_2 + \cdots + (a_{n-1} + a_n)\epsilon_{n-1} + a_n\epsilon_n.$$

This correspondence is shown in the following figure.

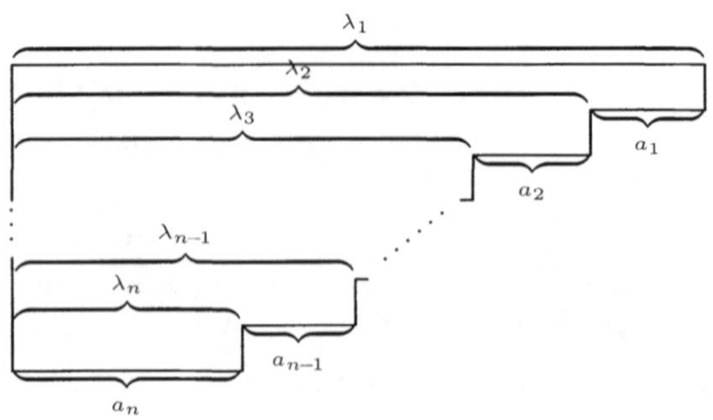

In the next section, we will investigate the connection between the crystal basis theory of $U_q(\mathfrak{gl}_n)$-modules in the category $\mathcal{O}^{\geq 0}_{\text{int}}$ and combinatorics of Young diagrams and tableaux. As is the case with quantum groups for Kac-Moody algebras, the crystal graphs of $U_q(\mathfrak{gl}_n)$-modules in $\mathcal{O}^{\geq 0}_{\text{int}}$ have a structure of $U_q(\mathfrak{gl}_n)$-crystal associated with the Cartan datum $(A, \Pi, \Pi^\vee, P, P^\vee)$, where

$$A = (a_{ij})_{i,j\in I} = \begin{pmatrix} 2 & -1 & 0 & 0 & \cdots & 0 \\ -1 & 2 & -1 & 0 & \cdots & 0 \\ 0 & -1 & 2 & 0 & \cdots & 0 \\ \vdots & \vdots & & \ddots & & \vdots \\ 0 & & \cdots & & 2 & -1 \\ 0 & & \cdots & & -1 & 2 \end{pmatrix},$$

$$\begin{aligned} \Pi &= \{\alpha_i = \epsilon_i - \epsilon_{i+1} \mid i = 1, \ldots, n-1\}, \\ \Pi^\vee &= \{h_i = E_{ii} - E_{i+1,i+1} \mid i = 1, \ldots, n-1\}, \\ P &= \mathbf{Z}\epsilon_1 \oplus \cdots \oplus \mathbf{Z}\epsilon_n, \\ P^\vee &= \mathbf{Z}E_{11} \oplus \cdots \oplus \mathbf{Z}E_{nn}. \end{aligned}$$

7.3. Tableaux and crystals

Definition 7.3.1.

(1) A ***Young diagram*** is a collection of boxes arranged in left-justified rows with a weakly decreasing number of boxes in each row.

(2) A ***skew Young diagram*** is a diagram obtained by removing a smaller Young diagram from a larger one that contains it.

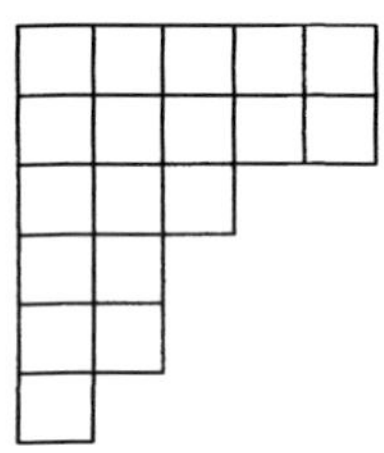

Young diagram

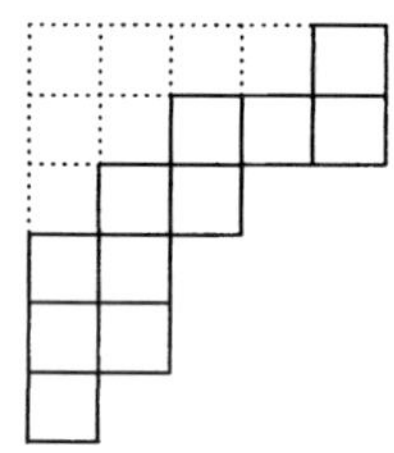

skew-Young diagram

Definition 7.3.2.

(1) A box in a skew Young diagram is called a ***removable corner*** if there is no box in the diagram to its right or beneath it.

(2) A place where a box can be added to a skew Young diagram to create a removable corner of a larger skew Young diagram is called an ***indent corner***.

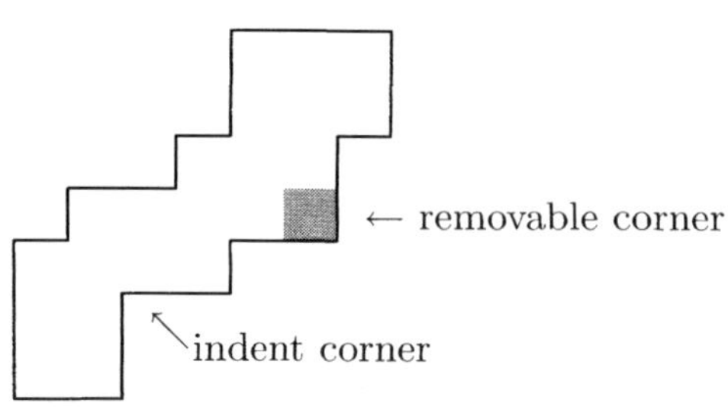

Definition 7.3.3. A ***tableau*** is a (skew) Young diagram filled with numbers, one for each box. A ***semistandard tableau*** is a tableau obtained from a (skew) Young diagram by filling the boxes with the numbers $1, 2, \ldots, n$ subject to the conditions:

(i) the entries in each row are weakly increasing,

(ii) the entries in each column are strictly increasing.

For a tableau T, we define its ***weight*** to be

$$\mathrm{wt}(T) = k_1\epsilon_1 + \cdots + k_n\epsilon_n,$$

where k_i denotes the number of i's appearing in T.

The shape and weight of a tableau is illustrated in the following figure.

$$\begin{array}{|c|c|c|c|}\hline 1 & 1 & 2 & 4 \\ \hline 2 & 3 & 3 \\ \cline{1-3} 4 \\ \cline{1-1}\end{array}$$

semistandard tableau of shape $\lambda = 4\epsilon_1 + 3\epsilon_2 + \epsilon_3$ with weight $\mu = 2\epsilon_1 + 2\epsilon_2 + 2\epsilon_3 + 2\epsilon_4$.

Let Y be a (skew) Young diagram and let N be the number of boxes in Y. We denote by $\mathcal{B}(Y)$ the set of all semistandard tableaux of shape Y. We would like to define an embedding of $\mathcal{B}(Y)$ into $\mathbf{B}^{\otimes N}$, where $\mathbf{B}$ is the crystal graph of the vector representation, so that $\mathcal{B}(Y)$ can be given a $U_q(\mathfrak{gl}_n)$-crystal structure. Roughly speaking, we would like to do the following:

(i) we first list the boxes in a semistandard tableau $T \in \mathcal{B}(Y)$ in some order $b_1, b_2, \ldots, b_N$;

(ii) we then identify the tableau T with the vector $b_1 \otimes \cdots \otimes b_N \in \mathbf{B}^{\otimes N}$.

Then the set $\mathcal{B}(Y)$ would be given a structure of $U_q(\mathfrak{gl}_n)$-crystal induced by that of $\mathcal{B}^{\otimes N}$. Such an embedding of $\mathcal{B}(Y)$ into $\mathbf{B}^{\otimes N}$ is called a ***reading*** of tableaux in $\mathcal{B}(Y)$. Among others, we single out two special cases.

Definition 7.3.4.

(1) The ***Far-Eastern reading*** of a semistandard tableau $T \in \mathcal{B}(Y)$ proceeds down columns from top to bottom and from right to left.

(2) The ***Middle-Eastern reading*** of a semistandard tableau $T \in \mathcal{B}(Y)$ moves across the rows from right to left and from top to bottom.

$$\begin{array}{cccc} \cline{4-4} & & & \multicolumn{1}{|c|}{2} \\ \cline{2-4} & \multicolumn{1}{|c|}{1} & \multicolumn{1}{c|}{2} & \multicolumn{1}{c|}{3} \\ \cline{1-4} \multicolumn{1}{|c|}{1} & \multicolumn{1}{c|}{2} & \multicolumn{1}{c|}{4} & \\ \cline{1-3} \multicolumn{1}{|c|}{3} & \multicolumn{1}{c|}{3} & & \\ \cline{1-2} \multicolumn{1}{|c|}{4} & & & \\ \cline{1-1} \end{array} = \boxed{2}\otimes\boxed{3}\otimes\boxed{2}\otimes\boxed{4}\otimes\boxed{1}\otimes\boxed{2}\otimes\boxed{3}\otimes\boxed{1}\otimes\boxed{3}\otimes\boxed{4}$$

Far-Eastern reading

$$\begin{array}{cccc} \cline{4-4} & & & \multicolumn{1}{|c|}{2} \\ \cline{2-4} & \multicolumn{1}{|c|}{1} & \multicolumn{1}{c|}{2} & \multicolumn{1}{c|}{3} \\ \cline{1-4} \multicolumn{1}{|c|}{1} & \multicolumn{1}{c|}{2} & \multicolumn{1}{c|}{4} & \\ \cline{1-3} \multicolumn{1}{|c|}{3} & \multicolumn{1}{c|}{3} & & \\ \cline{1-2} \multicolumn{1}{|c|}{4} & & & \\ \cline{1-1} \end{array} = \boxed{2}\otimes\boxed{3}\otimes\boxed{2}\otimes\boxed{1}\otimes\boxed{4}\otimes\boxed{2}\otimes\boxed{1}\otimes\boxed{3}\otimes\boxed{3}\otimes\boxed{4}$$

Middle-Eastern reading

More generally, we can define the notion of *admissible reading*. Let β and β' be two boxes in T at the sites (i, j) and (i', j'), respectively. We say that β is ***strictly higher than*** β' if $\beta \neq \beta'$, $i \leq i'$ and $j \geq j'$; i.e., if β lies in the northeast of β'. We also say that β and β' lie in ***comparable position*** if either of them is strictly higher than the other.

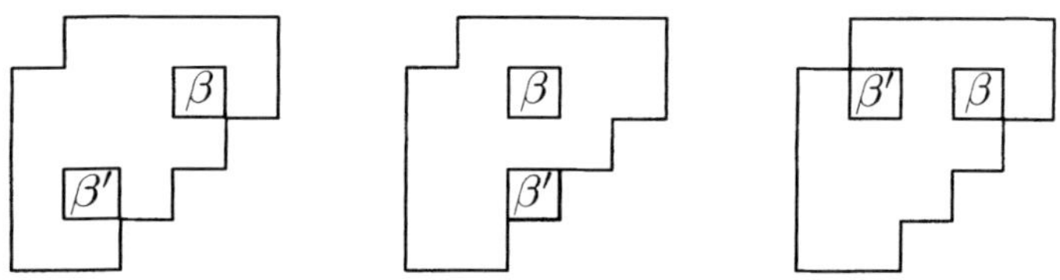

β is strictly higher than β'.

Definition 7.3.5. A reading $\Psi : \mathcal{B}(Y) \to \mathbf{B}^{\otimes N}$ is said to be ***admissible*** if β is read before β' whenever β is strictly higher than β'.

For instance, the Far-Eastern reading and the Middle-Eastern reading are admissible.

Theorem 7.3.6. *Let Y be a (skew) Young diagram and let $\mathcal{B}(Y)$ be the set of all semistandard tableaux of shape Y.*

(1) *For any admissible reading $\Psi : \mathcal{B}(Y) \to \mathbf{B}^{\otimes N}$, $\Psi(\mathcal{B}(Y)) \cup \{0\}$ is stable under the Kashiwara operators $\tilde{e}_i$ and $\tilde{f}_i$ $(i = 1, 2, \ldots, n-1)$. Hence an admissible reading defines a $U_q(\mathfrak{gl}_n)$-crystal structure on $\mathcal{B}(Y)$.*

(2) *The induced $U_q(\mathfrak{gl}_n)$-crystal structure on $\mathcal{B}(Y)$ does not depend on the choice of admissible reading.*

Before giving a proof, we illustrate the meaning of Theorem 7.3.6 in the following example.

Example 7.3.7. Let $n = 3$ and $Y = \begin{array}{|c|c|}\hline \ & \ \\ \hline \ & \ \\ \hline\end{array}$. Then the crystal $\mathbf{B}$ is equal to

$$\mathbf{B} : \quad \boxed{1} \xrightarrow{\ 1\ } \boxed{2} \xrightarrow{\ 2\ } \boxed{3}$$

and

$$\mathcal{B}(Y) = \left\{ \begin{array}{|c|c|}\hline 1 & 1 \\ \hline 2 & 2 \\ \hline\end{array}, \begin{array}{|c|c|}\hline 1 & 1 \\ \hline 2 & 3 \\ \hline\end{array}, \begin{array}{|c|c|}\hline 1 & 2 \\ \hline 2 & 3 \\ \hline\end{array}, \begin{array}{|c|c|}\hline 1 & 1 \\ \hline 3 & 3 \\ \hline\end{array}, \begin{array}{|c|c|}\hline 1 & 2 \\ \hline 3 & 3 \\ \hline\end{array}, \begin{array}{|c|c|}\hline 2 & 2 \\ \hline 3 & 3 \\ \hline\end{array} \right\}.$$

By the Far-Eastern reading, the $U_q(\mathfrak{gl}_3)$-crystal structure on $B(Y)$ is given as follows.

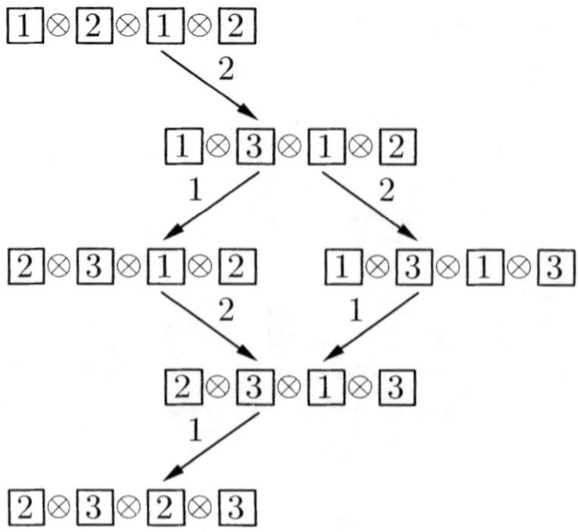

On the other hand, by the Middle-Eastern reading, we obtain the following.

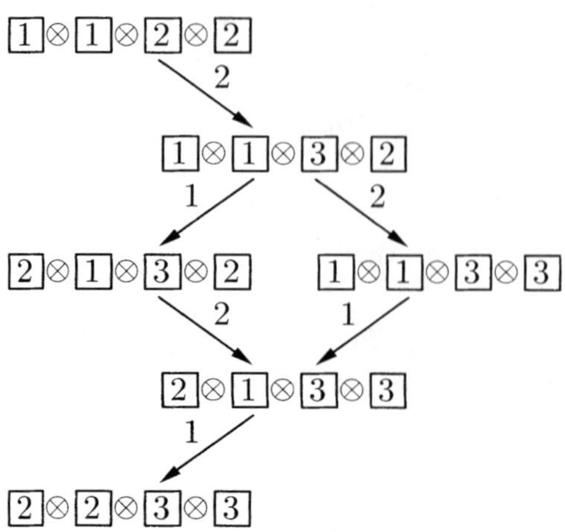

But, if we express each vector in $\mathbf{B}^{\otimes 4}$ as a semistandard tableau, both crystals coincide.

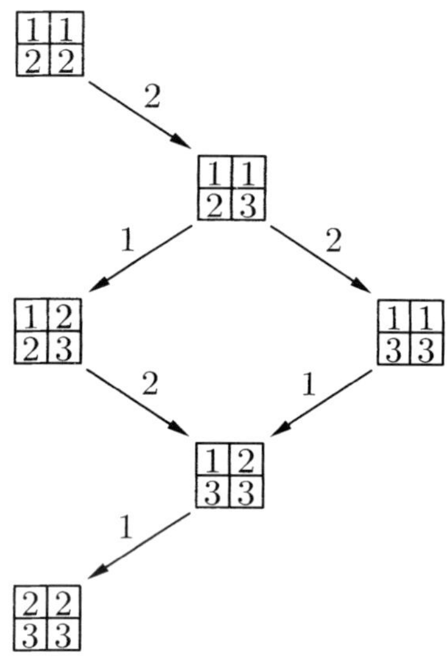

Proof of Theorem 7.3.6. Let $\Psi : \mathcal{B}(Y) \to \mathbf{B}^{\otimes N}$ be an admissible reading. Fix $i \in I = \{1, 2, \ldots, n-1\}$ and a semistandard tableau $T \in \mathcal{B}(Y)$. We first show that the tableaux $\tilde{e}_i T$ and $\tilde{f}_i T$ are independent of the choice of Ψ.

Suppose that T contains a rectangular subtableau

$$T_0 = \begin{array}{|c|c|c|}\hline i & \cdots & i \\ \hline i{+}1 & \cdots & i{+}1 \\ \hline \end{array}$$

with two rows such that the top row consists of $\boxed{i}$ and the bottom row consists of $\boxed{i{+}1}$. Such a rectangle is called an *i-trivial rectangle*. Indeed, we have $\tilde{e}_i T_0 = \tilde{f}_i T_0 = 0$. We assume that T_0 has maximal size among such rectangles.

Let T_1 be the subtableau of T consisting of the boxes that are strictly higher than the box $\boxed{i}$ that lies in the upper-right corner of T_0, and let T_2 be the subtableau of T consisting of the boxes that are strictly lower than the box $\boxed{i{+}1}$ that lies in the lower-left corner of T_0.

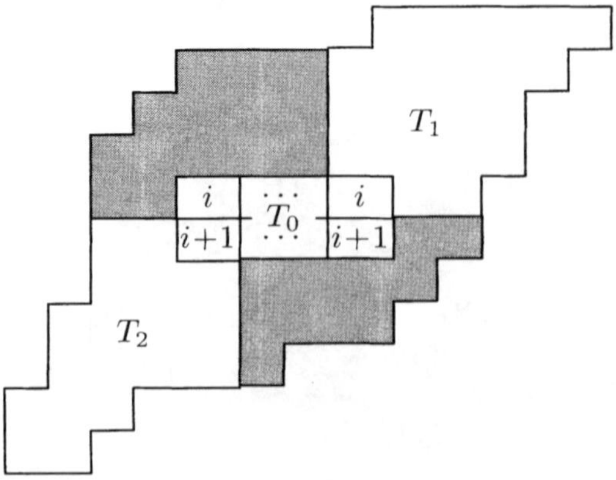

Since T is semistandard and T_0 is maximal, there are no boxes $\boxed{i}$ and $\boxed{i+1}$ in the shaded region of T. Hence for any admissible reading of T, T can be regarded as $T_1 \otimes T_0 \otimes T_2$ as a $U_q(\mathfrak{g}_{(i)})$-crystal, where $U_q(\mathfrak{g}_{(i)})$ is the subalgebra of $U_q(\mathfrak{gl}_n)$ generated by e_i, f_i and $K_i^{\pm 1}$. Moreover, since $\tilde{e}_i T_0 = \tilde{f}_i T_0 = 0$, T can be viewed as $T_1 \otimes T_2$ as a $U_q(\mathfrak{g}_{(i)})$-crystal.

By repeating the above argument, we may assume that the boxes $\boxed{i}$ and $\boxed{i+1}$ appear only in comparable positions except in i-trivial rectangles.

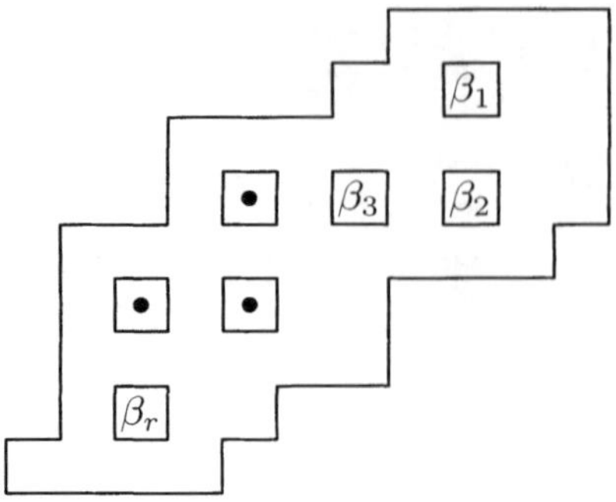

Therefore, for any admissible reading, T can be viewed as $\beta_1 \otimes \cdots \otimes \beta_r$, where $\beta_k = \boxed{i}$ or $\boxed{i+1}$, all of which lie in comparable positions, and hence $\tilde{e}_i T$ and $\tilde{f}_i T$ do not depend on the choice of an admissible reading Ψ.

It remains to show that $\tilde{e}_i T, \tilde{f}_i T \in \mathcal{B}(Y) \cup \{0\}$. Suppose that $\tilde{f}_i T \neq 0$. In this case, $\tilde{f}_i$ changes some box $b = \boxed{i}$ in T into $\boxed{i+1}$.

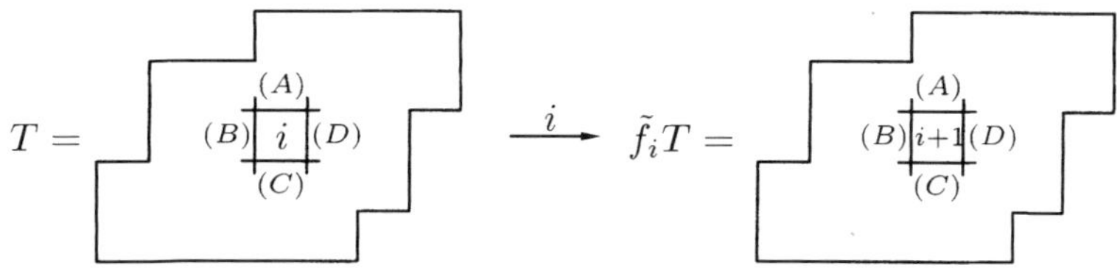

Let $b' = \boxed{j}$ be the box lying in one of the four sites (A), (B), (C), (D) around $b = \boxed{i}$. If b' lies in (A) or (B), then $j \leq i$ since T is semistandard. Hence there is nothing to prove.

If b' lies in (C), we already know $j \geq i+1$. If $j = i+1$, then by the tensor product rule, $\tilde{f}_i$ cannot act on b. Hence $j \geq i+2$. If b' lies in (D), then we already know $j \geq i$. If $j = i$, then $\tilde{f}_i$ would have acted on b', not on b, by the tensor product rule. Hence we must have $j \geq i+1$. Therefore, $\tilde{f}_i T$ is also semistandard.

Similarly, one can verify that $\tilde{e}_i T \in \mathcal{B}(Y) \cup \{0\}$ (Exercise 7.4), which completes the proof. □

7.4. Crystal graphs for $U_q(\mathfrak{gl}_n)$-modules

Recall that every finite dimensional irreducible $U_q(\mathfrak{gl}_n)$-module in the category $\mathcal{O}_{\text{int}}^{\geq 0}$ is of the form $V(\lambda)$ for some dominant integral weight $\lambda = \lambda_1\epsilon_1 + \cdots + \lambda_n\epsilon_n \in P_{\geq 0}^{+}$ (Theorem 7.2.3). We identify λ with the partition $(\lambda_1 \geq \lambda_2 \geq \cdots \geq \lambda_n \geq 0)$ and let Y be the corresponding Young diagram having λ_i boxes in the ith row. Let $\mathcal{B}(Y)$ denote the set of all semistandard tableaux of shape Y, which is given a $U_q(\mathfrak{gl}_n)$-crystal structure by an admissible reading. We first prove:

Theorem 7.4.1. *Let $\lambda = \lambda_1\epsilon_1 + \cdots + \lambda_n\epsilon_n \in P_{\geq 0}^{+}$ be a dominant integral weight and $V(\lambda)$ be the finite dimensional irreducible $U_q(\mathfrak{gl}_n)$-module in the category $\mathcal{O}_{\text{int}}^{\geq 0}$ with highest weight λ. Then the crystal graph $\mathcal{B}(\lambda)$ of $V(\lambda)$ is isomorphic to the $U_q(\mathfrak{gl}_n)$-crystal $\mathcal{B}(Y)$ consisting of all semistandard tableaux of shape Y.*

Proof. Let $N = \lambda_1 + \cdots + \lambda_n$. From the complete reducibility theorem, we know that $\mathbf{B}^{\otimes N}$ is a disjoint union of connected components, each of which is isomorphic to $\mathcal{B}(\mu)$ for some $\mu \in P_{\geq 0}^{+}$, a partition of N. Hence, to prove our theorem, it suffices to show that $\mathcal{B}(Y)$ is isomorphic to one of these connected components containing a maximal vector of weight λ.

We define T_Y to be the following tableau.

$T_Y =$ 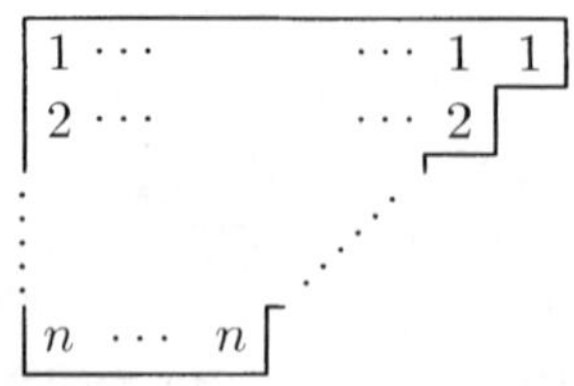

We claim that $\mathcal{B}(Y)$ is isomorphic to the connected component of $\mathbf{B}^{\otimes N}$ containing T_Y. Clearly, the weight of T_Y is equal to λ. By the Middle-Eastern reading, T_Y can be viewed as

$$\underbrace{\boxed{1 \ \cdots \ 1}}_{\lambda_1} \otimes \underbrace{\boxed{2 \ \cdots \ 2}}_{\lambda_2} \otimes \cdots \otimes \underbrace{\boxed{n \ \cdots \ n}}_{\lambda_n}.$$

Since $\lambda_i \geq \lambda_{i+1}$ for all $i \in I$, by the tensor product rule, we have $\tilde{e}_i T_Y = 0$ for all $i \in I$. Thus it is a maximal vector of weight λ.

It remains to show that for every semistandard tableau $T \in \mathcal{B}(Y)$, there exist $i_1, \ldots, i_r \in I$ such that $\tilde{e}_{i_1} \cdots \tilde{e}_{i_r} T = T_Y$. Equivalently, it suffices to show that if $\tilde{e}_i T = 0$ for all $i \in I$, then $T = T_Y$ (Exercise 7.5).

Let $T \in \mathcal{B}(Y)$ and assume that $\tilde{e}_i T = 0$ for all $i \in I$. Let k be the largest nonnegative integer such that the first k rows of T are the same as those of T_Y.

Suppose that $k < n$. We denote by T_0 the subtableau of T consisting of the first k rows of T and T_1 the subtableau of T consisting of the boxes below T_0. Let $b = \boxed{j}$ be the box lying in the upper-right corner of T_1, let T_2 be the column consisting of the boxes directly below $b = \boxed{j}$, and let T_3 denote the subtableau of T consisting of the columns of T_1 left to T_2. Hence T can be viewed as follows.

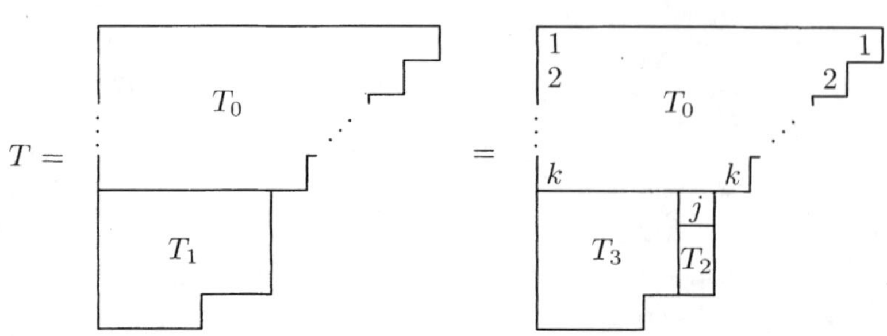

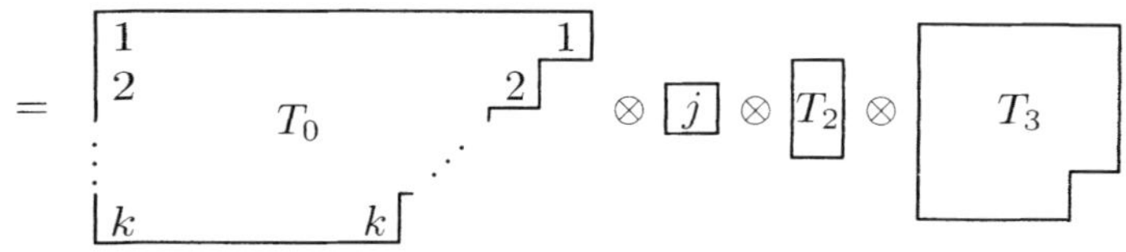

By the maximality of k, we must have $j > k+1$. On the other hand, since $\tilde{e}_i T = 0$ for all $i \in I$, by the tensor product rule, we must have $\tilde{e}_i(T_0 \otimes \boxed{j}) = 0$ for all $i \in I$, which implies $\tilde{e}_i T_0 = 0$ and $\varphi_i(T_0) \geq \varepsilon_i(\boxed{j})$ for all $i \in I$. In particular, $\varepsilon_i(\boxed{j}) = 0$ for all $i \geq k+1$. Hence we conclude that $j \leq k+1$, which is a contradiction. Therefore $k = n$ and $T = T_Y$. □

Remark 7.4.2. It is a well known fact that the basis vectors of the $\mathfrak{gl}_n$-module $V(\lambda)$ can be parameterized by semistandard tableaux of shape λ (see, for example, [**55**]). Theorem 7.4.1 gives a new interpretation of this fact in the language of crystal basis theory.

Example 7.4.3. (a) If $\lambda = \epsilon_1$, we have $\mathcal{B}(\lambda) = \mathbf{B}$, the crystal graph for the vector representation.

(b) When $n = 3$, the crystal graphs $\mathcal{B}(\lambda)$ can be realized as follows.

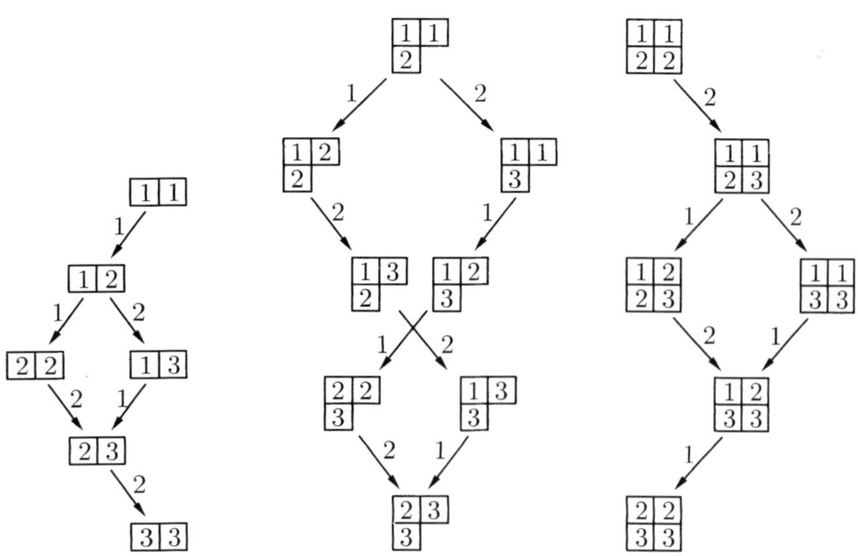

$\lambda = 2\epsilon_1$ $\lambda = 2\epsilon_1 + \epsilon_2$ $\lambda = 2\epsilon_1 + 2\epsilon_2$

Now we will describe a combinatorial rule of decomposing the tensor product of finite dimensional $U_q(\mathfrak{gl}_n)$-modules in the category $\mathcal{O}_{\text{int}}^{\geq 0}$ into a direct sum of irreducible submodules. The combinatorial algorithm we are going to describe here is exactly the same as the classical *Littlewood-Richardson rule* which has played a very important role in combinatorial representation theory for many years. Here, we will deduce the Littlewood-Richardson rule using the crystal basis theory. Since the structure of $U_q(\mathfrak{gl}_n)$-modules in the category $\mathcal{O}_{\text{int}}^{\geq 0}$ is reflected upon their crystal graphs (Theorem 4.2.10), we have only to decompose the tensor product of their crystal graphs into a disjoint union of connected components.

Let $\lambda = \lambda_1\epsilon_1 + \cdots + \lambda_n\epsilon_n$ ($\lambda_1 \geq \cdots \geq \lambda_n \geq 0$) and $\mu = \mu_1\epsilon_1 + \cdots + \mu_n\epsilon_n$ ($\mu_1 \geq \cdots \geq \mu_n \geq 0$) be dominant integral weights, and let Y and Y' denote the corresponding Young diagrams. Note that decomposing the tensor product $\mathcal{B}(Y) \otimes \mathcal{B}(Y')$ of $U_q(\mathfrak{gl}_n)$-crystals into connected components is equivalent to finding all the maximal vectors in $\mathcal{B}(Y) \otimes \mathcal{B}(Y')$.

We first consider the case when $\mu = \epsilon_1$. In this case, we have $\mathcal{B}(\mu) = \mathbf{B}$, the crystal graph of the vector representation.

By Corollary 4.4.4, the maximal vectors in $\mathcal{B}(Y) \otimes \mathbf{B}$ have the form $T_Y \otimes \boxed{j}$ with weight $\lambda + \epsilon_j$, where T_Y is the highest vector in $\mathcal{B}(Y)$ and $\boxed{j}$ satisfies $\varepsilon_i(\boxed{j}) \leq \lambda(h_i) = \lambda_i - \lambda_{i+1}$ for all $i = 1, \ldots, n-1$. Since

$$\varepsilon_i(\boxed{j}) = \begin{cases} 1 & \text{if } j = i+1, \\ 0 & \text{otherwise,} \end{cases}$$

we must have $\lambda_{j-1} - \lambda_j > 0$, which implies that there is an indent corner at the jth row of Y. Therefore, to each indent corner of Y, we get a connected component $\mathcal{B}(Y[j])$, where $Y[j]$ denotes the Young diagram obtained from Y by adding a box at the jth row. More generally, let $Y[j]$ denote the diagram obtained from Y by adding a box at the jth row and if $Y[j]$ is not a Young diagram, define $\mathcal{B}(Y[j])$ to be the null crystal; i.e., the empty set. Then we obtain:

Theorem 7.4.4. *Let $\lambda = \lambda_1\epsilon_1 + \cdots + \lambda_n\epsilon_n$ ($\lambda_1 \geq \cdots \geq \lambda_n \geq 0$) be a dominant integral weight and let Y be the Young diagram associated with λ. Then there is a $U_q(\mathfrak{gl}_n)$-crystal isomorphism*

$$\mathcal{B}(Y) \otimes \mathbf{B} \cong \bigoplus_{\boxed{j} \in \mathbf{B}} \mathcal{B}(Y[j]), \tag{7.9}$$

where $\bigoplus$ denotes the direct sum of crystals.

Example 7.4.5. Let $n = 3$ and $\lambda = 2\epsilon_1 + \epsilon_2 + \epsilon_3$. Then $Y =$ 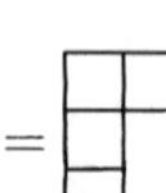 and we have

$Y[1] =$ 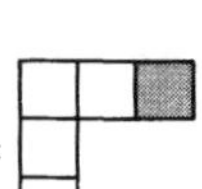, $Y[2] =$ 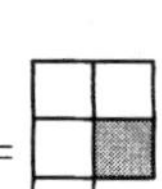, $Y[3] =$ 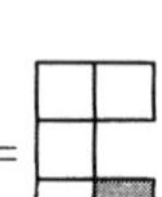.

Hence we obtain

$$\mathcal{B}(Y) \otimes \mathbf{B} \cong \mathcal{B}\left(\begin{array}{|c|c|c|}\hline \ & \ & \ \\ \hline \multicolumn{1}{|c|}{\ } \\ \cline{1-1} \multicolumn{1}{|c|}{\ } \\ \cline{1-1}\end{array}\right) \oplus \mathcal{B}\left(\begin{array}{|c|c|}\hline \ & \ \\ \hline \ & \ \\ \hline \multicolumn{1}{|c|}{\ } \\ \cline{1-1}\end{array}\right). \tag{7.10}$$

For the general case, let $\mu = \mu_1\epsilon_1 + \cdots + \mu_n\epsilon_n$ ($\mu_1 \geq \cdots \geq \mu_n$) be a dominant integral weight and Y' be the corresponding Young diagram. Then the $U_q(\mathfrak{gl}_n)$-crystal $\mathcal{B}(Y')$ can be embedded in $\mathbf{B}^{\otimes N}$ by an admissible reading, where $N = \mu_1 + \cdots + \mu_n$. Let $T \otimes T' \in \mathcal{B}(Y) \otimes \mathcal{B}(Y')$ be a maximal vector and write $T' = \boxed{j_1} \otimes \cdots \otimes \boxed{j_N} \in \mathbf{B}^{\otimes N}$. By Corollary 4.4.4, we must have $T = T_Y$ and $T_Y \otimes \boxed{j_1} \otimes \cdots \otimes \boxed{j_r}$ is a maximal vector for all $r = 1, 2, \ldots, N$. We define $Y[j_1, \ldots, j_r]$ to be the diagram obtained from $Y[j_1, \ldots, j_{r-1}]$ by adding a box at the j_rth row and set $\mathcal{B}(Y[j_1, \ldots, j_r]) = \emptyset$ unless $Y[j_1, \ldots, j_k]$ is a Young diagram for all $k = 1, \ldots, r$.

As we have seen in the discussion preceding Theorem 7.4.4, $T_Y \otimes \boxed{j_1} \otimes \cdots \otimes \boxed{j_N}$ is a maximal vector in $\mathcal{B}(Y) \otimes \mathcal{B}(Y') \subset \mathcal{B}(Y) \otimes \mathbf{B}^{\otimes N}$ if and only if $Y[j_1, \ldots, j_k]$ is a Young diagram for all $k = 1, \ldots, N$. Therefore we obtain the following combinatorial rule (*Littlewood-Richardson rule*) of decomposing the tensor product of $U_q(\mathfrak{gl}_n)$-crystals into a disjoint union of connected components.

Theorem 7.4.6. *Let* $\lambda = \lambda_1\epsilon_1 + \cdots + \lambda_n\epsilon_n$ ($\lambda_1 \geq \cdots \geq \lambda_n \geq 0$), $\mu = \mu_1\epsilon_1 + \cdots + \mu_n\epsilon_n$ ($\mu_1 \geq \cdots \geq \mu_n$) *be dominant integral weights and let* Y, Y' *denote the associated Young diagrams. Then there is a* $U_q(\mathfrak{gl}_n)$*-crystal isomorphism*

$$\mathcal{B}(Y) \otimes \mathcal{B}(Y') \cong \bigoplus_{\boxed{j_1} \otimes \cdots \otimes \boxed{j_N} \in \mathcal{B}(Y')} \mathcal{B}(Y[j_1, j_2, \ldots, j_N]), \tag{7.11}$$

where $N = \mu_1 + \cdots + \mu_n$.

Example 7.4.7. Let $n = 3$, $\lambda = 2\epsilon_1 + \epsilon_2$, and $\mu = 2\epsilon_1 + 2\epsilon_2$. Then

$$Y = \begin{array}{|c|c|}\hline \ & \ \\ \hline \multicolumn{1}{|c|}{\ } \\ \cline{1-1}\end{array} \quad \text{and} \quad Y' = \begin{array}{|c|c|}\hline \ & \ \\ \hline \ & \ \\ \hline\end{array},$$

and by the Far-Eastern reading, the crystal $\mathcal{B}(Y')$ consists of the following vectors:

$$\begin{array}{|c|c|}\hline 1 & 1 \\ \hline 2 & 2 \\ \hline\end{array} = \boxed{1} \otimes \boxed{2} \otimes \boxed{1} \otimes \boxed{2},$$

$$\begin{array}{|c|c|}\hline 1 & 1 \\ \hline 2 & 3 \\ \hline\end{array} = \boxed{1} \otimes \boxed{3} \otimes \boxed{1} \otimes \boxed{2},$$

$$\begin{array}{|c|c|}\hline 1 & 2 \\ \hline 2 & 3 \\ \hline\end{array} = \boxed{2} \otimes \boxed{3} \otimes \boxed{1} \otimes \boxed{2},$$

$$\begin{array}{|c|c|}\hline 1 & 1 \\ \hline 3 & 3 \\ \hline\end{array} = \boxed{1} \otimes \boxed{3} \otimes \boxed{1} \otimes \boxed{3},$$

$$\begin{array}{|c|c|}\hline 1 & 2 \\ \hline 3 & 3 \\ \hline \end{array} = \boxed{2} \otimes \boxed{3} \otimes \boxed{1} \otimes \boxed{3},$$

$$\begin{array}{|c|c|}\hline 2 & 2 \\ \hline 3 & 3 \\ \hline \end{array} = \boxed{2} \otimes \boxed{3} \otimes \boxed{2} \otimes \boxed{3}.$$

Hence we have the following.

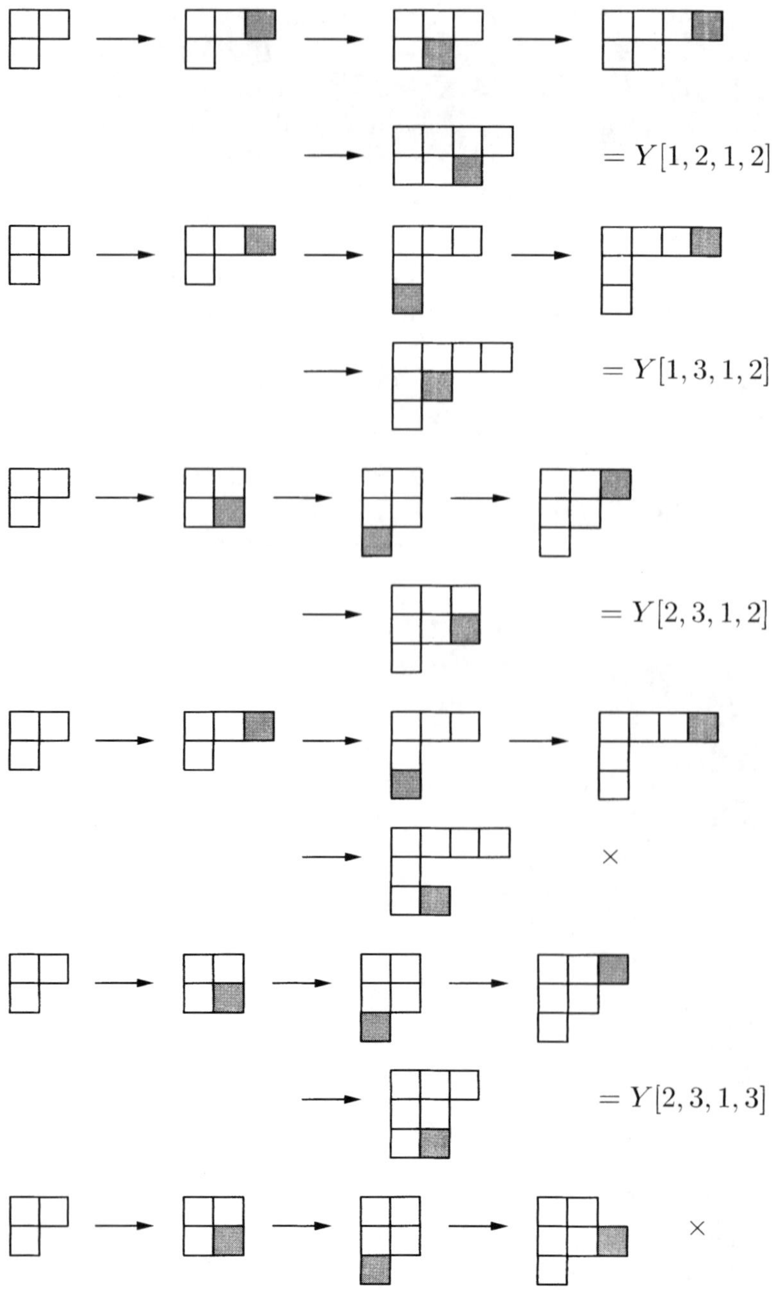

Therefore, by Theorem 7.4.6, we obtain

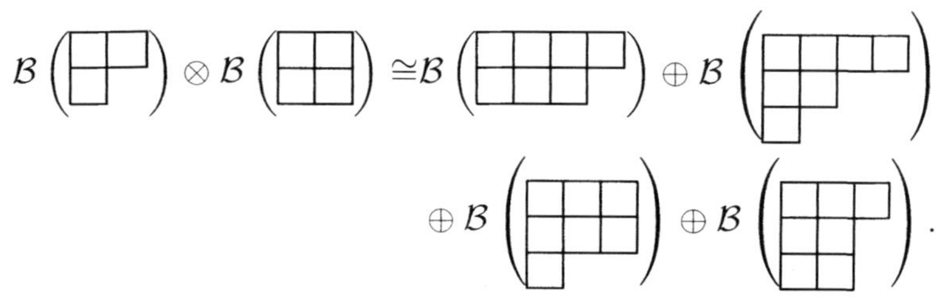

Exercises

7.1. Show that the equation (7.6) defines a $U_q(\mathfrak{gl}_n)$-module structure on the n-dimensional vector space $\mathbf{V} = \mathbf{C}(q)v_1 \oplus \cdots \oplus \mathbf{C}(q)v_n$ and $(\mathbf{L}, \mathbf{B})$ is a crystal basis of $\mathbf{V}$.

7.2. Verify the statements in Example 7.2.2.

7.3. Let $\lambda = (\lambda_1, \ldots, \lambda_n \geq 0)$ and $\mu = (\mu_1, \ldots, \mu_n)$ be two elements in $P_{\geq 0}$ such that $\lambda_1 + \cdots + \lambda_n = \mu_1 + \cdots + \mu_n$. Verify that $\lambda_1 + \cdots + \lambda_k \geq \mu_1 + \cdots + \mu_k$ for all $k \geq 1$ if and only if $\lambda - \mu \in Q_+$.

7.4. Let Y be a Young diagram with at most n rows and $\mathcal{B}(Y)$ be the set of all semistandard Young tableaux with shape Y. Show that if $T \in \mathcal{B}(Y)$, then $\tilde{e}_i T \in \mathcal{B}(Y) \cup \{0\}$ for all $i = 1, \ldots, n-1$.

7.5. Show that the following two statements are equivalent.

(a) For each $T \in \mathcal{B}(Y)$, there exist indices $i_1, \ldots, i_r \in I$ such that $\tilde{e}_{i_1} \cdots \tilde{e}_{i_r} T = T_Y$.

(b) If $\tilde{e}_i T = 0$ for all $i \in I$, then $T = T_Y$.

7.6. Verify the crystal graphs given in Example 7.4.3.

7.7. (a) Draw the $U_q(\mathfrak{gl}_3)$-crystal $\mathcal{B}(Y)$ for

$Y =$ 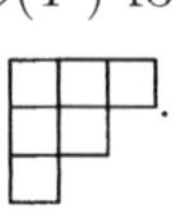.

(b) Draw the $U_q(\mathfrak{gl}_3)$-crystal $\mathcal{B}(Y)$ for

$Y =$ 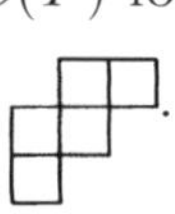.

(c) Verify that the $U_q(\mathfrak{gl}_3)$-crystal structures on $\mathcal{B}(Y)$ given by the Far-Eastern reading and the Middle-Eastern reading coincide with each other.

7.8. Decompose the tensor product $\mathcal{B}(Y) \otimes \mathcal{B}(Y')$ of $U_q(\mathfrak{gl}_3)$-crystals into a disjoint union of connected components in the following two cases.

(a)

$Y =$ 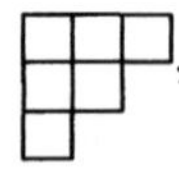, $\quad Y' =$ 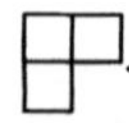.

(b)

$Y =$ 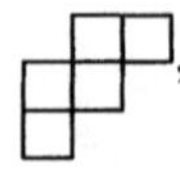, $\quad Y' =$ 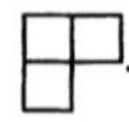.

Chapter 8

Crystal Graphs for Classical Lie Algebras

In this chapter, we further investigate the connection between crystal basis theory and the combinatorics of Young diagrams and Young tableaux. The main result of this section is a realization of crystal graphs for finite dimensional irreducible modules over the quantum groups corresponding to classical Lie algebras. The realization will be given in terms of semistandard Young tableaux satisfying certain additional conditions, depending on the type of Lie algebras.

Our approach is based on the works by Kashiwara and Nakashima [**41, 50**], and is essentially the same as the one given in Chapter 7. The only difference is that we will stick to the Far-Eastern reading because the crystal structure *does* depend on the choice of admissible reading except in the A_{n-1} case.

We will first take B_3 as an example to give the readers a feel for what the crystals would look like, and then present the explicit description of crystal graphs. The description of crystal graphs for all four cases are quite similar, with the complexity rising in the sequence A_{n-1}, C_n, B_n, and D_n.

In the last section, we will discuss the *generalized Littlewood-Richardson rule*—the tensor product decomposition of crystal graphs into a disjoint union of connected components.

8.1. Example: $U_q(B_3)$-crystals

In this section, we will mainly give a lot of examples of $U_q(B_3)$-crystals. Let $\mathfrak{g}$ be the finite dimensional simple Lie algebra of type B_3. Then $\mathfrak{g}$ is realized as the *special orthogonal Lie algebra* (see, for example, [**15**, Part III], [**17**, Ch.I], or [**19**, Ch.IV])

$$(8.1)\quad \begin{aligned}\mathfrak{g} &= \mathfrak{so}(7,\mathbf{C}) = \mathfrak{so}_7\\ &= \left\{ T = \begin{pmatrix} A & B & a\\ C & D & b\\ c & d & 0\end{pmatrix} \in M_{7\times 7}(\mathbf{C}) \;\middle|\; \begin{array}{l} A,B,C,D\in M_{3\times 3}(\mathbf{C}),\quad a,b\in M_{3\times 1}(\mathbf{C}),\quad c,d\in M_{1\times 3}(\mathbf{C}),\\ A^t=-D,\ B^t=-B,\ C^t=-C,\quad a^t=-d,\ b^t=-c\end{array}\right\}.\end{aligned}$$

Let E_{ij} denote the 7×7 elementary matrix having 1 at the (i,j)-entry and 0 elsewhere, and set

$$(8.2)\quad \begin{aligned} &e_1 = E_{12}-E_{54}, \qquad e_2 = E_{23}-E_{65}, \qquad e_3 = 2(E_{37}-E_{76}),\\ &f_1 = E_{21}-E_{45}, \qquad f_2 = E_{32}-E_{56}, \qquad f_3 = E_{73}-E_{67},\\ &h_1 = E_{11}-E_{22}-E_{44}+E_{55}, \qquad h_2 = E_{22}-E_{33}-E_{55}+E_{66},\\ &h_3 = 2(E_{33}-E_{66}).\end{aligned}$$

Then, as a Lie algebra, $\mathfrak{g}=\mathfrak{so}(7,\mathbf{C})$ is generated by e_i, f_i, h_i $(i=1,2,3)$.

Consider the linear functionals $\epsilon_i : M_{7\times 7}(\mathbf{C}) \longrightarrow \mathbf{C}$ defined by

$$\epsilon_i(T) = t_{ii}, \qquad T=(t_{ij}), \quad i,j=1,2,3.$$

Then, the *simple roots* and the *fundamental weights* are expressed as

$$(8.3)\quad \begin{aligned} &\alpha_1 = \epsilon_1-\epsilon_2, \qquad \alpha_2=\epsilon_2-\epsilon_3, \qquad \alpha_3=\epsilon_3,\\ &\omega_1=\epsilon_1, \qquad \omega_2=\epsilon_1+\epsilon_2, \qquad \omega_3=\frac{1}{2}(\epsilon_1+\epsilon_2+\epsilon_3).\end{aligned}$$

Using this notation, the *Cartan datum* for the special orthogonal Lie algebra $\mathfrak{g}=\mathfrak{so}(7,\mathbf{C})$ is given as follows:

$$(8.4)\quad \begin{aligned} &A = \begin{pmatrix} 2 & -1 & 0\\ -1 & 2 & -1\\ 0 & -2 & 2\end{pmatrix},\\ &\Pi = \{\alpha_1,\alpha_2,\alpha_3\}, \qquad \Pi^\vee = \{h_1,h_2,h_3\},\\ &P = \mathbf{Z}\omega_1\oplus\mathbf{Z}\omega_2\oplus\mathbf{Z}\omega_3, \qquad P^\vee = \mathbf{Z}h_1\oplus\mathbf{Z}h_2\oplus\mathbf{Z}h_3.\end{aligned}$$

The ***quantum special orthogonal algebra*** $U_q(\mathfrak{so}_7)$ is defined to be the quantum group associated with the Cartan datum given above.

Let us fix a seven-dimensional vector space

$$\mathbf{V} = \left(\bigoplus_{j=1}^{3} \mathbf{C}(q)v_j\right) \oplus \mathbf{C}(q)v_0 \oplus \left(\bigoplus_{j=1}^{3} \mathbf{C}(q)v_{\bar{j}}\right)$$

and set $\mathbf{N} = \{1, 2, 3, 0, \bar{3}, \bar{2}, \bar{1}\}$ to be the index set for the basis vectors of $\mathbf{V}$ with a linear ordering

$$1 \prec 2 \prec 3 \prec 0 \prec \bar{3} \prec \bar{2} \prec \bar{1}.$$

Then $\mathbf{V}$ becomes a $U_q(\mathfrak{so}_7)$-module with $U_q(\mathfrak{so}_7)$-action given by

$$q^h v_j = q^{\langle h, \mathrm{wt}(v_j)\rangle} v_j \quad \text{for } h \in P^\vee,\ j \in \mathbf{N},$$

$$e_i v_j = \begin{cases} v_i & \text{if } j = i+1,\ i = 1, 2, \\ v_{\overline{i+1}} & \text{if } j = \bar{i},\ i = 1, 2, \\ v_0 & \text{if } j = \bar{3},\ i = 3, \\ [2]_q v_3 & \text{if } j = 0,\ i = 3, \\ 0 & \text{otherwise,} \end{cases}$$

$$f_i v_j = \begin{cases} v_{i+1} & \text{if } j = i,\ i = 1, 2, \\ v_{\bar{i}} & \text{if } j = \overline{i+1},\ i = 1, 2, \\ v_0 & \text{if } j = 3,\ i = 3, \\ [2]_q v_{\bar{3}} & \text{if } j = 0,\ i = 3, \\ 0 & \text{otherwise,} \end{cases}$$

where

$$\mathrm{wt}(v_j) = \begin{cases} \epsilon_j & \text{for } j = 1, 2, 3, \\ 0 & \text{for } j = 0, \\ -\epsilon_j & \text{for } j = \bar{1}, \bar{2}, \bar{3}. \end{cases}$$

We call $\mathbf{V}$ the ***vector representation*** of $U_q(\mathfrak{so}_7)$.

Let $\mathbf{L} = \bigoplus_{j \in \mathbf{N}} \mathbf{A}_0 v_j$ and $\mathbf{B} = \{\boxed{j} = v_j + q\mathbf{L} \mid j \in \mathbf{N}\}$. Then $(\mathbf{L}, \mathbf{B})$ is a crystal basis of $\mathbf{V}$ with crystal graph

$$\boxed{1} \xrightarrow{1} \boxed{2} \xrightarrow{2} \boxed{3} \xrightarrow{3} \boxed{0} \xrightarrow{3} \boxed{\bar{3}} \xrightarrow{2} \boxed{\bar{2}} \xrightarrow{1} \boxed{\bar{1}}.$$

Note that $\mathbf{B} = \mathcal{B}(\omega_1) = \mathcal{B}(\epsilon_1)$ and that

$$\text{(8.5)} \qquad \mathrm{wt}(\boxed{j}) = \epsilon_j, \qquad \mathrm{wt}(\boxed{0}) = 0, \qquad \mathrm{wt}(\boxed{\bar{j}}) = -\epsilon_j \quad \text{for } j = 1, 2, 3.$$

Consider the tensor product of crystals $\mathbf{B} \otimes \mathbf{B}$ as is given in Figure 8.1. Rather than labelling the arrows with their colors, we chose to use different arrowheads in distinguishing them. The same convention will be used for

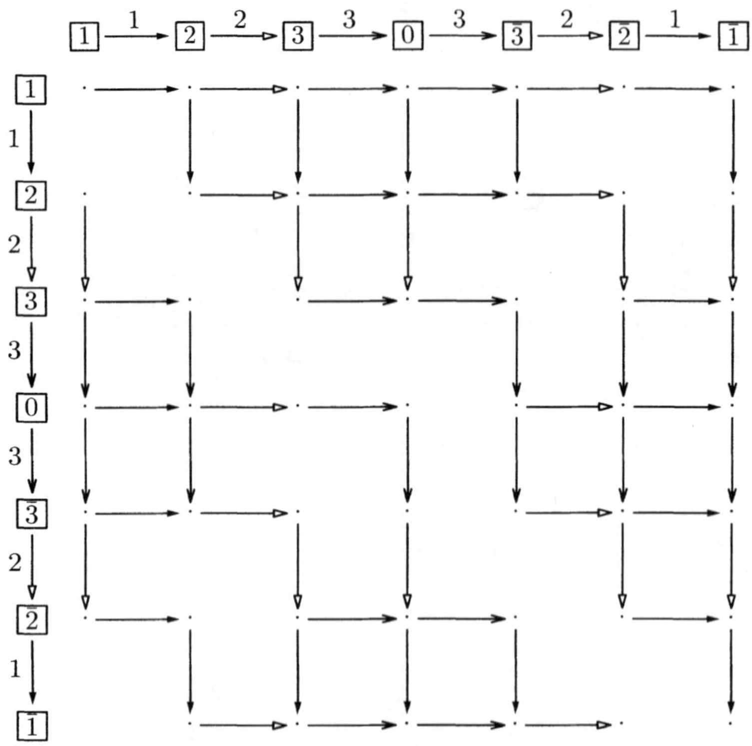

Figure 8.1. $\mathbf{B} \otimes \mathbf{B}$ for $U_q(\mathfrak{so}_7)$.

other crystal graphs appearing in this section. From Figure 8.1, we can see that

$$\mathbf{B} \otimes \mathbf{B} = \mathcal{B}(\omega_1) \otimes \mathcal{B}(\omega_1) = \mathcal{B}(2\omega_1) \oplus \mathcal{B}(\omega_2) \oplus \mathcal{B}(0),$$

where $\mathcal{B}(2\omega_1)$ corresponds to the upper triangular part of Figure 8.1, $\mathcal{B}(\omega_2)$ is the connected component filling in most of the lower triangular part, and $\mathcal{B}(0)$ is the isolated dot in the bottom left corner.

Note that the highest weight $2\omega_1 = 2\epsilon_1$ corresponds to the Young diagram $\boxed{}\boxed{}$ and $\omega_2 = \epsilon_1 + \epsilon_2$ corresponds to $\begin{array}{|c|}\hline \\ \hline \\ \hline\end{array}$. Hence, from Figure 8.1, we can conclude

$$\begin{aligned} &\mathcal{B}(2\omega_1) = \{\boxed{b}\boxed{a} \mid a \succeq b,\ (a,b) \neq (0,0)\}, \\ &\mathcal{B}(\omega_2) = \left\{ \begin{array}{|c|}\hline a \\ \hline b \\ \hline\end{array} \,\middle|\, \begin{array}{l} a \prec b \text{ with equality allowed} \\ \text{for zeros and } (a,b) \neq (1,\bar{1}) \end{array} \right\}. \end{aligned} \tag{8.6}$$

Next, consider the tensor product decomposition of crystals

$$\mathcal{B}(\omega_2) \otimes \mathbf{B} = \mathcal{B}(\omega_1 + \omega_2) \oplus \mathcal{B}(2\omega_3) \oplus \mathcal{B}(\omega_1)$$

as is shown in Figure 8.2.

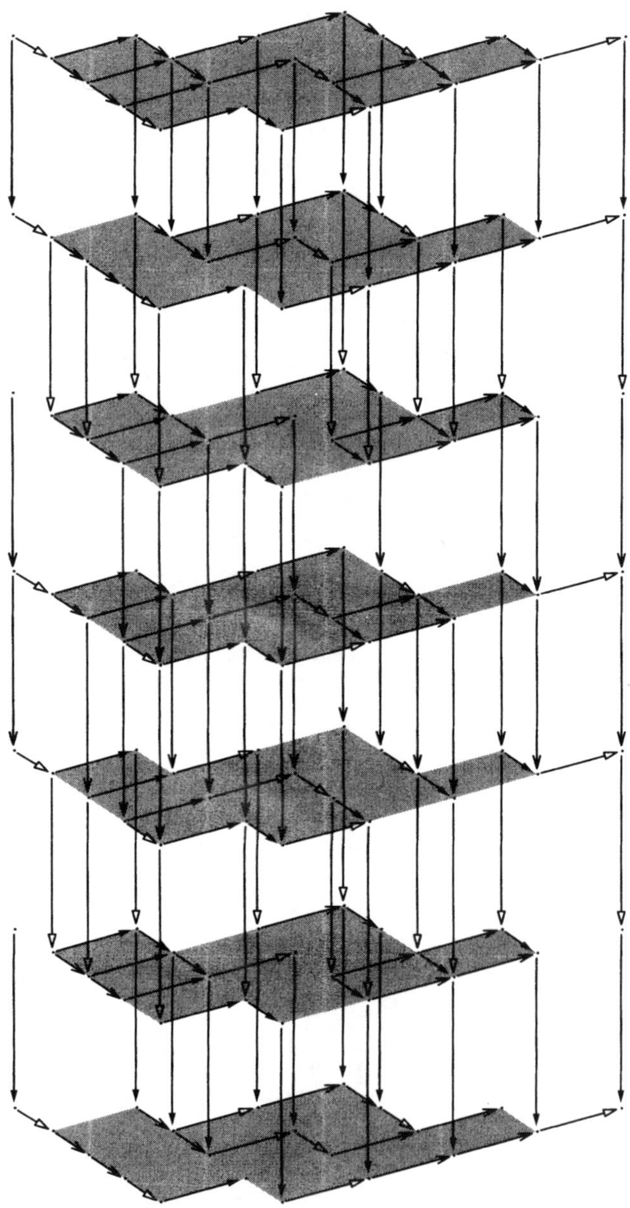

Figure 8.2. $\mathcal{B}(\omega_2) \otimes \mathbf{B}$ for $U_q(\mathfrak{so}_7)$.

We single out the connected component containing $\begin{array}{|c|}\hline 1 \\ \hline 2 \\ \hline \end{array} \otimes \begin{array}{|c|}\hline 3 \\ \hline \end{array}$ which is isomorphic to $\mathcal{B}(2\omega_3) = \mathcal{B}(\epsilon_1 + \epsilon_2 + \epsilon_3)$ (see Figure 8.3). Again, note that the highest weight $2\omega_3 = \epsilon_1 + \epsilon_2 + \epsilon_3$ corresponds to the Young diagram $\begin{array}{|c|}\hline \ \\ \hline \ \\ \hline \ \\ \hline \end{array}$.

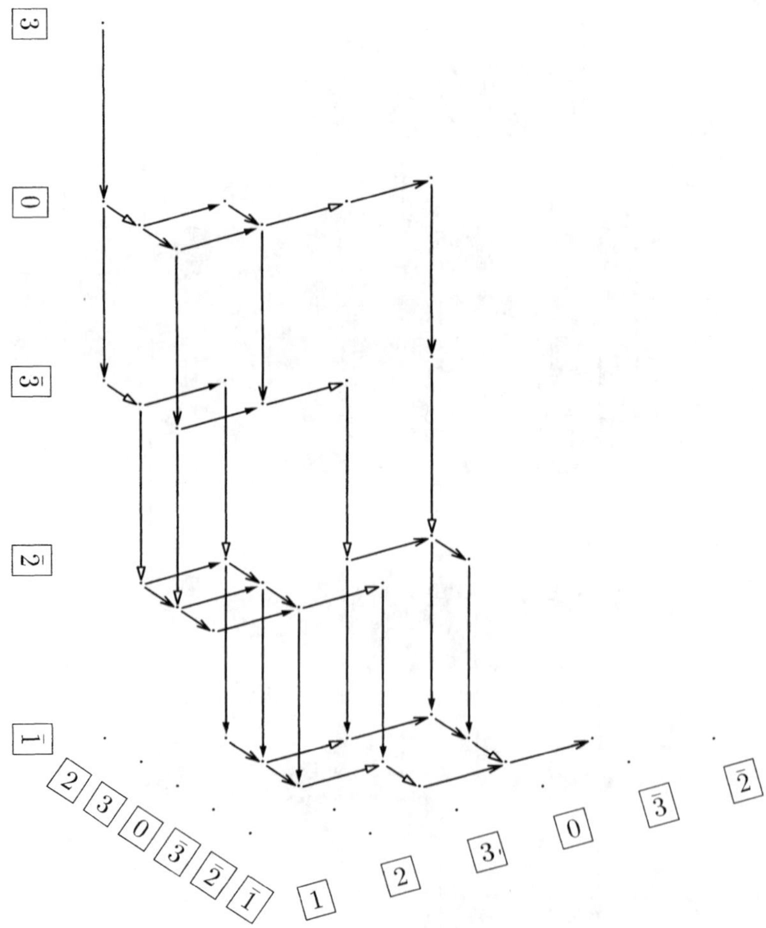

Figure 8.3. $\mathcal{B}(2\omega_3)$ for $U_q(\mathfrak{so}_7)$.

Thus, with the help of Figure 8.3, which has been singled out from Figure 8.2 for an easier view, we see that

$$(8.7)\quad \mathcal{B}(2\omega_3) = \left\{ \begin{array}{|c|}\hline a \\ \hline b \\ \hline c \\ \hline \end{array} \;\middle|\; \begin{array}{l} a \prec b \prec c \text{ with equalities allowed for zeros,} \\ (a,c) \neq (1,\bar{1}),\ (a,b) \neq (2,\bar{2}), \text{ and } (b,c) \neq (2,\bar{2}) \end{array} \right\}.$$

By now, we hope our scheme of realization of crystal graphs is more or less clear to the readers. To be more precise, given a dominant integral weight of the form $\lambda = a\omega_1 + b\omega_2 + 2c\omega_3$ $(a, b, c \in \mathbf{Z}_{\geq 0})$, we associate the Young diagram

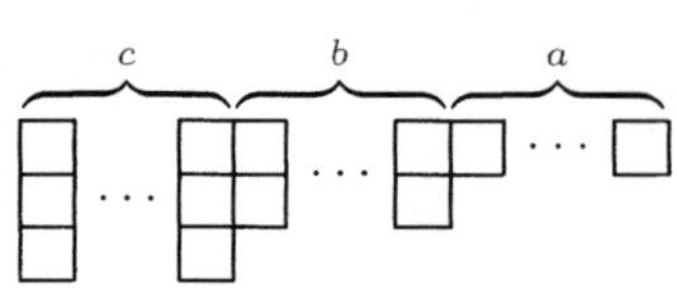

and we would like to realize the crystal graph $\mathcal{B}(\lambda)$ as the set $\mathcal{B}(Y)$ of Young tableaux of shape Y with entries in $\mathbf{N} = \{1, 2, 3, 0, \bar{3}, \bar{2}, \bar{1}\}$ satisfying certain conditions. Here, the crystal structure is given by the *Far-Eastern reading*. That is, for a Young tableau $T \in \mathcal{B}(Y)$, by the Far-Eastern reading, we view T as an element of

$$\mathcal{B}(\omega_1)^{\otimes a} \otimes \mathcal{B}(\omega_2)^{\otimes b} \otimes \mathcal{B}(2\omega_3)^{\otimes c} \subset \mathbf{B}^{\otimes N},$$

where $N = a + 2b + 3c$, and we define the action of Kashiwara operators by the tensor product rule. Then, the crystal graph $\mathcal{B}(\lambda)$ will be identified with the connected component of a maximal vector of weight λ in

$$\mathcal{B}(\omega_1)^{\otimes a} \otimes \mathcal{B}(\omega_2)^{\otimes b} \otimes \mathcal{B}(2\omega_3)^{\otimes c} \subset \mathbf{B}^{\otimes N}.$$

Thus, at least one of the conditions that T must satisfy is clear: every column of T must belong to $\mathcal{B}(\omega_1)$, $\mathcal{B}(\omega_2)$, or $B(2\omega_3)$. Hence the main task is to find the remaining conditions that can characterize the set $\mathcal{B}(Y)$.

Example 8.1.1. Consider the tensor product decomposition of $U_q(\mathfrak{so}_7)$-crystals

$$\mathcal{B}(2\omega_1) \otimes \mathbf{B} = \mathcal{B}(3\omega_1) \oplus \mathcal{B}(\omega_1 + \omega_2) \oplus \mathcal{B}(\omega_1)$$

as is shown in Figure 8.4. Then the crystal graphs $\mathcal{B}(3\omega_1)$ and $\mathcal{B}(\omega_1 + \omega_2)$ are realized as follows:

$$\mathcal{B}(3\omega_1) = \{\begin{array}{|c|c|c|}\hline c & b & a \\ \hline\end{array} \mid c \preccurlyeq b \preccurlyeq a,\ (a, b) \neq (0, 0),\ \text{and } (b, c) \neq (0, 0)\},$$

$$\mathcal{B}(\omega_1 + \omega_2) = \left\{ \begin{array}{|c|c|}\hline b & a \\ \hline c \\ \cline{1-1}\end{array} \;\middle|\; \begin{array}{|c|c|}\hline b & a \\ \hline\end{array} \in \mathcal{B}(2\omega_1),\ \begin{array}{|c|}\hline b \\ \hline c \\ \hline\end{array} \in \mathcal{B}(\omega_2) \right\}.$$

Note that we use the notation

$$\begin{aligned} \begin{array}{|c|c|c|}\hline c & b & a \\ \hline\end{array} &= \boxed{a} \otimes \boxed{b} \otimes \boxed{c} \in \mathcal{B}(\omega_1)^{\otimes 3} = \mathbf{B}^{\otimes 3}, \\ \begin{array}{|c|c|}\hline b & a \\ \hline c \\ \cline{1-1}\end{array} &= \boxed{a} \otimes \begin{array}{|c|}\hline b \\ \hline c \\ \hline\end{array} \qquad \in \mathcal{B}(\omega_1) \otimes \mathcal{B}(\omega_2) \subset \mathbf{B}^{\otimes 3}. \end{aligned}$$

However, this approach is not yet complete, because we still have not obtained a realization for $\mathcal{B}(\omega_3) = \mathcal{B}(\frac{1}{2}(\epsilon_1 + \epsilon_2 + \epsilon_3))$. For this purpose, we introduce the ***spin representation*** $\mathbf{V}_{\mathrm{sp}}$ of $U_q(\mathfrak{so}_7)$.

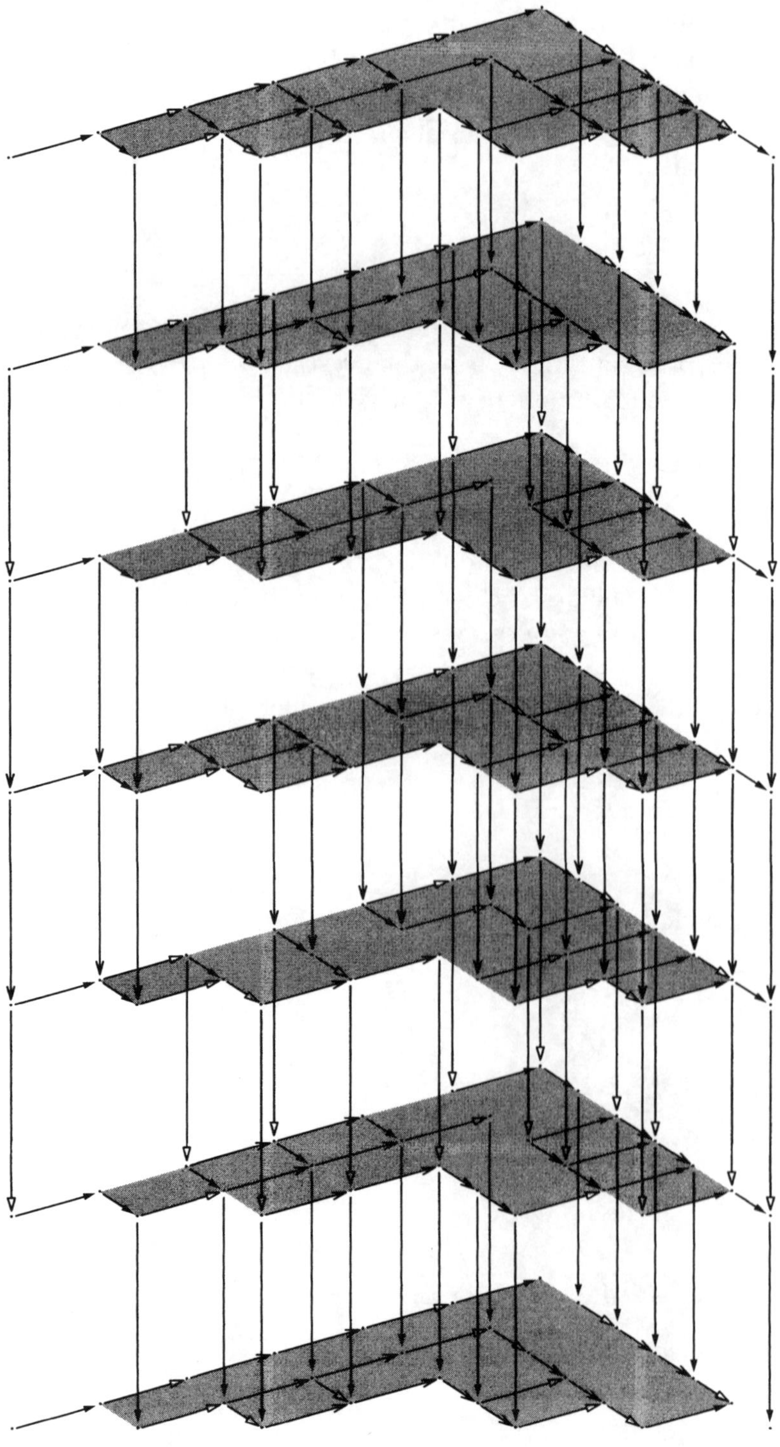

Figure 8.4. $\mathcal{B}(2\omega_1) \otimes \mathbf{B}$ for $U_q(\mathfrak{so}_7)$.

Let $\mathbf{V}_{\rm sp} = \bigoplus_{s_i=\pm} \mathbf{C}(q)(s_1, s_2, s_3)$ be the eight-dimensional vector space with the $U_q(\mathfrak{so}_7)$-action given by

$$q^h(s_1,s_2,s_3) = q^{\langle h, {\rm wt}(s_1,s_2,s_3)\rangle}(s_1,s_2,s_3) \quad \text{for } h \in P^\vee,$$

$$e_1(s_1,s_2,s_3) = \begin{cases} (+,-,s_3) & \text{if } (s_1,s_2) = (-,+), \\ 0 & \text{otherwise,} \end{cases}$$

$$e_2(s_1,s_2,s_3) = \begin{cases} (s_1,+,-) & \text{if } (s_2,s_3) = (-,+), \\ 0 & \text{otherwise,} \end{cases}$$

$$e_3(s_1,s_2,s_3) = \begin{cases} (s_1,s_2,+) & \text{if } s_3 = -, \\ 0 & \text{otherwise,} \end{cases}$$

$$f_1(s_1,s_2,s_3) = \begin{cases} (-,+,s_3) & \text{if } (s_1,s_2) = (+,-), \\ 0 & \text{otherwise,} \end{cases}$$

$$f_2(s_1,s_2,s_3) = \begin{cases} (s_1,-,+) & \text{if } (s_2,s_3) = (+,-), \\ 0 & \text{otherwise,} \end{cases}$$

$$f_3(s_1,s_2,s_3) = \begin{cases} (s_1,s_2,-) & \text{if } s_3 = +, \\ 0 & \text{otherwise,} \end{cases}$$

where the weight of each element is given by

$${\rm wt}(s_1,s_2,s_3) = \frac{1}{2}(s_1\epsilon_1 + s_2\epsilon_2 + s_3\epsilon_3).$$

For example, the weight of $(+,-,+)$ is $\frac{1}{2}(\epsilon_1 - \epsilon_2 + \epsilon_3)$. Then the spin representation $\mathbf{V}_{\rm sp}$ is isomorphic to $V(\omega_3)$. It has a crystal basis $(\mathbf{L}_{\rm sp}, \mathbf{B}_{\rm sp})$, where

$$\mathbf{L}_{\rm sp} = \bigoplus_{s_i=\pm} \mathbf{A}_0(s_1,s_2,s_3), \qquad \mathbf{B}_{\rm sp} = \{(s_1,s_2,s_3) \mid s_i\pm\},$$

and the crystal graph $\mathbf{B}_{\rm sp}$ is given in Figure 8.5.

Now, take the (generalized) Young diagram $Y_{\rm sp} = \begin{array}{|c|}\hline \\ \hline \\ \hline \vdots \\ \hline \\ \hline \end{array}$ consisting of *half-boxes* (in the sense that they are of half-unit width) and consider the (generalized) Young tableaux of shape $Y_{\rm sp}$ whose entries are taken from $\mathbf{N}_{\rm sp} = \{1,2,3,\bar{3},\bar{2},\bar{1}\}$ with a linear ordering

$$1 \prec 2 \prec 3 \prec \bar{3} \prec \bar{2} \prec \bar{1}.$$

We identify the element (s_1,s_2,s_3) of $\mathbf{B}_{\rm sp}$ with a tableau of shape $Y_{\rm sp}$ as follows: if the ith component is $+$ (respectively, $-$), we fill the ith box of

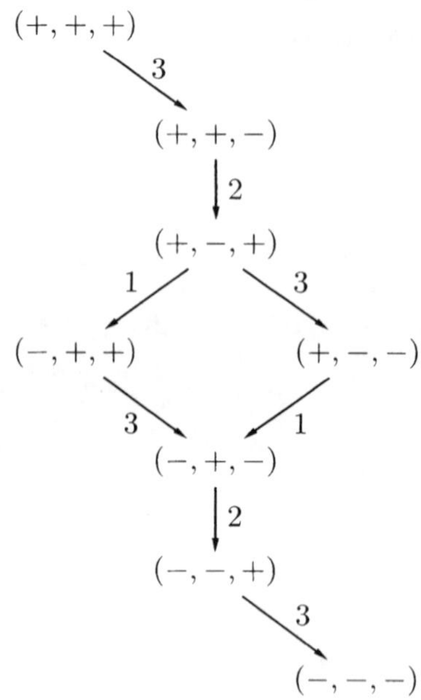

Figure 8.5. $\mathbf{B}_{\mathrm{sp}}$ for $U_q(\mathfrak{so}_7)$.

Y_{sp} with i (respectively, $\bar{i}$), and then reshuffle them so that they are linearly ordered. For example, $(-,+,-)$ is identified with the tableau $\begin{array}{|c|}\hline 2 \\ \hline \bar{3} \\ \hline \bar{1} \\ \hline \end{array}$.

Then the crystal graph $\mathbf{B}_{\mathrm{sp}} = \mathcal{B}(\omega_3)$ is characterized as

$$\mathbf{B}_{\mathrm{sp}} = \mathcal{B}(\omega_3) = \left\{ \begin{array}{|c|}\hline a \\ \hline b \\ \hline c \\ \hline \end{array} \;\middle|\; \begin{array}{l} a, b, c \in \mathbf{N}_{\mathrm{sp}} \text{ with } a \prec b \prec c, \\ i \text{ and } \bar{i} \text{ do not appear simultaneously} \end{array} \right\}.$$

We leave it as an exercise to the readers to draw the crystal graph $\mathbf{B}_{\mathrm{sp}}$ using this description (Exercise 8.2).

Therefore, if $\lambda = a\omega_1 + b\omega_2 + (2c+1)\omega_3$, we will identify the crystal graph $\mathcal{B}(\lambda)$ with the connected component containing a maximal vector of weight λ inside

$$\mathcal{B}(\omega_1)^{\otimes a} \otimes \mathcal{B}(\omega_2)^{\otimes b} \mathcal{B}(2\omega_3)^{\otimes c} \otimes \mathbf{B}_{\mathrm{sp}} \subset \mathbf{B}^{\otimes N} \otimes \mathbf{B}_{\mathrm{sp}},$$

where $N = a + 2b + 3c$. Hence, $\mathcal{B}(\lambda)$ will be characterized as the set $\mathcal{B}(Y)$ of (generalized) Young tableaux of shape

$Y =$ 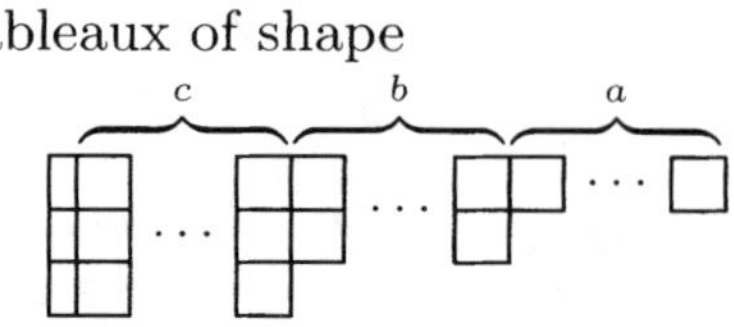

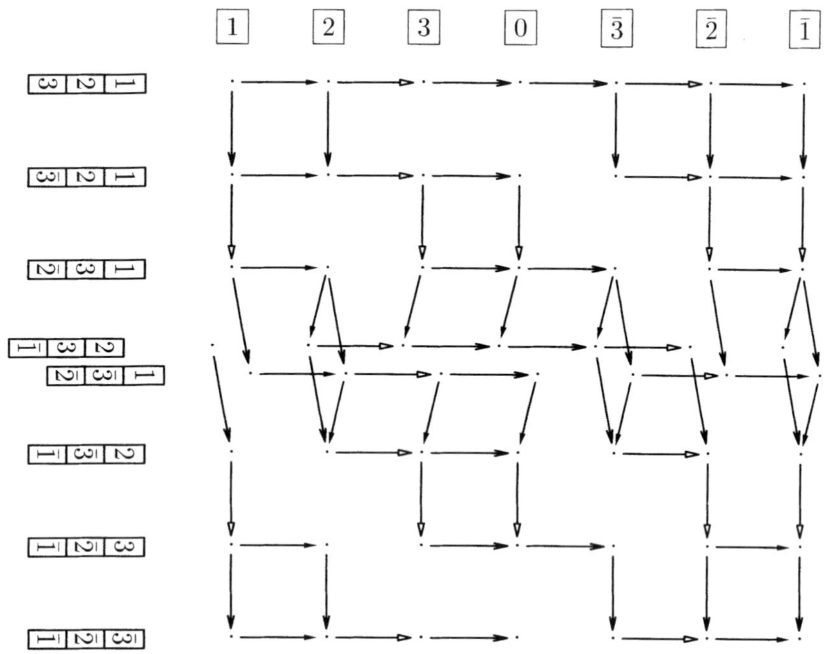

Figure 8.6. $\mathbf{B} \otimes \mathbf{B}_{\mathrm{sp}}$ for $U_q(\mathfrak{so}_7)$.

satisfying certain conditions. Here, the crystal structure on $\mathcal{B}(Y)$ is given by the Far-Eastern reading.

Example 8.1.2. Consider the tensor product of crystals

$$\mathcal{B}(\omega_1) \otimes \mathcal{B}(\omega_3) = \mathbf{B} \otimes \mathbf{B}_{\mathrm{sp}} = \mathcal{B}(\omega_1 + \omega_3) \oplus \mathcal{B}(\omega_3),$$

which is given in Figure 8.6. Then we have

$$\mathcal{B}(\omega_1 + \omega_3) = \left\{ \begin{array}{|c|c|} \hline b & a \\ \hline c & \\ \cline{1-1} d & \\ \cline{1-1} \end{array} \;\middle|\; \begin{array}{|c|} \hline b \\ \hline c \\ \hline d \\ \hline \end{array} \in \mathbf{B}_{\mathrm{sp}},\ b \preccurlyeq a \right\}.$$

In the following sections, for each type of classical Lie algebra, we will make explicit and precise these *certain conditions* that characterize the crystal graph $\mathcal{B}(\lambda)$.

8.2. Realization of $U_q(A_{n-1})$-crystals

Let $\mathfrak{g}$ be the finite dimensional simple Lie algebra of type A_{n-1}. Recall that $\mathfrak{g}$ is realized as the *special linear Lie algebra* (see, for example, [**15**, Part III], [**17**, Ch.I], or [**19**, Ch.IV])

$$\mathfrak{g} = \mathfrak{sl}(n, \mathbf{C}) = \mathfrak{sl}_n = \{T \in M_{n\times n}(\mathbf{C}) \mid \operatorname{tr} T = 0\}. \tag{8.8}$$

Let E_{ij} denote the $n \times n$ elementary matrix having 1 at the (i, j)-entry and 0 elsewhere, and set

$$e_i = E_{i,i+1}, \qquad f_i = E_{i+1,i}, \qquad h_i = E_{ii} - E_{i+1,i+1} \tag{8.9}$$

for $i = 1, 2, \ldots, n-1$. Then, as a Lie algebra, $\mathfrak{g} = \mathfrak{sl}(n, \mathbf{C})$ is generated by e_i, f_i, h_i ($i \in I = \{1, 2, \ldots, n-1\}$).

Consider the linear functionals $\epsilon_i : M_{n\times n}(\mathbf{C}) \longrightarrow \mathbf{C}$ defined by

$$\epsilon_i(T) = t_{ii}, \qquad T = (t_{ij}) \in M_{n\times n}(\mathbf{C}), \quad i, j = 1, 2, \ldots, n.$$

Then the *simple roots* and the *fundamental weights* are expressed as

$$\begin{aligned} &\alpha_i = \epsilon_i - \epsilon_{i+1}, \\ &\omega_i = \epsilon_1 + \epsilon_2 + \cdots + \epsilon_i \quad \text{for } i \in I. \end{aligned} \tag{8.10}$$

Note that $\epsilon_1 + \cdots + \epsilon_n = 0$ on $\mathfrak{g} = \mathfrak{sl}(n, \mathbf{C})$.

Using this notation, the *Cartan datum* for the special linear Lie algebra $\mathfrak{g} = \mathfrak{sl}(n, \mathbf{C})$ is given as follows:

$$\begin{aligned} A &= \begin{pmatrix} 2 & -1 & \cdots & 0 & 0 \\ -1 & 2 & -1 & & 0 \\ 0 & & \ddots & & 0 \\ 0 & \cdots & -1 & 2 & -1 \\ 0 & \cdots & & -1 & 2 \end{pmatrix}, \\ \Pi &= \{\alpha_1, \alpha_2, \ldots, \alpha_{n-1}\}, \qquad \Pi^\vee = \{h_1, h_2, \ldots, h_{n-1}\}, \\ P &= \mathbf{Z}\omega_1 \oplus \cdots \oplus \mathbf{Z}\omega_{n-1}, \qquad P^\vee = \mathbf{Z}h_1 \oplus \cdots \oplus \mathbf{Z}h_{n-1}. \end{aligned} \tag{8.11}$$

The ***quantum special linear algebra*** $U_q(\mathfrak{sl}_n)$ is defined to be the quantum group associated with the Cartan datum given above.

Let $\mathbf{V} = \bigoplus_{j=1}^n \mathbf{C}(q)v_j$ be an n-dimensional vector space with $U_q(\mathfrak{sl}_n)$-action defined by

$$\begin{aligned} K_i v_j &= \begin{cases} q\, v_j & \text{if } j = i, \\ q^{-1} v_j & \text{if } j = i+1, \\ v_j & \text{otherwise}, \end{cases} \\ e_i v_j &= \begin{cases} v_i & \text{if } j = i+1, \\ 0 & \text{otherwise}, \end{cases} \\ f_i v_j &= \begin{cases} v_{i+1} & \text{if } j = i, \\ 0 & \text{otherwise}. \end{cases} \end{aligned} \tag{8.12}$$

The $U_q(\mathfrak{sl}_n)$-module $\mathbf{V}$ is called the ***vector representation*** and it has a crystal basis $(\mathbf{L}, \mathbf{B})$, where

$$\mathbf{L} = \bigoplus_{j=1}^{n} \mathbf{A}_0 v_j, \qquad \mathbf{B} = \{\boxed{j} = v_j + q\mathbf{L} \mid j = 1, \ldots, n\}.$$

The crystal graph $\mathbf{B}$ of the *vector representation* is given below.

$$\boxed{1} \xrightarrow{\ 1\ } \boxed{2} \xrightarrow{\ 2\ } \cdots \xrightarrow{\ n-2\ } \boxed{n-1} \xrightarrow{\ n-1\ } \boxed{n}$$

Note that $\mathbf{B} = \mathcal{B}(\omega_1) = \mathcal{B}(\epsilon_1)$ and

$$\mathrm{wt}(\boxed{j}) = \epsilon_j \quad \text{for } j = 1, \ldots, n.$$

Let $\lambda = a_1\omega_1 + \cdots + a_{n-1}\omega_{n-1}$ ($a_i \in \mathbf{Z}_{\geq 0}$) be a dominant integral weight. Using the relation (8.10), λ can be identified with the partition $\lambda = (\lambda_1 \geq \lambda_2 \geq \cdots \geq \lambda_{n-1} \geq 0)$, where

$$\begin{aligned} \lambda_1 &= a_1 + a_2 + \cdots + a_{n-1}, \\ \lambda_2 &= a_2 + \cdots + a_{n-1}, \\ &\vdots \\ \lambda_{n-1} &= a_{n-1}. \end{aligned} \tag{8.13}$$

Hence the crystal graph $\mathcal{B}(\lambda)$ of the $U_q(\mathfrak{sl}_n)$-module $V(\lambda)$ can be identified with the connected component of $\mathbf{B}^{\otimes N}$ containing a maximal vector of weight λ, where $N = \lambda_1 + \lambda_2 + \cdots + \lambda_{n-1} = a_1 + 2a_2 + \cdots + (n-1)a_{n-1}$.

Let Y be the Young diagram associated with λ and let $\mathcal{B}(Y)$ be the set of all semistandard tableaux of shape Y with the $U_q(\mathfrak{sl}_n)$-crystal structure given by the Far-Eastern reading. Then, as is proved in Theorem 7.4.1, we have:

Theorem 8.2.1. *Let $\lambda = a_1\omega_1 + \cdots + a_{n-1}\omega_{n-1}$ be a dominant integral weight and let $V(\lambda)$ be the finite dimensional irreducible $U_q(\mathfrak{sl}_n)$-module with highest weight λ. Then the crystal graph $\mathcal{B}(\lambda)$ of $V(\lambda)$ is isomorphic to the $U_q(\mathfrak{sl}_n)$-crystal $\mathcal{B}(Y)$ consisting of all semistandard tableaux of shape Y.*

8.3. Realization of $U_q(C_n)$-crystals

Let $\mathfrak{g}$ be the finite dimensional simple Lie algebra of type C_n ($n \geq 2$). Then $\mathfrak{g}$ is realized as the *symplectic Lie algebra* (see, for example, [**15**, Part III],

[**17**, Ch.I], or [**19**, Ch.IV])

$$\begin{aligned} \mathfrak{g} &= \mathfrak{sp}(2n, \mathbf{C}) = \mathfrak{sp}_{2n} \\ &= \left\{ T = \begin{pmatrix} A & B \\ C & D \end{pmatrix} \in M_{2n\times 2n}(\mathbf{C}) \;\middle|\; \begin{array}{l} A, B, C, D \in M_{n\times n}(\mathbf{C}), \\ A^t = -D,\ B^t = B,\ C^t = C \end{array} \right\}. \end{aligned} \tag{8.14}$$

Let E_{ij} denote the $2n \times 2n$ elementary matrix having 1 at the (i,j)-entry and 0 elsewhere, and set

$$\begin{aligned} &e_i = E_{i,i+1} - E_{n+i+1,n+i}, \qquad f_i = E_{i+1,i} - E_{n+i,n+i+1}, \\ &h_i = E_{ii} - E_{i+1,i+1} - E_{n+i,n+i} + E_{n+i+1,n+i+1}, \\ &e_n = E_{n,2n}, \qquad f_n = E_{2n,n}, \qquad h_n = E_{n,n} - E_{2n,2n} \end{aligned} \tag{8.15}$$

for $i = 1, 2, \ldots, n-1$. Then, as a Lie algebra, $\mathfrak{g} = \mathfrak{sp}(2n, \mathbf{C})$ is generated by e_i, f_i, h_i $(i = 1, 2, \ldots, n)$.

Consider the linear functionals $\epsilon_i : M_{2n\times 2n}(\mathbf{C}) \longrightarrow \mathbf{C}$ defined by

$$\epsilon_i(T) = t_{ii}, \qquad T = (t_{ij}) \in M_{2n\times 2n}(\mathbf{C}), \quad i, j = 1, 2, \ldots, 2n.$$

Then, the *simple roots* and the *fundamental weights* are expressed as

$$\begin{aligned} &\alpha_i = \epsilon_i - \epsilon_{i+1} \quad \text{for } i = 1, 2, \ldots, n-1, \\ &\alpha_n = 2\epsilon_n, \\ &\omega_i = \epsilon_1 + \epsilon_2 + \cdots + \epsilon_i \quad \text{for } i = 1, 2, \ldots, n. \end{aligned} \tag{8.16}$$

Using these notations, the *Cartan datum* for the symplectic Lie algebra $\mathfrak{g} = \mathfrak{sp}(2n, \mathbf{C})$ is given as follows:

$$\begin{aligned} &A = \begin{pmatrix} 2 & -1 & \cdots & & 0 \\ -1 & 2 & -1 & & 0 \\ 0 & & \ddots & & 0 \\ 0 & & \cdots & 2 & -2 \\ 0 & & \cdots & -1 & 2 \end{pmatrix}, \\ &\Pi = \{\alpha_1, \alpha_2, \ldots, \alpha_n\}, \qquad \Pi^\vee = \{h_1, h_2, \ldots, h_n\}, \\ &P = \mathbf{Z}\omega_1 \oplus \cdots \oplus \mathbf{Z}\omega_n, \qquad P^\vee = \mathbf{Z}h_1 \oplus \cdots \oplus \mathbf{Z}h_n. \end{aligned} \tag{8.17}$$

The ***quantum symplectic algebra*** $U_q(\mathfrak{sp}_{2n})$ is defined to be the quantum group associated with the Cartan datum given above.

Let $\mathbf{V} = \left(\bigoplus_{j=1}^n \mathbf{C}(q)v_j\right) \oplus \left(\bigoplus_{j=1}^n \mathbf{C}(q)v_{\bar{j}}\right)$ be a $2n$-dimensional vector space and let $\mathbf{N} = \{1, 2, \ldots, n, \bar{n}, \ldots, \bar{2}, \bar{1}\}$ be the index set for the basis vectors of $\mathbf{V}$ with a linear ordering given by

$$1 \prec 2 \prec \cdots \prec n \prec \bar{n} \prec \cdots \prec \bar{2} \prec \bar{1}.$$

We define the $U_q(\mathfrak{sp}_{2n})$-action on $\mathbf{V}$ as follows:

$$q^h v_j = q^{\langle h, \mathrm{wt}(v_j)\rangle} v_j \quad \text{for } h \in P^\vee,\ j \in \mathbf{N},$$

(8.18)
$$e_i v_j = \begin{cases} v_i & \text{if } j = i+1, \\ v_{\overline{i+1}} & \text{if } j = \bar{i}, \\ 0 & \text{otherwise}, \end{cases}$$
$$f_i v_j = \begin{cases} v_{i+1} & \text{if } j = i, \\ v_{\bar{i}} & \text{if } j = \overline{i+1}, \\ 0 & \text{otherwise}, \end{cases}$$

where

$$\mathrm{wt}(v_j) = \epsilon_j, \qquad \mathrm{wt}(v_{\bar{j}}) = -\epsilon_j \quad \text{for } j = 1, 2, \ldots, n.$$

Then $\mathbf{V}$ becomes a $U_q(\mathfrak{sp}_{2n})$-module called the ***vector representation***, and it has a crystal basis $(\mathbf{L}, \mathbf{B})$, where

$$\mathbf{L} = \bigoplus_{j \in \mathbf{N}} \mathbf{A}_0 v_j, \qquad \mathbf{B} = \{\boxed{j} = v_j + q\mathbf{L} \mid j \in \mathbf{N}\}.$$

The $U_q(\mathfrak{sp}_{2n})$-crystal structure on $\mathbf{B}$ is given below.

$$\boxed{1} \xrightarrow{1} \boxed{2} \xrightarrow{2} \cdots \xrightarrow{n-1} \boxed{n} \xrightarrow{n} \boxed{\bar{n}} \xrightarrow{n-1} \cdots \xrightarrow{2} \boxed{\bar{2}} \xrightarrow{1} \boxed{\bar{1}}$$

Note that $\mathbf{B} = \mathcal{B}(\omega_1) = \mathcal{B}(\epsilon_1)$ and

$$\mathrm{wt}(\boxed{j}) = \epsilon_j, \qquad \mathrm{wt}(\boxed{\bar{j}}) = -\epsilon_j \quad \text{for } j = 1, 2, \ldots, n.$$

Let $\lambda = a_1\omega_1 + \cdots + a_n\omega_n$ ($a_i \in \mathbf{Z}_{\geq 0}$) be a dominant integral weight. Using the relation (8.16), λ can be identified with the partition $\lambda = (\lambda_1 \geq \lambda_2 \geq \cdots \geq \lambda_n \geq 0)$, where

(8.19)
$$\begin{aligned} \lambda_1 &= a_1 + a_2 + \cdots + a_n, \\ \lambda_2 &= a_2 + \cdots + a_n, \\ &\ \vdots \\ \lambda_n &= a_n. \end{aligned}$$

Thus the crystal graph $\mathcal{B}(\lambda)$ of $V(\lambda)$ can be embedded into

$$\mathcal{B}(\lambda) \hookrightarrow \mathcal{B}(\omega_1)^{\otimes a_1} \otimes \cdots \otimes \mathcal{B}(\omega_n)^{\otimes a_n} \hookrightarrow \mathbf{B}^{\otimes N},$$

where $N = \lambda_1 + \lambda_2 + \cdots + \lambda_n = a_1 + 2a_2 + \cdots + na_n$. Therefore, we can identify $\mathcal{B}(\lambda)$ with the connected component of $\mathbf{B}^{\otimes N}$ containing a maximal vector of weight λ.

Let Y be the Young diagram associated with λ. As in Section 8.1, we wish to characterize $\mathcal{B}(\lambda)$ as the set of semistandard tableaux of shape Y satisfying certain additional conditions. For this purpose, we introduce the notion of *semistandard C_n-tableaux*.

Definition 8.3.1. Let Y be a Young diagram with at most n rows.

(1) A C_n-***tableau*** of shape Y is a tableau obtained from Y by filling the boxes with entries from $\mathbf{N}$.

(2) A C_n-tableau is said to be ***semistandard*** if
  (a) the entries in each row are weakly increasing as we proceed to the right, and
  (b) the entries in each column are strictly increasing as we go down.

For a C_n-tableau T, we define its *weight* to be

$$\mathrm{wt}(T) = \sum_{i=1}^{n} (k_i - \overline{k_i})\epsilon_i, \tag{8.20}$$

where k_i (respectively, $\overline{k_i}$) is the number of i's (respectively, $\bar{i}$'s) appearing in T.

As in Chapter 7, one may expect the set of all semistandard C_n-tableaux to become a $U_q(\mathfrak{sp}_{2n})$-crystal by the Far-Eastern reading. However this set is *not* closed under the action of Kashiwara operators. Moreover the $U_q(\mathfrak{sp}_{2n})$-crystal containing all semistandard C_n-tableaux is *not* connected. Thus we need more conditions to characterize the connected $U_q(\mathfrak{sp}_{2n})$-crystal containing a maximal vector of weight λ.

To be more precise, we define $\mathcal{B}(Y)$ to be the set of all semistandard C_n-tableaux T satisfying the following conditions.

(C1) If T has a column of the form

$$\begin{array}{r|c|} \cline{2-2} & \\ \cline{2-2} p \rightarrow & i \\ \cline{2-2} & \\ \cline{2-2} q \rightarrow & \bar{i} \\ \cline{2-2} & \\ \cline{2-2} \end{array},$$

then we have

$$(q - p) + i > N,$$

where N is the length of the column. Here we mean to say that i is at the pth position from the top and that $\bar{j}$ is at the qth position from the top.

(C2) If T has a pair of adjacent columns having one of the following configurations with $p \le q < r \le s$ and $a \le b$:

$$\begin{array}{l} p \rightarrow \\ q \rightarrow \\ \\ r \rightarrow \\ s \rightarrow \end{array}
\begin{array}{c|c} a & \\ & b \\ & \\ & \bar{b} \\ & \bar{a} \end{array}, \quad
\begin{array}{c|c} a & \\ b & \\ & \\ \bar{b} & \\ & \bar{a} \end{array}, \quad
\begin{array}{c|c} a & a \\ & \\ & \\ & \\ & \bar{a} \end{array}, \quad
\begin{array}{c|c} a & \\ & \\ & \\ & \\ \bar{a} & \bar{a} \end{array},$$

then we have

$$(q-p)+(s-r) < b-a.$$

Thus any tableau containing one of the last two configurations cannot be an element of $\mathcal{B}(Y)$.

Example 8.3.2.

(1) Let Y be a Young diagram and set

$T_Y =$ 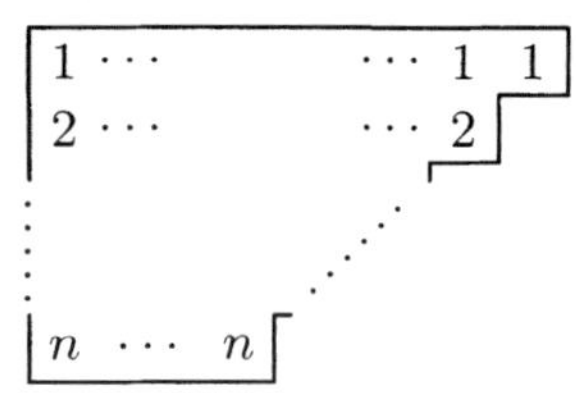

Then $T_Y \in \mathcal{B}(Y)$.

(2) When $n = 3$ and $\lambda = \omega_3$, we have $Y = \begin{array}{|c|}\hline \ \\ \hline \ \\ \hline \ \\ \hline\end{array}$, and the tableau $\begin{array}{|c|}\hline 2 \\ \hline 3 \\ \hline \bar{2} \\ \hline\end{array}$ belongs to $\mathcal{B}(Y)$ because

$$(q-p)+i = (3-1)+2 = 4 > 3.$$

On the other hand, the tableau $\begin{array}{|c|}\hline 1 \\ \hline 2 \\ \hline \bar{1} \\ \hline\end{array}$ does *not* belong to $\mathcal{B}(Y)$ because

$$(q-p)+i = (3-1)+1 = 3 \not> 3.$$

(3) When $n = 3$ and $\lambda = 2\omega_3$, we have $Y = \begin{array}{|c|c|}\hline \ & \ \\ \hline \ & \ \\ \hline \ & \ \\ \hline\end{array}$, and the tableau $\begin{array}{|c|c|}\hline 1 & 3 \\ \hline 2 & \bar{3} \\ \hline 3 & \bar{1} \\ \hline\end{array}$ belongs to $\mathcal{B}(Y)$, because

(i) $(q-p)+i = (2-1)+3 = 4 > 3$,

(ii) $(q-p)+(s-r) = (1-1)+(3-2) = 1 < b-a = 3-1 = 2$.

On the other hand, the tableau $\begin{array}{|c|c|}\hline 2 & 3 \\ \hline 3 & \bar{3} \\ \hline \bar{3} & \bar{2} \\ \hline\end{array}$ does *not* belong to $\mathcal{B}(Y)$, because

$$(q-p)+(s-r) = (1-1)+(3-2) = 1 \not< 1 = b-a = 3-2.$$

Theorem 8.3.3. *Let $\lambda = a_1\omega_1 + \cdots + a_n\omega_n$ ($a_i \in \mathbf{Z}_{\geq 0}$) be a dominant integral weight and let $V(\lambda)$ be the finite dimensional irreducible module over $U_q(\mathfrak{sp}_{2n})$ with highest weight λ. Let Y be the Young diagram associated with λ and let $\mathcal{B}(Y)$ denote the set of all semistandard C_n-tableaux of shape Y satisfying the conditions* **(C1)** *and* **(C2)**. *Then, by the Far-Eastern reading,*

$\mathcal{B}(Y)$ becomes a connected $U_q(\mathfrak{sp}_{2n})$-crystal containing a maximal vector of weight λ. Hence $\mathcal{B}(Y)$ is isomorphic to the crystal graph $\mathcal{B}(\lambda)$ of $V(\lambda)$.

To prove this theorem, we need the following lemma.

Lemma 8.3.4.

(1) *If $T \in \mathcal{B}(Y)$ has a column of the form*

$$\begin{array}{r|c|} \cline{2-2} & \\ \cline{2-2} p \longrightarrow & a \\ \cline{2-2} & \\ \cline{2-2} q \longrightarrow & \bar{b} \\ \cline{2-2} & \\ \cline{2-2} \end{array}\ ,$$

then we have

$$(q-p) + \max(a,b) > N,$$

where N is the length of the column.

(2) *If $T \in \mathcal{B}(Y)$ has a pair of adjacent columns having one of the following configurations with $p \le q < r \le s$, $a \le a'$ and $b \le b'$:*

$$\begin{array}{r} p \rightarrow \\ q \rightarrow \\ \\ r \rightarrow \\ s \rightarrow \end{array} \quad \begin{array}{c|c} a & \\ & a' \\ & \\ & \bar{b}' \\ & \bar{b} \end{array}\ , \qquad \begin{array}{c|c} a & \\ a' & \\ & \\ \bar{b}' & \\ & \bar{b} \end{array}\ ,$$

then we have

$$(q-p) + (s-r) < \max(a', b') - \min(a,b).$$

Proof. To prove (1), let j be the largest entry less than a such that both j and $\bar{j}$ appear in the column that contains a and $\bar{b}$. Assume that j (respectively, $\bar{j}$) lies in the kth row (respectively, lth row). If there is no such entry, we take $j = 0$, $k = 0$, $l = N + 1$.

$$\begin{array}{r|c|} \cline{2-2} & \\ \cline{2-2} k \longrightarrow & j \\ \cline{2-2} & A \\ \cline{2-2} p \longrightarrow & a \\ \cline{2-2} & \\ \cline{2-2} q \longrightarrow & \bar{b} \\ \cline{2-2} & B \\ \cline{2-2} l \longrightarrow & \bar{j} \\ \cline{2-2} & \\ \cline{2-2} \end{array}$$

Then there is no entry x in A for which $\bar{x}$ appears in B. Hence we must have

$$(p-k) + (l-q) \le \max(a,b) - j,$$

which implies

$$(l-k)+j \le (q-p)+\max(a,b).$$

Since T satisfies the condition **(C1)**, we must have $(l-k)+j > N$. (If $j=0$, $k=0$, $l=N+1$, this inequality is trivially satisfied.) Therefore, we obtain

$$(q-p)+\max(a,b) \ge (l-k)+j > N.$$

The proof of (2) is similar (Exercise 8.3). □

Proof of Theorem 8.3.3. As in the proof of Theorem 7.3.6 and Theorem 7.4.1, we need to prove:

(1) $\tilde{e}_i\mathcal{B}(Y) \subset \mathcal{B}(Y) \cup \{0\}$ for all $i \in I$,

(2) $\tilde{f}_i\mathcal{B}(Y) \subset \mathcal{B}(Y) \cup \{0\}$ for all $i \in I$,

(3) if $T \in \mathcal{B}(Y)$ satisfies $\tilde{e}_iT = 0$ for all $i \in I$, then $T = T_Y$.

We will first prove (2). The assertion (1) can be proved in a similar manner (Exercise 8.4).

Let $T \in \mathcal{B}(Y)$ be a semistandard C_n-tableau of shape Y satisfying the conditions **(C1)** and **(C2)**. Suppose $\tilde{f}_iT = T' \neq 0$. In this case, $\tilde{f}_i$ changes some box $b = \boxed{i}$ into $\boxed{i+1}$ or $b = \boxed{\overline{i+1}}$ into $\boxed{\overline{i}}$.

We will consider the first case only. Assume that $\tilde{f}_i$ changes $b = \boxed{i}$ into $\boxed{i+1}$. It is easy to verify that T' is semistandard (Exercise 8.4). Suppose further that $\boxed{\overline{i+1}}$ appears in the same column of T' as $\boxed{i+1}$ does, as is shown in the figure below.

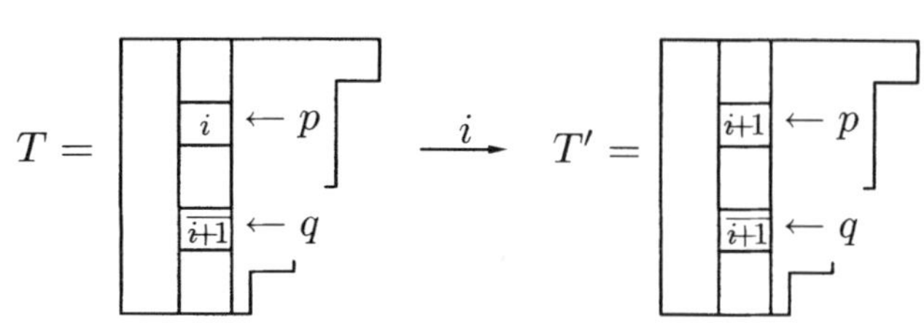

Then, $\boxed{\overline{i+1}}$ must appear at the same place in T, say, qth row of T. By Lemma 8.3.4 (1), we have

$$(q-p)+(i+1) > N,$$

which implies T' satisfies the condition **(C1)**.

Next, suppose T' has a configuration

$T' =$ 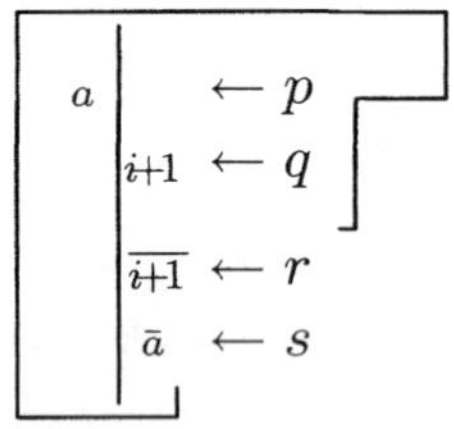 or $T' =$ 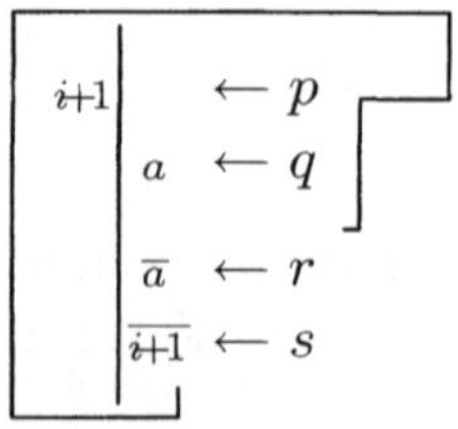.

Then T must have the form

$T =$ 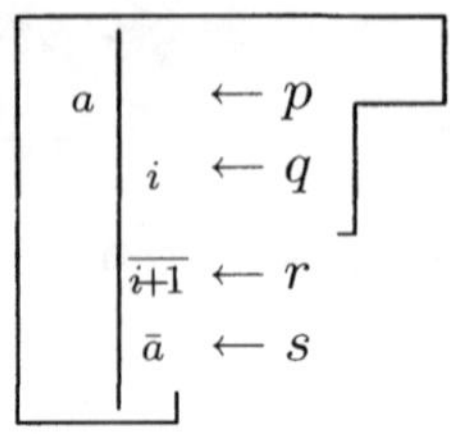 or $T =$ 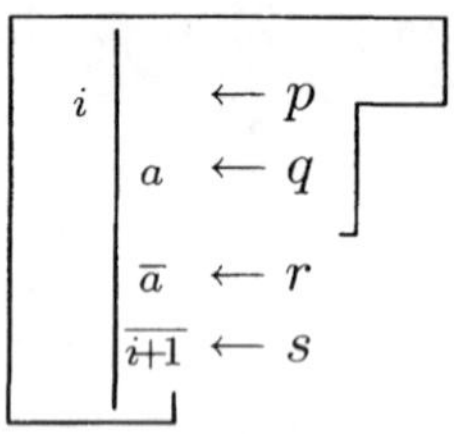.

By Lemma 8.3.4 (2), we have

$$(q-p)+(s-r) < (i+1)-a.$$

Hence, T' satisfies the condition **(C2)**.

Similarly, when $\tilde{f}_i$ changes the box $b = \boxed{i+1}$ into $\boxed{\bar{i}}$, one can show that $\tilde{f}_i T$ satisfies the conditions **(C1)** and **(C2)** (Exercise 8.4).

Now, it remains to prove the assertion (3). Let $T \in \mathcal{B}(Y)$ and assume that $\tilde{e}_i T = 0$ for all $i \in I$. Let k be the largest nonnegative integer such that the first k rows of T are the same as those of T_Y.

Suppose that $k < n$ and let $b = \boxed{j}$ be the rightmost box in the $(k+1)$th row. We denote by T_0 the subtableau of T consisting of the boxes lying in the northeast of $b = \boxed{j}$, T_1 the column consisting of the boxes directly below $b = \boxed{j}$, and T_2 the subtableau consisting of the columns left to $b = \boxed{j}$. Hence T can be viewed as

$T =$ 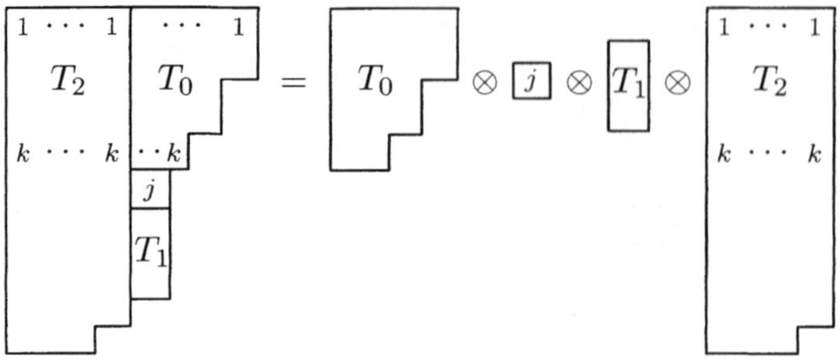 $= T_0 \otimes \boxed{j} \otimes T_1 \otimes T_2$.

Since $\tilde{e}_i T = 0$ for all $i \in I$, by the tensor product rule, we must have

$$\tilde{e}_i(T_0 \otimes \boxed{j}) = 0 \quad \text{for all } i \in I,$$

which implies $\tilde{e}_i T_0 = 0$ and $\varphi_i(T_0) \geq \varepsilon_i(\boxed{j})$ for all $i \in I$. In particular, $\varepsilon_i(\boxed{j}) = 0$ for all $i \geq k+1$. Hence we conclude that $j \preccurlyeq k+1$ or $j \succeq \bar{k}$.

On the other hand, by the maximality of k, we must have $j \succ k+1$. It follows that $j \succeq \bar{k}$. If $j \succ \bar{k}$, then $\tilde{e}_j(T_0 \otimes \boxed{j}) \neq 0$. If $j = \bar{k}$, then by the condition **(C1)**, we must have $(k+1) - k + k = k+1 > N$, which is impossible. (Here, N is the length of the column containing b.) Therefore, we have $k = n$ and $T = T_Y$, which completes the proof. $\square$

Example 8.3.5.

(1) If $\lambda = \omega_1$, then we have $\mathcal{B}(\lambda) = \mathbf{B}$, the crystal graph for the vector representation.

(2) If $n = 3$ and $\lambda = \omega_3$, then the crystal graph $\mathcal{B}(\lambda)$ can be realized as follows. Note that the tableau $\begin{array}{|c|}\hline 1\\\hline 2\\\hline \bar{1}\\\hline\end{array}$ does not appear in the crystal graph as we have seen in Example 8.3.2.

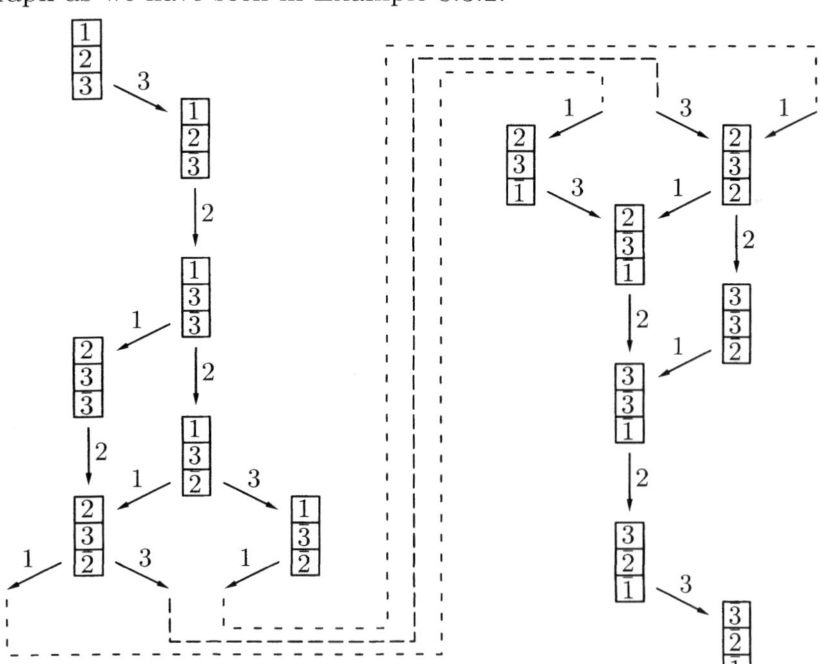

(3) If $n = 3$ and $\lambda = 2\omega_3$, we will illustrate the top and bottom parts of the crystal graph $\mathcal{B}(\lambda)$ in the following figure. It is an interesting exercise to verify that the tableau $\begin{array}{|c|c|}\hline 1 & 3\\\hline 2 & \bar{3}\\\hline 3 & \bar{1}\\\hline\end{array}$ appears in $\mathcal{B}(\lambda)$, whereas the tableau $\begin{array}{|c|c|}\hline 2 & 3\\\hline 3 & \bar{3}\\\hline \bar{3} & \bar{2}\\\hline\end{array}$ does not (see Example 8.3.2).

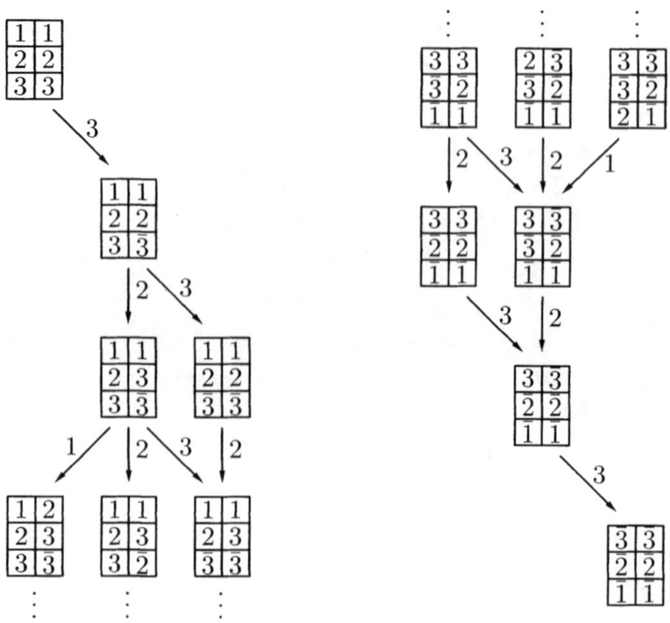

8.4. Realization of $U_q(B_n)$-crystals

Let $\mathfrak{g}$ be the finite dimensional simple Lie algebra of type B_n $(n \geq 3)$. Then $\mathfrak{g}$ is realized as the (*odd*) *special orthogonal Lie algebra* (see, for example, [**15**, Part III], [**17**, Ch.I], or [**19**, Ch.IV])

$$
\begin{aligned}
\mathfrak{g} &= \mathfrak{so}(2n+1, \mathbf{C}) = \mathfrak{so}_{2n+1} \\
&= \left\{ T = \begin{pmatrix} A & B & a \\ C & D & b \\ c & d & 0 \end{pmatrix} \in M_{(2n+1)\times(2n+1)}(\mathbf{C}) \;\middle|\; \begin{array}{l} A, B, C, D \in M_{n\times n}(\mathbf{C}), \quad a, b \in M_{n\times 1}(\mathbf{C}), \quad c, d \in M_{1\times n}(\mathbf{C}), \\ A^t = -D,\ B^t = -B,\ C^t = -C, \quad a^t = -d,\ b^t = -c \end{array} \right\}.
\end{aligned} \tag{8.21}
$$

Let E_{ij} denote the $(2n+1) \times (2n+1)$ elementary matrix having 1 at the (i, j)-entry and 0 elsewhere, and set

$$
\begin{aligned}
e_i &= E_{i,i+1} - E_{n+i+1,n+i}, \qquad f_i = E_{i+1,i} - E_{n+i,n+i+1}, \\
h_i &= E_{ii} - E_{i+1,i+1} - E_{n+i,n+i} + E_{n+i+1,n+i+1}, \\
e_n &= 2(E_{n,2n+1} - E_{2n+1,2n}), \qquad f_n = E_{2n+1,n} - E_{2n,2n+1}, \\
h_n &= 2(E_{n,n} - E_{2n,2n})
\end{aligned} \tag{8.22}
$$

for $i = 1, 2, \ldots, n-1$. Then, as a Lie algebra, $\mathfrak{g} = \mathfrak{so}(2n+1, \mathbf{C})$ is generated by e_i, f_i, h_i $(i = 1, 2, \ldots, n)$.

As usual, let $\epsilon_i : M_{(2n+1)\times(2n+1)}(\mathbf{C}) \longrightarrow \mathbf{C}$ denote the linear functional defined by

$$\epsilon_i(T) = t_{ii}, \qquad T = (t_{ij}) \in M_{2n+1}(\mathbf{C}), \quad i, j = 1, 2, \ldots, 2n+1.$$

Then, the *simple roots* and the *fundamental weights* are expressed as

$$\begin{aligned} \alpha_i &= \epsilon_i - \epsilon_{i+1} \quad \text{for } i = 1, 2, \ldots, n-1, \\ \alpha_n &= \epsilon_n, \\ \omega_i &= \epsilon_1 + \epsilon_2 + \cdots + \epsilon_i \quad \text{for } i = 1, 2, \ldots, n-1, \\ \omega_n &= \frac{1}{2}(\epsilon_1 + \cdots + \epsilon_n). \end{aligned} \tag{8.23}$$

Using this notation, the *Cartan datum* for the special orthogonal Lie algebra $\mathfrak{g} = \mathfrak{so}(2n+1, \mathbf{C})$ is given as follows:

$$\begin{aligned} A &= \begin{pmatrix} 2 & -1 & \cdots & & 0 \\ -1 & 2 & -1 & & 0 \\ 0 & & \ddots & & 0 \\ 0 & & \cdots & 2 & -1 \\ 0 & & \cdots & -2 & 2 \end{pmatrix}, \\ \Pi &= \{\alpha_1, \alpha_2, \ldots, \alpha_n\}, \qquad \Pi^\vee = \{h_1, h_2, \ldots, h_n\}, \\ P &= \mathbf{Z}\omega_1 \oplus \cdots \oplus \mathbf{Z}\omega_n, \qquad P^\vee = \mathbf{Z}h_1 \oplus \cdots \oplus \mathbf{Z}h_n. \end{aligned} \tag{8.24}$$

The ***quantum special orthogonal algebra*** $U_q(\mathfrak{so}_{2n+1})$ is defined to be the quantum group associated with the Cartan datum given above.

Let $\mathbf{V} = \left(\bigoplus_{j=1}^n \mathbf{C}(q)v_j\right) \oplus \mathbf{C}(q)v_0 \oplus \left(\bigoplus_{j=1}^n \mathbf{C}(q)v_{\bar{j}}\right)$ be a $(2n+1)$-dimensional vector space and let $\mathbf{N} = \{1, 2, \ldots, n, 0, \bar{n}, \ldots, \bar{2}, \bar{1}\}$ be the index set for the basis vectors of $\mathbf{V}$ with a linear ordering given by

$$1 \prec 2 \prec \cdots \prec n \prec 0 \prec \bar{n} \prec \cdots \prec \bar{2} \prec \bar{1}.$$

We define the $U_q(\mathfrak{so}_{2n+1})$-module action on $\mathbf{V}$ as follows:

$$(8.25)\qquad \begin{aligned} q^h v_j &= q^{\langle h, \mathrm{wt}(v_j)\rangle} v_j \quad \text{for } h \in P^\vee,\ j \in \mathbf{N}, \\ e_i v_j &= \begin{cases} v_i & \text{if } j = i+1,\ i \neq n, \\ v_{\overline{i+1}} & \text{if } j = \bar{i},\ i \neq n, \\ v_0 & \text{if } j = \bar{n},\ i = n, \\ [2]_q v_n & \text{if } j = 0,\ i = n, \\ 0 & \text{otherwise}, \end{cases} \\ f_i v_j &= \begin{cases} v_{i+1} & \text{if } j = i,\ i \neq n, \\ v_{\bar{i}} & \text{if } j = \overline{i+1},\ i \neq n, \\ v_0 & \text{if } j = n,\ i = n, \\ [2]_q v_{\bar{n}} & \text{if } j = 0,\ i = n, \\ 0 & \text{otherwise}, \end{cases} \end{aligned}$$

where

$$\mathrm{wt}(v_j) = \epsilon_j, \qquad \mathrm{wt}(v_0) = 0, \qquad \mathrm{wt}(v_{\bar{j}}) = -\epsilon_j \quad \text{for } j = 1, 2, \ldots, n.$$

Then $\mathbf{V}$ becomes a $U_q(\mathfrak{so}_{2n+1})$-module called the ***vector representation***, and it has a crystal basis $(\mathbf{L}, \mathbf{B})$, where

$$\mathbf{L} = \bigoplus_{j \in \mathbf{N}} \mathbf{A}_0 v_j, \qquad \mathbf{B} = \{\boxed{j} = v_j + q\mathbf{L} \mid j \in \mathbf{N}\}.$$

The $U_q(\mathfrak{so}_{2n+1})$-crystal structure on $\mathbf{B}$ is given below.

$$\boxed{1} \xrightarrow{1} \boxed{2} \xrightarrow{2} \cdots \xrightarrow{n-1} \boxed{n} \xrightarrow{n} \boxed{0} \xrightarrow{n} \boxed{\bar{n}} \xrightarrow{n-1} \cdots \xrightarrow{2} \boxed{\bar{2}} \xrightarrow{1} \boxed{\bar{1}}$$

Note that $\mathbf{B} = \mathcal{B}(\omega_1) = \mathcal{B}(\epsilon_1)$ and

$$\mathrm{wt}(\boxed{j}) = \epsilon_j, \qquad \mathrm{wt}(\boxed{0}) = 0, \qquad \mathrm{wt}(\boxed{\bar{j}}) = -\epsilon_j \quad \text{for } j = 1, 2, \ldots, n.$$

As we have seen in Section 8.1, we need the *spin representation* and its crystal graph to get a realization of all $\mathcal{B}(\lambda)$ with $\lambda \in P^+$.

Let $\mathbf{V}_{\mathrm{sp}} = \bigoplus_{s_i=\pm}(s_1,\dots,s_n)$ be the $2n$-dimensional vector space and define the $U_q(\mathfrak{so}_{2n+1})$-action on $\mathbf{V}_{\mathrm{sp}}$ by

(8.26)

$$
\begin{aligned}
q^h(s_1,\dots,s_n) &= q^{\langle h,\mathrm{wt}(s_1,\dots,s_n)\rangle}(s_1,\dots,s_n),\\
e_i(s_1,\dots,s_n) &= \begin{cases} (s_1,\dots,+,-,\dots,s_n) & \text{if } i\neq n,\ (s_i,s_{i+1})=(-,+),\\ (s_1,\dots,s_{n-1},+) & \text{if } i=n,\ s_n=-,\\ 0 & \text{otherwise,}\end{cases}\\
f_i(s_1,\dots,s_n) &= \begin{cases} (s_1,\dots,-,+,\dots,s_n) & \text{if } i\neq n,\ (s_i,s_{i+1})=(+,-),\\ (s_1,\dots,s_{n-1},-) & \text{if } i=n,\ s_n=+,\\ 0 & \text{otherwise,}\end{cases}
\end{aligned}
$$

where the weight of each element is given by

$$\mathrm{wt}(s_1,\dots,s_n) = \frac{1}{2}(s_1\epsilon_1+\cdots+s_n\epsilon_n).$$

We call $\mathbf{V}_{\mathrm{sp}}$ the ***spin representation*** of $U_q(\mathfrak{so}_{2n+1})$ and it is isomorphic to $V(\omega_n)=V(\frac{1}{2}(\epsilon_1+\cdots+\epsilon_n))$. The crystal graph $\mathbf{B}_{\mathrm{sp}}$ of $\mathbf{V}_{\mathrm{sp}}$ can be expressed as

$$\mathbf{B}_{\mathrm{sp}} = \{(s_1,s_2,\dots,s_n) \mid s_i=\pm\},$$

where action of the Kashiwara operators is given below.

$$(s_1,\dots,\overset{i}{+},\overset{i+1}{-},\dots,s_n) \xrightarrow{\ i\ } (s_1,\dots,\overset{i}{-},\overset{i+1}{+},\dots,s_n) \qquad (1\le i<n)$$

$$(s_1,\dots,s_{n-1},+) \xrightarrow{\ n\ } (s_1,\dots,s_{n-1},-)\ .$$

Note that $\mathbf{B}_{\mathrm{sp}} = \mathcal{B}(\omega_n) = \mathcal{B}(\frac{1}{2}(\epsilon_1+\cdots+\epsilon_n))$.

Now, take the (generalized) Young diagram $Y_{\mathrm{sp}} = \begin{array}{|c|}\hline \ \\ \hline \ \\ \hline \vdots \\ \hline \ \\ \hline\end{array}$ consisting of n *half-boxes* (in the sense that they are of half-unit width) and consider the (generalized) Young tableaux of shape Y_{sp} whose entries are taken from $\mathbf{N}_{\mathrm{sp}} = \{1,2,\dots,n,\bar{n},\dots,\bar{2},\bar{1}\}$ with a linear ordering

$$1\prec 2\prec\cdots\prec n\prec\bar{n}\prec\cdots\prec\bar{2}\prec\bar{1}.$$

We identify the element $(s_1,\dots,s_n)$ of $\mathbf{B}_{\mathrm{sp}}$ with a tableau of shape Y_{sp} as follows: if the ith component is $+$ (respectively, $-$), we fill the ith box of Y_{sp} with i (respectively, $\bar{i}$), and then reshuffle them so that they are linearly ordered. Then the crystal graph $\mathbf{B}_{\text{sp}} = \mathcal{B}(\omega_n)$ can be characterized as

$$\mathbf{B}_{\text{sp}} = \mathcal{B}(\omega_n) = \left\{ \begin{array}{|c|} \hline t_1 \\ \hline \vdots \\ \hline t_n \\ \hline \end{array} \;\middle|\; \begin{array}{l} t_i \in \mathbf{N}_{\text{sp}} \text{ with } t_1 \prec \cdots \prec t_n, \\ i \text{ and } \bar{i} \text{ do not appear simultaneously} \end{array} \right\},$$

where action of the Kashiwara operators is

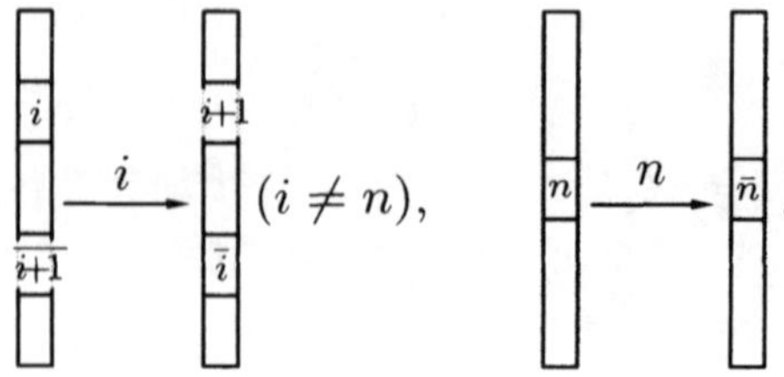

Let $\lambda = a_1\omega_1 + \cdots + a_n\omega_n$ ($a_i \in \mathbf{Z}_{\geq 0}$) be a dominant integral weight. Using the relation (8.23), λ can be written as $\lambda = \lambda_1\epsilon_1 + \cdots + \lambda_n\epsilon_n$, where

$$\begin{aligned} \lambda_1 &= a_1 + a_2 + \cdots + a_{n-1} + \frac{1}{2}a_n, \\ \lambda_2 &= a_2 + \cdots + a_{n-1} + \frac{1}{2}a_n, \\ &\vdots \\ \lambda_{n-1} &= a_{n-1} + \frac{1}{2}a_n, \\ \lambda_n &= \frac{1}{2}a_n. \end{aligned} \tag{8.27}$$

Hence we can associate a (generalized) Young diagram Y to λ as follows.

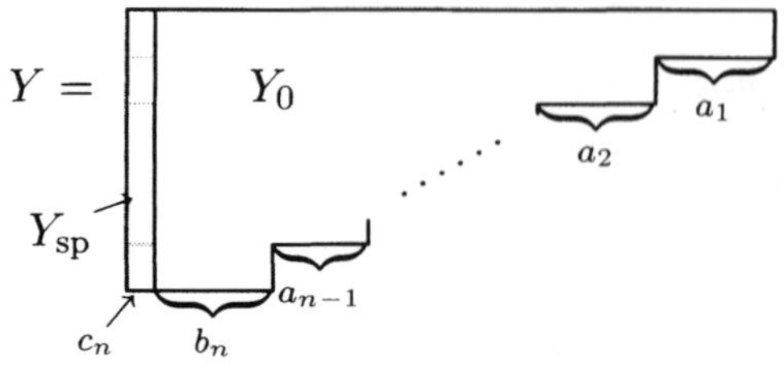

Here $a_n = 2b_n + c_n$, $c_n = 0$ or 1, and the column Y_{sp} consists of half-boxes. We will identify the (generalized) Young diagram with the sequence of half-integers $Y = (\lambda_1, \ldots, \lambda_n)$.

Definition 8.4.1.

(1) Let Y be a (generalized) Young diagram with at most n rows. A ***B_n-tableau of shape*** Y is a tableau obtained from Y by filling the boxes in Y_0 with the entries from $\mathbf{N}$ and the boxes in Y_{sp} with the entries from $\mathbf{N}_{\text{sp}}$.

(2) A B_n-tableau is said to be ***semistandard*** if

(a) the entries in each row are weakly increasing, but zeros cannot be repeated;
(b) the entries in each column in Y_0 are strictly increasing, but zeros can be repeated;
(c) the entries in the column $Y_{\rm sp}$ are strictly increasing, and i and $\bar{i}$ do not appear simultaneously.

For a B_n-tableau T, we write

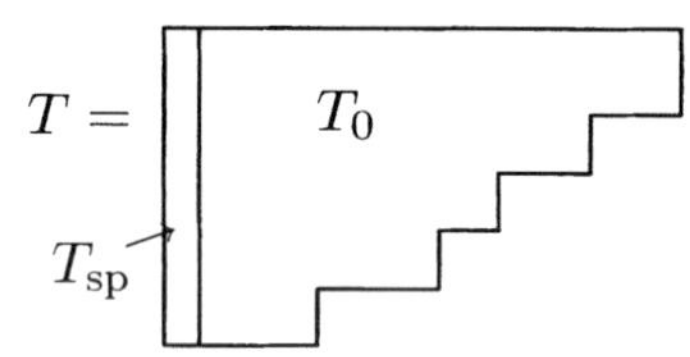

and define its *weight* to be

$$\mathrm{wt}(T) = \sum_{i=1}^{n}(k_i - \overline{k_i})\epsilon_i + \frac{1}{2}\sum_{i=1}^{n}(l_i - \overline{l_i})\epsilon_i,$$

where k_i (respectively, $\overline{k_i}$) is the number of i's (respectively, $\bar{i}$'s) appearing in T_0, and l_i (respectively, $\overline{l_i}$) is the number of i's (respectively, $\bar{i}$'s) appearing in $T_{\rm sp}$. Note that $l_i, \overline{l_i} = 0$ or 1 and they cannot be 1 at the same time.

As in Section 8.3, the set of all semistandard B_n-tableaux does not form a $U_q(\mathfrak{so}_{2n+1})$-crystal. Thus, to characterize the crystal graph $\mathcal{B}(\lambda)$, we need to introduce additional conditions.

We define $\mathcal{B}(Y)$ to be the set of all semistandard B_n-tableaux T satisfying the following conditions.

(B1) If T has a column of the form

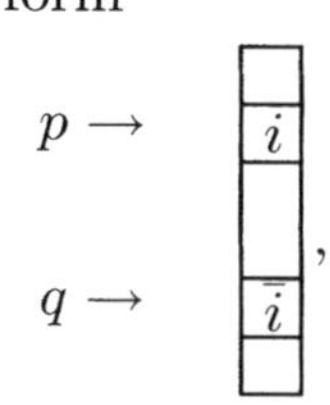

then we have

$$(q - p) + i > N,$$

where N is the length of the column.

(B2) If T has a pair of adjacent columns having one of the following configurations with $p \leq q < r \leq s$ and $a \leq b < n$:

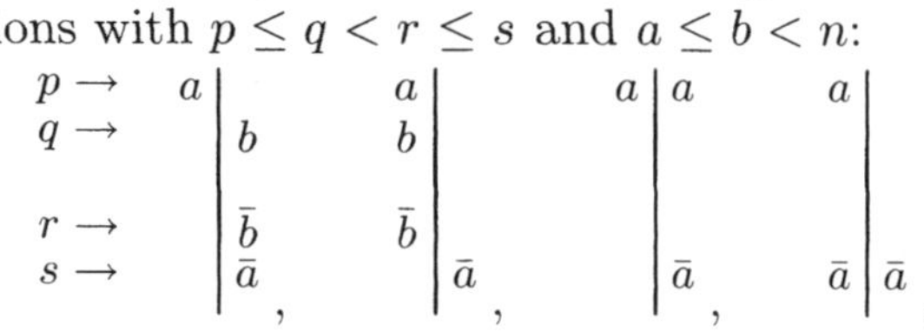

then we have

$$(q-p)+(s-r)<b-a.$$

Thus the tableau with one of the last two configurations cannot be an element of $\mathcal{B}(Y)$.

(B3) If T has a pair of adjacent columns having one of the following configurations with $p \le q < r = q+1 \le s$ and $a < n$:

$$\begin{array}{l} p\rightarrow \\ \\ q\rightarrow \\ r\rightarrow \\ \\ s\rightarrow \end{array}\quad \begin{array}{c|c} a & \\ & \\ & n \\ & \bar{n} \\ & \\ & \bar{a} \end{array}\,,\quad \begin{array}{c|c} a & \\ & \\ & n \\ & 0 \\ & \\ & \bar{a} \end{array}\,,\quad \begin{array}{c|c} a & \\ & \\ & 0 \\ & 0 \\ & \\ & \bar{a} \end{array}\,,\quad \begin{array}{c|c} a & \\ & \\ & 0 \\ & \bar{n} \\ & \\ & \bar{a} \end{array}\,,$$

$$\begin{array}{l} p\rightarrow \\ \\ q\rightarrow \\ r\rightarrow \\ \\ s\rightarrow \end{array}\quad \begin{array}{c|c} a & \\ & \\ n & \\ \bar{n} & \\ & \\ & \bar{a} \end{array}\,,\quad \begin{array}{c|c} a & \\ & \\ n & \\ 0 & \\ & \\ & \bar{a} \end{array}\,,\quad \begin{array}{c|c} a & \\ & \\ 0 & \\ 0 & \\ & \\ & \bar{a} \end{array}\,,\quad \begin{array}{c|c} a & \\ & \\ 0 & \\ \bar{n} & \\ & \\ & \bar{a} \end{array}\,,$$

then we have

$$(q-p)+(s-r)=s-p-1<n-a.$$

(B4) The tableau T cannot have a pair of adjacent columns having one of the following configurations with $p < s$:

$$\begin{array}{l} p\rightarrow \\ \\ \\ s\rightarrow \end{array}\quad \begin{array}{c|c} n & \\ & \\ & \\ & \bar{n} \end{array}\,,\quad \begin{array}{c|c} n & \\ & \\ & \\ & 0 \end{array}\,,\quad \begin{array}{c|c} 0 & \\ & \\ & \\ & 0 \end{array}\,,\quad \begin{array}{c|c} 0 & \\ & \\ & \\ & \bar{n} \end{array}\,.$$

Example 8.4.2. When $n=3$ and $\lambda = \omega_1 + \omega_2 + 3\omega_3$, we have

$$Y = \begin{array}{|c|c|c|c|} \hline \; & & & \\ \hline \; & & \\ \cline{1-3} \; & \\ \cline{1-2} \end{array}\,,$$

and the tableau $T = \begin{array}{|c|c|c|c|} \hline 1 & 2 & 3 & 0 \\ \hline 2 & \bar{3} & \bar{3} \\ \cline{1-3} 3 & \bar{2} \\ \cline{1-2} \end{array}$ belongs to $\mathcal{B}(\lambda)$. Moreover we have

$$\tilde{e}_3 T = \begin{array}{|c|c|c|c|} \hline 1 & 2 & 3 & 0 \\ \hline 2 & 0 & \bar{3} \\ \cline{1-3} 3 & \bar{2} \\ \cline{1-2} \end{array} \quad \in \mathcal{B}(\lambda),$$

$$\tilde{f}_3 T = \begin{array}{|c|c|c|c|} \hline 1 & 2 & 3 & 0 \\ \hline 2 & \bar{3} & \bar{3} \\ \cline{1-3} \bar{3} & \bar{2} \\ \cline{1-2} \end{array} \quad \in \mathcal{B}(\lambda).$$

Theorem 8.4.3. *Let $\lambda = a_1\omega_1 + \cdots + a_n\omega_n \in P^+$ be a dominant integral weight and let $V(\lambda)$ be the finite dimensional irreducible module over $U_q(\mathfrak{so}_{2n+1})$ with highest weight λ. Let Y be the (generalized) Young diagram* *associated with* λ *and let* $\mathcal{B}(Y)$ *denote the set of all semistandard* B_n*-tableaux of shape* Y *satisfying the conditions* **(B1)–(B4)**. *Then, by the Far-Eastern reading,* $\mathcal{B}(Y)$ *becomes a connected* $U_q(\mathfrak{so}_{2n+1})$*-crystal containing a maximal vector of weight* λ. *Hence* $\mathcal{B}(Y)$ *is isomorphic to the crystal graph* $\mathcal{B}(\lambda)$ *of* $V(\lambda)$.

Proof. We leave it to the readers as an exercise (Exercise 8.8). □

8.5. Realization of $U_q(D_n)$-crystals

Let $\mathfrak{g}$ be the finite dimensional simple Lie algebra of type D_n $(n \geq 4)$. Then $\mathfrak{g}$ is realized as the *(even) special orthogonal Lie algebra* (see, for example, [**15**, Part III], [**17**, Ch.I], or [**19**, Ch.IV])

$$\begin{aligned} \mathfrak{g} &= \mathfrak{so}(2n, \mathbf{C}) = \mathfrak{so}_{2n} \\ &= \left\{ T = \begin{pmatrix} A & B \\ C & D \end{pmatrix} \in M_{2n\times 2n}(\mathbf{C}) \,\middle|\, A, B, C, D \in M_{n\times n}(\mathbf{C}),\ \ A^t = -D,\ B^t = -B,\ C^t = -C \right\}. \end{aligned} \tag{8.28}$$

Let E_{ij} denote the $2n \times 2n$ elementary matrix having 1 at the (i, j)-entry and 0 elsewhere, and set

$$\begin{aligned} e_i &= E_{i,i+1} - E_{n+i+1,n+i}, \qquad f_i = E_{i+1,i} - E_{n+i,n+i+1}, \\ h_i &= E_{ii} - E_{i+1,i+1} - E_{n+i,n+i} + E_{n+i+1,n+i+1}, \\ e_n &= E_{n-1,2n} - E_{n,2n-1}, \qquad f_n = E_{2n,n-1} - E_{2n-1,n}, \\ h_n &= E_{n-1,n-1} + E_{n,n} - E_{2n-1,2n-1} - E_{2n,2n} \end{aligned} \tag{8.29}$$

for $i = 1, 2, \ldots, n-1$, Then, as a Lie algebra, $\mathfrak{g} = \mathfrak{so}(2n, \mathbf{C})$ is generated by e_i, f_i, h_i $(i = 1, 2, \ldots, n)$.

Let $\epsilon_i : M_{2n\times 2n}(\mathbf{C}) \longrightarrow \mathbf{C}$ denote the linear functional defined by

$$\epsilon_i(T) = t_{ii}, \qquad T = (t_{ij}) \in M_{2n\times 2n}(\mathbf{C}), \quad i, j = 1, 2, \ldots, 2n.$$

Then, the *simple roots* and the *fundamental weights* are expressed as

$$\begin{aligned} \alpha_i &= \epsilon_i - \epsilon_{i+1} \quad \text{for } i = 1, 2, \ldots, n-1, \\ \alpha_n &= \epsilon_{n-1} + \epsilon_n, \\ \omega_i &= \epsilon_1 + \epsilon_2 + \cdots + \epsilon_i \quad \text{for } i = 1, 2, \ldots, n-2, \\ \omega_{n-1} &= \frac{1}{2}(\epsilon_1 + \cdots + \epsilon_{n-1} - \epsilon_n), \\ \omega_n &= \frac{1}{2}(\epsilon_1 + \cdots + \epsilon_{n-1} + \epsilon_n). \end{aligned} \tag{8.30}$$

Using this notation, the *Cartan datum* for the special orthogonal Lie algebra $\mathfrak{g} = \mathfrak{so}(2n, \mathbf{C})$ is given as follows:

$$(8.31)\qquad \begin{aligned} A &= \begin{pmatrix} 2 & -1 & \cdots & & 0 & 0 \\ -1 & 2 & -1 & & 0 & 0 \\ & & & & & \\ 0 & & \ddots & & 0 & 0 \\ 0 & & \cdots & 2 & -1 & -1 \\ 0 & & \cdots & -1 & 2 & 0 \\ 0 & & \cdots & -1 & 0 & 2 \end{pmatrix}, \\ \Pi &= \{\alpha_1, \alpha_2, \ldots, \alpha_n\}, \qquad \Pi^\vee = \{h_1, h_2, \ldots, h_n\}, \\ P &= \mathbf{Z}\omega_1 \oplus \cdots \oplus \mathbf{Z}\omega_n, \qquad P^\vee = \mathbf{Z}h_1 \oplus \cdots \oplus \mathbf{Z}h_n. \end{aligned}$$

The ***quantum special orthogonal algebra*** $U_q(\mathfrak{so}_{2n})$ is defined to be the quantum group associated with the Cartan datum given above.

Let $\mathbf{V} = \left(\bigoplus_{j=1}^n \mathbf{C}(q)v_j\right) \oplus \left(\bigoplus_{j=1}^n \mathbf{C}(q)v_{\bar{j}}\right)$ be a $2n$-dimensional vector space and let $\mathbf{N} = \{1, 2, \ldots, n, \bar{n}, \ldots, \bar{2}, \bar{1}\}$ be the index set for the basis vectors of $\mathbf{V}$ with a linear ordering given by

$$1 \prec 2 \prec \cdots \prec \begin{matrix} n \\ \bar{n} \end{matrix} \prec \cdots \prec \bar{2} \prec \bar{1}.$$

Notice that the order between n and $\bar{n}$ is not defined.

We define the $U_q(\mathfrak{so}_{2n})$-module action on $\mathbf{V}$ as follows:

$$(8.32)\qquad \begin{aligned} q^h v_j &= q^{\langle h, \mathrm{wt}(v_j)\rangle} v_j \quad \text{for } h \in P^\vee,\ j \in \mathbf{N}, \\ e_i v_j &= \begin{cases} v_i & \text{if } j = i+1,\ i \neq n, \\ v_{\overline{i+1}} & \text{if } j = \bar{i},\ i \neq n, \\ v_n & \text{if } j = \overline{n-1},\ i = n, \\ v_{n-1} & \text{if } j = \bar{n},\ i = n, \\ 0 & \text{otherwise}, \end{cases} \\ f_i v_j &= \begin{cases} v_{i+1} & \text{if } j = i,\ i \neq n, \\ v_{\bar{i}} & \text{if } j = \overline{i+1},\ i \neq n, \\ v_{\bar{n}} & \text{if } j = n-1,\ i = n, \\ v_{\overline{n-1}} & \text{if } j = n,\ i = n, \\ 0 & \text{otherwise}, \end{cases} \end{aligned}$$

where

$$\mathrm{wt}(v_j) = \epsilon_j, \qquad \mathrm{wt}(v_{\bar{j}}) = -\epsilon_j \quad \text{for } j = 1, 2, \ldots, n.$$

Then $\mathbf{V}$ becomes a $U_q(\mathfrak{so}_{2n})$-module called the ***vector representation***, and it has a crystal basis $(\mathbf{L}, \mathbf{B})$, where

$$\mathbf{L} = \bigoplus_{j \in \mathbf{N}} \mathbf{A}_0 v_j, \qquad \mathbf{B} = \{\boxed{j} = v_j + q\mathbf{L} \mid j \in \mathbf{N}\}.$$

The $U_q(\mathfrak{so}_{2n})$-crystal structure on $\mathbf{B}$ is given below.

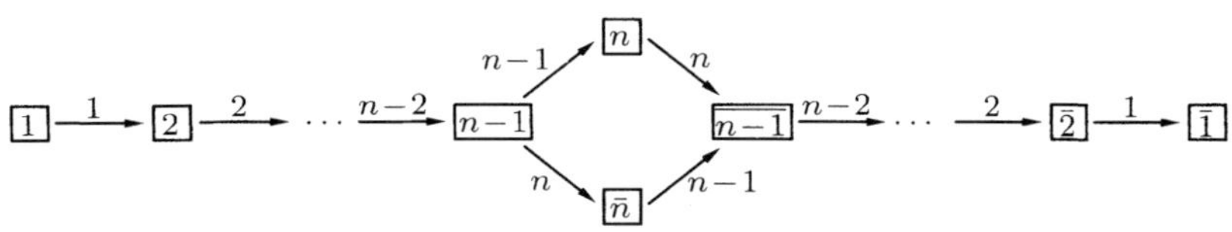

Note that $\mathbf{B} = \mathcal{B}(\omega_1) = \mathcal{B}(\epsilon_1)$ and

$$\mathrm{wt}(\boxed{j}) = \epsilon_j, \qquad \mathrm{wt}(\boxed{\bar{j}}) = -\epsilon_j \quad \text{for } j = 1, 2, \ldots, n.$$

As in the case of $U_q(\mathfrak{so}_{n+1})$-crystals, we need the *spin representations* and their crystal graphs to get a realization of all $\mathcal{B}(\lambda)$ with $\lambda \in P^+$. The ***spin representations*** $\mathbf{V}_{\mathrm{sp}}^{\pm}$ are $2n$-dimensional vector spaces

$$\mathbf{V}_{\mathrm{sp}}^{+} = \bigoplus_{\substack{s_i = \pm \\ s_1 \cdots s_n = +}} (s_1, \ldots, s_n),$$

$$\mathbf{V}_{\mathrm{sp}}^{-} = \bigoplus_{\substack{s_i = \pm \\ s_1 \cdots s_n = -}} (s_1, \ldots, s_n)$$

with the $U_q(\mathfrak{so}_{2n})$-action given by
(8.33)

$$q^h(s_1, \ldots, s_n) = q^{\langle h, \mathrm{wt}(s_1, \ldots, s_n) \rangle}(s_1, \ldots, s_n),$$

$$e_i(s_1, \ldots, s_n) = \begin{cases} (s_1, \ldots, +, -, \ldots, s_n) & \text{if } i \neq n,\ (s_i, s_{i+1}) = (-, +), \\ (s_1, \ldots, s_{n-2}, +, +) & \text{if } i = n,\ (s_{n-1}, s_n) = (-, -), \\ 0 & \text{otherwise}, \end{cases}$$

$$f_i(s_1, \ldots, s_n) = \begin{cases} (s_1, \ldots, -, +, \ldots, s_n) & \text{if } i \neq n,\ (s_i, s_{i+1}) = (+, -), \\ (s_1, \ldots, s_{n-2}, -, -) & \text{if } i = n,\ (s_{n-1}, s_n) = (+, +), \\ 0 & \text{otherwise}, \end{cases}$$

where the weight of each element is given by

$$\mathrm{wt}(s_1, \ldots, s_n) = \frac{1}{2}(s_1\epsilon_1 + \cdots + s_n\epsilon_n).$$

The spin representations $\mathbf{V}_{\mathrm{sp}}^{+}$ and $\mathbf{V}_{\mathrm{sp}}^{-}$ are isomorphic to the finite dimensional irreducible modules $V(\omega_n)$ and $V(\omega_{n-1})$, respectively. The crystal graphs $\mathbf{B}_{\mathrm{sp}}^{\pm}$ can be expressed as

$$\mathbf{B}_{\mathrm{sp}}^{+} = \{(s_1, s_2, \dots, s_n) \mid s_j = \pm, \quad s_1 \cdots s_n = +\},$$
$$\mathbf{B}_{\mathrm{sp}}^{-} = \{(s_1, s_2, \dots, s_n) \mid s_j = \pm, \quad s_1 \cdots s_n = -\},$$

where action of the Kashiwara operators is given below.

$$(s_1, \dots, \overset{i}{+}, \overset{i+1}{-}, \dots, s_n) \xrightarrow{\ i\ } (s_1, \dots, \overset{i}{-}, \overset{i+1}{+}, \dots, s_n) \qquad (1 \le i < n)$$
$$(s_1, \dots, s_{n-2}, +, +) \xrightarrow{\ n\ } (s_1, \dots, s_{n-2}, -, -)\ .$$

The crystal graphs $\mathbf{B}_{\mathrm{sp}}^{\pm}$ can be realized using the (generalized) Young tableaux consisting of half-boxes:

$$\mathbf{B}_{\mathrm{sp}}^{+} = \mathcal{B}(\omega_n) = \left\{ \begin{array}{|c|}\hline t_1 \\ \vdots \\ t_n \\ \hline\end{array} \;\middle|\; \begin{array}{l} t_i \in \mathbf{N}_{\mathrm{sp}} \text{ with } t_1 \prec \cdots \prec t_n, \\ i \text{ and } \bar{i} \text{ do not appear simultaneously}, \\ \text{if } t_k = n, \text{ then } n-k \text{ is even}, \\ \text{if } t_k = \bar{n}, \text{ then } n-k \text{ is odd}. \end{array} \right\},$$

$$\mathbf{B}_{\mathrm{sp}}^{-} = \mathcal{B}(\omega_{n-1}) = \left\{ \begin{array}{|c|}\hline t_1 \\ \vdots \\ t_n \\ \hline\end{array} \;\middle|\; \begin{array}{l} t_i \in \mathbf{N}_{\mathrm{sp}} \text{ with } t_1 \prec \cdots \prec t_n, \\ i \text{ and } \bar{i} \text{ do not appear simultaneously}, \\ \text{if } t_k = n, \text{ then } n-k \text{ is odd}, \\ \text{if } t_k = \bar{n}, \text{ then } n-k \text{ is even}. \end{array} \right\}.$$

Here the action of Kashiwara operators is given below.

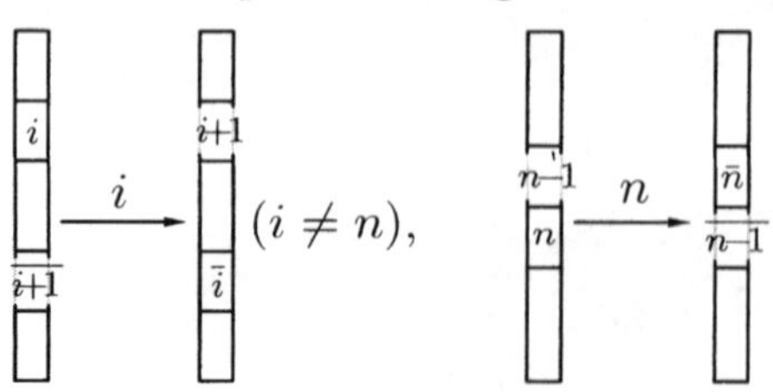

Let $\lambda = a_1\omega_1 + \cdots + a_n\omega_n$ $(a_i \in \mathbf{Z}_{\ge 0})$ be a dominant integral weight. Using the relation (8.30), λ can be written as $\lambda = \lambda_1\epsilon_1 + \cdots + \lambda_n\epsilon_n$, where

$$\begin{aligned}
\lambda_1 &= a_1 + a_2 + \cdots + a_{n-2} + \frac{1}{2}(a_{n-1} + a_n), \\
\lambda_2 &= a_2 + \cdots + a_{n-1} + \frac{1}{2}(a_{n-1} + a_n), \\
&\ \vdots \\
\lambda_{n-1} &= \frac{1}{2}(a_{n-1} + a_n), \\
\lambda_n &= \frac{1}{2}(a_n - a_{n-1}).
\end{aligned} \tag{8.34}$$

Hence we can associate a (generalized) Young diagram Y to λ as follows.

If $a_n \ge a_{n-1}$, write $a_n - a_{n-1} = 2b_n + c_n$ with $b_n \in \mathbf{Z}_{\ge 0}$, $c_n = 0$ or 1, and set

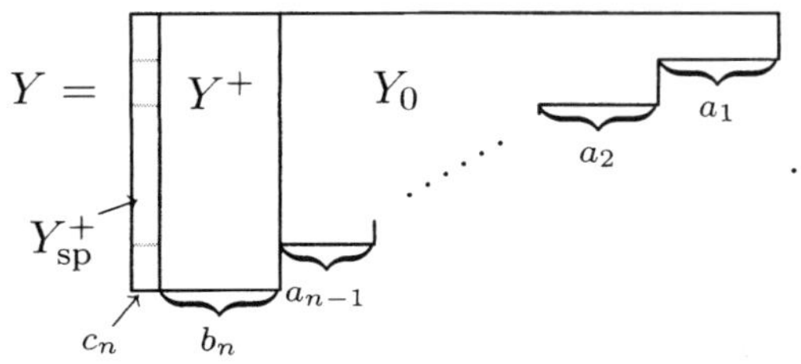

If $a_n \le a_{n-1}$, write $a_{n-1} - a_n = 2\bar{b}_n + \bar{c}_n$ with $\bar{b}_n \in \mathbf{Z}_{\ge 0}$, $\bar{c}_n = 0$ or 1, and set

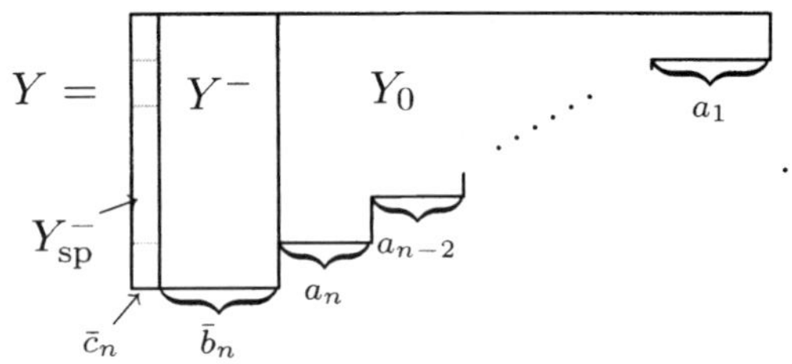

Here, the column $Y_{\rm sp}^{\pm}$ consists of half-boxes. We will identify the (generalized) Young diagram with the sequence of half-integers $Y = (\lambda_1, \dots, \lambda_n)$. When $a_n \le a_{n-1}$, we color the parts Y^- and $Y_{\rm sp}^-$ black to distinguish it from the one with $a_n \ge a_{n-1}$.

Definition 8.5.1.

(1) Let Y be a (generalized) Young diagram with at most n rows. A ***D_n-tableau of shape*** Y is a tableau obtained from Y by filling the boxes in Y with the entries from $\mathbf{N}$.

(2) A D_n-tableau is said to be ***semistandard*** if
   (a) the entries in each row are weakly increasing, hence n and $\bar{n}$ do not appear simultaneously;
   (b) the entries in each column of Y_0 and $Y^{\pm}$ are strictly increasing, but n and $\bar{n}$ can appear successively;
   (c) the entries in the column $Y_{\rm sp}^{\pm}$ are strictly increasing, and i and $\bar{i}$ do not appear simultaneously.

For a D_n-tableau T, we write

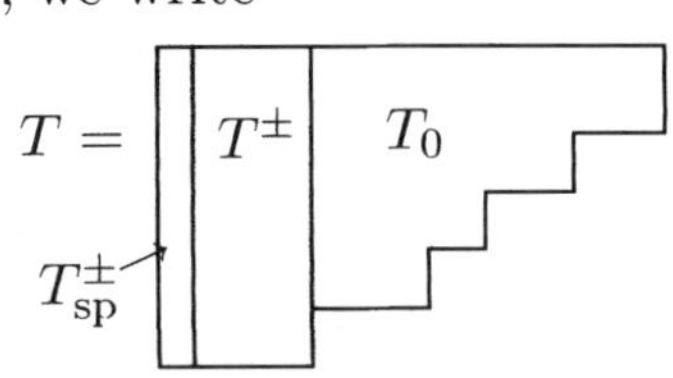

and define its *weight* to be

$$\mathrm{wt}(T) = \sum_{i=1}^{n} (k_i - \overline{k_i})\epsilon_i + \frac{1}{2}\sum_{i=1}^{n} (l_i - \overline{l_i})\epsilon_i,$$

where k_i (respectively, $\overline{k_i}$) is the number of i's (respectively, $\bar{i}$'s) appearing in T_0 and $T^{\pm}$, and l_i (respectively, $\overline{l_i}$) is the number of i's (respectively, $\bar{i}$'s) appearing in $T^{\pm}_{\mathrm{sp}}$. Note that $l_i, \overline{l_i} = 0$ or 1 and they cannot be 1 at the same time.

We define $\mathcal{B}(Y)$ to be the set of all semistandard D_n-tableaux T satisfying the following conditions.

(D1) If T has a column of the form

$$\begin{array}{r|c|} \cline{2-2} & \\ \cline{2-2} p \rightarrow & i \\ \cline{2-2} & \\ & \\ \cline{2-2} q \rightarrow & \bar{i} \\ \cline{2-2} & \\ \cline{2-2} \end{array},$$

then we have

$$(q-p)+i>N,$$

where N is the length of the column.

(D2) If T^+ or T^+_{sp} has a column whose kth entry is n (respectively, $\bar{n}$), then $n-k$ is even (respectively, odd).

(D3) If T^- or T^-_{sp} has a column whose kth entry is n (respectively, $\bar{n}$), then $n-k$ is odd (respectively, even).

(D4) If T has a pair of adjacent columns having one of the following configurations with $p \le q < r \le s$ and $a \le b < n$:

$$\begin{array}{r} p \rightarrow \\ q \rightarrow \\ \\ r \rightarrow \\ s \rightarrow \end{array}
\begin{array}{c|c} a & \\ & b \\ & \\ & \bar{b} \\ & \bar{a} \end{array}, \quad
\begin{array}{c|c} a & \\ b & \\ & \\ \bar{b} & \\ & \bar{a} \end{array}, \quad
\begin{array}{c|c} a & a \\ & \\ & \\ & \\ & \bar{a} \end{array}, \quad
\begin{array}{c|c} a & \\ & \\ & \\ & \\ \bar{a} & \bar{a} \end{array},$$

then we have

$$(q-p)+(s-r)<b-a.$$

Thus the tableau with one of the last two configurations cannot be an element of $\mathcal{B}(Y)$.

(D5) If T has a pair of adjacent columns having one of the following configurations with $p \le q < r = q+1 \le s$ and $a < n$:

$$\begin{array}{r} p \rightarrow \\ \\ q \rightarrow \\ r \rightarrow \\ \\ s \rightarrow \end{array}
\begin{array}{c|c} a & \\ & \\ & n \\ & \bar{n} \\ & \\ & \bar{a} \end{array}, \quad
\begin{array}{c|c} a & \\ & \\ & \bar{n} \\ & n \\ & \\ & \bar{a} \end{array}, \quad
\begin{array}{c|c} a & \\ & \\ n & \\ \bar{n} & \\ & \\ & \bar{a} \end{array}, \quad
\begin{array}{c|c} a & \\ & \\ \bar{n} & \\ n & \\ & \\ & \bar{a} \end{array},$$

then we have

$$(q-p)+(s-r)=s-p-1<n-a.$$

(D6) T cannot have a pair of adjacent columns having one of the following configurations with $p < s$:

$$\begin{array}{l} p \rightarrow \\ \\ s \rightarrow \end{array} \quad \begin{array}{c|c} n & \\ & \\ & n \end{array}\;, \quad \begin{array}{c|c} n & \\ & \\ & \bar{n} \end{array}\;, \quad \begin{array}{c|c} \bar{n} & \\ & \\ & n \end{array}\;, \quad \begin{array}{c|c} \bar{n} & \\ & \\ & \bar{n} \end{array}\;.$$

(D7) If T has a pair of adjacent columns having one of the following configurations with $p \leq q < r \leq s$ and $a < n$:

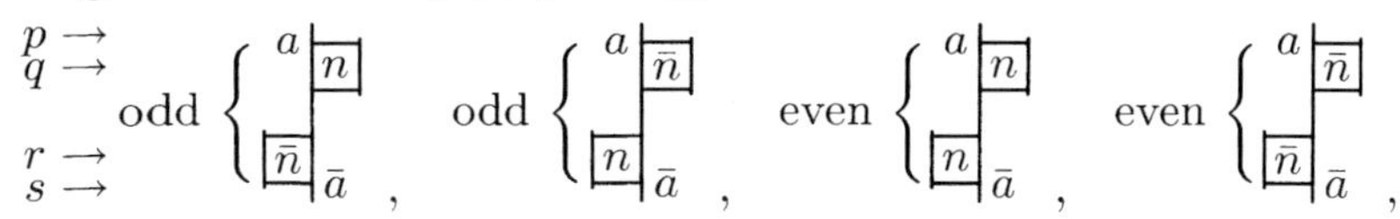

then we have $s - p < n - a$.

Theorem 8.5.2. *Let $\lambda = a_1\omega_1 + \cdots + a_n\omega_n$ $(a_i \in \mathbf{Z}_{\geq 0})$ be a dominant integral weight and let $V(\lambda)$ be the finite dimensional irreducible module over $U_q(\mathfrak{so}_{2n})$ with highest weight λ. Let Y be the (generalized) Young diagram associated with λ and let $\mathcal{B}(Y)$ denote the set of all semistandard D_n-tableaux of shape Y satisfying the conditions* **(D1)**–**(D7)**. *Then, by the Far-Eastern reading, $\mathcal{B}(Y)$ becomes a connected $U_q(\mathfrak{so}_{2n})$-crystal containing a maximal vector of weight λ. Hence $\mathcal{B}(Y)$ is isomorphic to the crystal graph $\mathcal{B}(\lambda)$ of $V(\lambda)$.*

Proof. We leave it to the readers as an exercise (Exercise 8.12). □

8.6. Tensor product decomposition of crystals

In this section, we will present the *generalized Littlewood-Richardson rule* of decomposing the tensor product of crystal graphs for classical Lie algebras. The proofs will be left to the readers as exercises because they are quite similar to the one given in the last part of Chapter 7.

Recall that finite dimensional irreducible modules over classical Lie algebras are parameterized by their highest weights that are dominant integral. In the previous sections, we have seen that there is a canonical correspondence between the set of dominant integral weights and the set of generalized Young diagrams. We will give a precise definition of generalized Young diagrams. Then it is easy to show that this canonical correspondence is indeed a bijection.

For a sequence of half integers $\lambda_j \in \frac{1}{2}\mathbf{Z}$ $(j = 1, \ldots, n)$, such that $\lambda_j - \lambda_{j+1} \in \mathbf{Z}_{\geq 0}$, we associate a diagram $Y = (\lambda_1, \ldots, \lambda_n)$ with n rows whose

jth has length λ_j (including the *negative length*). For example, we have

$$\left(\frac{5}{2}, \frac{1}{2}, -\frac{3}{2}\right) =$$ 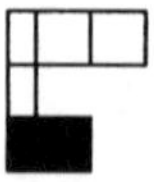 ,

where the black boxes mean the negative length. If all $\lambda_j \in \mathbf{Z}_{\geq 0}$, we get an ordinary Young diagram.

Definition 8.6.1. Let $Y = (\lambda_1, \lambda_2, \ldots, \lambda_n)$ be a diagram such that $\lambda_j \in \frac{1}{2}\mathbf{Z}$ for $j = 1, \ldots, n$ and $\lambda_j - \lambda_{j+1} \in \mathbf{Z}_{\geq 0}$ for $j = 1, \ldots, n-1$.

(1) Y is a ***generalized Young diagram of type*** A_n ***or*** C_n if $\lambda_j \in \mathbf{Z}_{\geq 0}$ for all j.

(2) Y is a ***generalized Young diagram of type*** B_n if $\lambda_j \in \frac{1}{2}\mathbf{Z}_{\geq 0}$ for all j.

(3) Y is a ***generalized Young diagram of type*** D_n if $\lambda_j \in \frac{1}{2}\mathbf{Z}_{\geq 0}$ for all $j = 1, \ldots, n-1$ and $\lambda_{n-1} \geq |\lambda_n|$.

The following proposition is straightforward.

Proposition 8.6.2. *For each type of classical Lie algebras, there is a canonical bijection between the set of generalized Young diagrams and the set of dominant integral weights given by*

$$Y = (\lambda_1, \lambda_2, \ldots, \lambda_n) \longmapsto \sum_{j=1}^{n} \lambda_j \epsilon_j$$

$$= \begin{cases} \sum_{j=1}^{n-1} (\lambda_j - \lambda_{j+1})\omega_j + \lambda_n \omega_n & \textit{if } \mathfrak{g} = A_n \textit{ and } C_n, \\ \sum_{j=1}^{n-1} (\lambda_j - \lambda_{j+1})\omega_j + 2\lambda_n \omega_n & \textit{if } \mathfrak{g} = B_n, \\ \sum_{j=1}^{n-1} (\lambda_j - \lambda_{j+1})\omega_j + (\lambda_{n-1} + \lambda_n)\omega_n & \textit{if } \mathfrak{g} = D_n. \end{cases}$$

Proof. We leave it to the readers as an exercise (Exercise 8.13). Note that the inverse of this map was given in the previous sections. □

Except for type D_n, it is quite clear that the realizations of $\mathcal{B}(\lambda)$ given in the previous sections for $\lambda = \sum_{j=1}^{n} \lambda_j \varepsilon_j$ was in terms of (generalized) Young tableaux of shape $Y = (\lambda_1, \ldots, \lambda_n)$. For type D_n, we can see that $\mathcal{B}(a\omega_{n-1} + b\omega_n)$ and $\mathcal{B}(b\omega_{n-1} + a\omega_n)$ are realized with the (generalized) Young tableaux of the same shape. We are distinguishing the two as generalized Young diagrams by coloring the nth row black if the coefficient of ω_{n-1} is greater than that of ω_n.

Let $Y = (\lambda_1, \lambda_2, \dots, \lambda_n)$ be a generalized Young diagram. Following the notation used in Chapter 7, we define

$$\begin{aligned} Y[\,j\,] &= (\lambda_1, \dots, \lambda_j + 1, \dots, \lambda_n) \quad \text{for } j = 1, \dots, n, \\ Y[\,\bar{j}\,] &= (\lambda_1, \dots, \lambda_j - 1, \dots, \lambda_n) \quad \text{for } j = 1, \dots, n, \\ Y[n+1] &= (\lambda_1 - 1, \lambda_2 - 1, \dots, \lambda_n - 1) \quad \text{for type } A_n, \\ Y[\,0\,] &= \begin{cases} Y & \text{if } \lambda_n > 0, \\ (\lambda_1, \dots, \lambda_{n-1}, -\infty) & \text{if } \lambda_n = 0. \end{cases} \end{aligned}$$

Note that the sequence $Y[\,j\,]$ defined above may not be a generalized Young diagram in many cases. In particular, $Y[\,0\,]$ is not a generalized Young diagram if $\lambda_n = 0$. For any $b = \boxed{*} \in \mathbf{B}$, we abuse the notation and write $Y[\,b\,] = Y[\boxed{*}]$ to denote $Y[\,*\,]$. If Y' is a diagram which is not a generalized Young diagram, $\mathcal{B}(Y')$ will denote the null crystal; i.e., the empty set. Then, by a similar argument as in the proof of Theorem 7.4.4, we obtain:

Proposition 8.6.3. *Let λ be a dominant integral weight for a classical Lie algebra $\mathfrak{g}$ and let Y be the generalized Young diagram associated with λ. Then there exists a $U_q(\mathfrak{g})$-crystal isomorphism*

$$\mathcal{B}(Y) \otimes \mathbf{B} \cong \bigoplus_{b \in \mathbf{B}} \mathcal{B}(Y[\,b\,]).$$

Proof. We leave it to the readers as an exercise (Exercise 8.14). □

Example 8.6.4. Let $\mathfrak{g} = B_3$ and $\lambda = \omega_1 + 2\omega_3$. Then $Y = (2,1,1) =$ 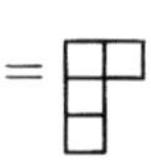

and we have

$$\mathbf{B} = \{\boxed{1}, \boxed{2}, \boxed{3}, \boxed{0}, \boxed{\bar{3}}, \boxed{\bar{2}}, \boxed{\bar{1}}\}.$$

Thus the various generalized Young diagrams obtained by attaching elements of $\mathbf{B}$ to Y are given as follows:

$Y[1] =$ 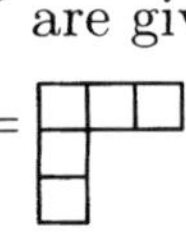, $Y[2] =$ 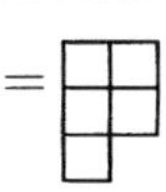, $Y[3] = \times$, $Y[0] =$ 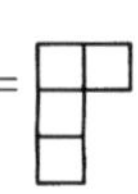,

$Y[\bar{3}] =$ 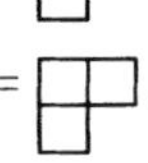, $Y[\bar{2}] = \times$, $Y[\bar{1}] =$ 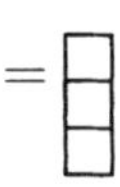.

Hence we obtain

$\mathcal{B}(Y) \otimes \mathbf{B} \cong \mathcal{B}\Big($ 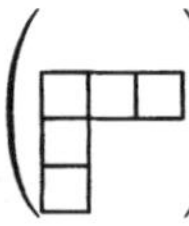 $\Big) \oplus \mathcal{B}\Big($ 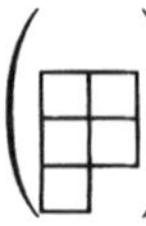 $\Big)$

$\oplus \mathcal{B}\Big($  $\Big) \oplus \mathcal{B}\Big($ 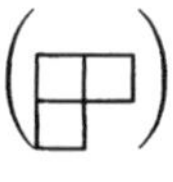 $\Big) \oplus \mathcal{B}\Big($  $\Big)$, which yields

$$\begin{aligned}\mathcal{B}(\omega_1+2\omega_3)\otimes\mathcal{B}(\omega_1)\cong&\mathcal{B}(2\omega_1+2\omega_3)\oplus\mathcal{B}(\omega_2+2\omega_3)\oplus\mathcal{B}(\omega_1+2\omega_3)\\&\oplus\mathcal{B}(\omega_1+\omega_2)\oplus\mathcal{B}(2\omega_3).\end{aligned}$$

We now consider the tensor product of $\mathcal{B}(Y)$ with the crystal graph(s) of the spin representation(s). For an element $b=(s_1,\ldots,s_n)\in\mathbf{B}_{\mathrm{sp}}$, $\mathbf{B}_{\mathrm{sp}}^{\pm}$ and a generalized Young diagram $Y=(\lambda_1,\ldots,\lambda_n)$, we define

$$Y[b]=\Big(\lambda_1+\frac{1}{2}s_1,\lambda_2+\frac{1}{2}s_2,\ldots,\lambda_n+\frac{1}{2}s_n\Big).$$

Proposition 8.6.5. *Let λ be a dominant integral weight for the classical Lie algebra $\mathfrak{g}=B_n$ or D_n, and let Y be the generalized Young diagram associated with λ.*

Then there exists a $U_q(\mathfrak{g})$-crystal isomorphism

$$\mathcal{B}(Y)\otimes\mathcal{B}\cong\bigoplus_{b\in\mathcal{B}}\mathcal{B}(Y[b]),$$

where $\mathcal{B}=\mathbf{B}_{\mathrm{sp}},\mathbf{B}_{\mathrm{sp}}^{\pm}$.

Proof. We leave it to the readers as an exercise (Exercise 8.15). □

For the general case, let μ be a dominant integral weight for the classical Lie algebra $\mathfrak{g}$ and Y' be the corresponding generalized Young diagram. Then the $U_q(\mathfrak{g})$-crystal $\mathcal{B}(Y')$ can be embedded in $\mathbf{B}^{\otimes k}$, $\mathbf{B}^{\otimes k}\otimes\mathbf{B}_{\mathrm{sp}}$, or $\mathbf{B}^{\otimes k}\otimes\mathbf{B}_{\mathrm{sp}}^{\pm}$ for some $k>0$, and every element of $\mathcal{B}(Y')$ can be identified with a sequence $b=(b_1,b_2,\ldots,b_N)$ with $b_i\in\mathbf{B},\mathbf{B}_{\mathrm{sp}},\mathbf{B}_{\mathrm{sp}}^{\pm}$. For such a sequence, we define

$$Y[b_1,\ldots,b_N]=Y[b_1,\ldots,b_{N-1}][b_N].$$

Then $Y[b_1,\ldots,b_N]$ is a generalized Young diagram if and only if $Y[b_1,\ldots,b_i]$ is a generalized Young diagram for all $i=1,2,\ldots,N$.

We now present the *generalized Littlewood-Richardson rule* for classical Lie algebras, the main theorem of this section.

Theorem 8.6.6. *Let λ and μ be dominant integral weights for a classical Lie algebra $\mathfrak{g}$ and let Y and Y' be the corresponding generalized Young diagrams. Then there exists a $U_q(\mathfrak{g})$-crystal isomorphism*

$$\mathcal{B}(Y)\otimes\mathcal{B}(Y')\cong\bigoplus_{b_1\otimes\ldots\otimes b_N\in\mathcal{B}(\mu)}\mathcal{B}(Y[b_1,\ldots,b_N]).$$

Proof. We leave it to the readers as an exercise (Exercise 8.16). □

Example 8.6.7. Let $\mathfrak{g} = D_4$, $\lambda = \omega_2 + \omega_3$, and $\mu = 2\omega_1$. Then the corresponding generalized Young diagrams are

$Y =$ 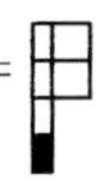, $Y' =$ 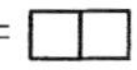,

and the crystal $\mathcal{B}(Y')$ is realized as

$$\mathcal{B}(Y') = \{ \boxed{b}\boxed{a} \mid b \preccurlyeq a \}.$$

Here, the ordering is given by

$$1 \prec 2 \prec 3 \prec \begin{matrix} 4 \\ \bar{4} \end{matrix} \prec \bar{3} \prec \bar{2} \prec \bar{1}.$$

Below, we list $\boxed{a} \otimes \boxed{b} = \boxed{b}\boxed{a} \in \mathcal{B}(Y')$ and the corresponding $Y[\,a, b\,]$ for the cases when $Y[\,a, b\,]$ is a generalized Young diagram.

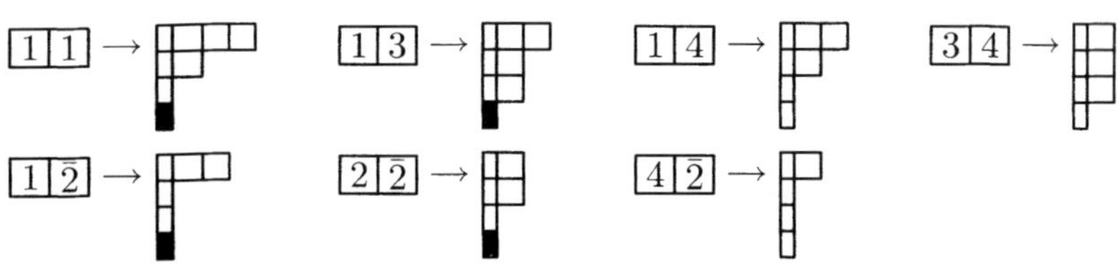

Hence we obtain the decomposition

$$\begin{aligned} \mathcal{B}(\omega_2 + \omega_3) \otimes \mathcal{B}(2\omega_1) \cong\ & \mathcal{B}(2\omega_1 + \omega_2 + \omega_3) \oplus \mathcal{B}(\omega_1 + 2\omega_3 + \omega_4) \\ & \oplus \mathcal{B}(\omega_1 + \omega_2 + \omega_4) \oplus \mathcal{B}(\omega_3 + 2\omega_4) \\ & \oplus \mathcal{B}(2\omega_1 + \omega_3) \oplus \mathcal{B}(\omega_2 + \omega_3) \oplus \mathcal{B}(\omega_1 + \omega_4). \end{aligned}$$

Exercises

8.1. Let $\mathfrak{g}$ be a classical Lie algebra.
  (a) Verify that the elements e_i, f_i, h_i $(i \in I)$ given in this chapter generate $\mathfrak{g}$ as a Kac-Moody algebra.
  (b) Verify that the vector representation $\mathbf{V}$ is indeed a $U_q(\mathfrak{g})$-module with a crystal basis $(\mathbf{L}, \mathbf{B})$.

8.2. Draw the crystal graph $\mathbf{B}_{\mathrm{sp}}$ over $U_q(\mathfrak{so}_7)$ using the generalized Young tableaux of shape Y_{sp}.

8.3. Complete the proof of Lemma 8.3.4.

8.4. Complete the proof of Theorem 8.3.3.

8.5. For $\mathfrak{g} = C_3$, draw the crystal graph $\mathcal{B}(2\omega_3)$ and find the dimension of $V(2\omega_3)$.

8.6. Let $\mathfrak{g} = B_3$.

(a) For $\lambda = \omega_1 + 2\omega_3$ and $\omega_1 + \omega_2 + 2\omega_3$, find the corresponding generalized Young diagram and give a description of the crystal graph $\mathcal{B}(Y)$.
(b) For $\lambda = \omega_1 + 3\omega_3$ and $\omega_1 + \omega_2 + 3\omega_3$, find the corresponding generalized Young diagram and give a description of the crystal graph $\mathcal{B}(Y)$.

8.7. Verify the statements in Example 8.4.2.

8.8. Prove Theorem 8.4.3.

8.9. Draw the $U_q(\mathfrak{so}_7)$-crystal $\mathcal{B}(\lambda)$ for $\lambda = 2\omega_3$ and $3\omega_3$.

8.10. (a) Verify that $\mathbf{V}_{\mathrm{sp}}^{\pm}$ is a module over $U_q(\mathfrak{so}_{2n})$ with a crystal basis $(\mathbf{L}_{\mathrm{sp}}^{\pm}, \mathbf{B}_{\mathrm{sp}}^{\pm})$.
(b) Draw the crystal graph $\mathbf{B}_{\mathrm{sp}}^{\pm}$ for $\mathfrak{g} = D_4$. What are the highest weight vectors for them?

8.11. Let $\mathfrak{g} = D_4$. For $\lambda = \omega_1+\omega_3+\omega_4$, $\omega_1+2\omega_3+\omega_4$, and $\omega_1+\omega_3+2\omega_4$, find the corresponding generalized Young diagram and give a description of the crystal graph $\mathcal{B}(Y)$.

8.12. Prove Theorem 8.5.2.

8.13. Prove Proposition 8.6.2.

8.14. (a) Prove Proposition 8.6.3.
(b) For $\mathfrak{g} = B_3$ and $\lambda = \omega_1 + 3\omega_3$, find the tensor product decomposition of $\mathcal{B}(\lambda) \otimes \mathbf{B}$.

8.15. (a) Prove Proposition 8.6.5.
(b) For $\mathfrak{g} = D_4$ and $\lambda = \omega_3 + 3\omega_4$, find the tensor product decomposition of $\mathcal{B}(\lambda) \otimes \mathbf{B}_{\mathrm{sp}}^{\pm}$.

8.16. Prove Theorem 8.6.6.

Chapter 9

Solvable Lattice Models

In this chapter, we review the very basic theory of solvable lattice models and discuss its connection with the representation theory of the quantum affine algebra $U_q(\widehat{\mathfrak{sl}}_2)$. In particular, we show that the one-point function for the 6-vertex model can be expressed as the quotient of the string function by the character of the basic representation of $U_q(\widehat{\mathfrak{sl}}_2)$.

9.1. The 6-vertex model

Consider a two-dimensional (infinite) lattice consisting of *vertices*, *edges*, and *faces*.

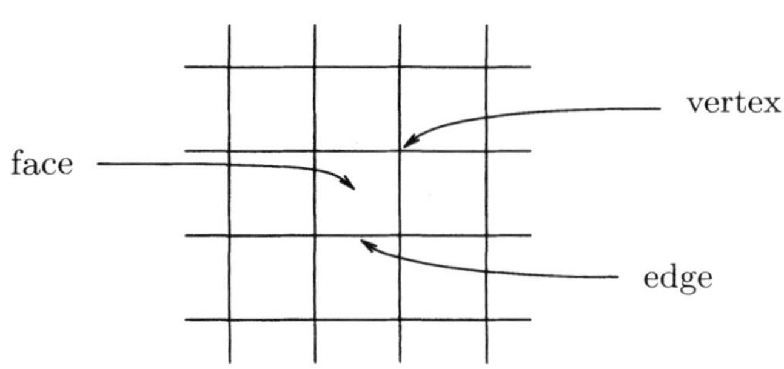

Definition 9.1.1.

(1) A ***configuration*** is an assignment of ***spins*** $\varepsilon = \pm$ to each edge.

(2) Fix a face of the lattice and denote it by $\star$. Among the configurations, there exist two distinguished configurations called the ***ground-state configurations*** given below.

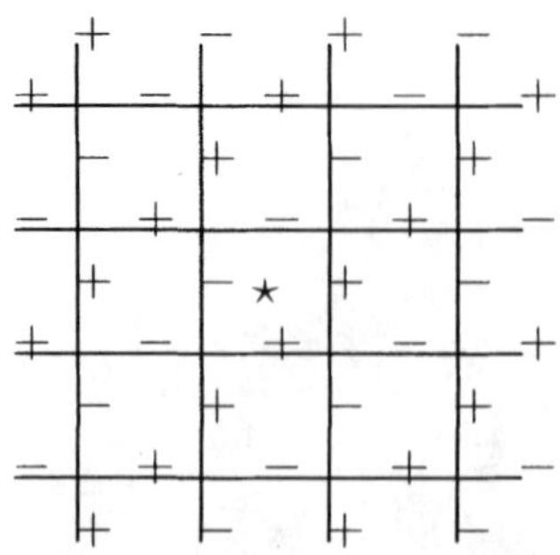

$C^{(+)}$

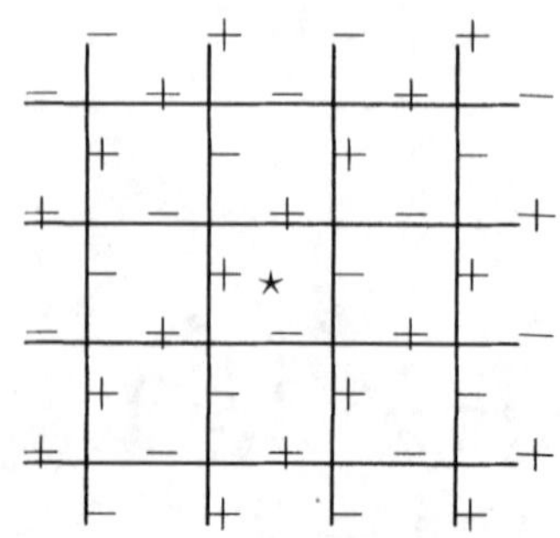

$C^{(-)}$

(3) A configuration $C = \{C(e)\}_{e:\,\mathrm{edge}}$, $C(e) = \pm$, is said to be in the **0-*sector*** (respectively, **1-*sector***) if $C(e) = C^{(+)}(e)$ (respectively, $C(e) = C^{(-)}(e)$) for all but finitely many edges e.

For simplicity, we mainly focus on the configurations in the 0-sector.

Definition 9.1.2. Let x and q be nonzero complex numbers. The ***6-vertex model*** is defined to be the two-dimensional lattice model that assigns the ***Boltzmann weight*** to each vertex as follows:

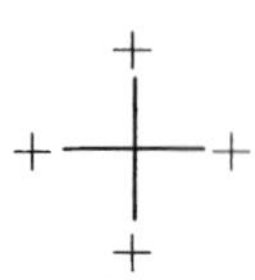 $= x - q^2$,  $= x - q^2$,

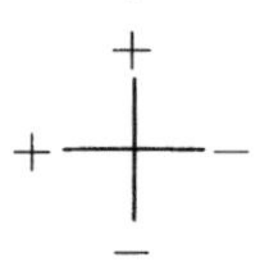 $= 1 - q^2$, 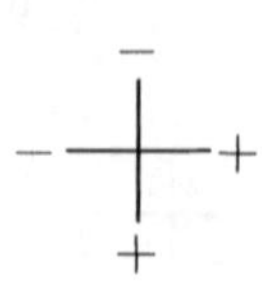 $= x(1 - q^2)$,

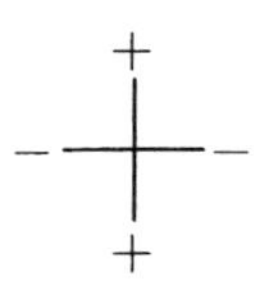 $= q(x - 1)$,  $= q(x - 1)$,

other vertices $= 0$.

We say that a configuration C satisfies the ***6-vertex condition*** if the configuration around each vertex in C is given by one of the above six configurations.

Let $V = \mathbf{C}v_+ \oplus \mathbf{C}v_-$ be a two-dimensional vector space and set

$$V^{\mathrm{aff}} = \mathbf{C}[z, z^{-1}] \otimes_{\mathbf{C}} V,$$

where z is an indeterminate. Choose a nonzero complex number ζ and let $\mathbf{J}_\zeta$ be the maximal ideal of $\mathbf{C}[z, z^{-1}]$ generated by $z - \zeta$. Then there is an isomorphism of fields $\mathbf{C}[z, z^{-1}]/\mathbf{J}_\zeta \xrightarrow{\sim} \mathbf{C}$ given by $z \longmapsto \zeta$. The ***evaluation space*** V_ζ of V at $z = \zeta$ is defined to be

$$V_\zeta = \mathbf{C} \otimes_{\mathbf{C}[z,z^{-1}]} V^{\mathrm{aff}} \cong V^{\mathrm{aff}}/\mathbf{J}_\zeta V^{\mathrm{aff}}. \tag{9.1}$$

Let ζ_i $(i = 1, 2)$ be complex numbers that are not roots of unity and define a $\mathbf{C}$-linear map $R : V_{\zeta_1} \otimes V_{\zeta_2} \longrightarrow V_{\zeta_2} \otimes V_{\zeta_1}$ by

$$R(v_{\varepsilon_1} \otimes v_{\varepsilon_2}) = \sum_{\varepsilon_1', \varepsilon_2' = \pm} R_{\varepsilon_1', \varepsilon_2'}^{\varepsilon_1, \varepsilon_2} v_{\varepsilon_1'} \otimes v_{\varepsilon_2'} = \sum_{\varepsilon_1', \varepsilon_2' = \pm} \begin{array}{ccc} & \varepsilon_1 & \\ \varepsilon_1' & + & \varepsilon_2 \\ & \varepsilon_2' & \end{array} v_{\varepsilon_1'} \otimes v_{\varepsilon_2'}, \tag{9.2}$$

where ε_j, ε_j' $(k = 1, 2)$ are the spins and $R_{\varepsilon_1', \varepsilon_2'}^{\varepsilon_1, \varepsilon_2} = \begin{array}{ccc} & \varepsilon_1 & \\ \varepsilon_1' & + & \varepsilon_2 \\ & \varepsilon_2' & \end{array}$ is the Boltzmann weight given by the 6-vertex condition with $x = \zeta_2/\zeta_1$. Then the linear map R can be expressed as the following 4×4 matrix:

$$R = R(x) = \begin{pmatrix} x - q^2 & 0 & 0 & 0 \\ 0 & 1 - q^2 & q(x-1) & 0 \\ 0 & q(x-1) & x(1-q^2) & 0 \\ 0 & 0 & 0 & x - q^2 \end{pmatrix}. \tag{9.3}$$

The linear map R is called the ***R-matrix*** for the 6-vertex model.

Let $\mathcal{B} = \{+, -\}$ and define a function $H : \mathcal{B} \times \mathcal{B} \to \mathbf{Z}$ by

$$H(\varepsilon, \varepsilon') = \begin{cases} 0 & \text{if } \varepsilon = +,\ \varepsilon' = -, \\ 1 & \text{otherwise.} \end{cases} \tag{9.4}$$

The function H is called the ***energy function*** for the 6-vertex model. We have taken q to be a nonzero complex number, but if we set $q = 0$, the R-matrix becomes a diagonal matrix:

$$R = \begin{pmatrix} x & 0 & 0 & 0 \\ 0 & 1 & 0 & 0 \\ 0 & 0 & x & 0 \\ 0 & 0 & 0 & x \end{pmatrix} = \operatorname{diag}\left(x^{H(\varepsilon, \varepsilon')}\right)_{\varepsilon,\, \varepsilon' = \pm}.$$

One can verify by a direct calculation that the R-matrix for the 6-vertex model satisfies the ***Yang-Baxter equation*** (Exercise 9.1):

$$R_{12}\left(\frac{\zeta_3}{\zeta_2}\right) R_{23}\left(\frac{\zeta_3}{\zeta_1}\right) R_{12}\left(\frac{\zeta_2}{\zeta_1}\right) = R_{23}\left(\frac{\zeta_2}{\zeta_1}\right) R_{12}\left(\frac{\zeta_3}{\zeta_1}\right) R_{23}\left(\frac{\zeta_3}{\zeta_2}\right), \tag{9.5}$$

where $R_{12} = R \otimes \mathrm{id}$ and $R_{23} = \mathrm{id} \otimes R$ on $V_{\zeta_1} \otimes V_{\zeta_2} \otimes V_{\zeta_3}$.

The Yang-Baxter equation is one of the main ingredients of the theory of solvable lattice models. In the next section, we will give a conceptual proof of the Yang-Baxter equation using the representation theory of the quantum affine algebra $U_q(\widehat{\mathfrak{sl}}_2)$. It can be viewed graphically as follows.

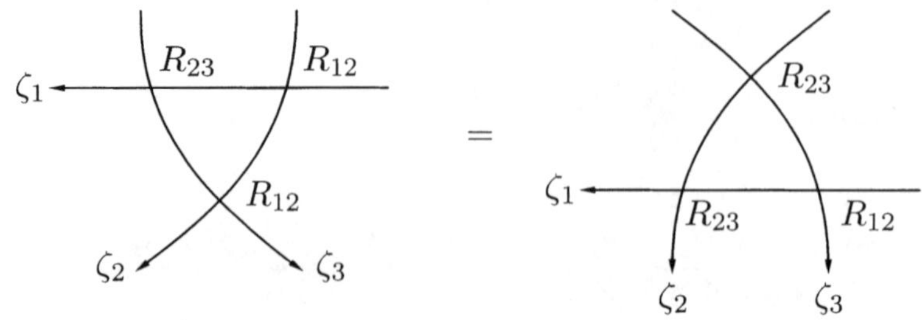

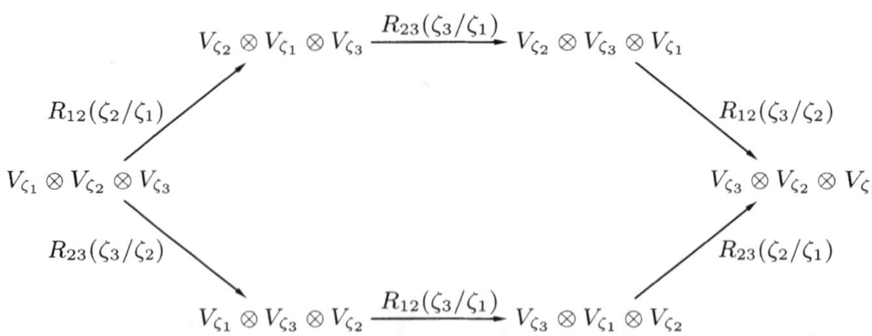

Let $C = \{C(e)\}_{e:\,\text{edge}}$ be a configuration in the 0-sector. We define the ***Boltzmann weight*** of the configuration C to be

$$B(C) = \prod_{\text{all vertices in } C} + .$$

Note that $B(C) \neq 0$ only if the configuration C satisfies the 6-vertex condition. The ***partition function*** Z of the 6-vertex model is defined to be the sum of all Boltzmann weights $B(C)$, where C runs over all the configurations in the 0-sector:

$$Z = \sum_{C\,:\,\substack{\text{all configurations}\\ \text{in the 0-sector}}} B(C).$$

The acute reader will complain of convergence problems. But the discussion of such matters is beyond the scope of this book. The interested readers may refer to [**24**].

For a configuration $C = \{C(e)\}_{e:\,\text{edge}}$ in the 0-sector satisfying the 6-vertex condition, we define its ***weight configuration***. This is a set of integers $W(C) = \{W(j, C)\}_{j:\,\text{faces}}$, indexed by the faces, satisfying the following condition. Let j be a face and j' be the face right below j or to the right of j. Then we must have

$$W(j', C) = \begin{cases} W(j, C) + 1 & \text{if } e = +, \\ W(j, C) - 1 & \text{if } e = -, \end{cases} \tag{9.6}$$

where e is the edge between j and j'.

Let $C^{(+)}$ be the ground-state configuration of the 0-sector and fix a face $\star$. By setting $C^{(+)}(\star) = 0$, we can determine the weight configuration $W(C^{(+)})$ of $C^{(+)}$:

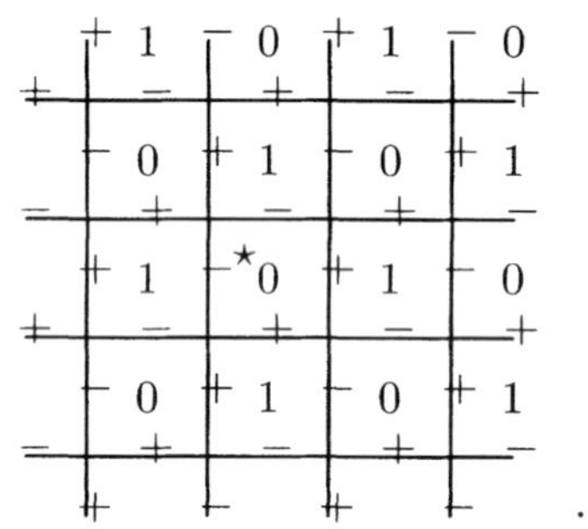

For any configuration $C = \{C(e)\}_{e:\,\text{edge}}$ in the 0-sector satisfying the 6-vertex condition, we choose the boundary condition as

$$W(j, C) = W(j, C^{(+)}) \qquad \text{for all but finitely many } j\text{'s}. \tag{9.7}$$

One can show that weight configuration $W(j, C)$ is uniquely determined by the 6-vertex condition and the boundary condition (Exercise 9.2).

Example 9.1.3. Let C be the configuration

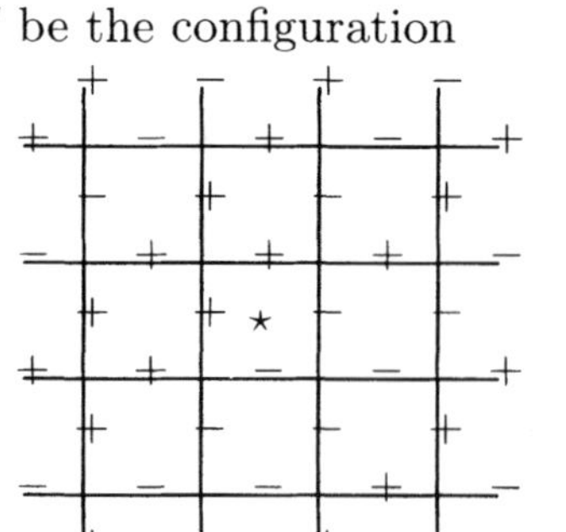

Then the weight configuration $W(C)$ is given by

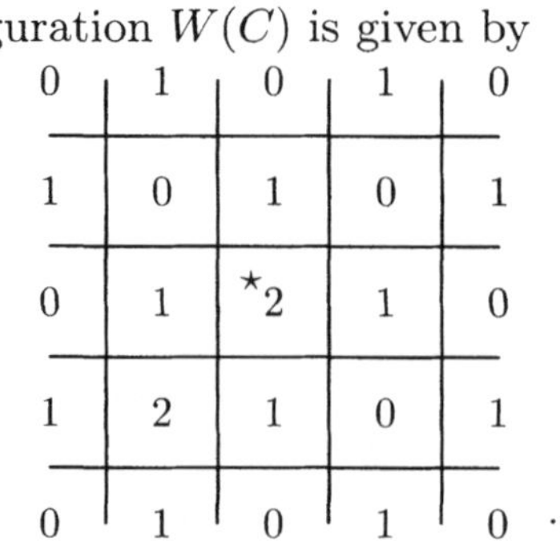

For an integer $a \in \mathbf{Z}$, set

$$G(a) = \sum_{C:\, W(\star,C)=a} B(C) \tag{9.8}$$

and define the ***one-point function*** for the 6-vertex model to be

$$F(a) = \frac{G(a)}{Z}. \tag{9.9}$$

Note that $Z = \sum_{a\in\mathbf{Z}} G(a)$. Hence the one-point function can be interpreted as the *probability* for a configuration to have $W(\star, C) = a$.

The *corner transfer matrix method* introduced by Baxter reduces the two-dimensional sum $G(a) = \sum_{C:\, W(\star,C)=a} B(C)$ to the one-dimensional sum over the *paths* [**3**]. More precisely, consider a sequence of spins $\varepsilon = \pm$:

$$\mathbf{p}_{\Lambda_0} = (\ldots, \mathbf{p}_{\Lambda_0}(k), \ldots, \mathbf{p}_{\Lambda_0}(1), \mathbf{p}_{\Lambda_0}(0)) = (\ldots, +, -, +, -, +, -). \tag{9.10}$$

The sequence $\mathbf{p}_{\Lambda_0}$ is called the ***ground-state path*** of the 0-sector. A sequence $\mathbf{p} = (\mathbf{p}(k))_{k=0}^{\infty} = (\ldots, \mathbf{p}(k), \ldots, \mathbf{p}(1), \mathbf{p}(0))$ of spins is called a Λ_0-***path*** if $\mathbf{p}(k) = \mathbf{p}_{\Lambda_0}(k) = (-)^{k+1}$ for all sufficiently large k. We denote by $\mathcal{P}(\Lambda_0)$ the set of all Λ_0-paths.

For a Λ_0-path $\mathbf{p} = (\mathbf{p}(k))_{k=0}^{\infty}$, we define its ***weight sequence*** $W(\mathbf{p}) = W(k, \mathbf{p})_{k=0}^{\infty}$ by

$$W(k,\mathbf{p}) = \begin{cases} W(k+1,\mathbf{p}) + 1 & \text{if } \mathbf{p}(k) = +, \\ W(k+1,\mathbf{p}) - 1 & \text{if } \mathbf{p}(k) = -. \end{cases}$$

For example, if $\mathbf{p} = \mathbf{p}_{\Lambda_0}$, we could set

$$W(\mathbf{p}_{\Lambda_0}) = (\ldots, 1, 0, 1, 0, 1, 0).$$

Given a Λ_0-path $\mathbf{p}$, the weight sequence is uniquely determined if we add the boundary condition that it should be equal to $W(k, \mathbf{p}_{\Lambda_0})$ for all sufficiently large k. So for $\mathbf{p} = (\ldots, +, -, +, +, -, -)$, the weight sequence is given by

$$W(\mathbf{p}) = (\ldots, 1, 0, 1, 2, 1, 0).$$

Through the *corner transfer matrix method* of Baxter, the sum (9.8) essentially reduces to the sum

$$G'(a) = \sum_{\substack{\mathbf{p}\in\mathcal{P}(\Lambda_0)\\ W(0,\mathbf{p})=a}} q^{2\sum_{k=0}^{\infty}(k+1)(H(\mathbf{p}(k+1),\mathbf{p}(k)) - H(\mathbf{p}_{\Lambda_0}(k+1),\mathbf{p}_{\Lambda_0}(k)))}, \tag{9.11}$$

where H denotes the energy function of the 6-vertex model (see [**3**] for more detail). Now one can evaluate this sum (Exercise 9.3):

$$G'(a) = \begin{cases} \dfrac{q^{\frac{a^2}{2}}}{\prod_{n=1}^{\infty}(1-q^{2n})} & \text{if } a \text{ is even}, \\ 0 & \text{if } a \text{ is odd}. \end{cases} \tag{9.12}$$

9.2. The quantum affine algebra $U_q(\widehat{\mathfrak{sl}}_2)$

In this section, we will show that there exists a close relation between the one-point function for the 6-vertex model and the character for the basic representation of the quantum affine algebra $U_q(\widehat{\mathfrak{sl}}_2)$.

Let $I = \{0, 1\}$ be the index set and let $A = (a_{ij})_{i,j=0,1} = \begin{pmatrix} 2 & -2 \\ -2 & 2 \end{pmatrix}$ be the generalized Cartan matrix of affine $A_1^{(1)}$ type. The set of *simple roots* (respectively, *simple coroots*) is denoted by $\Pi = \{\alpha_0, \alpha_1\}$ (respectively, $\Pi^\vee = \{h_0, h_1\}$), and the *dual weight lattice* is the free abelian group $P^\vee = \mathbf{Z}h_0 \oplus \mathbf{Z}h_1 \oplus \mathbf{Z}d$, where d is the *grading element* satisfying $\alpha_0(d) = 1$, $\alpha_1(d) = 0$.

Define the ***fundamental weights*** Λ_i $(i = 0, 1)$ to be linear functionals on $\mathfrak{h} = \mathbf{C} \otimes_{\mathbf{Z}} P^\vee$ satisfying $\Lambda_i(h_j) = \delta_{ij}$, $\Lambda_i(d) = 0$ $(i, j = 0, 1)$, and let $\delta = \alpha_0 + \alpha_1$ be the *null root*. Then the *weight lattice* is defined to be $P = \mathbf{Z}\Lambda_0 \oplus \mathbf{Z}\Lambda_1 \oplus \mathbf{Z}\delta$.

Definition 9.2.1. Let q be a nonzero complex number which is not a root of unity. The ***quantum affine algebra*** $U_q(\widehat{\mathfrak{sl}}_2)$ is the quantum group associated with the Cartan datum $(A, \Pi, \Pi^\vee, P, P^\vee)$ of affine type $A_1^{(1)}$. That is, it is the associative algebra over $\mathbf{C}$ generated by the elements e_i, f_i $(i = 0, 1)$ and q^h $(h \in P^\vee)$ with defining relations given in Definition 3.1.1.

The quantum affine algebra $U_q(\widehat{\mathfrak{sl}}_2)$ has a Hopf algebra structure with the comultiplication, the counit, and the antipode defined as in Proposition 3.1.2.

The subalgebra of $U_q(\widehat{\mathfrak{sl}}_2)$ generated by e_i, f_i, $K_i^{\pm 1}$ $(i = 0, 1)$ will be denoted by $U_q'(\widehat{\mathfrak{sl}}_2)$ and will also be called the ***quantum affine algebra*** of type $A_1^{(1)}$. It can be regarded as the quantum group associated with the Cartan datum $(A, \Pi, \Pi^\vee, \overline{P}, \overline{P}^\vee)$, where $\overline{P} = \mathbf{Z}\Lambda_0 \oplus \mathbf{Z}\Lambda_1$, $\overline{P}^\vee = \mathbf{Z}h_0 \oplus \mathbf{Z}h_1$, and α_i, Λ_i $(i = 0, 1)$ are considered as linear functionals on $\overline{\mathfrak{h}} = \mathbf{C} \otimes_{\mathbf{Z}} \overline{P}^\vee$. In particular, $\delta = 0$ on $\overline{\mathfrak{h}}$. The elements of P (respectively, $\overline{P}$) are called ***affine weights*** (respectively, ***classical weights***). The image of the projection $\mathrm{cl} : P \to \overline{P}$ will be denoted with the bar notation. So $\mathrm{cl}(\lambda) = \overline{\lambda}$. We will fix an embedding $\mathrm{aff} : \overline{P} \to P$ such that $\mathrm{cl} \circ \mathrm{aff} = \mathrm{id}_{\overline{P}}$ and $\mathrm{aff} \circ \mathrm{cl}(\alpha_i) = \alpha_i$ for $i \neq 0$.

The main difference between $U_q(\widehat{\mathfrak{sl}}_2)$ and $U_q'(\widehat{\mathfrak{sl}}_2)$ is that the algebra $U_q'(\widehat{\mathfrak{sl}}_2)$ can have finite dimensional irreducible modules, while all the nontrivial irreducible $U_q(\widehat{\mathfrak{sl}}_2)$-modules are infinite dimensional. On the other hand, the infinite dimensional irreducible $U_q(\widehat{\mathfrak{sl}}_2)$-modules have finite dimensional weight spaces, whereas the weight spaces of infinite dimensional $U'_q(\widehat{\mathfrak{sl}}_2)$-modules are infinite dimensional.

Let V be a finite dimensional $U'_q(\widehat{\mathfrak{sl}}_2)$-module and let z be an indeterminate. Set

$$\begin{aligned} V^{\mathrm{aff}} &= \mathbf{C}[z, z^{-1}] \otimes_{\mathbf{C}} V \\ &= \mathrm{Span}_{\mathbf{C}}\{z^m \otimes v \mid v \in V, m \in \mathbf{Z}\} \end{aligned}$$

and define the $U_q(\widehat{\mathfrak{sl}}_2)$-module action on V^{aff} by

$$\begin{aligned} e_1(z^m \otimes v) &= z^m \otimes e_1 v, & e_0(z^m \otimes v) &= z^{m+1} \otimes e_0 v, \\ f_1(z^m \otimes v) &= z^m \otimes f_1 v, & f_0(z^m \otimes v) &= z^{m-1} \otimes f_0 v, \\ K_1(z^m \otimes v) &= z^m \otimes K_1 v, & K_0(z^m \otimes v) &= z^m \otimes K_0 v, \\ q^d(z^m \otimes v) &= q^m z^m \otimes v. \end{aligned}$$

The $U_q(\widehat{\mathfrak{sl}}_2)$-module V^{aff} thus defined is called the ***affinization*** of V. We will identify $v \in V$ with $1 \otimes v \in V^{\mathrm{aff}}$.

Let ζ be a nonzero complex number. As before, we set $\mathbf{J}_\zeta$ to be the maximal ideal of $\mathbf{C}[z, z^{-1}]$ generated by $z - \zeta$ so that $\mathbf{C} \cong \mathbf{C}[z, z^{-1}]/\mathbf{J}_\zeta$. The ***evaluation module*** V_ζ of V at $z = \zeta$ is defined to be the $U'_q(\widehat{\mathfrak{sl}}_2)$-module

$$V_\zeta = \mathbf{C} \otimes_{\mathbf{C}[z,z^{-1}]} V^{\mathrm{aff}} \cong V^{\mathrm{aff}}/\mathbf{J}_\zeta V^{\mathrm{aff}}.$$

Example 9.2.2. Let $V = \mathbf{C}v_+ \oplus \mathbf{C}v_-$ be the $U'_q(\widehat{\mathfrak{sl}}_2)$-module defined by

$$\begin{aligned} e_1 v_+ &= 0, & f_1 v_+ &= v_-, & K_1 v_+ &= q v_+, \\ e_1 v_- &= v_+, & f_1 v_- &= 0, & K_1 v_- &= q^{-1} v_-, \\ e_0 v_+ &= v_-, & f_0 v_+ &= 0, & K_0 v_+ &= q^{-1} v_+, \\ e_0 v_- &= 0, & f_0 v_- &= v_+, & K_0 v_- &= q v_-. \end{aligned}$$

Then the $U'_q(\widehat{\mathfrak{sl}}_2)$-module action on V^{aff} is given by

$$\begin{aligned} e_1(z^m \otimes v_+) &= 0, & e_1(z^m \otimes v_-) &= z^m \otimes v_+, \\ f_1(z^m \otimes v_+) &= z^m \otimes v_-, & f_1(z^m \otimes v_-) &= 0, \\ K_1(z^m \otimes v_+) &= q z^m \otimes v_+, & K_1(z^m \otimes v_-) &= q^{-1}(z^m \otimes v_-), \\ e_0(z^m \otimes v_+) &= z^{m+1} \otimes v_-, & e_0(z^m \otimes v_-) &= 0, \\ f_0(z^m \otimes v_+) &= 0, & f_0(z^m \otimes v_-) &= z^{m-1} \otimes v_+, \\ K_0(z^m \otimes v_+) &= q^{-1}(z^m \otimes v_+), & K_0(z^m \otimes v_-) &= q z^m \otimes v_-. \end{aligned}$$

Hence, for a nonzero complex number ζ, the $U_q'(\widehat{\mathfrak{sl}}_2)$-module action on the evaluation module V_ζ is given by

$$\begin{aligned} e_1 v_+ &= 0, & f_1 v_+ &= v_-, & K_1 v_+ &= q v_+, \\ e_1 v_- &= v_+, & f_1 v_- &= 0, & K_1 v_- &= q^{-1} v_-, \\ e_0 v_+ &= \zeta v_-, & f_0 v_+ &= 0, & K_0 v_+ &= q^{-1} v_+, \\ e_0 v_- &= 0, & f_0 v_- &= \zeta^{-1} v_+, & K_0 v_- &= q v_-. \end{aligned}$$

We now want to talk about crystal graphs. But since we have changed the choice of the base field in defining the quantum groups in this section, we should also change the definition of the crystal basis. For suitable choices of $\mathbf{F}$ and q (for example, $\mathbf{F} = \mathbf{Q}$ and $q \in \mathbf{C} \setminus \mathbf{Q}$), the crystal graph of V is equal to

$\mathcal{B}$: 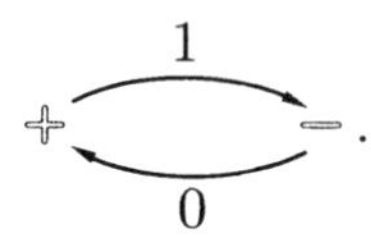

Hence the crystal graph $\mathcal{B}^{\text{aff}}$ of V^{aff} would look like the following.

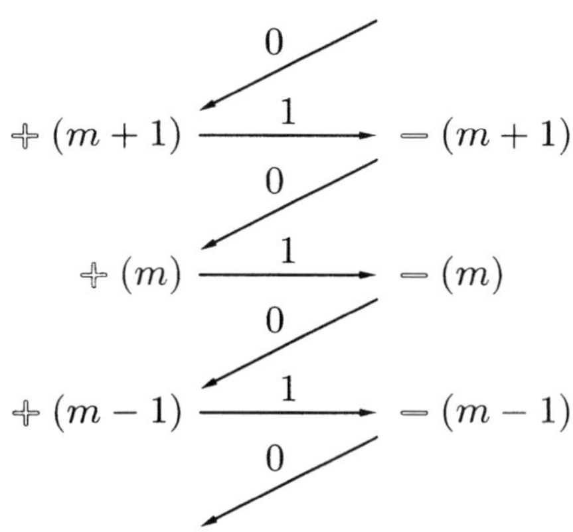

The evaluation module V_ζ has a crystal graph when $\zeta = 1$, and it is the same as $\mathcal{B}$ given above.

We recall some of the fundamental results on finite dimensional $U_q'(\widehat{\mathfrak{sl}}_2)$-modules developed in [**6**]. A nonempty finite set of nonzero complex numbers is called a ***q-string*** if it is of the form $\{\omega, q^{-2}\omega, \ldots, q^{-2r}\omega\}$ for some nonzero complex number ω and a nonnegative integer r. For example, the set

$$S_n(\zeta) = \{q^{n-1}\zeta, q^{n-3}\zeta, \ldots, q^{-n+3}\zeta, q^{-n+1}\zeta\}$$

is a q-string. Two q-strings S_1 and S_2 are said to be in ***general position*** if either $S_1 \cup S_2$ is not a q-string or $S_1 \subset S_2$ or $S_2 \subset S_1$. A set of more than two q-strings is in ***general position*** if every pair of q-strings in them are in general position.

We denote by $V(m)$ the $(m+1)$-dimensional irreducible $U_q(\mathfrak{sl}_2)$-module with basis $\{u, f_1, \ldots, f_1^m u\}$, where $e_1 u = 0$ and $K_1 u = q^m u$. Then $V(m)$ becomes a $U_q'(\widehat{\mathfrak{sl}}_2)$-module by defining

$$\begin{aligned}
K_0 f_1^j u &= q^{-m+2j} f_1^j u, \\
e_0 f_1^j u &= f_1^{j+1} u, \\
f_0 f_1^j u &= j(m-j+1) f_1^{j-1} u.
\end{aligned} \tag{9.13}$$

For a nonzero complex number ζ, let $V(m)_\zeta$ denote the evaluation module of $V(m)$ at $z = \zeta$. In [**6**], Chari and Pressley proved the precise condition under which the tensor products of evaluation modules $V(m)_\zeta$'s are irreducible.

Proposition 9.2.3 ([**6**]). *Let $V(m_i)$ be the (m_i+1)-dimensional $U_q'(\widehat{\mathfrak{sl}}_2)$-module defined by* (9.13) *and let ζ_i be nonzero complex numbers* ($i = 1, \ldots, r$). *Then the tensor product of evaluation modules $V(m_1)_{\zeta_1} \otimes \cdots \otimes V(m_r)_{\zeta_r}$ is irreducible if and only if the q-strings $S_{m_1}(\zeta_1), \ldots, S_{m_r}(\zeta_r)$ are in general position.*

For $i = 1, 2, 3$, let $V(m_i)$ be the (m_i+1)-dimensional $U_q'(\widehat{\mathfrak{sl}}_2)$-module defined by (9.13) and let ζ_i be nonzero complex numbers such that the q-strings $S_{m_i}(\zeta_i)$ are in general position. Consider the evaluation module $V(m_i)_{\zeta_i}$ ($i = 1, 2, 3$). Observe that the $\mathbf{C}$-linear isomorphism

$$(V(m_1)_{\zeta_1} \otimes V(m_2)_{\zeta_2}) \otimes V(m_3)_{\zeta_3} \xrightarrow{\sim} V(m_1)_{\zeta_1} \otimes (V(m_2)_{\zeta_2} \otimes V(m_3)_{\zeta_3})$$

defined by $(v_1 \otimes v_2) \otimes v_3 \mapsto v_1 \otimes (v_2 \otimes v_3)$ is a $U_q'(\widehat{\mathfrak{sl}}_2)$-module isomorphism due to the coassociativity of the comultiplication on $U_q'(\widehat{\mathfrak{sl}}_2)$. However, the $\mathbf{C}$-linear isomorphism $V(m_i)_{\zeta_i} \otimes V(m_j)_{\zeta_j} \longrightarrow V(m_j)_{\zeta_j} \otimes V(m_i)_{\zeta_i}$ given by $v_i \otimes v_j \mapsto v_j \otimes v_i$ ($i, j = 1, 2, 3$) is *not* a $U_q'(\widehat{\mathfrak{sl}}_2)$-module isomorphism.

A $\mathbf{C}$-linear isomorphism $R : V(m_i)_{\zeta_i} \otimes V(m_j)_{\zeta_j} \to V(m_j)_{\zeta_j} \otimes V(m_i)_{\zeta_i}$ is called an ***R-matrix*** if it is an *intertwiner* between $U_q'(\widehat{\mathfrak{sl}}_2)$-modules; that is, if it satisfies

$$\begin{aligned}
&R \circ \Delta(u) = \Delta(u) \circ R \quad \text{for all } u \in U_q'(\widehat{\mathfrak{sl}}_2), \\
&R \circ (\zeta_1 \otimes 1) = (1 \otimes \zeta_1) \circ R, \\
&R \circ (1 \otimes \zeta_2) = (\zeta_2 \otimes 1) \circ R.
\end{aligned}$$

For $i = 1, 2, 3$, let u_i denote the maximal vector of the $U_q(\mathfrak{sl}_2)$-module $V(m_i)$; i.e., $e_1 u_i = 0$, $K_1 u_i = q^{m_i} u_i$. We normalize the R-matrix by

$$R(u_i \otimes u_j) = u_j \otimes u_i \quad \text{for } i, j = 1, 2, 3.$$

Then the R-matrix satisfies the *Yang-Baxter equation*:

Proposition 9.2.4. *As linear maps from* $V(m_1)_{\zeta_1} \otimes V(m_2)_{\zeta_2} \otimes V(m_3)_{\zeta_3}$ *to* $V(m_3)_{\zeta_3} \otimes V(m_2)_{\zeta_2} \otimes V(m_1)\zeta_1$*, we have*

$$R_{12}\left(\frac{\zeta_3}{\zeta_2}\right) R_{23}\left(\frac{\zeta_3}{\zeta_1}\right) R_{12}\left(\frac{\zeta_2}{\zeta_1}\right) = R_{23}\left(\frac{\zeta_2}{\zeta_1}\right) R_{12}\left(\frac{\zeta_3}{\zeta_1}\right) R_{23}\left(\frac{\zeta_3}{\zeta_2}\right),$$

where $R_{12} = R \otimes \mathrm{id}$ *and* $R_{23} = \mathrm{id} \otimes R$.

Proof. Consider the *hexagon diagram*:

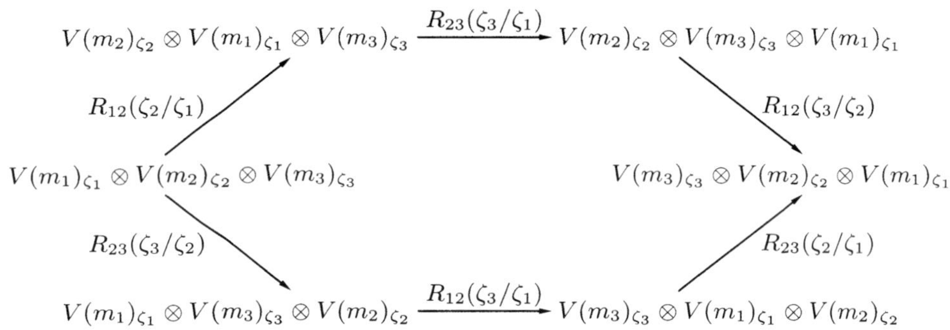

Let

$$F = R_{12}\left(\frac{\zeta_3}{\zeta_2}\right) R_{23}\left(\frac{\zeta_3}{\zeta_1}\right) R_{12}\left(\frac{\zeta_2}{\zeta_1}\right),$$

$$G = R_{23}\left(\frac{\zeta_2}{\zeta_1}\right) R_{12}\left(\frac{\zeta_3}{\zeta_1}\right) R_{23}\left(\frac{\zeta_3}{\zeta_2}\right).$$

Since the $U_q'(\widehat{\mathfrak{sl}}_2)$-modules $V(m_1)_{\zeta_1} \otimes V(m_2)_{\zeta_2} \otimes V(m_3)_{\zeta_3}$ and $V(m_3)_{\zeta_3} \otimes V(m_2)_{\zeta_2} \otimes V(m_1)_{\zeta_1}$ are irreducible, by Schur's Lemma, F is a scalar multiple of G. Since $F(u_1 \otimes u_2 \otimes u_3) = u_3 \otimes u_2 \otimes u_1 = G(u_1 \otimes u_2 \otimes u_3)$, we have $F = G$. □

Consider the two-dimensional $U_q'(\widehat{\mathfrak{sl}}_2)$-module $V = \mathbf{C}v_+ \oplus \mathbf{C}v_-$ defined in Example 9.2.2. Let ζ_i $(i = 1, 2)$ be nonzero complex numbers such that $S_1(\zeta_i) = \{\zeta_i\}$ $(i = 1, 2)$ are in general position, and let V_{ζ_i} denote the evaluation module of V at $z = \zeta_i$. We will calculate the R-matrix $R : V_{\zeta_1} \otimes V_{\zeta_2} \to V_{\zeta_2} \otimes V_{\zeta_1}$ and show that it is a scalar multiple of the R-matrix for the 6-vertex model.

We normalize the R-matrix by $R(v_+ \otimes v_+) = v_+ \otimes v_+$. Since $v_- \otimes v_- = f_1^{(2)}(v_+ \otimes v_+)$, the intertwining property of R yields

$$R(v_- \otimes v_-) = v_- \otimes v_-.$$

Since R preserves the weights, we may write

$$R(v_+ \otimes v_-) = Av_+ \otimes v_- + Bv_- \otimes v_+,$$
$$R(v_- \otimes v_+) = Cv_+ \otimes v_- + Dv_- \otimes v_+$$

for some $A, B, C, D \in \mathbf{C}$. By the intertwining property of R, we have the following commutative diagram.

$$\begin{array}{ccc} V_{\zeta_1} \otimes V_{\zeta_2} & \xrightarrow{\quad R \quad} & V_{\zeta_2} \otimes V_{\zeta_1} \\ {\scriptstyle \Delta(x)} \downarrow & & \downarrow {\scriptstyle \Delta(x)} \\ V_{\zeta_1} \otimes V_{\zeta_2} & \xrightarrow{\quad R \quad} & V_{\zeta_2} \otimes V_{\zeta_1} \end{array}$$

Thus we have

$$\begin{aligned} \Delta(f_1)R(v_+ \otimes v_-) &= (f_1 \otimes 1 + K_1 \otimes f_1)(Av_+ \otimes v_- + Bv_- \otimes v_+) \\ &= Av_- \otimes v_- + q^{-1}Bv_- \otimes v_- = (A + q^{-1}B)(v_- \otimes v_-) \end{aligned}$$

and

$$\begin{aligned} R\Delta(f_1)(v_+ \otimes v_-) &= R(f_1 \otimes 1 + K_1 \otimes f_1)(v_+ \otimes v_-) \\ &= R(v_- \otimes v_-) = v_- \otimes v_-, \end{aligned}$$

which implies

$$A + q^{-1}B = 1. \tag{9.14}$$

On the other hand,

$$\begin{aligned} \Delta(f_0)R(v_+ \otimes v_-) &= (f_0 \otimes 1 + K_0 \otimes f_0)(Av_+ \otimes v_- + Bv_- \otimes v_+) \\ &= Aq^{-1}\zeta_1^{-1}v_+ \otimes v_+ + B\zeta_2^{-1}v_+ \otimes v_+ \\ &= (Aq^{-1}\zeta_1^{-1} + B\zeta_2^{-1})(v_+ \otimes v_+) \end{aligned}$$

and

$$\begin{aligned} R\Delta(f_0)(v_+ \otimes v_-) &= R(f_0 \otimes 1 + K_0 \otimes f_0)(v_+ \otimes v_-) \\ &= R(q^{-1}\zeta_2^{-1}v_+ \otimes v_+) \\ &= q^{-1}\zeta_2^{-1}v_+ \otimes v_+. \end{aligned}$$

It follows that

$$Aq^{-1}\zeta_1^{-1} + B\zeta_2^{-1} = q^{-1}\zeta_2^{-1}. \tag{9.15}$$

Combining (9.14) and (9.15) yields

$$A = \frac{1 - q^2}{x - q^2}, \qquad B = \frac{q(x - 1)}{x - q^2}, \quad \text{where } x = \frac{\zeta_2}{\zeta_1}.$$

Similarly, we have

$$C = \frac{q(x - 1)}{x - q^2}, \qquad D = \frac{x(1 - q^2)}{x - q^2}. \tag{9.16}$$

Therefore, we obtain

$$R = \begin{pmatrix} 1 & 0 & 0 & 0 \\ 0 & \frac{1-q^2}{x-q^2} & \frac{q(x-1)}{x-q^2} & 0 \\ 0 & \frac{q(x-1)}{x-q^2} & \frac{x(1-q^2)}{x-q^2} & 0 \\ 0 & 0 & 0 & 1 \end{pmatrix}.$$

Multiplying each entry by $x-q^2$, we get the R-matrix for the 6-vertex model. Hence the Boltzmann weights of the 6-vertex model can be interpreted as the entries of the $U_q'(\widehat{\mathfrak{sl}}_2)$-module isomorphism $R : V_{\zeta_1} \otimes V_{\zeta_2} \to V_{\zeta_2} \otimes V_{\zeta_1}$ and the Yang-Baxter equation follows immediately from the intertwining properties of R.

Now, we further investigate the connection between the theory of the 6-vertex model and the representation theory of the quantum affine algebra $U_q(\widehat{\mathfrak{sl}}_2)$. We will show that the *one-point function* for the 6-vertex model is essentially the same as the quotient of the *string function* by the *character* for the basic representation of $U_q(\widehat{\mathfrak{sl}}_2)$.

Recall that the fundamental weight Λ_0 is the linear functional on $\mathfrak{h} = \mathbf{C} \otimes_{\mathbf{Z}} P^\vee$ defined by

$$\Lambda_0(h_0) = 1, \qquad \Lambda_0(h_1) = 0, \qquad \Lambda_0(d) = 0.$$

The ***basic representation*** of the quantum affine algebra $U_q(\widehat{\mathfrak{sl}}_2)$ is the irreducible highest weight module $V(\Lambda_0)$ over $U_q(\widehat{\mathfrak{sl}}_2)$ with highest weight Λ_0. Let v_{Λ_0} denote the highest weight vector of $V(\Lambda_0)$. Then as a $U_q(\widehat{\mathfrak{sl}}_2)$-module, $V(\Lambda_0)$ is generated by v_{Λ_0} with defining relations

$$\begin{aligned} &q^h v_{\Lambda_0} = q^{\Lambda_0(h)} v_{\Lambda_0} \quad (h \in P^\vee), \\ &e_i v_{\Lambda_0} = 0 \quad (i = 0, 1), \\ &f_0^2 v_{\Lambda_0} = 0, \qquad f_1 v_{\Lambda_0} = 0. \end{aligned}$$

The weight multiplicities of $V(\Lambda_0)$ are determined by the formula

$$\dim V(\Lambda_0)_\lambda = p\left(1 - \frac{(\lambda|\lambda)}{2}\right),$$

where $p(n)$ denotes the number of partitions of n into a sum of positive integers and $(\ |\ )$ is the nondegenerate symmetric bilinear form on $\mathfrak{h}^*$ satisfying

$$\begin{aligned} &(\Lambda_0|\Lambda_0) = 2, \qquad (\Lambda_0|\alpha_0) = 1, \qquad (\Lambda_0|\alpha_1) = 0, \\ &(\alpha_0|\alpha_0) = 2, \qquad (\alpha_0|\alpha_1) = -2, \qquad (\alpha_1|\alpha_1) = 2 \end{aligned}$$

(see, for example, [**13, 14, 28**]).

Let λ be a weight of $V(\Lambda_0)$. Then it can be written in the form

$$\begin{aligned}\lambda &= \Lambda_0 - a_0\alpha_0 - a_1\alpha_1 = \Lambda_0 - a_0(\alpha_0+\alpha_1) + (a_0 - a_1)\alpha_1 \\ &= \Lambda_0 + \frac{a}{2}\alpha_1 - n\delta,\end{aligned}$$

where $a_0, a_1 \in \mathbf{Z}_{\geq 0}$, $a = 2(a_0 - a_1) \in 2\mathbf{Z}$, $n = a_0 \in \mathbf{Z}_{\geq 0}$. We define the function $S'(a)$ $(a \in 2\mathbf{Z})$ to be

$$S'(a) = \sum_{n=0}^{\infty} \dim V(\Lambda_0)_{\Lambda_0 + \frac{a}{2}\alpha_1 - n\delta} q^{2n}.$$

Notice that this is part of the character for the irreducible representation $V(\Lambda_0)$ with some specializations. We will see in the next section that $S'(a)$ is a modified version of a *string function*. One can easily verify that

$$\begin{aligned}S'(a) &= \sum_{n=0}^{\infty} p\left(n - \frac{a^2}{4}\right) q^{2n} = q^{\frac{a^2}{2}} \sum_{n=0}^{\infty} p(n) q^{2n} \\ &= \frac{q^{\frac{a^2}{2}}}{\prod_{n=1}^{\infty}(1 - q^{2n})} = G'(a).\end{aligned}$$

Recall that

$$Z = \sum_{a \in 2\mathbf{Z}} G(a),$$

and notice that $\operatorname{ch} V(\Lambda_0)$ looks almost the same after substituting q in place of $e^{-\alpha_i}$ $(i = 0, 1)$.

$$\operatorname{ch} V(\Lambda_0)|_{(e^{-\alpha_i} = q)} = e^{\Lambda_0} \sum_{a \in 2\mathbf{Z}} q^{-\frac{a}{2}} S'(a) = e^{\Lambda_0} \sum_{a \in 2\mathbf{Z}} q^{-\frac{a}{2}} G'(a).$$

We have stated that $G(a)$ is *essentially* the same as $G'(a) = S'(a)$. So we may expect the partition function Z to be *essentially* equal to the character $\operatorname{ch} V(\Lambda_0)$.

Actually, in a sense which we shall make precise in the next section, the one-point function $F(a)$ may be written as

$$F(a) = \frac{G(a)}{Z} \sim \frac{\text{string function}}{\operatorname{ch} V(\Lambda_0)}.$$

9.3. Crystals and paths

In the previous section, we have seen some evidence that the one-point function for the 6-vertex model coincides with the quotient of the string function by the character for the basic representation of the quantum affine algebra $U_q(\widehat{\mathfrak{sl}}_2)$. In this section, we will give an explanation of this phenomenon using the *crystal basis theory.* Most of the proofs will be postponed to Chapter 10, where we will develop the general theory of *perfect crystals* for quantum affine algebras.

Let $V = \mathbf{C}v_+ \oplus \mathbf{C}v_-$ be the two-dimensional $U_q'(\widehat{\mathfrak{sl}}_2)$-module defined in Example 9.2.2 with crystal graph

$\mathcal{B}$:

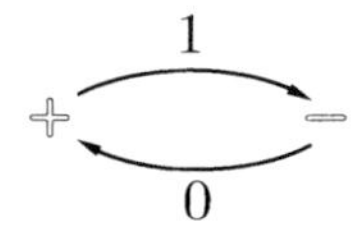

and let $\mathcal{B}(\Lambda_i)$ denote the crystal graph of the basic representation $V(\Lambda_i)$ of $U_q(\widehat{\mathfrak{sl}}_2)$ $(i = 0, 1)$. Recall that the evaluation module V_1 of V at $z = 1$ coincides with V and its crystal graph is the same as $\mathcal{B}$ given above.

Consider the tensor product of crystals $\mathcal{B}(\Lambda_0) \otimes \mathcal{B}$. It is easy to check that $u_{\Lambda_0} \otimes +$ is a maximal vector of weight Λ_1, where u_{Λ_0} is the highest weight vector in $\mathcal{B}(\Lambda_0)$. Moreover $u_{\Lambda_0} \otimes +$ is the *only* maximal vector in $\mathcal{B}(\Lambda_0) \otimes \mathcal{B}$. Indeed by the tensor product rule for Kashiwara operators, the only possible maximal vectors in $\mathcal{B}(\Lambda_0) \otimes \mathcal{B}$ are $u_{\Lambda_0} \otimes +$ and $u_{\Lambda_0} \otimes -$. But, since $\varphi_1(u_{\Lambda_0}) = 0$ and $\varepsilon_1(-) = 1$,

$$\tilde{e}_1(u_{\Lambda_0} \otimes -) = u_{\Lambda_0} \otimes \tilde{e}_1(-) = u_{\Lambda_0} \otimes + \neq 0.$$

We now show that every vector in $\mathcal{B}(\Lambda_0) \otimes \mathcal{B}$ is connected to $u_{\Lambda_0} \otimes +$ by applying $\tilde{e}_0$ and $\tilde{e}_1$. Let $u \otimes \varepsilon \in \mathcal{B}(\Lambda_0) \otimes \mathcal{B}$, where $\varepsilon = \pm$. The tensor product rule for Kashiwara operators implies that, for any $i \in I = \{0, 1\}$, the operator $\tilde{e}_i$ would act on the second factor first if possible and then it would act on the first factor. Hence there exist $j_1, \ldots, j_s \in I$ with r sufficeintly large such that

$$\tilde{e}_{j_s} \cdots \tilde{e}_{j_1}(u \otimes \varepsilon) = u_{\Lambda_0} \otimes \varepsilon', \quad \text{where } \varepsilon, \varepsilon' = \pm.$$

If $\varepsilon' = +$, we are done. If $\varepsilon' = -$, then applying $\tilde{e}_1$ would give $u_{\Lambda_0} \otimes +$.

This structure allows us to expect with a high possibility that there exists an isomorphism of crystals

$$\mathcal{B}(\Lambda_0) \otimes \mathcal{B} \xrightarrow{\sim} \mathcal{B}(\Lambda_1) \quad \text{given by} \quad u_{\Lambda_0} \otimes + \longmapsto u_{\Lambda_1}.$$

This will be proved in the next chapter. Similarly, one can verify that $u_{\Lambda_1} \otimes -$ is the only maximal vector in $\mathcal{B}(\Lambda_1) \otimes \mathcal{B}$ and every vector in $\mathcal{B}(\Lambda_1) \otimes \mathcal{B}$ is connected to $u_{\Lambda_1} \otimes -$ by $\tilde{e}_0$ and $\tilde{e}_1$. Therefore, we may also conclude that there is an isomorphism of crystals

$$\mathcal{B}(\Lambda_1) \otimes \mathcal{B} \xrightarrow{\sim} \mathcal{B}(\Lambda_0) \quad \text{given by} \quad u_{\Lambda_1} \otimes - \longmapsto u_{\Lambda_0}.$$

In Chapter 10, we will give a rigorous proof of these statements in a more general context using the theory of perfect crystals.

By taking the composition of the above crystal isomorphisms repeatedly, we observe that there is a crystal isomorphism

$$\mathcal{B}(\Lambda_0) \xrightarrow{\sim} \mathcal{B}(\Lambda_1) \otimes \mathcal{B} \xrightarrow{\sim} \mathcal{B}(\Lambda_0) \otimes \mathcal{B} \otimes \mathcal{B} \xrightarrow{\sim} \cdots$$

given by

$$u_{\Lambda_0} \longmapsto u_{\Lambda_1} \otimes - \longmapsto u_{\Lambda_0} \otimes + \otimes - \longmapsto \cdots .$$

Thus we can identify the highest weight vector u_{Λ_0} in $\mathcal{B}(\Lambda_0)$ with the infinite sequence

$$\begin{aligned} \mathbf{p}_{\Lambda_0} &= (\mathbf{p}_{\Lambda_0}(k))_{k=0}^{\infty} = (\dots, +, -, +, -, +, -) \\ &= \cdots \otimes + \otimes - \otimes + \otimes - \in \mathcal{B}^{\otimes\infty}. \end{aligned}$$

Similarly, the highest weight vector u_{Λ_1} in $\mathcal{B}(\Lambda_1)$ can be identified with the infinite sequence

$$\begin{aligned} \mathbf{p}_{\Lambda_1} &= (\mathbf{p}_{\Lambda_1}(k))_{k=0}^{\infty} = (\dots, -, +, -, +, -, +) \\ &= \cdots \otimes - \otimes + \otimes - \otimes + \in \mathcal{B}^{\otimes\infty}. \end{aligned}$$

Recall that

$$\mathcal{B}(\Lambda_0) = \{\tilde{f}_{i_r} \cdots \tilde{f}_{i_1} u_{\Lambda_0} \mid r \geq 0, i_k = 0, 1\} \setminus \{0\}.$$

Let $u = \tilde{f}_{i_r} \cdots \tilde{f}_{i_1} u_{\Lambda_0} \in \mathcal{B}(\Lambda_0)$. Observe that, under the crystal isomorphism given above, we have

$$\begin{aligned} \tilde{f}_{i_1} u_{\Lambda_0} &\longmapsto \tilde{f}_{i_1}(u_{\Lambda_1} \otimes -) = u_{\Lambda_1} \otimes \tilde{f}_{i_1}(-) \doteq u_{\Lambda_1} \otimes \varepsilon_1 \\ &\longmapsto \cdots \otimes + \otimes - \otimes + \otimes \varepsilon_1 \qquad (\varepsilon_1 = \pm), \\ \tilde{f}_{i_2}\tilde{f}_{i_1} u_{\Lambda_0} &\longmapsto \tilde{f}_{i_2}(u_{\Lambda_1} \otimes \varepsilon_1) \longmapsto \tilde{f}_{i_2}(u_{\Lambda_0} \otimes + \otimes \varepsilon_1) \\ &= \begin{cases} \tilde{f}_{i_2}(u_{\Lambda_0} \otimes +) \otimes \varepsilon_1, \\ u_{\Lambda_0} \otimes + \otimes \tilde{f}_{i_2}\varepsilon_1 \end{cases} \\ &= u_{\Lambda_0} \otimes \varepsilon_2' \otimes \varepsilon_1' \longmapsto \cdots \otimes + \otimes - \otimes \varepsilon_2' \otimes \varepsilon_1' \qquad (\varepsilon_1', \varepsilon_2' = \pm), \\ &\ \ \vdots \\ \tilde{f}_{i_r} \cdots \tilde{f}_{i_1} u_{\Lambda_0} &\longmapsto \cdots \otimes + \otimes - \otimes \cdots \otimes \varepsilon_r'' \otimes \cdots \varepsilon_2'' \otimes \varepsilon_1'' \qquad (\varepsilon_j'' = \pm). \end{aligned}$$

Hence every vector $u \in \mathcal{B}(\Lambda_0)$ can be identified with an infinite sequence

$$\mathbf{p} = (\mathbf{p}(k))_{k=0}^{\infty} = \cdots \otimes \mathbf{p}(k) \otimes \cdots \otimes \mathbf{p}(1) \otimes \mathbf{p}(0)$$

such that $\mathbf{p}(k) = \mathbf{p}_{\Lambda_0}(k) = (-)^{k+1}$ for all sufficiently large k. Recall that such an infinite sequence is called a Λ_0-*path*. The set $\mathcal{P}(\Lambda_0)$ of all Λ_0-paths is given a $U_q'(\widehat{\mathfrak{sl}}_2)$-crystal structure by the above identification of paths with vectors in $\mathcal{B}^{\otimes\infty}$, and we obtain a crystal isomorphism

$$\mathcal{P}(\Lambda_0) \cong \mathcal{B}(\Lambda_0). \tag{9.17}$$

In the following example, we illustrate the crystal graph structure of $\mathcal{P}(\Lambda_0)$.

Example 9.3.1. The crystal structure of $\mathcal{P}(\Lambda_0)$ is shown below.

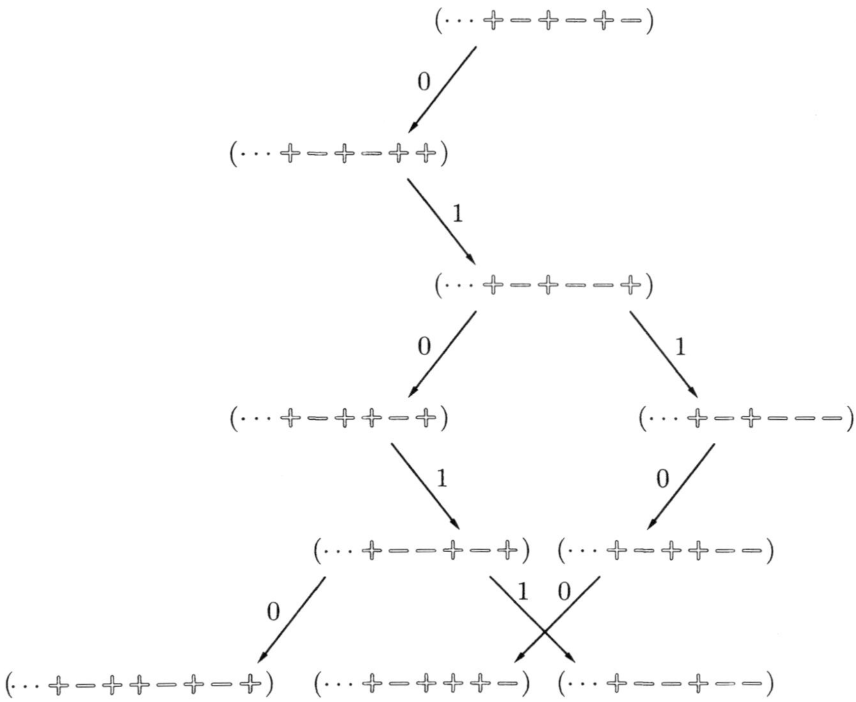

To express the character and the string function of the basic representation $V(\Lambda_0)$ of $U_q(\widehat{\mathfrak{sl}}_2)$, we need to calculate the *affine weight* of a Λ_0-path $\mathbf{p} = (\mathbf{p}(k))_{k=0}^{\infty} \in \mathcal{P}(\Lambda_0)$. We denote by $\operatorname{wt}\mathbf{p}$ (respectively, $\overline{\operatorname{wt}}\mathbf{p}$) the affine weight (respectively, the classical weight) of a Λ_0-path $\mathbf{p}$.

Since $\overline{\operatorname{wt}}(+) = \Lambda_1 - \Lambda_0$, $\overline{\operatorname{wt}}(-) = \Lambda_0 - \Lambda_1$, and $\mathbf{p}(k) = \mathbf{p}_{\Lambda_0}(k) = (-)^{k+1}$ for all sufficiently large k, the *classical weight* of $\mathbf{p}$ can be calculated easily:

$$\overline{\operatorname{wt}}\mathbf{p} = \Lambda_0 + \sum_{k=0}^{\infty}(\overline{\operatorname{wt}}\mathbf{p}(k) - \overline{\operatorname{wt}}\mathbf{p}_{\Lambda_0}(k)) = \Lambda_0 + a(\Lambda_1 - \Lambda_0)$$

for some $a \in 2\mathbf{Z}$. Note that $\overline{\operatorname{wt}}\mathbf{p} = \Lambda_0 + a(\Lambda_1 - \Lambda_0)$ if and only if $W(0, \mathbf{p}) = a$ (Exercise 9.7).

In Section 10.6, we will show that the affine weight of $\mathbf{p}$ is determined by the formula

$$\begin{aligned}\operatorname{wt}\mathbf{p} = \Lambda_0 &+ \sum_{k=0}^{\infty}(\overline{\operatorname{wt}}\mathbf{p}(k) - \overline{\operatorname{wt}}\mathbf{p}_{\Lambda_0}(k)) \\ &- \left(\sum_{k=0}^{\infty}(k+1)\left(H(\mathbf{p}(k+1),\mathbf{p}(k)) - H(\mathbf{p}_{\Lambda_0}(k+1),\mathbf{p}_{\Lambda_0}(k))\right)\right)\delta,\end{aligned}$$

where H is the energy function for the 6-vertex model. Thus we may write

$$\operatorname{wt}\mathbf{p} = \Lambda_0 + \frac{a}{2}\alpha_1 - n(\mathbf{p})\delta,$$

where $W(0,\mathbf{p}) = a$ and

$$n(\mathbf{p}) = \sum_{k=0}^{\infty}(k+1)\left(H(\mathbf{p}(k+1),\mathbf{p}(k)) - H(\mathbf{p}_{\Lambda_0}(k+1),\mathbf{p}_{\Lambda_0}(k))\right).$$

Set

$$\mathcal{P}(\Lambda_0;a) = \{\mathbf{p} \in \mathcal{P}(\Lambda_0) \mid \overline{\operatorname{wt}}\mathbf{p} = \Lambda_0 + a(\Lambda_1 - \Lambda_0)\}.$$

Then the *corner transfer matrix method* of Baxter [**3**] tells us that

$$G(a) = q^{-4(\rho\,|\,\Lambda_0 + \frac{a}{2}\alpha_1)} \sum_{\mathbf{p}\in\mathcal{P}(\Lambda_0;a)} q^{4(\rho\,|\,\delta)n(\mathbf{p})}.$$

We define the string function $S(a)$ to be

$$S(a) = \sum_{\mathrm{cl}(\mu)=\Lambda_0+\frac{a}{2}\alpha_1} \dim V(\Lambda_0)_\mu e^\mu.$$

The sum is over all weights μ of $V(\Lambda_0)$ such that $\mathrm{cl}(\mu) = \Lambda_0 + \frac{a}{2}\alpha_1$. We also define the operator $(q^{-4\rho}\,|\quad)$ on the group algebra $\mathbf{C}[P]$ by

$$(q^{-4\rho}\,|\,e^\mu) = q^{-4(\rho\,|\,\mu)}.$$

Then from the crystal isomorphism (9.17) and our discussion on the weight of a path, we may easily conclude

$$G(a) = (q^{-4\rho}\,|\,S(a)).$$

We can also notice that

$$Z = \sum_{a\in\mathbf{Z}} G(a) = (q^{-4\rho}\,|\,\operatorname{ch}V(\Lambda_0)).$$

Hence the one-point function is given by

$$F(a) = \frac{G(a)}{Z} = \left(q^{-4\rho}\,\Big|\,\frac{S(a)}{\operatorname{ch}V(\Lambda_0)}\right),$$

which is a specialization of the quotient of the string function by the character of a basic representation. (See [**36**] for more detail.)

Exercises

9.1. Show by a direct calculation that the R-matrix given by (9.3) satisfies the Yang-Baxter equation.

9.2. Show that equation (9.6) defines a weight configuration on any configuration satisfying the 6-vertex condition. Show that it is uniquely defined for configurations in the 0-sector by the boundary condition (9.7).

9.3. Compute the sum (9.11) to obtain (9.12)

9.4. Verify that the equation (9.13) defines a $U'_q(\widehat{\mathfrak{sl}}_2)$-module structure on $V(m)$.

9.5. Using the intertwining properties of the R-matrix, verify the equation (9.16).

9.6. Using the tensor product rule for the Kashiwara operators, verify the crystal structure of $\mathcal{P}(\Lambda_0)$ given in Example 9.3.1.

9.7. Show that for a Λ_0-path $\mathbf{p}$, we have $\overline{\mathrm{wt}\,\mathbf{p}} = \Lambda_0 + a(\Lambda_0 - \Lambda_1)$ if and only if $W(\star, \mathbf{p}) = a$.

Chapter 10

Perfect Crystals

In this chapter, we develop the theory of *perfect crystals* for quantum affine algebras. We will first study the properties of *vertex operators* and then prove a fundamental crystal isomorphism theorem. Using this crystal isomorphism, the crystal graph $\mathcal{B}(\lambda)$ of the irreducible highest weight $U_q(\mathfrak{g})$-module $V(\lambda)$ will be realized in terms of certain *paths*.

10.1. Quantum affine algebras

For the reader's convenience, we recall the definition of quantum affine algebras. Let $I = \{0, 1, \ldots, n\}$ be an index set and let $A = (a_{i,j})_{i,j \in I}$ be a generalized Cartan matrix of affine type. Consider a free abelian group of rank $n + 2$

$$P^{\vee} = \mathbf{Z}h_0 \oplus \mathbf{Z}h_1 \oplus \cdots \oplus \mathbf{Z}h_n \oplus \mathbf{Z}d \tag{10.1}$$

and let $\mathfrak{h} = \mathbf{C} \otimes_{\mathbf{Z}} P^{\vee}$ be its complexification. The free abelian group $P^{\vee}$ is called the ***dual weight lattice*** and the complex vector space $\mathfrak{h}$ is called the ***Cartan subalgebra***.

We define the linear functionals α_i and Λ_i $(i \in I)$ on $\mathfrak{h}$ by

$$\begin{aligned} &\alpha_i(h_j) = a_{ji}, \qquad \alpha_i(d) = \delta_{0,i}, \\ &\Lambda_i(h_j) = \delta_{ij}, \qquad \Lambda_i(d) = 0 \quad (i, j \in I). \end{aligned} \tag{10.2}$$

As in Chapter 2, the α_i (respectively, h_i) are called the ***simple roots*** (respectively, ***simple coroots***) and the Λ_i are called the ***fundamental weights***. We denote by $\Pi = \{\alpha_i \mid i \in I\}$ and $\Pi^{\vee} = \{h_i \mid i \in I\}$ the set of simple roots and simple coroots, respectively.

The ***affine weight lattice*** is defined to be

$$P = \{\lambda \in \mathfrak{h}^* \mid \lambda(P^\vee) \subset \mathbf{Z}\}. \tag{10.3}$$

The quintuple $(A, \Pi, \Pi^\vee, P, P^\vee)$ is called an ***affine Cartan datum***. To each affine Cartan datum, we can associate the ***affine Kac-Moody algebra*** $\mathfrak{g}$ [**28**, Ch.1]. The center of $\mathfrak{g}$ is one-dimensional and is generated by the ***canonical central element***

$$c = c_0 h_0 + c_1 h_1 + \cdots + c_n h_n. \tag{10.4}$$

Moreover the imaginary roots of $\mathfrak{g}$ are nonzero integral multiples of the ***null root***

$$\delta = d_0 \alpha_0 + d_1 \alpha_1 + \cdots + d_n \alpha_n. \tag{10.5}$$

Here, c_i and d_i $(i \in I)$ are the nonnegative integers given in [**28**, Ch.4]. Using the fundamental weights and the null root, the affine weight lattice can be written as

$$P = \mathbf{Z}\Lambda_0 \oplus \mathbf{Z}\Lambda_1 \oplus \cdots \oplus \mathbf{Z}\Lambda_n \oplus \mathbf{Z}\frac{1}{d_0}\delta. \tag{10.6}$$

In fact, $A_{2n}^{(2)}$ is the only case for which $d_0 \neq 1$. The elements of P are called the ***affine weights***.

Set

$$P^+ = \{\lambda \in P \mid \lambda(h_i) \in \mathbf{Z}_{\geq 0} \text{ for all } i \in I\}. \tag{10.7}$$

The elements of P^+ are called the ***affine dominant integral weights*** and the ***level*** of an affine dominant integral weight $\lambda \in P^+$ is defined to be the nonnegative integer $\lambda(c)$.

Example 10.1.1. In this example, we present affine Dynkin diagrams, the canonical central elements, and the null roots for affine Cartan data of classical type.

(1) $A_1^{(1)}$

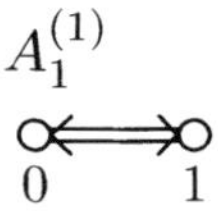

$c = h_0 + h_1,$
$\delta = \alpha_0 + \alpha_1.$

(2) $A_n^{(1)}$ $(n \geq 2)$

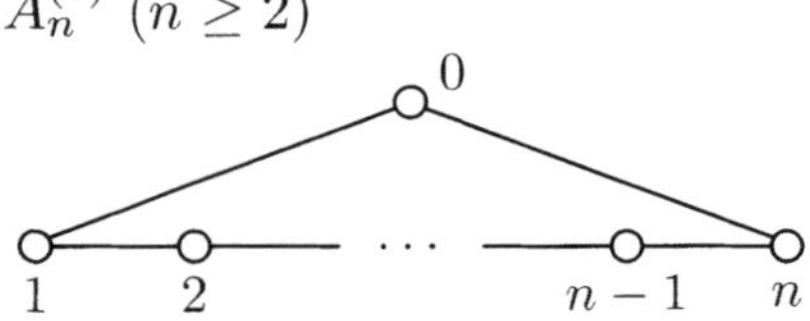

$c = h_0 + h_1 + \cdots + h_n,$
$\delta = \alpha_0 + \alpha_1 + \cdots + \alpha_n.$

(3) $A_{2n-1}^{(2)}$ $(n \geq 3)$

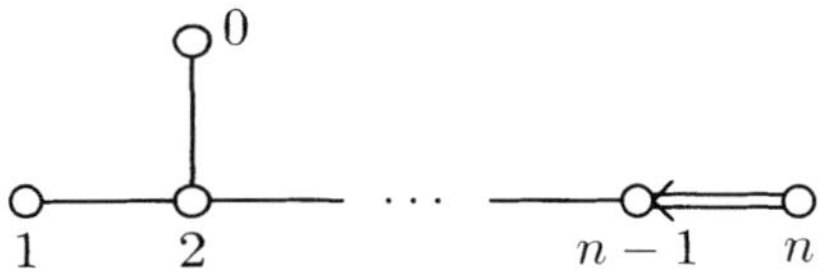

$$c = h_0 + h_1 + 2h_2 + \cdots + 2h_{n-1} + 2h_n,$$
$$\delta = \alpha_0 + \alpha_1 + 2\alpha_2 + \cdots + 2\alpha_{n-1} + \alpha_n.$$

(4) $D_n^{(1)}$ $(n \geq 4)$

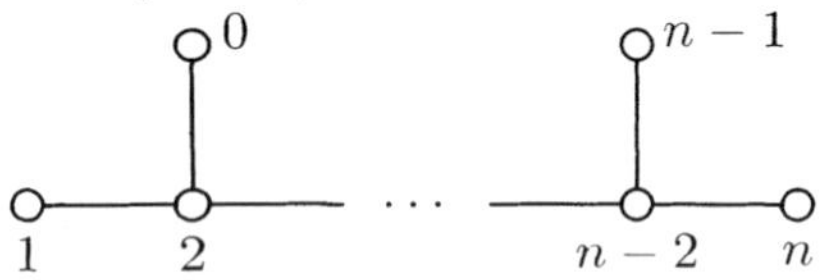

$$c = h_0 + h_1 + 2h_2 + \cdots + 2h_{n-2} + h_{n-1} + h_n,$$
$$\delta = \alpha_0 + \alpha_1 + 2\alpha_2 + \cdots + 2\alpha_{n-2} + \alpha_{n-1} + \alpha_n.$$

(5) $A_{2n}^{(2)}$ $(n \geq 2)$

$$c = h_0 + 2h_1 + \cdots + 2h_{n-1} + 2h_n,$$
$$\delta = 2\alpha_0 + 2\alpha_1 + \cdots + 2\alpha_{n-1} + \alpha_n.$$

(6) $D_{n+1}^{(2)}$ $(n \geq 2)$

$$c = h_0 + 2h_1 + \cdots + 2h_{n-1} + h_n,$$
$$\delta = \alpha_0 + \alpha_1 + \cdots + \alpha_{n-1} + \alpha_n.$$

(7) $B_n^{(1)}$ $(n \geq 3)$

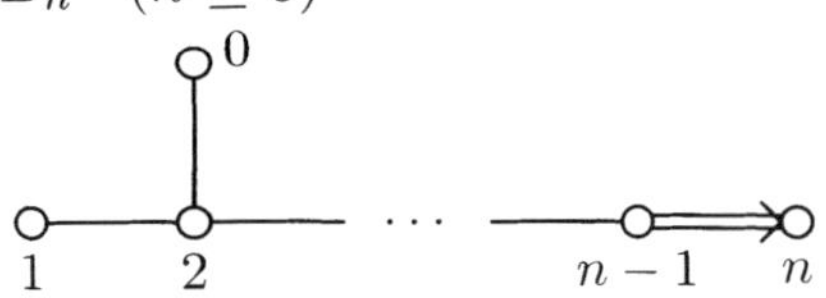

$$c = h_0 + h_1 + 2h_2 + \cdots + 2h_{n-1} + h_n,$$
$$\delta = \alpha_0 + \alpha_1 + 2\alpha_2 + \cdots + 2\alpha_{n-1} + 2\alpha_n.$$

(8) $C_n^{(1)}$ $(n \geq 2)$

$$c = h_0 + h_1 + \cdots + h_{n-1} + h_n,$$
$$\delta = \alpha_0 + 2\alpha_1 + \cdots + 2\alpha_{n-1} + \alpha_n.$$

Definition 10.1.2. The ***quantum affine algebra*** $U_q(\mathfrak{g})$ is the quantum group associated with the affine Cartan datum $(A, \Pi, \Pi^\vee, P, P^\vee)$. That is, it is the associative algebra over $\mathbf{C}(q)$ with unity generated by e_i, f_i $(i \in I)$, and q^h $(h \in P^\vee)$ satisfying the defining relations in Definition 3.1.1.

The quantum affine algebra $U_q(\mathfrak{g})$ has a Hopf algebra structure with comultiplication, counit, and antipode as given in Proposition 3.1.2.

The subalgebra of $U_q(\mathfrak{g})$ generated by e_i, f_i, $K_i^{\pm 1}$ $(i \in I)$ is denoted by $U_q'(\mathfrak{g})$ and is also called the ***quantum affine algebra***. Let

$$\bar{P}^\vee = \mathbf{Z}h_0 \oplus \mathbf{Z}h_1 \oplus \cdots \oplus \mathbf{Z}h_n \quad \text{and} \quad \bar{\mathfrak{h}} = \mathbf{C} \otimes_{\mathbf{Z}} \bar{P}^\vee. \tag{10.8}$$

Consider α_i and Λ_i as linear functionals on $\bar{\mathfrak{h}}$ and set

$$\bar{P} = \mathbf{Z}\Lambda_0 \oplus \mathbf{Z}\Lambda_1 \oplus \cdots \oplus \mathbf{Z}\Lambda_n. \tag{10.9}$$

For example, $\delta = 0$ as an element in $\bar{P}$. The elements of $\bar{P}$ are called the ***classical weights***. The quintuple $(A, \Pi, \Pi^\vee, \bar{P}, \bar{P}^\vee)$ is called a ***classical Cartan datum*** and the quantum affine algebra $U_q'(\mathfrak{g})$ can be regarded as the quantum group associated with the classical Cartan datum.

The projection $\mathrm{cl} : P \to \bar{P}$ will be denoted by $\lambda \mapsto \bar{\lambda}$. We will fix an embedding $\mathrm{aff} : \bar{P} \to P$ such that $\mathrm{cl} \circ \mathrm{aff} = \mathrm{id}$, $\mathrm{aff} \circ \mathrm{cl}(\alpha_i) = \alpha_i$ for $i \neq 0$. We define

$$\bar{P}^+ = \mathrm{cl}(P^+) = \{\lambda \in \bar{P} \mid \lambda(h_i) \geq 0 \quad \text{for all } i \in I\}. \tag{10.10}$$

The elements of $\bar{P}^+$ are called the ***classical dominant integral weights***. For simplicity, we will often omit the notations for the projections and the embedding. We will state explicitly whether a linear functional is an affine weight or a classical weight whenever it could cause a confusion. A classical dominant integral weight is said to have ***level*** $l \geq 0$ if $\lambda(c) = l$. Note that it has the same level as its affine counterpart.

The main difference between the algebras $U_q(\mathfrak{g})$ and $U_q'(\mathfrak{g})$ lies in the fact that $U_q'(\mathfrak{g})$ can have finite dimensional irreducible modules while all the nontrivial irreducible $U_q(\mathfrak{g})$-modules are infinite dimensional. On the other hand, the infinite dimensional irreducible $U_q(\mathfrak{g})$-modules in the category $\mathcal{O}^q_{\mathrm{int}}$ have finite dimensional weight spaces, whereas the weight spaces for infinite dimensional irreducible $U_q'(\mathfrak{g})$-modules in the category $\mathcal{O}^q_{\mathrm{int}}$ are infinite dimensional.

Let V be a finite dimensional $U_q'(\mathfrak{g})$-module with weight space decomposition $V = \oplus_{\lambda \in \bar{P}} V_\lambda$, where

$$V_\lambda = \{v \in V \mid K_i v = q_i^{\lambda(h_i)} v \quad \text{for } i \in I\}.$$

For an indeterminate z, set

$$V^{\mathrm{aff}} = \mathbf{C}(q)[z, z^{-1}] \otimes_{\mathbf{C}(q)} V \tag{10.11}$$

and define the $U_q(\mathfrak{g})$-module action on V^{aff} by

$$\begin{aligned}
e_0(z^m \otimes v) &= z^{m+1} \otimes (e_0 v),\\
e_i(z^m \otimes v) &= z^m \otimes (e_i v) \quad \text{for } i \neq 0,\\
f_0(z^m \otimes v) &= z^{m-1} \otimes (f_0 v),\\
f_i(z^m \otimes v) &= z^m \otimes (f_i v) \quad \text{for } i \neq 0,\\
K_i(z^m \otimes v) &= z^m \otimes (K_i v),\\
q^d(z^m \otimes v) &= q^{md_0} z^m \otimes v.
\end{aligned} \tag{10.12}$$

The $U_q(\mathfrak{g})$-module V^{aff} thus defined is called the ***affinization*** of V. For $v \in V_\lambda$ and $m \in \mathbf{Z}$, the affine weight of $z^m \otimes v$ is $\text{aff}(\lambda) + m\delta \in P$.

Let $\zeta \in \mathbf{C}(q)$ be a nonzero scalar and let $\mathbf{J}_\zeta$ denote the maximal ideal of $\mathbf{C}(q)[z, z^{-1}]$ generated by $z - \zeta$. Then, there exists an isomorphism of fields

$$\mathbf{C}(q)[z, z^{-1}]/\mathbf{J}_\zeta \xrightarrow{\sim} \mathbf{C}(q)$$

given by $z \mapsto \zeta$. By specializing V^{aff} at $z = \zeta$, we obtain a $U_q'(\mathfrak{g})$-module V_ζ, which is called the ***evaluation module*** of V at ζ:

$$V_\zeta = \mathbf{C}(q) \otimes_{\mathbf{C}(q)[z,z^{-1}]} V^{\text{aff}} \xrightarrow{\sim} V^{\text{aff}}/\mathbf{J}_\zeta V^{\text{aff}}. \tag{10.13}$$

Note that $\mathbf{J}_\zeta V^{\text{aff}}$ is just a $U_q'(\mathfrak{g})$-submodule of V^{aff}, not a $U_q(\mathfrak{g})$-submodule. Hence, for a nonzero scalar $\zeta \in \mathbf{C}(q)$, we can take $V_\zeta = V$ and define the $U_q'(\mathfrak{g})$-module action on V_ζ using the homomorphism $\text{ev} : U_q'(\mathfrak{g}) \longrightarrow \text{End}_{\mathbf{C}(q)} V$ given by

$$\begin{aligned}
\text{ev}(e_i) &= \begin{cases} e_i & \text{if } i \neq 0,\\ \zeta e_0 & \text{if } i = 0,\end{cases}\\
\text{ev}(f_i) &= \begin{cases} f_i & \text{if } i \neq 0,\\ \zeta^{-1} f_0 & \text{if } i = 0,\end{cases}\\
\text{ev}(K_i) &= K_i \quad \text{for } i \in I.
\end{aligned} \tag{10.14}$$

Suppose that V has a crystal basis $(\mathcal{L}, \mathcal{B})$. Then its affinization V^{aff} has a crystal basis $(\mathcal{L}^{\text{aff}}, \mathcal{B}^{\text{aff}})$, where

$$\begin{aligned}
\mathcal{L}^{\text{aff}} &= \mathbf{A}_0[z, z^{-1}] \otimes_{\mathbf{A}_0} \mathcal{L},\\
\mathcal{B}^{\text{aff}} &= \{b(m) \mid m \in \mathbf{Z}, b \in \mathcal{B}\},
\end{aligned}$$

and the action of Kashiwara operators on $\mathcal{B}^{\text{aff}}$ is given by

$$\begin{aligned}
\tilde{e}_0(b(m)) &= (\tilde{e}_0 b)(m+1), & \tilde{f}_0(b(m)) &= (\tilde{f}_0 b)(m-1),\\
\tilde{e}_i(b(m)) &= (\tilde{e}_i b)(m), & \tilde{f}_i(b(m)) &= (\tilde{f}_i b)(m) \quad \text{for } i \neq 0.
\end{aligned}$$

The evaluation module V_ζ has a crystal basis when $\zeta = 1$ and its crystal graph is the same as $\mathcal{B}$.

Example 10.1.3. Consider the quantum affine algebra $U_q(\widehat{\mathfrak{sl}}_3) = U_q(A_2^{(1)})$ and let $V = \mathbf{C}(q)v_1 \oplus \mathbf{C}(q)v_2 \oplus \mathbf{C}(q)v_3$ be the three-dimensional $U'_q(\widehat{\mathfrak{sl}}_3)$-module defined by

$$
\begin{aligned}
e_i v_j &= \begin{cases} v_{j-1} & \text{if } j \equiv i+1 \pmod 3, \\ 0 & \text{otherwise}, \end{cases} \\
f_i v_j &= \begin{cases} v_{j+1} & \text{if } j \equiv i \pmod 3, \\ 0 & \text{otherwise}, \end{cases} \\
K_i v_j &= \begin{cases} q v_j & \text{if } j \equiv i \pmod 3, \\ q^{-1} v_j & \text{if } j \equiv i+1 \pmod 3, \\ v_j & \text{otherwise}. \end{cases}
\end{aligned} \tag{10.15}
$$

Then, $\mathcal{L} = \mathbf{A}_0 v_1 \oplus \mathbf{A}_0 v_2 \oplus \mathbf{A}_0 v_3$ is the crystal lattice of V and its crystal graph $\mathcal{B}$ is given below.

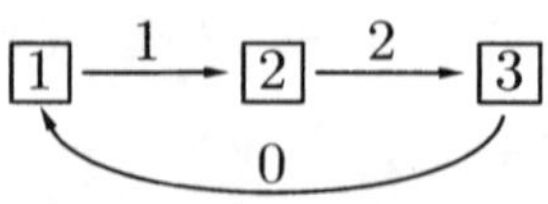

Hence the crystal graph $\mathcal{B}^{\text{aff}}$ is given as follows.

$$\cdots \xrightarrow{2} \boxed{3}(m+1) \xrightarrow{0} \boxed{1}(m) \xrightarrow{1} \boxed{2}(m) \xrightarrow{2} \boxed{3}(m) \xrightarrow{0} \boxed{1}(m-1) \longrightarrow \cdots$$

Moreover the $U'_q(\widehat{\mathfrak{sl}}_3)$-module action on the evaluation module V_ζ is given by

$$
\begin{aligned}
e_i v_j &= \begin{cases} \zeta v_3 & \text{if } i = 0,\ j \equiv 1 \pmod 3, \\ v_{j-1} & \text{if } i \neq 0,\ j \equiv i+1 \pmod 3, \\ 0 & \text{otherwise}, \end{cases} \\
f_i v_j &= \begin{cases} \zeta^{-1} v_1 & \text{if } i = 0,\ j \equiv 0 \pmod 3, \\ v_{j+1} & \text{if } i \neq 0,\ j \equiv i \pmod 3, \\ 0 & \text{otherwise}, \end{cases} \\
K_i v_j &= \begin{cases} q v_j & \text{if } j \equiv i \pmod 3, \\ q^{-1} v_j & \text{if } j \equiv i+1 \pmod 3, \\ v_j & \text{otherwise}. \end{cases}
\end{aligned}
$$

As we have seen in Section 9.2, a $\mathbf{C}(q)$-linear isomorphism

$$R : V_{\zeta_1} \otimes V_{\zeta_2} \longrightarrow V_{\zeta_2} \otimes V_{\zeta_1} \quad (\zeta_1, \zeta_2 \in \mathbf{C}(q)^\times) \tag{10.16}$$

is called an ***R*-matrix** if it is an *intertwiner* between $U_q'(\mathfrak{g})$-modules $V_{\zeta_1} \otimes V_{\zeta_2}$ and $V_{\zeta_2} \otimes V_{\zeta_1}$; i.e., if it satisfies

$$\begin{aligned} R \circ \Delta(u) &= \Delta(u) \circ R \quad \text{for all } u \in U_q'(\mathfrak{g}), \\ R \circ (\zeta_1 \otimes \mathrm{id}) &= (\mathrm{id} \otimes \zeta_1) \circ R, \\ R \circ (\mathrm{id} \otimes \zeta_2) &= (\zeta_2 \otimes \mathrm{id}) \circ R. \end{aligned} \tag{10.17}$$

Using the argument in the proof of Proposition 9.2.4, one can show that the R-matrix R satisfies the *Yang Baxter equation.*

Proposition 10.1.4. *Let $\zeta_i \in \mathbf{C}(q)^\times$ $(i = 1, 2, 3)$ be nonzero scalars such that $V_{\zeta_{i_1}} \otimes V_{\zeta_{i_2}} \otimes V_{\zeta_{i_3}}$ is irreducible for all $i_k = 1, 2, 3$. Then, as $\mathbf{C}(q)$-linear maps $V_{\zeta_1} \otimes V_{\zeta_2} \otimes V_{\zeta_3} \longrightarrow V_{\zeta_3} \otimes V_{\zeta_2} \otimes V_{\zeta_1}$, we have*

$$\begin{aligned} &R_{12}(\zeta_3/\zeta_2)R_{23}(\zeta_3/\zeta_1)R_{12}(\zeta_2/\zeta_1) \\ &\qquad = R_{23}(\zeta_3/\zeta_2)R_{12}(\zeta_3/\zeta_1)R_{23}(\zeta_2/\zeta_1), \end{aligned} \tag{10.18}$$

where $R_{12} = R \otimes \mathrm{id}$ and $R_{23} = \mathrm{id} \otimes R$.

Proof. The proof is left to the readers as an exercise (Exercise 10.2). □

10.2. Energy functions and combinatorial R-matrices

Let V be a finite dimensional $U_q'(\mathfrak{g})$-module with crystal basis $(\mathcal{L}, \mathcal{B})$, and let V^{aff} denote its affinization with crystal basis $(\mathcal{L}^{\mathrm{aff}}, \mathcal{B}^{\mathrm{aff}})$.

Definition 10.2.1. An ***energy function*** on $\mathcal{B}$ is a $\mathbf{Z}$-valued function $H : \mathcal{B} \otimes \mathcal{B} \to \mathbf{Z}$ satisfying the following conditions:

$$H(\tilde{e}_i(b_1 \otimes b_2)) = \begin{cases} H(b_1 \otimes b_2) & \text{if } i \neq 0, \\ H(b_1 \otimes b_2) + 1 & \text{if } i = 0,\ \varphi_0(b_1) \geq \varepsilon_0(b_2), \\ H(b_1 \otimes b_2) - 1 & \text{if } i = 0,\ \varphi_0(b_1) < \varepsilon_0(b_2) \end{cases} \tag{10.19}$$

for all $i \in I$, $b_1 \otimes b_2 \in \mathcal{B} \otimes \mathcal{B}$ with $\tilde{e}_i(b_1 \otimes b_2) \in \mathcal{B} \otimes \mathcal{B}$.

For an energy function H on $\mathcal{B}$, we define the corresponding ***affine energy function***

$$H^{\mathrm{aff}} : \mathcal{B}^{\mathrm{aff}} \otimes \mathcal{B}^{\mathrm{aff}} \to \mathbf{Z} \tag{10.20}$$

by

$$H^{\mathrm{aff}}(b_1(m) \otimes b_2(n)) = m - n - H(b_1 \otimes b_2) \tag{10.21}$$

for $b_1, b_2 \in \mathcal{B}$, $m, n \in \mathbf{Z}$.

Example 10.2.2. Let V be the three-dimensional $U_q'(\widehat{\mathfrak{sl}}_3)$-module defined in Example 10.1.3 and let $\mathcal{B}$ be its crystal graph. By the tensor product rule, the crystal structure on $\mathcal{B} \otimes \mathcal{B}$ is given as follows.

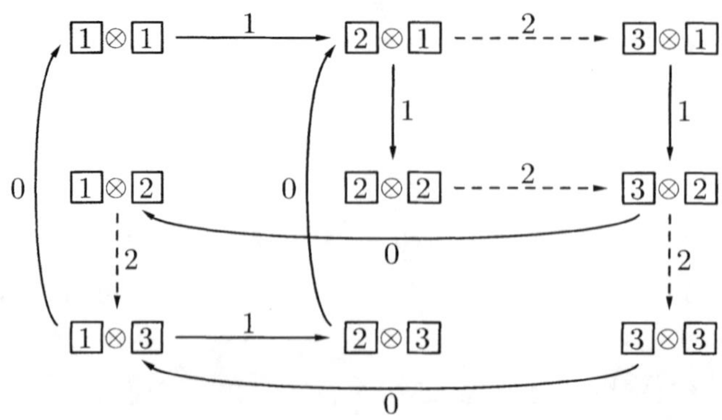

Define a map $H : \mathcal{B} \otimes \mathcal{B} \to \mathbf{Z}$ by

$$H(\boxed{i} \otimes \boxed{j}) = \begin{cases} 1 & \text{if } i \geq j, \\ 0 & \text{if } i < j. \end{cases} \tag{10.22}$$

Then, H is an energy function on $\mathcal{B}$. Hence the corresponding affine energy function $H^{\text{aff}} : \mathcal{B}^{\text{aff}} \otimes \mathcal{B}^{\text{aff}} \to \mathbf{Z}$ is given by

$$H^{\text{aff}}(\boxed{i}(m) \otimes \boxed{j}(n)) = \begin{cases} m - n - 1 & \text{if } i \geq j, \\ m - n & \text{if } i < j. \end{cases} \tag{10.23}$$

Next we define the notion of *R-matrices on crystals* as the intertwiner between the tensor products of affine crystals. An ***affine crystal*** is an abstract crystal associated with an affine Cartan datum $(A, \Pi, \Pi^\vee, P, P^\vee)$. Similarly, a ***classical crystal*** is an abstract crystal associated with a classical Cartan datum $(A, \Pi, \Pi^\vee, \bar{P}, \bar{P}^\vee)$. An affine crystal (respectively, classical crystal) is often referred to as a $U_q(\mathfrak{g})$-crystal (respectively, $U_q'(\mathfrak{g})$-crystal).

Definition 10.2.3. A ***combinatorial R-matrix*** on $\mathcal{B}^{\text{aff}}$ is an endomorphism of affine crystals $R : \mathcal{B}^{\text{aff}} \otimes \mathcal{B}^{\text{aff}} \to \mathcal{B}^{\text{aff}} \otimes \mathcal{B}^{\text{aff}}$ such that

$$\begin{aligned} (T \otimes \text{id}) \circ R &= R \circ (\text{id} \otimes T), \\ (\text{id} \otimes T) \circ R &= R \circ (T \otimes \text{id}), \end{aligned} \tag{10.24}$$

where $T : \mathcal{B}^{\text{aff}} \to \mathcal{B}^{\text{aff}}$ is the shift operator defined by

$$T(b(m)) = b(m+1). \tag{10.25}$$

Pictorially, the definition of combinatorial R-matrices can be explained using the following commutative diagram.

$$\begin{array}{ccc} \mathcal{B}^{\text{aff}} \otimes \mathcal{B}^{\text{aff}} & \xrightarrow{\quad R \quad} & \mathcal{B}^{\text{aff}} \otimes \mathcal{B}^{\text{aff}} \\ {\scriptstyle \tilde{e}_i,\, \tilde{f}_i}\Big\downarrow {\scriptstyle \begin{matrix} T \otimes \text{id} \\ \text{id} \otimes T \end{matrix}} & & {\scriptstyle \tilde{e}_i,\, \tilde{f}_i}\Big\downarrow {\scriptstyle \begin{matrix} \text{id} \otimes T \\ T \otimes \text{id} \end{matrix}} \\ \mathcal{B}^{\text{aff}} \otimes \mathcal{B}^{\text{aff}} & \xrightarrow{\quad R \quad} & \mathcal{B}^{\text{aff}} \otimes \mathcal{B}^{\text{aff}} \end{array}$$

Example 10.2.4. Let V be the three-dimensional $U_q'(\widehat{\mathfrak{sl}}_3)$-module defined in Example 10.1.3 and let $\mathcal{B}$ be its crystal graph. Define a map $R : \mathcal{B}^{\mathrm{aff}} \otimes \mathcal{B}^{\mathrm{aff}} \to \mathcal{B}^{\mathrm{aff}} \otimes \mathcal{B}^{\mathrm{aff}}$ by

$$R(\boxed{i}(m) \otimes \boxed{j}(n)) = \begin{cases} \boxed{i}(n+1) \otimes \boxed{j}(m-1) & \text{if } i \geq j, \\ \boxed{i}(n) \otimes \boxed{j}(m) & \text{if } i < j. \end{cases} \tag{10.26}$$

Then one can verify that R is a combinatorial R-matrix on $\mathcal{B}^{\mathrm{aff}}$ (Exercise 10.4).

The following lemma provides a generic way of constructing combinatorial R-matrices.

Lemma 10.2.5. *Let V be a finite dimensional $U_q'(\mathfrak{g})$-module with crystal basis $(\mathcal{L}, \mathcal{B})$ and let $H : \mathcal{B} \otimes \mathcal{B} \to \mathbf{Z}$ be an energy function on $\mathcal{B}$. Define a map $R : \mathcal{B}^{\mathrm{aff}} \otimes \mathcal{B}^{\mathrm{aff}} \to \mathcal{B}^{\mathrm{aff}} \otimes \mathcal{B}^{\mathrm{aff}}$ by*

$$R(b_1(m) \otimes b_2(n)) = b_1(n + H(b_1 \otimes b_2)) \otimes b_2(m - H(b_1 \otimes b_2)) \tag{10.27}$$

for $b_1, b_2 \in \mathcal{B}$, $m, n \in \mathbf{Z}$. Then R is a combinatorial R-matrix.

Proof. It is easy to verify that R satisfies the following conditions:

$$(T \otimes \mathrm{id}) \circ R = R \circ (\mathrm{id} \otimes T), \qquad (\mathrm{id} \otimes T) \circ R = R \circ (T \otimes \mathrm{id}).$$

We will show that R commutes with the action of Kashiwara operators.

For $i = 0$, $b_1, b_2 \in \mathcal{B}$, $m, n \in \mathbf{Z}$, by the definition of energy functions, we have

$$\begin{aligned}
&R(\tilde{e}_0(b_1(m) \otimes b_2(n))) \\
&\quad = \begin{cases} R((\tilde{e}_0 b_1)(m+1) \otimes b_2(n)) & \text{if } \varphi_0(b_1) \geq \varepsilon_0(b_2), \\ R(b_1(m) \otimes (\tilde{e}_0 b_2)(n+1)) & \text{if } \varphi_0(b_1) < \varepsilon_0(b_2), \end{cases} \\
&\quad = \begin{cases} (\tilde{e}_0 b_1)(n + H(\tilde{e}_0 b_1 \otimes b_2)) \otimes b_2(m + 1 - H(\tilde{e}_0 b_1 \otimes b_2)) \\ \qquad\qquad\qquad\qquad \text{if } \varphi_0(b_1) \geq \varepsilon_0(b_2), \\ b_1(n + 1 + H(b_1 \otimes \tilde{e}_0 b_2)) \otimes (\tilde{e}_0 b_2)(m - H(b_1 \otimes \tilde{e}_0 b_2)) \\ \qquad\qquad\qquad\qquad \text{if } \varphi_0(b_1) < \varepsilon_0(b_2), \end{cases} \\
&\quad = \begin{cases} (\tilde{e}_0 b_1)(n + H(b_1 \otimes b_2) + 1) \otimes b_2(m - H(b_1 \otimes b_2)) \\ \qquad\qquad\qquad\qquad \text{if } \varphi_0(b_1) \geq \varepsilon_0(b_2), \\ b_1(n + H(b_1 \otimes b_2)) \otimes (\tilde{e}_0 b_2)(m - H(b_1 \otimes b_2) + 1) \\ \qquad\qquad\qquad\qquad \text{if } \varphi_0(b_1) < \varepsilon_0(b_2), \end{cases} \\
&\quad = \tilde{e}_0(b_1(n + H(b_1 \otimes b_2)) \otimes b_2(m - H(b_1 \otimes b_2))) \\
&\quad = \tilde{e}_0 R(b_1(m) \otimes b_2(n)),
\end{aligned}$$

as desired.

Similarly, one can prove

$$R(\tilde{e}_i(b_1(m) \otimes b_2(n))) = \tilde{e}_i R(b_1(m) \otimes b_2(n)) \tag{10.28}$$

for $i \neq 0$, $b_1, b_2 \in \mathcal{B}$, $m, n \in \mathbf{Z}$ (Exercise 10.5). □

Example 10.2.6. Let $\mathcal{B}$ be the crystal graph of the three-dimensional $U_q'(\widehat{\mathfrak{sl}}_3)$-module $V = \mathbf{C}(q)v_1 \oplus \mathbf{C}(q)v_2 \oplus \mathbf{C}(q)v_3$ defined in Example 10.1.3 and let $H : \mathcal{B} \otimes \mathcal{B} \to \mathbf{Z}$ be the energy function on $\mathcal{B}$ given by (10.22). Then the combinatorial R-matrix on $\mathcal{B}^{\mathrm{aff}}$ given by Lemma 10.2.5 coincides with the one given in Example 10.2.4.

The following lemma shows that the connected components of $\mathcal{B}^{\mathrm{aff}} \otimes \mathcal{B}^{\mathrm{aff}}$ can be characterized in terms of the affine energy function. This fact will be used in a crucial way in determining the affine weights of *paths* (Section 10.6).

Lemma 10.2.7. *Let V be a finite dimensional $U_q'(\mathfrak{g})$-module with crystal basis $(\mathcal{L}, \mathcal{B})$ and let $H : \mathcal{B} \otimes \mathcal{B} \to \mathbf{Z}$ be an energy function on $\mathcal{B}$. Consider the affine energy function $H^{\mathrm{aff}} : \mathcal{B}^{\mathrm{aff}} \otimes \mathcal{B}^{\mathrm{aff}} \to \mathbf{Z}$ defined by*

$$H^{\mathrm{aff}}(b_1(m) \otimes b_2(n)) = m - n - H(b_1 \otimes b_2)$$

for $b_1, b_2 \in \mathcal{B}$, $m, n \in \mathbf{Z}$. Then H^{aff} is constant on each connected component of $\mathcal{B}^{\mathrm{aff}} \otimes \mathcal{B}^{\mathrm{aff}}$.

Proof. Let $b_1(m) \otimes b_2(n) \in \mathcal{B}^{\mathrm{aff}} \otimes \mathcal{B}^{\mathrm{aff}}$ ($b_1, b_2 \in \mathcal{B}$, $m, n \in \mathbf{Z}$) and consider the combinatorial R-matrix $R : \mathcal{B}^{\mathrm{aff}} \otimes \mathcal{B}^{\mathrm{aff}} \to \mathcal{B}^{\mathrm{aff}} \otimes \mathcal{B}^{\mathrm{aff}}$ defined by (10.27). Then we have

$$\begin{aligned}
&R(b_1(m) \otimes b_2(n)) \\
&\quad = b_1(n + H(b_1 \otimes b_2)) \otimes b_2(m - H(b_1 \otimes b_2)) \\
&\quad = b_1(-H^{\mathrm{aff}}(b_1(m) \otimes b_2(n)) + m) \otimes b_2(H^{\mathrm{aff}}(b_1(m) \otimes b_2(n)) + n) \\
&\quad = (T^{-H^{\mathrm{aff}}(b_1(m)\otimes b_2(n))} \otimes T^{H^{\mathrm{aff}}(b_1(m)\otimes b_2(n))})(b_1(m) \otimes b_2(n)).
\end{aligned}$$

Observe that

$$\begin{aligned}
&R \circ \tilde{e}_i(b_1(m) \otimes b_2(n)) \\
&\quad = (T^{-H^{\mathrm{aff}}(\tilde{e}_i(b_1(m)\otimes b_2(n)))} \otimes T^{H^{\mathrm{aff}}(\tilde{e}_i(b_1(m)\otimes b_2(n)))})(\tilde{e}_i(b_1(m) \otimes b_2(n)))
\end{aligned}$$

and that

$$\begin{aligned}
&\tilde{e}_i \circ R(b_1(m) \otimes b_2(n)) \\
&\quad = \tilde{e}_i \left(T^{-H^{\mathrm{aff}}(b_1(m)\otimes b_2(n))} \otimes T^{H^{\mathrm{aff}}(b_1(m)\otimes b_2(n))}\right)(b_1(m) \otimes b_2(n)) \\
&\quad = (T^{-H^{\mathrm{aff}}(b_1(m)\otimes b_2(n))} \otimes T^{H^{\mathrm{aff}}(b_1(m)\otimes b_2(n))})(\tilde{e}_i(b_1(m) \otimes b_2(n))).
\end{aligned}$$

Since $R \circ \tilde{e}_i = \tilde{e}_i \circ R$ for all $i \in I$, we must have

$$H^{\text{aff}}(\tilde{e}_i(b_1(m) \otimes b_2(n))) = H^{\text{aff}}(b_1(m) \otimes b_2(n))$$

for all $i \in I$.

Similarly, we can verify

$$H^{\text{aff}}(\tilde{f}_i(b_1(m) \otimes b_2(n))) = H^{\text{aff}}(b_1(m) \otimes b_2(n))$$

for all $i \in I$, which completes the proof. □

Example 10.2.8. Recall the three-dimensional $U_q'(\widehat{\mathfrak{sl}}_3)$-module V defined in Example 10.1.3 with crystal basis $(\mathcal{L}, \mathcal{B})$. Then all the connected components of $\mathcal{B}^{\text{aff}} \otimes \mathcal{B}^{\text{aff}}$ are isomorphic to each other as is shown below.

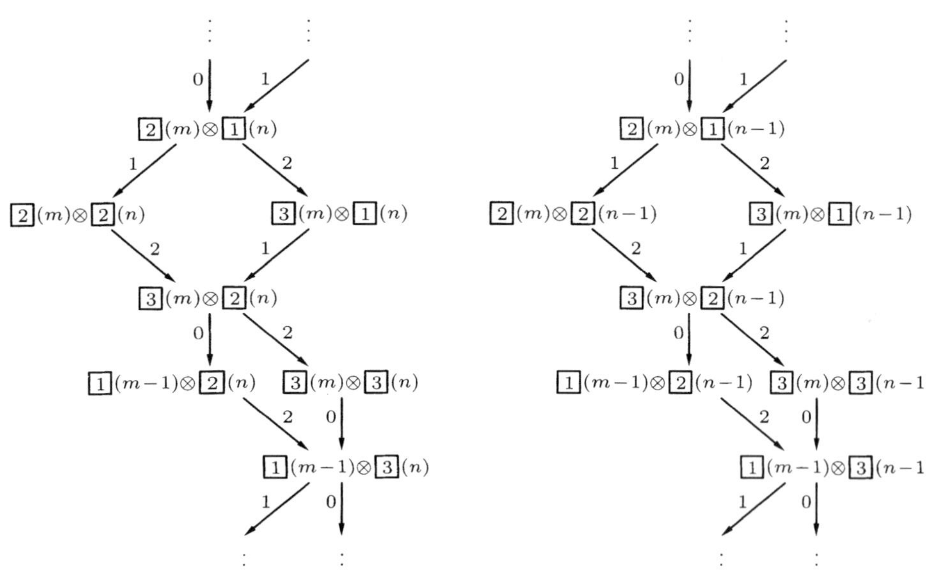

Let $H : \mathcal{B} \otimes \mathcal{B} \to \mathbf{Z}$ be the energy function on $\mathcal{B}$ defined by (10.22) and let $H^{\text{aff}} : \mathcal{B}^{\text{aff}} \otimes \mathcal{B}^{\text{aff}} \to \mathbf{Z}$ be the affine energy function defined by (10.23). Then it is straightforward to verify that H^{aff} is constant on each component of $\mathcal{B}^{\text{aff}} \otimes \mathcal{B}^{\text{aff}}$.

10.3. Vertex operators for $U_q(\mathfrak{sl}_2)$-modules

Let W be a finite dimensional $U_q(\mathfrak{sl}_2)$-module and consider a $U_q(\mathfrak{sl}_2)$-module homomorphism

$$\Phi : V(l) \longrightarrow V(m) \otimes W,$$

where $V(l)$ and $V(m)$ are irreducible $U_q(\mathfrak{sl}_2)$-modules of dimension $l+1$ and $m+1$, respectively. Such a $U_q(\mathfrak{sl}_2)$-module homomorphism is called a ***vertex operator***. In this section, we will study the basic properties of vertex operators for finite dimensional $U_q(\mathfrak{sl}_2)$-modules (see [**9**]).

Proposition 10.3.1. *Let W be a finite dimensional $U_q(\mathfrak{sl}_2)$-module with crystal basis $(\mathcal{L}, \mathcal{B})$, and let*

$$\Phi : V(l) \longrightarrow V(m) \otimes W \quad (l, m \in \mathbf{Z}_{\geq 0})$$

be a vertex operator, where $V(l)$ and $V(m)$ are irreducible $U_q(\mathfrak{sl}_2)$-modules with highest weight vectors u_0 and v_0, respectively.

(1) *Write $\Phi(u_0) = v_0 \otimes w_\star + \sum v_j \otimes w_j$, where $v_0, v_j \in V(m)$, $w_\star, w_j \in W$ and $\mathrm{wt}(v_j) \neq m$. If $w_\star \in \mathcal{L}$, then Φ preserves the crystal lattice; i.e., $\Phi(\mathcal{L}(l)) \subset \mathcal{L}(m) \otimes \mathcal{L}$.*

(2) *If $w_\star + q\mathcal{L} \in \mathcal{B}$, then the vertex operator Φ induces a strict crystal morphism $\bar{\Phi} : \mathcal{B}(l) \longrightarrow \mathcal{B}(m) \otimes \mathcal{B}$.*

Proof. (1) We may assume that $W = V(n)$ $(n \in \mathbf{Z}_{\geq 0})$ and $l = m + n - 2s$ with $0 \leq s \leq \min(m, n)$. Let w_0 denote the highest weight vector of $W = V(n)$. Using the condition $w_\star \in \mathcal{L}$, we may write $w_\star = a f^{(s)} w_0 \in \mathcal{L}(n)$ for some $a \in \mathbf{A}_0$. Recall the element

$$E_l = v_0 \otimes f^{(s)} w_0 + \sum_{k=1}^{s} (-1)^k q^{k(m-s+1)} \left(\prod_{t=1}^{k} \frac{q^{2(n-s+t)} - 1}{q^{2(m-t+1)} - 1} \right) f^{(k)} v_0 \otimes f^{(s-k)} w_0$$

and its properties from Exercise 4.6. This is the unique, up to scalar multiple, highest weight vector in $V(m) \otimes V(n)$ of highest weight l. So we must have $\Phi(u_0) = a E_l$. Then, the same exercise shows $\Phi(f^{(k)} u_0) = a f^{(k)}(E_l) \in \mathcal{L}(m) \otimes \mathcal{L}(n)$.

(2) We already know that $\bar{\Phi}$ commutes with $\tilde{e}$ and $\tilde{f}$. The condition $w_\star + q\mathcal{L} \in \mathcal{B}$ tells us that $a \equiv 1 \pmod{q\mathbf{A}_0}$. So that

$$\Phi(f^{(k)} u_0) \equiv f^{(k)}(E_l) \equiv \tilde{f}^k (v_0 \otimes f^{(s)} w_0) \mod q\mathcal{L}(m) \otimes \mathcal{L}(n).$$

Hence, by the tensor product rule, we obtain

$$\bar{\Phi}(f^{(k)} u_0) = \tilde{f}^k (v_0 \otimes f^{(s)} w_0) \in \mathcal{B}(m) \otimes \mathcal{B}(n).$$

□

Suppose that a finite dimensional $U_q(\mathfrak{sl}_2)$-module W has a crystal basis $(\mathcal{L}, \mathcal{B})$ and let $\{w_j \mid j = 1, \ldots, \dim W\}$ be an $\mathbf{A}_0$-basis of $\mathcal{L}$ such that

$$\mathcal{B} = \{b_j = w_j + q\mathcal{L} \mid j = 1, \ldots, \dim W\}.$$

Given a vertex operator $\Phi : V(l) \longrightarrow V(m)\otimes W$, consider the *jth component* $\Phi_j : V(l) \longrightarrow V(m)$ of Φ defined by

$$\Phi(u) = \sum_j \Phi_j(u) \otimes w_j \quad \text{for } u \in V(l),\ j = 1, \dots, \dim W. \tag{10.29}$$

Using the components Φ_j, we define a linear map $\Phi^\vee : V(m) \otimes V(l) \longrightarrow W$ by

$$\Phi^\vee(v \otimes u) = \sum_j (\Phi_j(u), v) w_j \quad (u \in V(l), v \in V(m)), \tag{10.30}$$

where (,) denotes the symmetric bilinear form on $V(m)$ satisfying (5.2) (see Section 5.1).

Proposition 10.3.2. *The linear map $\Phi^\vee : V(m)\otimes V(l) \longrightarrow W$ satisfies the following conditions:*

$$\begin{aligned} &\mathrm{wt}\, \Phi^\vee(v \otimes u) = \mathrm{wt}(u) - \mathrm{wt}(v), \\ &\Phi^\vee(v \otimes e\, u) = e\, \Phi^\vee(v \otimes u) + q^{-\langle h, \mathrm{wt}(u)\rangle - 1} \Phi^\vee(fv \otimes u), \\ &\Phi^\vee(v \otimes fu) = q^{\langle h, \mathrm{wt}(v)\rangle} f \Phi^\vee(v \otimes u) + q^{\langle h, \mathrm{wt}(v)\rangle + 1} \Phi^\vee(e\, v \otimes u) \end{aligned} \tag{10.31}$$

for $u \in V(l)$, $v \in V(m)$.

Proof. Recall that $(\Phi_j(u), v) \neq 0$ only if $\mathrm{wt}(\Phi_j(u)) = \mathrm{wt}(v)$. Now our assertion follows directly from the intertwining properties of the vertex operator Φ (Exercise 10.6). □

Conversely, suppose that we have a linear map $\Phi^\vee : V(m)\otimes V(l) \longrightarrow W$ satisfying the conditions in (10.31). For $u \in V(l)$, $v \in V(m)$, we write

$$\Phi^\vee(v \otimes u) = \sum_j \phi_j^\vee(u, v) w_j \quad \text{for some } \phi_j^\vee(u,v) \in \mathbf{C}(q).$$

Since $\Phi^\vee$ is linear, there is a natural linear map $\phi_j^\vee(u) : V(m) \longrightarrow \mathbf{C}(q)$ defined by

$$\phi_j^\vee(u)(v) = \phi_j^\vee(u, v) \quad \text{for all } u \in V(l),\ v \in V(m).$$

Recall that the symmetric bilinear form (,) is nondegenerate on $V(m)$. Hence there exists a unique element $\Phi_j(u) \in V(m)$ such that

$$\phi_j^\vee(u)(v) = \phi_j^\vee(u, v) = (\Phi_j(u), v),$$

which induces a $\mathbf{C}(q)$-linear map $\Phi : V(l) \longrightarrow V(m) \otimes W$ given by

$$\Phi(u) = \sum_j \Phi_j(u) \otimes w_j.$$

Then the conditions in (10.31) imply that Φ is a $U_q(\mathfrak{sl}_2)$-module homomorphism (Exercise 10.7) and we can write

$$\Phi^\vee(v \otimes u) = \sum_j (\Phi_j(u), v) w_j.$$

Proposition 10.3.3. *If $\Phi^\vee : V(m) \otimes V(l) \longrightarrow W$ is a $\mathbf{C}(q)$-linear map satisfying the conditions in* (10.31) *and if $\Phi^\vee(v_0 \otimes u_0) \in \mathcal{L}$, then we have*

$$\Phi^\vee(\mathcal{L}(m) \otimes \mathcal{L}(l)) \subset \mathcal{L}.$$

Proof. As we have seen in the preceding paragraph, there exists a vertex operator $\Phi : V(l) \longrightarrow V(m) \otimes W$ such that

$$\Phi^\vee(v \otimes u) = \sum_j (\Phi_j(u), v) w_j \quad (u \in V(l), v \in V(m)),$$

where Φ_j is the jth component of Φ. We would like to apply Proposition 10.3.1 to deduce our claim. Write

$$\sum_j \Phi_j(u_0) \otimes w_j = \Phi(u_0) = v_0 \otimes w_\star + \sum_i v_i \otimes w_i'$$

with $w_\star, w_i' \in W$ and $\mathrm{wt}(v_i) \neq m$. Applying $(\ \ , v_0) \otimes \mathrm{id}$ to the left-hand side, we get

$$\sum_j (\Phi_j(u_0), v_0) w_j = \Phi^\vee(v_0 \otimes u_0),$$

and applying it to the right-hand side, we get $w_\star$. Hence $w_\star = \Phi^\vee(v_0 \otimes u_0) \in \mathcal{L}$. Since we have taken w_j to satisfy $b_j = w_j + q\mathcal{L}$, Proposition 10.3.1 now implies $\Phi_j(u) \in \mathcal{L}(m)$ for all $u \in \mathcal{L}(l)$, $j = 1, \ldots, \dim W$. We already know $(\mathcal{L}(m), \mathcal{L}(m)) \subset \mathbf{A}_0$, which gives $(\Phi_j(u), v) \in \mathbf{A}_0$ for all $u \in \mathcal{L}(l)$ and $v \in \mathcal{L}(m)$. Therefore, we have $\Phi^\vee(v \otimes u) \in \mathcal{L}$ for all $u \in \mathcal{L}(l)$ and $v \in \mathcal{L}(m)$. □

10.4. Vertex operators for quantum affine algebras

In this section, we will study the *vertex operators* for integrable modules over quantum affine algebras. As in the previous section, the main reference for this section is [**9**]. Let $\mathfrak{g}$ be an affine Kac-Moody algebra and let $U_q(\mathfrak{g})$ denote the corresponding quantum affine algebra. Recall that a $U_q(\mathfrak{g})$-module V is *integrable* if

(i) $V = \bigoplus_{\lambda \in P} V_\lambda$ with $\dim V_\lambda < \infty$ for all $\lambda \in P$,

(ii) e_i and f_i $(i \in I)$ are locally nilpotent on V.

For an integrable $U_q(\mathfrak{g})$-module $V = \bigoplus_{\lambda \in P} V_\lambda$, we define its ***formal completion*** to be $\widehat{V} = \prod_{\lambda \in P} V_\lambda$. Thus the formal completion of the tensor product, which we denote by $V\widehat{\otimes}W$, of $U_q(\mathfrak{g})$-modules V and W can be written as

$$V\widehat{\otimes}W = \Big\{ \sum_{\text{infinite}} v_i \otimes w_i \,\Big|\, v_i \in V, w_i \in W \Big\}.$$

Definition 10.4.1. Let W be a finite dimensional $U_q'(\mathfrak{g})$-module and let $W^{\text{aff}} = \mathbf{C}(q)[z, z^{-1}] \otimes_{\mathbf{C}(q)} W$ be its affinization. For dominant integral weights $\lambda, \mu \in P^+$, a $U_q(\mathfrak{g})$-module homomorphism of the form

$$\widehat{\Phi}(z) : V(\lambda) \longrightarrow V(\mu)\widehat{\otimes}W^{\text{aff}} \tag{10.32}$$

is called a ***vertex operator***.

Fix a basis $\{w_j \,|\, j = 1, \ldots, \dim W\}$ of W and write $w_j z^{-n} = z^{-n} \otimes w_j$ $(n \in \mathbf{Z}, j = 1, \ldots, \dim W)$. Given a vertex operator $\widehat{\Phi}(z) : V(\lambda) \to V(\mu)\widehat{\otimes}W^{\text{aff}}$, we define the linear maps $\Phi_{j,n} : V(\lambda) \to V(\mu)$ by

$$\widehat{\Phi}(z)(u) = \sum_j \sum_{n\in\mathbf{Z}} \Phi_{j,n}(u) \otimes w_j z^{-n} \tag{10.33}$$

for $u \in V(\lambda)$, $n \in \mathbf{Z}$, $j = 1, \ldots, \dim W$. The linear map $\Phi_{j,n}$ is called the (j,n)*-th component* of $\widehat{\Phi}(z)$. Note that

$$\begin{aligned} &\Phi_{j,n}(V(\lambda)_\tau) \subset V(\mu)_{\tau - \text{wt}(w_j) + n\delta}, \\ &\Phi_{j,n}(u) = 0 \quad \text{for each } u \in V(\lambda), \text{ if } n \gg 0. \end{aligned} \tag{10.34}$$

Using the linear maps $\Phi_{j,n}$, we define a linear map $\Phi : V(\lambda) \longrightarrow \widehat{V}(\mu) \otimes W$ by

$$\Phi(u) = \sum_j \left(\sum_{n\in\mathbf{Z}} \Phi_{j,n}(u) \right) \otimes w_j \quad \text{for } u \in V(\lambda). \tag{10.35}$$

Then it is straightforward to verify that Φ is a $U_q'(\mathfrak{g})$-module homomorphism (Exercise 10.8).

We may also go the other way around. Given a $U_q'(\mathfrak{g})$-module homomorphism $\Phi : V(\lambda) \longrightarrow \widehat{V}(\mu) \otimes W$, we may write

$$\Phi(u) = \sum_j \Phi_j(u) \otimes w_j = \sum_j \left(\sum_{n\in\mathbf{Z}} \Phi_{j,n}(u) \right) \otimes w_j \quad (u \in V(\lambda)) \tag{10.36}$$

so that (10.34) is satisfied. Define a $\mathbf{C}(q)$-linear map $\widehat{\Phi}(z) : V(\lambda) \to V(\mu)\widehat{\otimes}W^{\text{aff}}$ by

$$\widehat{\Phi}(z)(u) = \sum_j \sum_{n\in\mathbf{Z}} \Phi_{j,n}(u) \otimes w_j z^{-n}. \tag{10.37}$$

Then one can verify that $\widehat{\Phi}(z)$ is a $U_q(\mathfrak{g})$-module homomorphism (Exercise 10.9).

This shows that there is a one-to-one correspondence between the $U_q(\mathfrak{g})$-module homomorphisms $V(\lambda) \to V(\mu)\widehat{\otimes}W^{\mathrm{aff}}$ and the $U_q'(\mathfrak{g})$-module homomorphisms $V(\lambda) \to \widehat{V}(\mu) \otimes W$ given by $\widehat{\Phi}(z) \longmapsto \Phi$. Thus, a $U_q'(\mathfrak{g})$-module homomorphism $\Phi : V(\lambda) \to \widehat{V}(\mu)\otimes W$ will also be called a ***vertex operator***. From now on, we will focus on the vertex operators $\Phi : V(\lambda) \to \widehat{V}(\mu) \otimes W$, because they are more convenient to deal with in developing the theory of perfect crystals.

Suppose that the finite dimensional $U_q'(\mathfrak{g})$-module W has a crystal basis $(\mathcal{L}, \mathcal{B})$. Then the pair $(\widehat{\mathcal{L}}(\mu) \otimes \mathcal{L}, \mathcal{B}(\mu) \otimes \mathcal{B})$ is a crystal basis of the $U_q'(\mathfrak{g})$-module $\widehat{V}(\mu) \otimes W$.

Remark 10.4.2. This is a crystal basis in a slightly weaker sense. We need infinite sums of elements from $\mathcal{B}(\mu) \otimes \mathcal{B}$ to represent elements from $(\widehat{\mathcal{L}}(\mu) \otimes \mathcal{L})/q(\widehat{\mathcal{L}}(\mu) \otimes \mathcal{L})$.

Let v_λ and v_μ denote the highest weight vectors of $V(\lambda)$ and $V(\mu)$, respectively. Write

$$\Phi(v_\lambda) = v_\mu \otimes w_\star + \sum v_j \otimes w_j, \tag{10.38}$$

where $v_\mu, v_j \in V(\mu)$, $w_\star, w_j \in W$ and $\mathrm{wt}(v_j) \neq \mu$. The vector $w_\star$ is called the ***leading term*** of the vertex operator Φ. Given dominant integral weights $\lambda, \mu \in P^+$, set

$$W^{(\lambda,\mu)} = \{w \in W \mid \mathrm{wt}(w) = \lambda - \mu,\ e_i^{\mu(h_i)+1} w = 0 \quad \text{for all } i \in I\}. \tag{10.39}$$

Then the following theorem was proved in [**9**].

Theorem 10.4.3. *The map* $\Phi \longmapsto w_\star =$ (leading term of Φ) *defines a* $\mathbf{C}(q)$*-linear isomorphism*

$$\mathrm{Hom}_{U_q'(\mathfrak{g})}(V(\lambda), \widehat{V}(\mu) \otimes W) \xrightarrow{\ \sim\ } W^{(\lambda,\mu)}.$$

Proof. Let $U_q'(\mathfrak{b}^+)$ be the subalgebra of $U_q'(\mathfrak{g})$ generated by e_i and $K_i^{\pm 1}$ $(i \in I)$. Then the highest weight vector v_λ generates a one-dimensional $U_q'(\mathfrak{b}^+)$-submodule $\mathbf{C}(q)v_\lambda$ with defining relations:

$$e_i v_\lambda = 0, \qquad K_i^{\pm 1} v_\lambda = q_i^{\pm\lambda(h_i)} v_\lambda.$$

We first claim that there is a $\mathbf{C}(q)$-linear isomorphism

$$\mathrm{Hom}_{U_q'(\mathfrak{g})}(V(\lambda), \widehat{V}(\mu) \otimes W) \xrightarrow{\ \sim\ } \mathrm{Hom}_{U_q'(\mathfrak{b}^+)}(\mathbf{C}(q)v_\lambda, \widehat{V}(\mu) \otimes W) \tag{10.40}$$

given by the natural restriction map $\Phi \longmapsto \Phi|_{\mathbf{C}(q)v_\lambda}$. It is easy to see that the $\mathbf{C}(q)$-linear map $\Phi \longmapsto \Phi|_{\mathbf{C}(q)v_\lambda}$ is injective. So it remains to show that this map is surjective.

Let $\phi : \mathbf{C}(q)v_\lambda \to \widehat{V}(\mu) \otimes W$ be a $U'_q(\mathfrak{b}^+)$-module homomorphism and set

$$w_\lambda = \phi(v_\lambda) = \sum v_j \otimes w_j,$$

where $\{w_j \mid j = 1, \ldots, \dim W\}$ is a $\mathbf{C}(q)$-linear basis of W. By definition, we have $e_i w_\lambda = 0$ and $K_i w_\lambda = q_i^{\lambda(h_i)} w_\lambda$ for all $i \in I$. Define a linear map $\Phi : V(\lambda) \longrightarrow \widehat{V}(\mu) \otimes W$ by

$$\Phi(Pv_\lambda) = Pw_\lambda = \sum \Delta(P)(v_j \otimes w_j),$$

where P is a homogeneous element of $U_q^-(\mathfrak{g})$. To prove Φ is a well defined $U'_q(\mathfrak{g})$-module homomorphism, it suffices to show that $f_i^{\lambda(h_i)+1} w_\lambda = 0$ for all $i \in I$. Recall that

$$\Delta(f_i^N) = f_i^N \otimes 1 + (\text{intermediate terms}) + K_i^N \otimes f_i^N.$$

For each $j = 1, \ldots, \dim W$, the vector $v_j \in \widehat{V}(\mu)$ may be written as an infinite sum of weight vectors in $V(\mu)_{\lambda - \mathrm{wt}(w_j)}$. Still, we have

$$f_i^N v_j \in \widehat{V}(\mu)_{\mu - \mathrm{wt}(w_j) - N\alpha_i} = \{0\} \quad \text{for some sufficiently large } N > 0.$$

That is, for each $i \in I$, there is a positive integer $N_i > 0$ such that $f_i^{N_i} v_j = 0$. On the other hand, since W is finite dimensional, we can choose a positive integer $N_i > 0$ such that $f_i^{N_i} w = 0$ for all $w \in W$. Hence for each $i \in I$, there exists a sufficiently large integer $N_i > 0$ such that

$$f_i^{N_i} w_\lambda = \sum_j \left(f_i^{N_i} v_j \otimes w_j + (\cdots) + K_i^{N_i} v_j \otimes f_i^{N_i} w_j \right) = 0.$$

Therefore, in $\widehat{V}(\mu) \otimes W$, $w_\lambda = \Phi(v_\lambda)$ generates an integrable highest weight $U'_q(\mathfrak{g})$-module with highest weight λ, which must be isomorphic to $V(\lambda)$. It follows that $f_i^{\lambda(h_i)+1} w_\lambda = 0$ for all $i \in I$.

Next, we claim that there is a $\mathbf{C}(q)$-linear isomorphism

$$\mathrm{Hom}_{U'_q(\mathfrak{b}^+)}(\mathbf{C}(q)v_\lambda, \widehat{V}(\mu) \otimes W) \xrightarrow{\sim} W^{(\lambda,\mu)}. \tag{10.41}$$

We leave construction of the canonical $\mathbf{C}(q)$-linear isomorphism

$$\mathrm{Hom}_{U'_q(\mathfrak{b}^+)}(\mathbf{C}(q)v_\lambda, \widehat{V}(\mu) \otimes W) \xrightarrow{\sim} \mathrm{Hom}_{U'_q(\mathfrak{b}^+)}(V^*(\mu) \otimes \mathbf{C}(q)v_\lambda, W)$$

to the readers (Exercise 10.10). The finite dual space $V^*(\mu)$ appearing in this isomorphism is isomorphic to the irreducible lowest weight $U'_q(\mathfrak{g})$-module with lowest weight $-\mu$. Hence we have

$$\begin{aligned} V^*(\mu) &= U_q'(\mathfrak{b}^+)v_\mu^*, \\ f_i v_\mu^* &= 0 \quad \text{for all } i \in I, \\ K_i v_\mu^* &= q_i^{-\mu(h_i)} v_\mu^*, \\ e_i^{\mu(h_i)+1} v_\mu^* &= 0 \quad \text{for all } i \in I. \end{aligned}$$

Note that any $U_q'(\mathfrak{b}^+)$-module homomorphism $\Psi : V^*(\mu) \otimes \mathbf{C}(q)v_\lambda \longrightarrow W$ is completely determined by $\Psi(v_\mu^* \otimes v_\lambda) = w_\star \in W$ and that $w_\star$ must satisfy the condition:

$$\mathrm{wt}(w_\star) = \lambda - \mu, \qquad e_i^{\mu(h_i)+1} w_\star = 0 \quad \text{for all } i \in I.$$

Hence $w_\star \in W^{(\lambda,\mu)}$ and our claim is proved.

Finally, it remains to show that $w_\star$ is actually the leading term of Φ, or its constant multiple not depending on Φ. This follows from the definitions of canonical $\mathbf{C}(q)$-linear isomorphisms

$$\begin{aligned} \mathrm{Hom}_{U_q'(\mathfrak{b}^+)}(\mathbf{C}(q)v_\lambda, \widehat{V}(\mu) \otimes W) &\xrightarrow{\sim} \mathrm{Hom}_{U_q'(\mathfrak{b}^+)}(V^*(\mu) \otimes \mathbf{C}(q)v_\lambda, W) \\ &\xrightarrow{\sim} W^{(\lambda,\mu)} \end{aligned}$$

(Exercise 10.10). □

Theorem 10.4.4. *Let W be a finite dimensional $U_q'(\mathfrak{g})$-module with crystal basis $(\mathcal{L}, \mathcal{B})$. For dominant integral weights $\lambda, \mu \in P^+$, suppose that there exists a vertex operator $\Phi : V(\lambda) \longrightarrow \widehat{V}(\mu) \otimes W$ with the leading term $w_\star \in W$.*

(1) *If $w_\star \in \mathcal{L}$, then Φ preserves the crystal lattice; i.e., $\Phi(\mathcal{L}(\lambda)) \subset \widehat{\mathcal{L}}(\mu) \otimes \mathcal{L}$.*

(2) *If $w_\star + q\mathcal{L} \in \mathcal{B}$, then Φ induces a strict crystal morphism $\overline{\Phi} : \mathcal{B}(\lambda) \longrightarrow \mathcal{B}(\mu) \otimes \mathcal{B}$.*

Proof. (1) Let $\{w_j \,|\, j = 1, \ldots, \dim W\}$ be an $\mathbf{A}_0$-basis of $\mathcal{L}$ such that

$$\mathcal{B} = \{b_j = w_j + q\mathcal{L} \,|\, j = 1, \ldots, \dim W\},$$

and let $\Phi_{j,n} : V(\lambda) \to V(\mu)$ be the (j,n)-th components of Φ ($n \in \mathbf{Z}, j = 1, \ldots, \dim W$). We would like to show that

$$\Phi(u) = \sum_j \sum_{n \in \mathbf{Z}} \Phi_{j,n}(u) \otimes w_j \in \widehat{\mathcal{L}}(\mu) \otimes \mathcal{L} \quad \text{for all } u \in \mathcal{L}(\lambda).$$

It suffices to show that $\Phi_{j,n}(u) \in \mathcal{L}(\mu)$ for all $u \in \mathcal{L}(\lambda)$.

Define a $\mathbf{C}(q)$-linear map $\Phi^\vee : V(\mu) \otimes V(\lambda) \longrightarrow W$ by

$$\Phi^\vee(v \otimes u) = \sum_j \sum_{n \in \mathbf{Z}} (\Phi_{j,n}(u), v) w_j \quad \text{for } u \in V(\lambda),\ v \in V(\mu),$$

where $(\ ,\ )$ is the symmetric bilinear form on $V(\mu)$ defined in Section 5.1. Using the intertwining properties of Φ, one can show that $\Phi^\vee$ satisfies the conditions in (10.31) (see Exercise 10.6). To prove our claim, it suffices to show that

$$(10.42)\quad \Phi^\vee(v \otimes u) \in \mathcal{L} \quad \text{for all } u \in \mathcal{L}(\lambda)_{\lambda-\alpha},\ v \in \mathcal{L}(\mu)_{\mu-\beta} \quad (\alpha, \beta \in Q_+).$$

Then we would have $(\Phi_{j,n}(u), v) \in \mathbf{A}_0$ for all $u \in \mathcal{L}(\lambda)_{\lambda-\alpha}, v \in \mathcal{L}(\mu)_{\mu-\beta}$ and by Proposition 5.1.7, we would conclude $\Phi_{j,n}(u) \in \mathcal{L}(\mu)$ for all $u \in \mathcal{L}(\lambda)$ as desired.

We will prove our assertion (10.42) by induction on $|\alpha|$ and $|\beta|$. If $|\alpha| = |\beta| = 0$, then $u = av_\lambda$, $v = bv_\mu$ for some $a, b \in \mathbf{A}_0$ and $\Phi^\vee(av_\mu \otimes bv_\lambda) = abw_\star \in \mathcal{L}$.

If $|\alpha| > 0$, we may assume $u = \tilde{f}_i u'$ for some $i \in I$ and $u' \in \mathcal{L}(\lambda)_{\lambda-\alpha+\alpha_i}$. Write $u' = \sum_{k \geq 0} f_i^{(k)} u'_k$ with $e_i u'_k = 0$ and $v = \sum_{l \geq 0} f_i^{(l)} v_l$ with $e_i v_l = 0$. Recall that Proposition 4.2.11 yields $u'_k \in \mathcal{L}(\lambda)$, $v_l \in \mathcal{L}(\mu)$ for all $k, l \geq 0$ and $u = \sum_{k \geq 0} f_i^{(k+1)} u'_k$. By the induction hypothesis, we have $\Phi^\vee(v_l \otimes u'_k) \in \mathcal{L}$. Hence, using Proposition 10.3.3, we conclude that

$$\Phi^\vee(v \otimes u) = \sum_{k,l \geq 0} \Phi^\vee(f_i^{(l)} v_l \otimes f_i^{(k+1)} u'_k) \in \mathcal{L}.$$

The case when $|\beta| > 0$ can be proved in a similar manner (Exercise 10.11).

(2) We already know that $\overline{\Phi}$ commutes with the Kashiwara operators $\tilde{e}_i$ and $\tilde{f}_i$ $(i \in I)$. To prove $\overline{\Phi}(\mathcal{B}(\lambda)) \subset \mathcal{B}(\mu) \otimes \mathcal{B}$, by induction, it suffices to show that if $b = u + q\mathcal{L}(\lambda) \in \mathcal{B}(\lambda)$, $\tilde{f}_i b \in \mathcal{B}(\lambda)$ and $\Phi(u) + q(\widehat{\mathcal{L}}(\mu) \otimes \mathcal{L}) \in \mathcal{B}(\mu) \otimes \mathcal{B}$, then $\Phi(\tilde{f}_i u) + q(\widehat{\mathcal{L}}(\mu) \otimes \mathcal{L}) \in \mathcal{B}(\mu) \otimes \mathcal{B}$. This follows directly from Proposition 10.3.1 (2). □

10.5. Perfect crystals

In this section, we define the notion of *perfect crystals* and prove a fundamental crystal isomorphism theorem which will be used to obtain *path realization* of crystal graphs for irreducible highest weight modules over quantum affine algebras.

Let $\mathcal{B}$ be a classical crystal. For $b \in \mathcal{B}$, we define

$$\varepsilon(b) = \sum_i \varepsilon_i(b)\Lambda_i, \qquad \varphi(b) = \sum_i \varphi_i(b)\Lambda_i. \tag{10.43}$$

Note that $\mathrm{wt}(b) = \varphi(b) - \varepsilon(b)$. For a positive integer $l > 0$, set

$$\bar{P}_l^+ = \{\lambda \in \bar{P}^+ \mid \langle c, \lambda \rangle = l\}. \tag{10.44}$$

Definition 10.5.1. For a positive integer $l > 0$, we say that a finite classical crystal $\mathcal{B}$ is a ***perfect crystal of level*** l if it satisfies the following conditions:

(1) there exists a finite dimensional $U_q'(\mathfrak{g})$-module with a crystal basis whose crystal graph is isomorphic to $\mathcal{B}$,

(2) $\mathcal{B} \otimes \mathcal{B}$ is connected,

(3) there exists a classical weight $\lambda_0 \in \bar{P}$ such that

$$\mathrm{wt}(\mathcal{B}) \subset \lambda_0 + \sum_{i \neq 0} \mathbf{Z}_{\leq 0}\alpha_i, \quad \#(\mathcal{B}_{\lambda_0}) = 1,$$

(4) for any $b \in \mathcal{B}$, we have $\langle c, \varepsilon(b) \rangle \geq l$,

(5) for each $\lambda \in \bar{P}_l^+$, there exist unique vectors $b^\lambda \in \mathcal{B}$ and $b_\lambda \in \mathcal{B}$ such that $\varepsilon(b^\lambda) = \lambda$, $\varphi(b_\lambda) = \lambda$.

For a perfect crystal $\mathcal{B}$, we define

$$\mathcal{B}^{\min} = \{b \in \mathcal{B} \mid \langle c, \varepsilon(b) \rangle = l\}. \tag{10.45}$$

Then the maps $\varepsilon, \varphi : \mathcal{B}^{\min} \longrightarrow \bar{P}_l^+$ are bijections.

A finite dimensional $U_q'(\mathfrak{g})$-module V is called a ***perfect representation of level*** l if it has a crystal basis whose crystal graph is a perfect crystal of level l.

In the following, we give some examples of perfect crystals.

Example 10.5.2.

(1) Let $\mathfrak{g} = A_1^{(1)}$ and let $\mathcal{B}$ be the $U_q'(\mathfrak{g})$-crystal given by

$$\mathcal{B}: \quad + \underset{0}{\overset{1}{\rightleftarrows}} - .$$

Then $\mathcal{B}$ is a perfect crystal of level 1 with

$$b^{\Lambda_0} = +, \quad b_{\Lambda_0} = -; \qquad b^{\Lambda_1} = -, \quad b_{\Lambda_1} = +.$$

(2) Let $\mathfrak{g} = A_1^{(1)}$ and, for a positive integer $l > 0$, let $\mathcal{B}_l$ be the $U_q'(\mathfrak{g})$-crystal given by

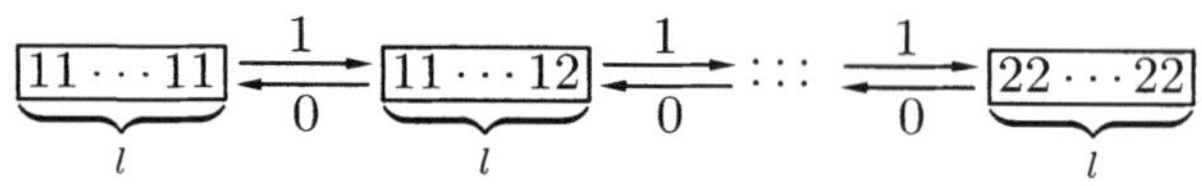

Then $\mathcal{B}_l$ is a perfect crystal of level l, where

$$b^\lambda = \underbrace{\boxed{1\cdots1}}_{a}\underbrace{\boxed{2\cdots2}}_{l-a} \quad \text{and} \quad b_\lambda = \underbrace{\boxed{1\cdots1}}_{l-a}\underbrace{\boxed{2\cdots2}}_{a}$$

for $\lambda = a\Lambda_0 + (l-a)\Lambda_1$.

(3) For $\mathfrak{g} = A_2^{(1)}$, the $U_q'(\mathfrak{g})$-crystal given below is a perfect crystal of level 1.

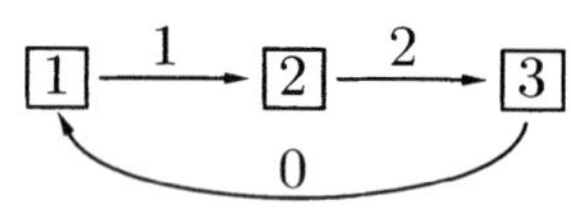

(4) Let $\mathfrak{g} = A_2^{(1)}$ and set

$$\mathcal{B}_l = \{b = (b_1, b_2, b_3) \in \mathbf{Z}^3 \mid 0 \le b_i \le l, b_1 + b_2 + b_3 = l\}.$$

We denote

$$b = (b_1, b_2, b_3) = \underbrace{\boxed{1\cdots1}}_{b_1}\underbrace{\boxed{2\cdots2}}_{b_2}\underbrace{\boxed{3\cdots3}}_{b_3}.$$

Then $\mathcal{B}_l$ is a perfect crystal of level l with $U_q'(\mathfrak{g})$-crystal structure given below.

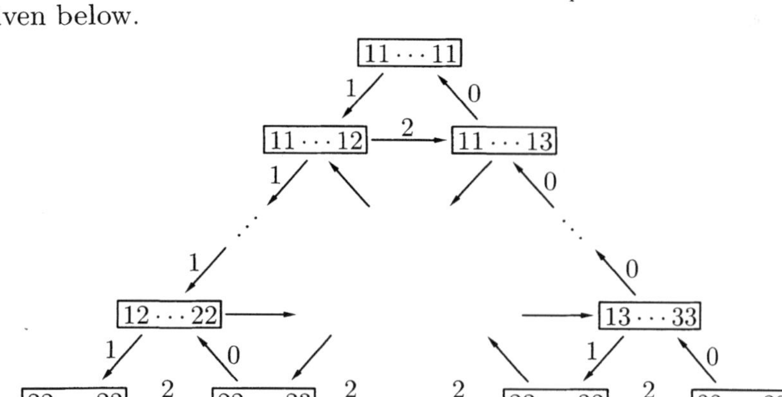

Here, for $\lambda = a_0\Lambda_0 + a_1\Lambda_1 + a_2\Lambda_2$, we have $b^\lambda = (a_0, a_1, a_2)$, $b_\lambda = (a_1, a_2, a_0)$.

(5) Let $\mathfrak{g} = C_2^{(1)}$ and consider the $U_q'(\mathfrak{g})$-crystal $\mathcal{B}$ given by

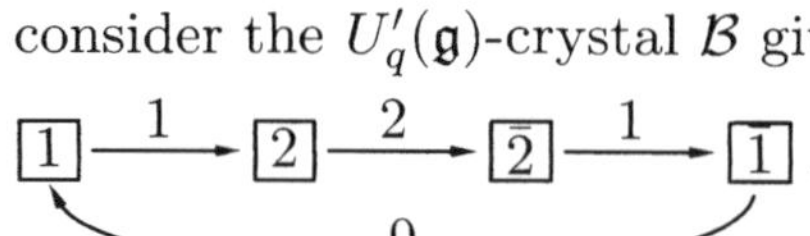

The crystal $\mathcal{B}$ is *not* perfect because for $\lambda = \Lambda_1$, we may have

$$b^{\Lambda_1} = \boxed{2} \text{ or } \boxed{\bar{1}}; \qquad b_{\Lambda_1} = \boxed{1} \text{ or } \boxed{\bar{2}}.$$

(6) Let $\mathfrak{g} = C_2^{(1)}$ and let $\mathcal{B}$ be the $U_q'(\mathfrak{g})$-crystal defined by

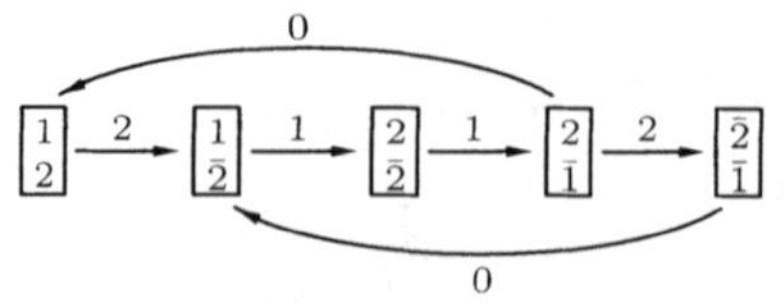

Then $\mathcal{B}$ is a perfect crystal of level 1. Indeed, we have

$$\begin{aligned} b^{\Lambda_0} &= \boxed{\begin{smallmatrix}1\\2\end{smallmatrix}}, & b_{\Lambda_0} &= \boxed{\begin{smallmatrix}\bar{2}\\\bar{1}\end{smallmatrix}}; \\ b^{\Lambda_1} &= \boxed{\begin{smallmatrix}2\\\bar{2}\end{smallmatrix}}, & b_{\Lambda_1} &= \boxed{\begin{smallmatrix}2\\\bar{2}\end{smallmatrix}}; \\ b^{\Lambda_2} &= \boxed{\begin{smallmatrix}\bar{2}\\\bar{1}\end{smallmatrix}}, & b_{\Lambda_2} &= \boxed{\begin{smallmatrix}1\\2\end{smallmatrix}}. \end{aligned}$$

Remark 10.5.3. In [**35, 37**], one can find an explicit description of *coherent* families of perfect crystals of arbitrary level $l > 0$ for quantum affine algebras of type $A_n^{(1)}$ $(n \geq 1)$, $B_n^{(1)}$ $(n \geq 3)$, $C_n^{(1)}$ $(n \geq 2)$, $D_n^{(1)}$ $(n \geq 4)$, $A_{2n-1}^{(2)}$ $(n \geq 3)$, $A_{2n}^{(2)}$ $(n \geq 2)$, and $D_{n+1}^{(2)}$. In [**56**], Yamane found a family of perfect crystals of arbitrary level for the quantum affine algebras of type $G_2^{(1)}$. Finding perfect crystals for exceptional quantum affine algebras other than $U_q(G_2^{(1)})$ is still an open problem.

Theorem 10.5.4. *Let $\mathcal{B}$ be a perfect crystal of level $l > 0$ and let $\lambda \in \bar{P}_l^+$ be a classical dominant integral weight.*

(1) *If $u \otimes b \in \mathcal{B}(\lambda) \otimes \mathcal{B}$ satisfies $\tilde{e}_i(u \otimes b) = 0$ for all $i \in I$, then $u = u_\lambda$ and $b = b^\lambda$, where u_λ is the highest weight vector of $\mathcal{B}(\lambda)$ and b^λ is the unique vector in $\mathcal{B}$ such that $\varepsilon(b^\lambda) = \lambda$. Hence $u_\lambda \otimes b^\lambda$ is the only maximal vector in $\mathcal{B}(\lambda) \otimes \mathcal{B}$ and its weight is $\lambda + \mathrm{wt}(b^\lambda) = \varphi(b^\lambda)$.*

(2) *For each $u \otimes b \in \mathcal{B}(\lambda) \otimes \mathcal{B}$, there exists a sequence of indices $i_1, \ldots, i_r \in I$ such that*

$$\tilde{e}_{i_r} \cdots \tilde{e}_{i_1}(u \otimes b) = u_\lambda \otimes b^\lambda.$$

In particular, $\mathcal{B}(\lambda) \otimes \mathcal{B}$ is connected.

Proof. (1) Suppose that $\tilde{e}_i(u \otimes b) = 0$ for all $i \in I$. By the tensor product rule, we must have $u = u_\lambda$ and $\varepsilon_i(b) \leq \lambda(h_i)$ for all $i \in I$.

Let $c = c_0 h_0 + c_1 h_1 + \cdots + c_n h_n$ be the canonical central element. Since $\mathcal{B}$ is perfect, we have

$$\begin{aligned} l &\leq \langle c, \varepsilon(b) \rangle = c_0 \varepsilon_0(b) + c_1 \varepsilon_1(b) + \cdots + c_n \varepsilon_n(b) \\ &\leq c_0 \lambda(h_0) + c_1 \lambda(h_1) + \cdots + c_n \lambda(h_n) = \langle c, \lambda \rangle = l. \end{aligned}$$

If follows that $\varepsilon_i(b) = \lambda(h_i)$ for all $i \in I$, which implies $b = b^\lambda$.

(2) Let $u \otimes b \in \mathcal{B}(\lambda) \otimes \mathcal{B}$. Recall that, by the tensor product rule, the Kashiwara operator $\tilde{e}_i$ "tends to" act on the second component first and then move on to act on the first component. Thus, for each $i \in I$, we have $\tilde{e}_i^N(u \otimes b) = \tilde{e}_i u \otimes \tilde{e}_i^{N-1} b$ for some sufficiently large $N > 0$. Therefore, there exists a sequence of indices $j_1, \dots, j_s \in I$ such that

$$\tilde{e}_{j_s} \cdots \tilde{e}_{j_1}(u \otimes b) = u_\lambda \otimes b' \quad \text{for some } b' \in \mathcal{B}.$$

Hence it suffices to prove our statement for $u \otimes b = u_\lambda \otimes b$.

Suppose that our assertion is false. Then there would exist an infinite sequence of indices $\{i_k\}_{k \geq 1}$ such that $\tilde{e}_{i_k} \cdots \tilde{e}_{i_1}(u_\lambda \otimes b) \neq 0$ for all $k \geq 1$. By the tensor product rule, we have

$$\tilde{e}_{i_k} \cdots \tilde{e}_{i_1}(u_\lambda \otimes b) = u_\lambda \otimes \tilde{e}_{i_k} \cdots \tilde{e}_{i_1} b \neq 0.$$

Since $\mathcal{B}$ is a finite set, there exist a vector $b' \in \mathcal{B}$ and positive integers $s \leq t$ such that

$$\tilde{e}_{i_t} \cdots \tilde{e}_{i_s} b' = b',$$

which implies $\alpha_{i_s} + \cdots + \alpha_{i_t} = 0$ in $\bar{P}$. Hence $\alpha_{i_s} + \cdots + \alpha_{i_t}$ must be a multiple of the null root δ and the sequence of indices $\{i_s, i_{s+1}, \dots, i_t\}$ contains all $i \in I$.

Set $b_s = b'$ and recursively set $b_{k+1} = \tilde{e}_{i_k} b_k$ $(s \leq k \leq t)$. Then we have $\tilde{e}_{i_k}(u_\lambda \otimes b_k) = u_\lambda \otimes b_{k+1}$ and $b_{t+1} = b' = b_s$. Thus the tensor product rule implies $\lambda(h_{i_k}) = \varphi_{i_k}(u_\lambda) < \varepsilon_{i_k}(b_k)$ for all k with $s \leq k \leq t$. Since $\mathcal{B}$ is perfect and since the sequence $\{i_s, i_{s+1}, \dots, i_t\}$ contains all $i \in I$, there exists a unique vector $b_\lambda \in \mathcal{B}$ such that

$$\varphi_{i_k}(b_\lambda) = \lambda(h_{i_k}) \quad \text{for all } i_k \in \{i_s, i_{s+1}, \dots, i_t\}.$$

It follows that $\tilde{e}_{i_k}(b_\lambda \otimes b_k) = b_\lambda \otimes \tilde{e}_{i_k} b_k = b_\lambda \otimes b_{k+1}$, which implies

$$H(b_\lambda \otimes b_{k+1}) = \begin{cases} H(b_\lambda \otimes b_k) & \text{if } i_k \neq 0, \\ H(b_\lambda \otimes b_k) - 1 & \text{if } i_k = 0. \end{cases}$$

Therefore we obtain

$$H(b_\lambda \otimes b_{t+1}) = H(b_\lambda \otimes b_s) - \#\{k \mid i_k = 0\}.$$

But, since $b_{t+1} = b_s$, there is no k such that $i_k = 0$, which is a contradiction. □

We are now ready to prove the main theorem of this section.

Theorem 10.5.5. *Let $\mathcal{B}$ be a perfect crystal of level $l > 0$. Then for any classical dominant integral weight $\lambda \in \bar{P}_l^+$, there exists a (strict) crystal isomorphism*

$$\Psi : \mathcal{B}(\lambda) \xrightarrow{\sim} \mathcal{B}(\mu) \otimes \mathcal{B} \quad \textit{given by } u_\lambda \longmapsto u_\mu \otimes b_\lambda,$$

where u_λ, u_μ are the highest weight vectors of $\mathcal{B}(\lambda)$ and $\mathcal{B}(\mu)$, respectively, b_λ is the unique vector in $\mathcal{B}$ such that $\varphi(b_\lambda) = \lambda$, and $\mu = \lambda - \mathrm{wt}(b_\lambda) = \varepsilon(b_\lambda)$.

Proof. Let W be a finite dimensional $U_q'(\mathfrak{g})$-module with a crystal basis whose crystal graph is isomorphic to $\mathcal{B}$. We know that the dimension of

$$W^{(\lambda,\mu)} = \{w \in W \mid \mathrm{wt}(w) = \lambda - \mu,\ e_i^{\mu(h_i)+1} w = 0 \quad \text{for all } i \in I\}$$

is equal to the number of elements in

$$\begin{aligned}
&\{b \in \mathcal{B} \mid \mathrm{wt}(b) = \lambda - \mu,\ \tilde{e}_i^{\mu(h_i)+1} b = 0 \quad \text{for all } i \in I\} \\
&\quad = \{b \in \mathcal{B} \mid \mathrm{wt}(b) = \lambda - \mu,\ \varepsilon_i(b) \le \mu(h_i) \quad \text{for all } i \in I\} \\
&\quad = \{b \in \mathcal{B} \mid \mathrm{wt}(b) = \lambda - \mu,\ \varepsilon(b) = \mu\} \\
&\quad = \{b \in \mathcal{B} \mid \varphi(b) = \lambda\} \\
&\quad = \{b_\lambda\}.
\end{aligned}$$

The last equality here is one of the requirements for $\mathcal{B}$ being perfect. Choose an element $w_\star \in W^{(\lambda,\mu)}$ such that $w_\star + q\mathcal{L} = b_\lambda \in \mathcal{B}$. By Theorem 10.4.3, there exists a unique vertex operator $\Phi : V(\lambda) \longrightarrow \widehat{V}(\mu) \otimes W$ satisfying

$$\Phi(v_\lambda) = v_\mu \otimes w_\star + \sum_j v_j \otimes w_j \quad (v_j \in V(\mu), w_j \in W).$$

Then, by Theorem 10.4.4, the vertex operator Φ preserves the crystal lattice and induces a strict crystal morphism

$$\Psi : \mathcal{B}(\lambda) \longrightarrow \mathcal{B}(\mu) \otimes \mathcal{B} \quad \text{given by} \quad u_\lambda \longmapsto u_\mu \otimes b_\lambda$$

such that $\Psi(\mathcal{B}(\lambda)) \subset \mathcal{B}(\mu) \otimes \mathcal{B}$. Since $\mathcal{B}(\mu) \otimes \mathcal{B}$ is connected, by Proposition 4.5.8, Ψ must be an isomorphism. □

10.6. Path realization of crystal graphs

Fix a positive integer $l > 0$ and let $\mathcal{B}$ be a perfect crystal of level l. By Theorem 10.5.5, for any classical dominant integral weight $\lambda \in \bar{P}_l^+$, there exists a unique crystal isomorphism

$$\Psi : \mathcal{B}(\lambda) \xrightarrow{\sim} \mathcal{B}(\varepsilon(b_\lambda)) \otimes \mathcal{B} \quad \text{given by} \quad u_\lambda \longmapsto u_{\varepsilon(b_\lambda)} \otimes b_\lambda, \tag{10.46}$$

where b_λ is the unique vector in $\mathcal{B}$ such that $\varphi(b_\lambda) = \lambda$. Set

$$\lambda_0 = \lambda, \quad \lambda_{k+1} = \varepsilon(b_{\lambda_k}); \qquad b_0 = b_\lambda, \quad b_{k+1} = b_{\lambda_{k+1}}.$$

Thus we have a crystal isomorphism

$$\Psi : \mathcal{B}(\lambda_j) \xrightarrow{\sim} \mathcal{B}(\lambda_{j+1}) \otimes \mathcal{B} \quad \text{given by} \quad u_{\lambda_j} \longmapsto u_{\lambda_{j+1}} \otimes b_j$$

such that $\varphi(b_j) = \lambda_j$, $\varepsilon(b_j) = \lambda_{j+1}$. By taking the composition of these crystal isomorphisms, we obtain a sequence of crystal isomorphisms

$$\mathcal{B}(\lambda) \xrightarrow{\sim} \mathcal{B}(\lambda_1) \otimes \mathcal{B} \xrightarrow{\sim} \mathcal{B}(\lambda_2) \otimes \mathcal{B} \otimes \mathcal{B} \xrightarrow{\sim} \cdots$$

given by

$$u_\lambda \longmapsto u_{\lambda_1} \otimes b_0 \longmapsto u_{\lambda_2} \otimes b_1 \otimes b_0 \longmapsto \cdots,$$

which yields two infinite sequences

$$\mathbf{w}_\lambda = (\lambda_k)_{k=0}^{\infty} = (\dots, \lambda_{k+1}, \lambda_k, \dots, \lambda_1, \lambda_0) \in (\bar{P}_l^+)^{\infty}$$

and

$$\mathbf{p}_\lambda = (b_k)_{k=0}^{\infty} = \cdots \otimes b_{k+1} \otimes b_k \otimes \cdots \otimes b_1 \otimes b_0 \in \mathcal{B}^{\otimes\infty}.$$

Thus, for each $k \geq 1$, we get a crystal isomorphism

$$\Psi_k : \mathcal{B}(\lambda) \xrightarrow{\sim} \mathcal{B}(\lambda_k) \otimes \mathcal{B}^{\otimes k} \tag{10.47}$$

given by

$$u_\lambda \longmapsto u_{\lambda_k} \otimes b_{k-1} \otimes \cdots \otimes b_1 \otimes b_0.$$

In the sequences $\mathbf{w}_\lambda$ and $\mathbf{p}_\lambda$, since there are only finitely many elements in $\bar{P}_l^+$ (and in $\mathcal{B}$), there exist $N > 0$ and $k \geq 0$ such that $\lambda_{k+N} = \lambda_k$ (hence $b_{k+N} = b_k$ since φ is bijective). Note that

$$\varepsilon(b_{k-1}) = \lambda_k = \lambda_{k+N} = \varepsilon(b_{k+N-1}).$$

Since ε is bijective, we have $b_{k-1} = b_{k+N-1}$, which in turn yields

$$\lambda_{k-1} = \varphi(b_{k-1}) = \varphi(b_{k+N-1}) = \lambda_{k+N-1}.$$

By repeating this procedure, we obtain

$$\lambda_{j+N} = \lambda_j, \qquad b_{j+N} = b_j \quad \text{for } j = 0, \dots, k.$$

Similarly, one can prove $\lambda_{j+N} = \lambda_j$, $b_{j+N} = b_j$ for $j \geq k+1$ (Exercise 10.12).

Hence the sequences $\mathbf{w}_\lambda$ and $\mathbf{p}_\lambda$ are periodic with the same period $N > 0$.

Definition 10.6.1.

(1) The sequence $\mathbf{p}_\lambda = (b_k)_{k=0}^{\infty} = \cdots \otimes b_{k+1} \otimes b_k \otimes \cdots \otimes b_1 \otimes b_0$ is called the ***ground-state path*** of weight λ.

(2) A ***λ-path*** in $\mathcal{B}$ is a sequence $\mathbf{p} = (\mathbf{p}_k)_{k=0}^{\infty} = \cdots \otimes \mathbf{p}_{k+1} \otimes \mathbf{p}_k \otimes \cdots \otimes \mathbf{p}_1 \otimes \mathbf{p}_0$ with $\mathbf{p}_k \in \mathcal{B}$ such that $\mathbf{p}_k = b_k$ for all $k \gg 0$.

Let $\mathcal{P}(\lambda)$ be the set of all λ-paths in $\mathcal{B}$. We would like to define a $U_q'(\mathfrak{g})$-crystal structure on $\mathcal{P}(\lambda)$. We first prove:

Lemma 10.6.2. *For any $u \in \mathcal{B}(\lambda)$, there exists a sufficiently large positive integer $N > 0$ such that $\Psi_N(u) \in u_{\lambda_N} \otimes \mathcal{B}^{\otimes N}$.*

Proof. Given $u \in \mathcal{B}(\lambda)$, it suffices to show that if $\Psi_N(u) \in u_{\lambda_N} \otimes \mathcal{B}^{\otimes N}$, then $\tilde{f}_i u = 0$ or $\Psi_{N+1}(\tilde{f}_i u) \in u_{\lambda_{N+1}} \otimes \mathcal{B}^{\otimes(N+1)}$.

If $\Psi_N(u) = u_{\lambda_N} \otimes b$ with $b \in \mathcal{B}^{\otimes N}$, then $\Psi_{N+1}(u) = u_{\lambda_{N+1}} \otimes b_N \otimes b$. It follows that

$$\begin{aligned}\Psi_{N+1}(\tilde{f}_i u) &= \tilde{f}_i(u_{\lambda_{N+1}} \otimes b_N \otimes b) \\ &= \begin{cases} \tilde{f}_i(u_{\lambda_{N+1}} \otimes b_N) \otimes b & \text{if } \varphi_i(u_{\lambda_{N+1}} \otimes b_N) > \varepsilon_i(b), \\ u_{\lambda_{N+1}} \otimes b_N \otimes \tilde{f}_i b & \text{if } \varphi_i(u_{\lambda_{N+1}} \otimes b_N) \le \varepsilon_i(b). \end{cases}\end{aligned}$$

Since $\varphi_i(u_{\lambda_{N+1}}) = \lambda_{N+1}(h_i) = \varepsilon_i(b_N)$, we always have $\tilde{f}_i(u_{\lambda_{N+1}} \otimes b_N) = u_{\lambda_{N+1}} \otimes \tilde{f}_i b_N$. To summarize,

$$\Psi_{N+1}(\tilde{f}_i u) = \begin{cases} u_{\lambda_{N+1}} \otimes \tilde{f}_i b_N \otimes b & \text{if } \varphi_i(u_{\lambda_{N+1}} \otimes b_N) > \varepsilon_i(b), \\ u_{\lambda_{N+1}} \otimes b_N \otimes \tilde{f}_i b & \text{if } \varphi_i(u_{\lambda_{N+1}} \otimes b_N) \le \varepsilon_i(b). \end{cases}$$

This proves our claim. □

Let $\mathbf{p} = (\mathbf{p}_k)_{k=0}^{\infty}$ be a λ-path in $\mathcal{B}$ and let $N > 0$ be the smallest positive integer such that $\mathbf{p}_k = b_k$ for all $k \ge N$. For each $i \in I$, we define

$$\begin{aligned} \overline{\mathrm{wt}}\mathbf{p} &= \lambda_N + \sum_{k=0}^{N-1} \overline{\mathrm{wt}}\mathbf{p}_k, \\ \tilde{e}_i\mathbf{p} &= \cdots \otimes \mathbf{p}_{N+1} \otimes \tilde{e}_i(\mathbf{p}_N \otimes \cdots \otimes \mathbf{p}_0), \\ \tilde{f}_i\mathbf{p} &= \cdots \otimes \mathbf{p}_{N+1} \otimes \tilde{f}_i(\mathbf{p}_N \otimes \cdots \otimes \mathbf{p}_0), \\ \varepsilon_i(\mathbf{p}) &= \max(\varepsilon_i(\mathbf{p}') - \varphi_i(b_N), 0), \\ \varphi_i(\mathbf{p}) &= \varphi_i(\mathbf{p}') + \max(\varphi_i(b_N) - \varepsilon_i(\mathbf{p}'), 0), \end{aligned} \tag{10.48}$$

where $\mathbf{p}' = \mathbf{p}_{N-1} \otimes \cdots \otimes \mathbf{p}_1 \otimes \mathbf{p}_0$ and $\overline{\mathrm{wt}}$ denotes the classical weights. Then it is straightforward to prove the following proposition (Exercise 10.13).

Proposition 10.6.3. *The maps* $\overline{\mathrm{wt}} : \mathcal{P}(\lambda) \longrightarrow \bar{P}$, $\tilde{e}_i, \tilde{f}_i : \mathcal{P}(\lambda) \longrightarrow \mathcal{P}(\lambda) \sqcup \{0\}$, $\varepsilon_i, \varphi_i : \mathcal{P}(\lambda) \longrightarrow \mathbf{Z}$ *given in* (10.48) *define a* $U_q'(\mathfrak{g})$*-crystal structure on* $\mathcal{P}(\lambda)$.

We now prove the main result of this section: the ***path realization*** of the $U_q'(\mathfrak{g})$-crystal $\mathcal{B}(\lambda)$.

Theorem 10.6.4. *There exists an isomorphism of* $U_q'(\mathfrak{g})$*-crystals*

$$\Psi : \mathcal{B}(\lambda) \xrightarrow{\sim} \mathcal{P}(\lambda) \quad \text{given by} \quad u_\lambda \longmapsto \mathbf{p}_\lambda.$$

Proof. Let $u \in \mathcal{B}(\lambda)$ and find the smallest positive integer $N > 0$ such that

$$\Psi_N(u) = u_{\lambda_N} \otimes \mathbf{p}_{N-1} \otimes \cdots \otimes \mathbf{p}_1 \otimes \mathbf{p}_0 \in u_{\lambda_N} \otimes \mathcal{B}^{\otimes N}.$$

Define the map $\Psi : \mathcal{B}(\lambda) \longrightarrow \mathcal{P}(\lambda)$ by

$$\begin{aligned}\Psi(u) &= \mathbf{p} = (\mathbf{p}_k)_{k=0}^{\infty} \\ &= \cdots \otimes b_{N+1} \otimes b_N \otimes \mathbf{p}_{N-1} \otimes \cdots \otimes \mathbf{p}_1 \otimes \mathbf{p}_0.\end{aligned}$$

Then we have

$$\begin{aligned}\tilde{f}_i\Psi(u) &= \tilde{f}_i(\cdots \otimes b_{N+1} \otimes b_N \otimes \mathbf{p}_{N-1} \otimes \cdots \otimes \mathbf{p}_1 \otimes \mathbf{p}_0) \\ &= \cdots \otimes b_{N+1} \otimes \tilde{f}_i(b_N \otimes \mathbf{p}_{N-1} \otimes \cdots \otimes \mathbf{p}_1 \otimes \mathbf{p}_0).\end{aligned}$$

On the other hand, recall that

$$\begin{aligned}\Psi_{N+1}(\tilde{f}_i u) = \tilde{f}_i\Psi_{N+1}(u) &= \tilde{f}_i(u_{\lambda_{N+1}} \otimes b_N \otimes \mathbf{p}_{N-1} \otimes \cdots \otimes \mathbf{p}_1 \otimes \mathbf{p}_0) \\ &= u_{\lambda_{N+1}} \otimes \tilde{f}_i(b_N \otimes \mathbf{p}_{N-1} \otimes \cdots \otimes \mathbf{p}_1 \otimes \mathbf{p}_0).\end{aligned}$$

Hence we obtain

$$\begin{aligned}\Psi(\tilde{f}_i u) &= \cdots \otimes b_{N+1} \otimes \tilde{f}_i(b_N \otimes \mathbf{p}_{N-1} \otimes \cdots \otimes \mathbf{p}_1 \otimes \mathbf{p}_0) \\ &= \tilde{f}_i\Psi(u).\end{aligned}$$

Similarly, one can prove the map Ψ commutes with $\tilde{e}_i$'s $(i \in I)$.

It is an easy exercise to verify that Ψ is a bijection (Exercise 10.14). □

Example 10.6.5.

(1) Let $\mathfrak{g} = A_1^{(1)}$ and $\mathcal{B}$ be the perfect crystal of level 1 given in Example 10.5.2 (1). The ground-state path $\mathbf{p}_{\Lambda_0}$ is the sequence $\mathbf{p}_{\Lambda_0} = (\ldots, 1, 2, 1, 2, 1, 2)$ and the path realization of the crystal graph $\mathcal{B}(\Lambda_0)$ is given in Example 9.3.1.

(2) Let $\mathfrak{g} = A_2^{(1)}$ and let $\mathcal{B}$ be the perfect crystal of level 1 given in Example 10.5.2 (3). The ground-state path $\mathbf{p}_{\Lambda_0}$ is the sequence $\mathbf{p}_{\Lambda_0} = (\ldots, 1, 2, 3, 1, 2, 3, 1, 2, 3)$ and the path realization of the crystal graph $\mathcal{B}(\Lambda_0)$ is given in the following figure.

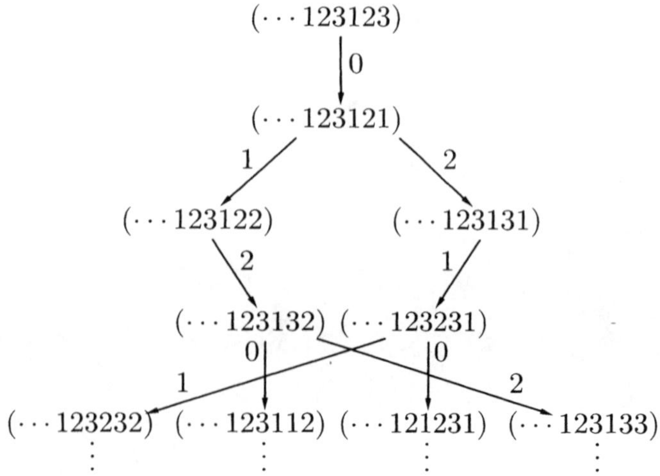

(3) Let $\mathfrak{g} = C_2^{(1)}$ and let $\mathcal{B}$ be the perfect crystal of level 1 given in Example 10.5.2 (6). The ground-state path $\mathbf{p}_{\Lambda_1}$ is the sequence

$$\mathbf{p}_{\Lambda_1} = \left(\cdots \begin{smallmatrix} 2 & 2 & 2 & 2 & 2 & 2 \\ \bar{2} & \bar{2} & \bar{2} & \bar{2} & \bar{2} & \bar{2} \end{smallmatrix}\right)$$

and the path realization of the crystal graph $\mathcal{B}(\Lambda_0)$ is given in the following figure.

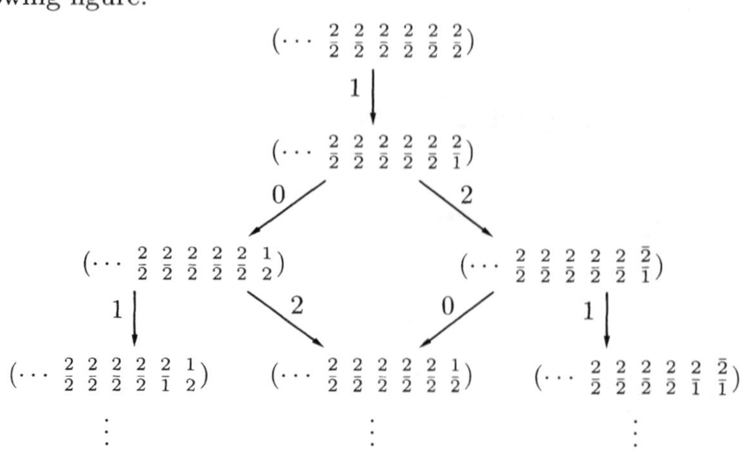

Using the path realization of crystal graphs, we would like to determine the *character* of the irreducible highest weight $U_q(\mathfrak{g})$-module $V(\lambda)$ with highest weight $\lambda \in P^+$. Since

$$\operatorname{ch} V(\lambda + \delta) = e^{\delta} \operatorname{ch} V(\lambda),$$

we will restrict our discussion to $\lambda \in P^+$ such that $\lambda(d) = 0$. We need to calculate the *affine weight* of a path $\mathbf{p} \in \mathcal{P}(\lambda)$. For this purpose, instead of the crystal isomorphism

$$\Psi : \mathcal{B}(\lambda) \xrightarrow{\sim} \mathcal{B}(\varepsilon(b_\lambda)) \otimes \mathcal{B},$$

mapping $u_\lambda \longmapsto u_{\varepsilon(b_\lambda)} \otimes b_\lambda$, we will make use of the crystal embedding

$$\Psi^{\text{aff}} : \mathcal{B}(\lambda) \hookrightarrow \mathcal{B}(\varepsilon(b_\lambda)) \otimes \mathcal{B}^{\text{aff}} \tag{10.49}$$

given by $u_\lambda \longmapsto u_{\varepsilon(b_\lambda)} \otimes b_\lambda(0)$. The reader may easily check the existence of such an embedding (Exercise 10.15). By taking the composition of these crystal embeddings repeatedly, we obtain a sequence of crystal embeddings

$$\mathcal{B}(\lambda) \hookrightarrow \mathcal{B}(\lambda_1) \otimes \mathcal{B}^{\text{aff}} \hookrightarrow \mathcal{B}(\lambda_2) \otimes \mathcal{B}^{\text{aff}} \otimes \mathcal{B}^{\text{aff}} \hookrightarrow \cdots$$

given by

$$u_\lambda \longmapsto u_{\lambda_1} \otimes b_0(0) \longmapsto u_{\lambda_2} \otimes b_1(0) \otimes b_0(0) \longmapsto \cdots .$$

Thus, for each $k \geq 1$, we get a crystal embedding

$$\Psi_k^{\text{aff}} : \mathcal{B}(\lambda) \hookrightarrow \mathcal{B}(\lambda_k) \otimes (\mathcal{B}^{\text{aff}})^{\otimes k}$$

given by

$$u_\lambda \longmapsto u_{\lambda_k} \otimes b_{k-1}(0) \otimes \cdots \otimes b_1(0) \otimes b_0(0).$$

Let $u \in \mathcal{B}(\lambda)$ and let $\mathbf{p} = (\mathbf{p}_k)_{k=0}^{\infty} = \Psi(u)$ be the corresponding λ-path in $\mathcal{P}(\lambda)$. Then for a positive integer $N > 0$ such that

$$\Psi_N(u) = u_{\lambda_N} \otimes \mathbf{p}_{N-1} \otimes \cdots \otimes \mathbf{p}_1 \otimes \mathbf{p}_0 \in u_{\lambda_N} \otimes \mathcal{B}^{\otimes N},$$

we have

$$\Psi_N^{\text{aff}}(u) = u_{\lambda_N} \otimes \mathbf{p}_{N-1}(c_{N-1}) \otimes \cdots \otimes \mathbf{p}_0(c_0) \in u_{\lambda_N} \otimes (\mathcal{B}^{\text{aff}})^{\otimes N}$$

for some $c_0, c_1, \ldots, c_{N-1} \in \mathbf{Z}$, which yields a crystal embedding

$$\Psi^{\text{aff}} : \mathcal{B}(\lambda) \hookrightarrow (\mathcal{B}^{\text{aff}})^{\otimes\infty}$$

given by

$$\Psi^{\text{aff}}(u) = \mathbf{p}^{\text{aff}} = \cdots \otimes b_{N+1}(0) \otimes b_N(0) \otimes \mathbf{p}_{N-1}(c_{N-1}) \otimes \cdots \otimes \mathbf{p}_0(c_0).$$

Hence the *affine weight* of the path $\mathbf{p}$ is determined by

$$\begin{aligned} \operatorname{wt}(\mathbf{p}) = \operatorname{wt}(\mathbf{p}^{\text{aff}}) &= \lambda_N + \sum_{k=0}^{N-1} \overline{\operatorname{wt}}\mathbf{p}_k + \left(\sum_{k=0}^{N-1} c_k\right)\delta \\ &= \lambda + \sum_{k=0}^{\infty} (\overline{\operatorname{wt}}\mathbf{p}_k - \overline{\operatorname{wt}} b_k) + \left(\sum_{k=0}^{N-1} c_k\right)\delta. \end{aligned} \tag{10.50}$$

Lemma 10.6.6. *For any $u \in \mathcal{B}(\lambda)$ and $N > 0$, write*

$$\Psi_N^{\mathrm{aff}}(u) = u' \otimes \mathbf{p}_{N-1}(c_{N-1}) \otimes \cdots \otimes \mathbf{p}_1(c_1) \otimes \mathbf{p}_0(c_0)$$

for some $u' \in \mathcal{B}(\lambda_N)$, $\mathbf{p}_0, \mathbf{p}_1, \ldots, \mathbf{p}_{N-1} \in \mathcal{B}$, $c_0, c_1, \ldots, c_{N-1} \in \mathbf{Z}$. Then we have

$$c_{k+1} - c_k = H(\mathbf{p}_{k+1} \otimes \mathbf{p}_k) - H(b_{k+1} \otimes b_k) \quad \text{for all } k \geq 0.$$

Proof. Since $\operatorname{Im} \Psi_N^{\mathrm{aff}}$ is connected, $\mathbf{p}_{k+1}(c_{k+1}) \otimes \mathbf{p}_k(c_k)$ and $b_{k+1}(0) \otimes b_k(0)$ belong to the same connected components of $\mathcal{B}^{\mathrm{aff}} \otimes \mathcal{B}^{\mathrm{aff}}$ for all $k \geq 0$. By Lemma 10.2.7, we have

$$H^{\mathrm{aff}}(\mathbf{p}_{k+1}(c_{k+1}) \otimes \mathbf{p}_k(c_k)) = H^{\mathrm{aff}}(b_{k+1}(0) \otimes b_k(0)),$$

which yields

$$c_{k+1} - c_k - H(\mathbf{p}_{k+1} \otimes \mathbf{p}_k) = -H(b_{k+1} \otimes b_k).$$

Hence we obtain

$$c_{k+1} - c_k = H(\mathbf{p}_{k+1} \otimes \mathbf{p}_k) - H(b_{k+1} \otimes b_k) \quad \text{for all } k \geq 0.$$

□

Now, we can determine the affine weight of a λ-path $\mathbf{p} \in \mathcal{P}(\lambda)$.

Theorem 10.6.7. *Let $\mathbf{p} = (\mathbf{p}_k)_{k=0}^{\infty}$ be a λ-path in $\mathcal{P}(\lambda)$. Then the affine weight of $\mathbf{p}$ is given by the formula*

$$\begin{aligned} \operatorname{wt}(\mathbf{p}) &= \lambda + \sum_{k=0}^{\infty} (\overline{\operatorname{wt}} \mathbf{p}_k - \overline{\operatorname{wt}} b_k) \\ &\quad - \left(\sum_{k=0}^{\infty} (k+1) \left(H(\mathbf{p}_{k+1} \otimes \mathbf{p}_k) - H(b_{k+1} \otimes b_k) \right) \right) \delta. \end{aligned} \tag{10.51}$$

Therefore we have

$$\operatorname{ch} V(\lambda) = \sum_{\mathbf{p} \in \mathcal{P}(\lambda)} e^{\operatorname{wt} \mathbf{p}}. \tag{10.52}$$

Proof. Let $\mathbf{p} = (\mathbf{p}_k)_{k=0}^{\infty}$ be a λ-path in $\mathcal{P}(\lambda)$ and let $u \in \mathcal{B}(\lambda)$ be such that $\Psi(u) = \mathbf{p}$. Take a sufficiently large $N > 0$ such that

$$\Psi_N(u) = u_{\lambda_N} \otimes \mathbf{p}_{N-1} \otimes \cdots \otimes \mathbf{p}_1 \otimes \mathbf{p}_0 \in u_{\lambda_N} \otimes \mathcal{B}^{\otimes N}.$$

Then we have $\mathbf{p}_k = b_k$ for all $k \geq N$ and

$$\Psi_N^{\mathrm{aff}}(u) = u_{\lambda_N} \otimes \mathbf{p}_{N-1}(c_{N-1}) \otimes \cdots \otimes \mathbf{p}_0(c_0) \in u_{\lambda_N} \otimes (\mathcal{B}^{\mathrm{aff}})^{\otimes N}$$

for some $c_0, c_1, \ldots, c_{N-1} \in \mathbf{Z}$. By (10.50), the affine weight of $\mathbf{p}$ is given by

$$\mathrm{wt}(\mathbf{p}) = \lambda + \sum_{k=0}^{\infty} (\overline{\mathrm{wt}}\mathbf{p}_k - \overline{\mathrm{wt}}b_k) + \left(\sum_{k=0}^{N-1} c_k\right)\delta.$$

Since $c_{k+1} - c_k = H(\mathbf{p}_{k+1} \otimes \mathbf{p}_k) - H(b_{k+1} \otimes b_k)$ by Lemma 10.6.6, we have

$$\begin{aligned}\sum_{k=0}^{\infty}(k+1)\,(H(\mathbf{p}_{k+1} \otimes \mathbf{p}_k) - H(b_{k+1} \otimes b_k)) &= \sum_{k=0}^{N-1}(k+1)(c_{k+1} - c_k)\\ &= (c_1 - c_0) + 2(c_2 - c_1) + \cdots + (N-1)(c_{N-1} - c_{N-2}) + N(-c_{N-1})\\ &= -(c_0 + c_1 + \cdots + c_{N-1}),\end{aligned}$$

which proves our formula. □

Example 10.6.8.

(1) Let $\mathfrak{g} = A_2^{(1)}$ and let $\mathcal{B}$ be the perfect crystal of level 1 given in Example 10.5.2 (3). Recall that the ground-state path of weight Λ_0 is

$$\mathbf{p}_{\Lambda_0} = (\ldots, 1, 2, 3, 1, 2, 3)$$

and consider the path

$$\mathbf{p} = (\ldots, 1, 2, 3, 1, 3, 2).$$

Since the energy function on $\mathcal{B}$ is given by (10.22), the affine weight formula (10.51) gives

$$\begin{aligned}\mathrm{wt}\,\mathbf{p} &= \Lambda_0 + ((\Lambda_0 - \Lambda_1) - (\Lambda_2 - \Lambda_1))\\ &\quad + ((\Lambda_2 - \Lambda_1) - (\Lambda_0 - \Lambda_1))\\ &\quad - ((1-0) + 2(0-0))\delta\\ &= \Lambda_0 - \delta.\end{aligned}$$

(2) Let $\mathfrak{g} = C_2^{(1)}$ and let $\mathcal{B}$ be the perfect crystal of level 1 given in Example 10.5.2 (6). Recall that ground-state path $\mathbf{p}_{\Lambda_1}$ is the sequence

$$\mathbf{p}_{\Lambda_1} = (\cdots \genfrac{}{}{0pt}{}{2}{\bar 2}\ \genfrac{}{}{0pt}{}{2}{\bar 2}\ \genfrac{}{}{0pt}{}{2}{\bar 2}\ \genfrac{}{}{0pt}{}{2}{\bar 2}\ \genfrac{}{}{0pt}{}{2}{\bar 2}\ \genfrac{}{}{0pt}{}{2}{\bar 2})$$

and consider the path

$$\mathbf{p} = (\cdots \genfrac{}{}{0pt}{}{2}{\bar 2}\ \genfrac{}{}{0pt}{}{2}{\bar 2}\ \genfrac{}{}{0pt}{}{2}{\bar 2}\ \genfrac{}{}{0pt}{}{2}{\bar 2}\ \genfrac{}{}{0pt}{}{2}{\bar 1}\ \genfrac{}{}{0pt}{}{1}{\bar 2}).$$

Then, using the tensor product rule, it is easy to verify that there is an energy function $H : \mathcal{B} \otimes \mathcal{B} \longrightarrow \mathbf{Z}$ such that

$$H(\genfrac{}{}{0pt}{}{2}{\bar 2} \otimes \genfrac{}{}{0pt}{}{2}{\bar 2}) = 1, \qquad H(\genfrac{}{}{0pt}{}{2}{\bar 2} \otimes \genfrac{}{}{0pt}{}{2}{\bar 1}) = 1, \qquad H(\genfrac{}{}{0pt}{}{2}{\bar 1} \otimes \genfrac{}{}{0pt}{}{1}{\bar 2}) = 2 \tag{10.53}$$

(Exercise 10.16). Hence by the weight formula (10.51), we have

$$\begin{aligned}\mathrm{wt}\,\mathbf{p} &= \Lambda_1 + ((\Lambda_1 - \Lambda_2) + (\Lambda_2 - \Lambda_1)) \\ &\quad - ((2-1) + 2(1-1))\delta \\ &= \Lambda_1 - \delta.\end{aligned}$$

Exercises

10.1. Verify that the equation (10.15) defines a $U_q'(\widehat{\mathfrak{sl}}_3)$-module structure on the space $V = \mathbf{C}(q)v_1 \oplus \mathbf{C}(q)v_2 \oplus \mathbf{C}(q)v_3$.

10.2. Prove Proposition 10.1.4.

10.3. Verify the statements in Example 10.2.2.

10.4. Verify that the map $R : \mathcal{B}^{\mathrm{aff}} \otimes \mathcal{B}^{\mathrm{aff}} \longrightarrow \mathcal{B}^{\mathrm{aff}} \otimes \mathcal{B}^{\mathrm{aff}}$ defined by (10.26) is a combinatorial R-matrix.

10.5. Complete the proof of Lemma 10.2.5.

10.6. Complete the proof of Proposition 10.3.2.

10.7. Let $\Phi^\vee : V(m) \otimes V(l) \longrightarrow W$ be a linear map satisfying the conditions in (10.31), and let $\Phi : V(l) \longrightarrow V(m) \otimes W$ be the linear map induced by $\Phi^\vee$ such that

$$\Phi^\vee(v \otimes u) = \sum_j (\Phi_j(u), v) w_j,$$

where $\Phi(u) = \sum_j \Phi_j(u) \otimes w_j$.
Show that Φ is a $U_q(\mathfrak{sl}_2)$-module homomorphism.

10.8. Given a vertex operator

$$\widehat{\Phi}(z)(u) = \sum_j \sum_{n \in \mathbf{Z}} \Phi_{j,n}(u) \otimes w_j z^{-n},$$

define a linear map $\Phi : V(\lambda) \longrightarrow \widehat{V}(\mu) \otimes W$ by

$$\phi(u) = \sum_j (\sum_{n \in \mathbf{Z}} \Phi_{j,n}(u)) \otimes w_j.$$

Show that Φ is a $U_q'(\mathfrak{g})$-module homomorphism.

10.9. Given a $U_q'(\mathfrak{g})$-module homomorphism $\Phi : V(\lambda) \longrightarrow \widehat{V}(\mu) \otimes W$, write

$$\Phi(u) = \sum_j \Phi_j(u) \otimes w_j = \sum_j \left(\sum_{n \in \mathbf{Z}} \Phi_{j,n}(u) \right) \otimes w_j \quad (u \in V(\lambda))$$

and define a $\mathbf{C}(q)$-linear map $\widehat{\Phi}(z) : V(\lambda) \to V(\mu)\widehat{\otimes}W^{\text{aff}}$ by

$$\widehat{\Phi}(z)(u) = \sum_j \sum_{n \in \mathbf{Z}} \Phi_{j,n}(u) \otimes w_j z^{-n}.$$

Show that $\widehat{\Phi}(z)$ is a $U_q(\mathfrak{g})$-module homomorphism.

10.10. Complete the proof of Theorem 10.4.3.

10.11. Complete the proof of Theorem 10.4.4.

10.12. Verify that the sequences $\mathbf{w}_\lambda$ and $\mathbf{p}_\lambda$ are periodic sequences with the same period.

10.13. Prove Proposition 10.6.3.

10.14. Show that the map $\Psi : \mathcal{B}(\lambda) \longrightarrow \mathcal{P}(\lambda)$ in Theorem 10.6.4 is a bijection.

10.15. Show that the map

$$\Psi^{\text{aff}} : \mathcal{B}(\lambda) \hookrightarrow \mathcal{B}(\varepsilon(b_\lambda)) \otimes \mathcal{B}^{\text{aff}}$$

given by $u_\lambda \longmapsto u_{\varepsilon(b_\lambda)} \otimes b_\lambda(0)$ defines a (strict) crystal embedding.

10.16. Let $\mathfrak{g} = C_2^{(1)}$ and let $\mathcal{B}$ be a perfect crystal of level 1 given in Example 10.5.2 (6). Show that there exists an energy function H on $\mathcal{B}$ satisfying (10.53).

Chapter 11

Combinatorics of Young Walls

In this chapter, we give a realization of crystal graphs for basic representations of classical quantum affine algebras of type $A_n^{(1)}$ $(n \geq 1)$, $A_{2n-1}^{(2)}$ $(n \geq 3)$, $D_n^{(1)}$ $(n \geq 4)$, $A_{2n}^{(2)}$, $D_{n+1}^{(2)}$ $(n \geq 2)$, and $B_n^{(1)}$ $(n \geq 3)$ using some new combinatorial objects which we call the *Young walls.* The Young walls consist of colored blocks with various shapes that are built on a given *ground-state wall* and can be viewed as generalizations of Young diagrams. The rules for building Young walls and the action of Kashiwara operators are given explicitly in terms of combinatorics of Young walls. They are quite similar to playing with LEGO® blocks or the Tetris® game. The crystal graphs for basic representations are characterized as the sets of all *reduced proper Young walls.* The characters of basic representations can be computed easily by counting the number of colored blocks in reduced proper Young walls that have been added to the ground-state wall.

11.1. Perfect crystals of level 1 and path realization

We briefly review the path realization of crystal graphs for irreducible highest weight representations of $U_q(\mathfrak{g})$. In this section, we will focus on the irreducible highest weight representations with dominant integral highest weights of level 1, which are called the ***basic representations***.

Let $\mathcal{B}$ be a perfect crystal of level 1 for the quantum affine algebra $U_q(\mathfrak{g})$ and let $\lambda \in \bar{P}_1^+$ be a dominant integral weight of level 1. The *ground-state path* $\mathbf{p}_\lambda = (\mathbf{p}_\lambda(k))_{k=0}^\infty = (b_k)_{k=0}^\infty$ is determined by the condition

$$\lambda_0 = \lambda, \quad b_0 = b_\lambda; \qquad \lambda_{k+1} = \varepsilon(b_k), \quad b_{k+1} = b_{\lambda_{k+1}} \quad (k \geq 0).$$

Here, b_λ is the unique element of $\mathcal{B}$ such that $\varphi(b_\lambda) = \lambda$, so $\varphi(b_{k+1}) = \lambda_{k+1}$. Then, by Theorem 10.6.4, there is a crystal isomorphism

$$\text{(11.1)} \quad \mathcal{B}(\lambda) \cong \mathcal{P}(\lambda) = \{\mathbf{p} = (\mathbf{p}(k))_{k=0}^\infty \mid \mathbf{p}(k) \in \mathcal{B},\ \mathbf{p}(k) = b_k, \forall k \gg 0\}.$$

In the following example, we illustrate the perfect crystals of level 1, the ground-state paths, and the top parts of crystal graphs for basic representations.

Example 11.1.1. $A_n^{(1)}$ $(n \geq 1)$.

(1) Perfect crystal of level 1:

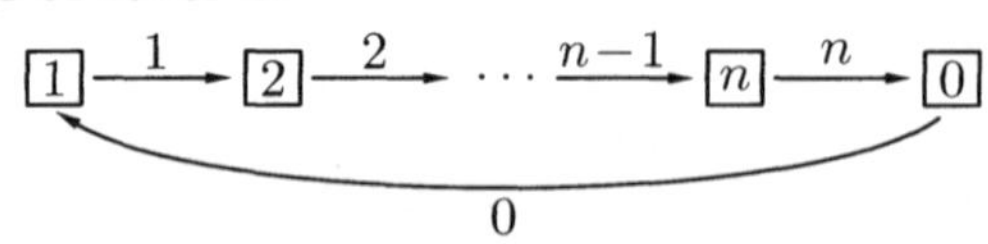

$$b_{\Lambda_i} = \boxed{i}, \quad b^{\Lambda_i} = \boxed{i{+}1} \quad \text{for } i = 0, 1, \ldots, n.$$

(2) Ground-state paths:

$$\mathbf{p}_{\Lambda_i} = (\mathbf{p}_{\Lambda_i}(k))_{k=0}^\infty = (\cdots 1\,2 \cdots n\,0\,1\,2 \cdots i).$$

(3) Crystal graph $\mathcal{B}(\Lambda_0)$ for $A_2^{(1)}$:

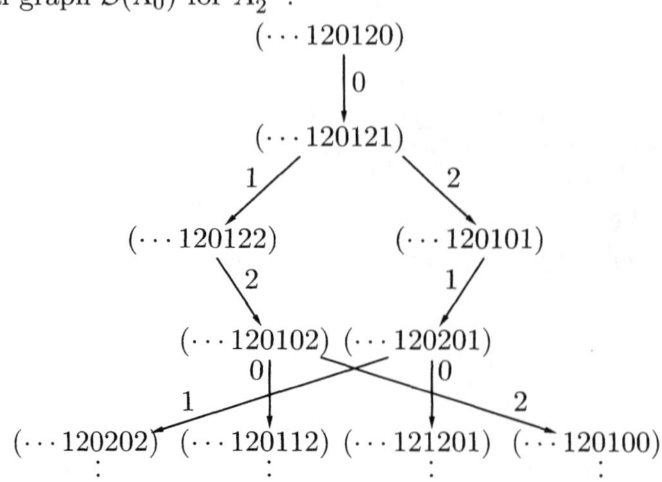

Example 11.1.2. $A_{2n-1}^{(2)}$ $(n \geq 3)$.

(1) Perfect crystal of level 1:

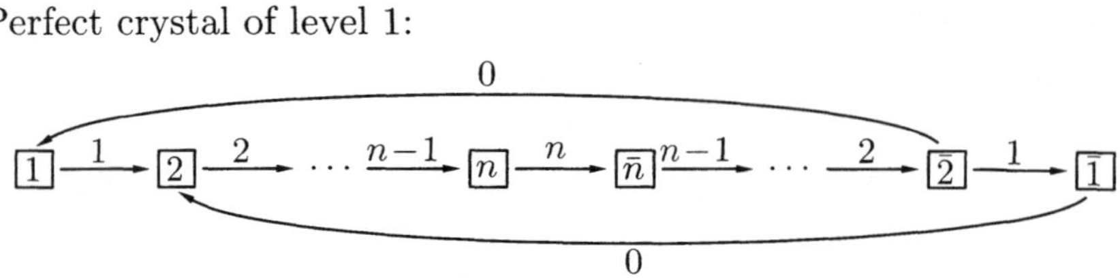

$$b_{\Lambda_0} = \boxed{\bar{1}}, \quad b^{\Lambda_0} = \boxed{1}; \qquad b_{\Lambda_1} = \boxed{1}, \quad b^{\Lambda_1} = \boxed{\bar{1}}.$$

(2) Ground-state paths:

$$\mathbf{p}_{\Lambda_0} = (\mathbf{p}_{\Lambda_0}(k))_{k=0}^{\infty} = (\cdots\ 1\,\bar{1}\,1\,\bar{1}\,1\,\bar{1}\,),$$
$$\mathbf{p}_{\Lambda_1} = (\mathbf{p}_{\Lambda_1}(k))_{k=0}^{\infty} = (\cdots\ \bar{1}\,1\,\bar{1}\,1\,\bar{1}\,1\,).$$

(3) Crystal graph $\mathcal{B}(\Lambda_0)$ for $A_5^{(2)}$:

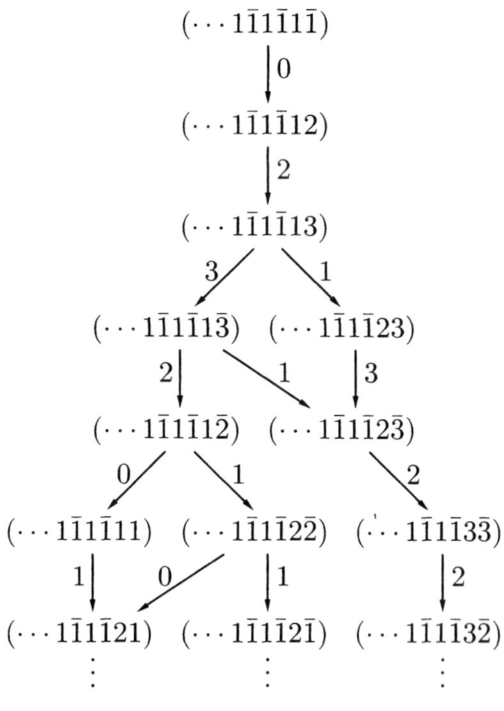

Example 11.1.3. $D_n^{(1)}$ $(n \geq 4)$.

(1) Perfect crystal of level 1:

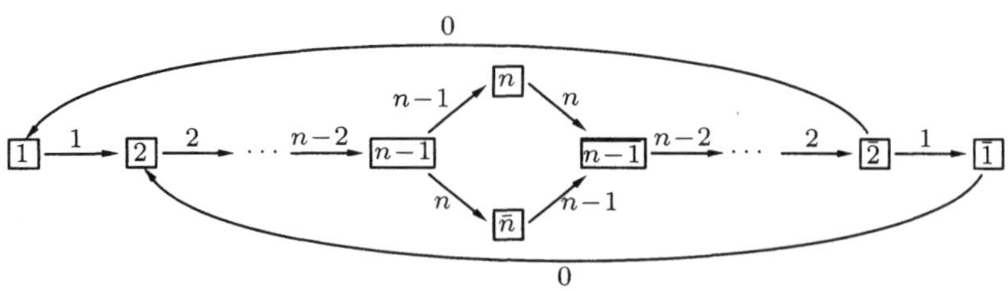

$$b_{\Lambda_0} = \boxed{\bar{1}}, \quad b^{\Lambda_0} = \boxed{1}; \qquad b_{\Lambda_1} = \boxed{1}, \quad b^{\Lambda_1} = \boxed{\bar{1}},$$

$$b_{\Lambda_{n-1}} = \boxed{\bar{n}}, \quad b^{\Lambda_{n-1}} = \boxed{n}; \qquad b_{\Lambda_n} = \boxed{n}, \quad b^{\Lambda_n} = \boxed{\bar{n}}.$$

(2) Ground-state paths:

$$\mathbf{p}_{\Lambda_0} = (\mathbf{p}_{\Lambda_0}(k))_{k=0}^{\infty} = (\cdots\ 1\,\bar{1}\,1\,\bar{1}\,1\,\bar{1}\,),$$
$$\mathbf{p}_{\Lambda_1} = (\mathbf{p}_{\Lambda_1}(k))_{k=0}^{\infty} = (\cdots\ \bar{1}\,1\,\bar{1}\,1\,\bar{1}\,1\,),$$
$$\mathbf{p}_{\Lambda_{n-1}} = (\mathbf{p}_{\Lambda_{n-1}}(k))_{k=0}^{\infty} = (\cdots\ n\,\bar{n}\,n\,\bar{n}\,n\,\bar{n}\,),$$
$$\mathbf{p}_{\Lambda_n} = (\mathbf{p}_{\Lambda_n}(k))_{k=0}^{\infty} = (\cdots\ \bar{n}\,n\,\bar{n}\,n\,\bar{n}\,n\,).$$

(3) Crystal graph $\mathcal{B}(\Lambda_0)$ for $D_4^{(1)}$:

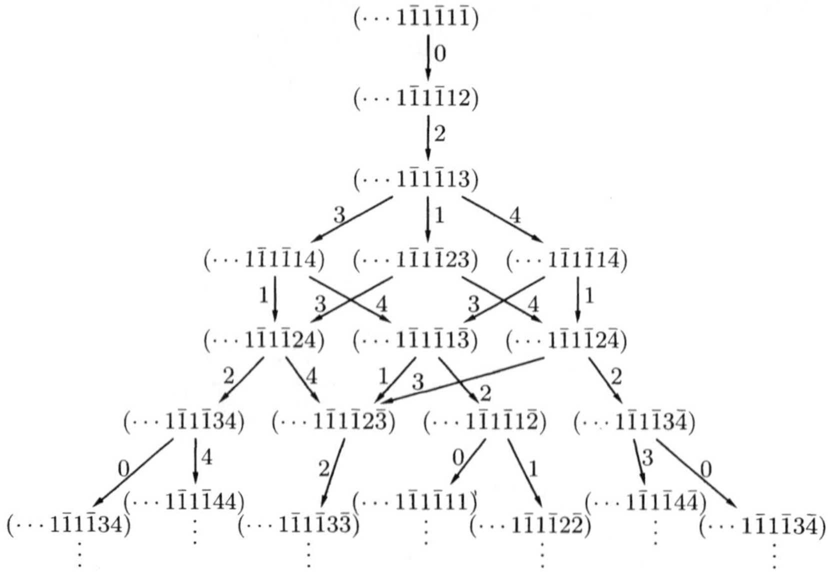

Example 11.1.4. $A_{2n}^{(2)}$ $(n \geq 2)$.

(1) Perfect crystal of level 1:

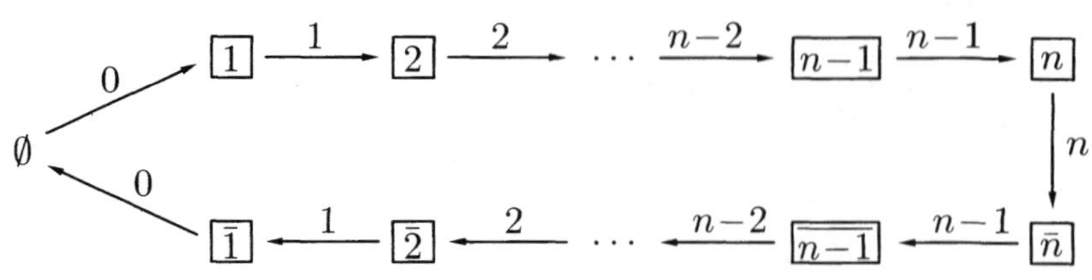

$$b_{\Lambda_0} = \emptyset, \qquad b^{\Lambda_0} = \emptyset.$$

(2) Ground-state path:

$$\mathbf{p}_{\Lambda_0} = (\mathbf{p}_{\Lambda_0}(k))_{k=0}^{\infty} = (\cdots\ \emptyset\,\emptyset\,\emptyset\,\emptyset\,\emptyset\,)$$

(3) Crystal graph $\mathcal{B}(\Lambda_0)$ for $A_4^{(2)}$:

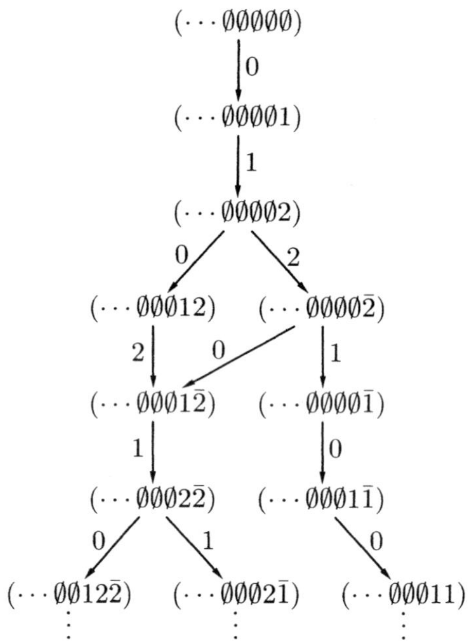

Example 11.1.5. $D_{n+1}^{(2)}$ $(n \geq 2)$.

(1) Perfect crystal of level 1:

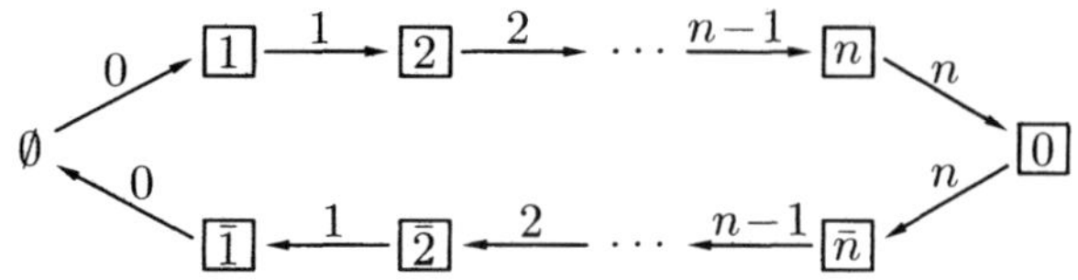

$$b_{\Lambda_0} = \emptyset, \quad b^{\Lambda_0} = \emptyset; \qquad b_{\Lambda_n} = \boxed{0}, \quad b^{\Lambda_n} = \boxed{\bar{0}}.$$

(2) Ground-state paths:

$$\mathbf{p}_{\Lambda_0} = (\mathbf{p}_{\Lambda_0}(k))_{k=0}^{\infty} = (\cdots \emptyset\emptyset\emptyset\emptyset\emptyset),$$
$$\mathbf{p}_{\Lambda_n} = (\mathbf{p}_{\Lambda_n}(k))_{k=0}^{\infty} = (\cdots 00000).$$

(3) Crystal graph $\mathcal{B}(\Lambda_0)$ for $D_3^{(2)}$:

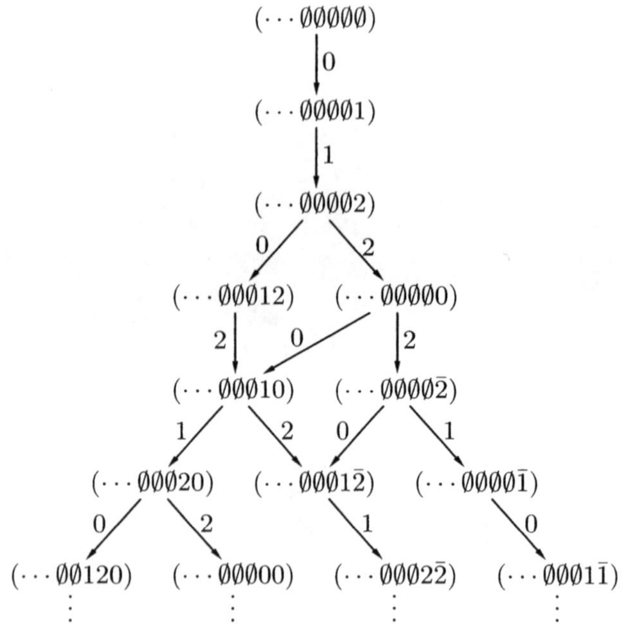

Example 11.1.6. $B_n^{(1)}$ $(n \geq 3)$.

(1) Perfect crystal of level 1:

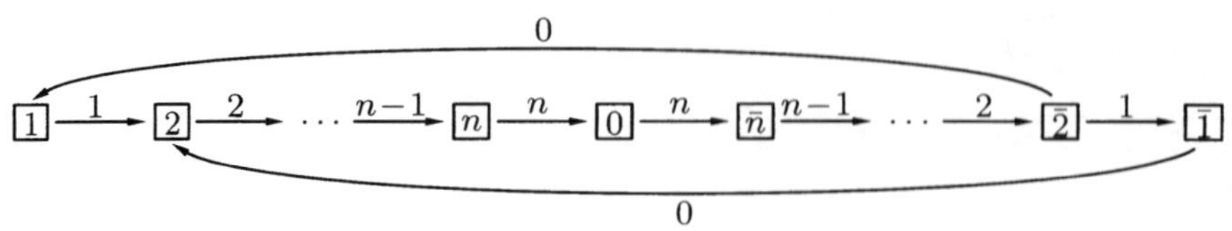

$$
\begin{aligned}
&b_{\Lambda_0} = \boxed{\bar{1}}, \qquad b_{\Lambda_1} = \boxed{1}, \qquad b_{\Lambda_n} = \boxed{0}, \\
&b^{\Lambda_0} = \boxed{1}, \qquad b^{\Lambda_1} = \boxed{\bar{1}}, \qquad b^{\Lambda_n} = \boxed{0}.
\end{aligned}
$$

(2) Ground-state paths:

$$
\begin{aligned}
\mathbf{p}_{\Lambda_0} &= (\mathbf{p}_{\Lambda_0}(k))_{k=0}^{\infty} = (\cdots 1\,\bar{1}\,1\,\bar{1}\,1\,\bar{1}), \\
\mathbf{p}_{\Lambda_1} &= (\mathbf{p}_{\Lambda_1}(k))_{k=0}^{\infty} = (\cdots \bar{1}\,1\,\bar{1}\,1\,\bar{1}\,1), \\
\mathbf{p}_{\Lambda_n} &= (\mathbf{p}_{\Lambda_n}(k))_{k=0}^{\infty} = (\cdots 0\,0\,0\,0\,0\,0).
\end{aligned}
$$

(3) Crystal graph $\mathcal{B}(\Lambda_0)$ for $B_3^{(1)}$:

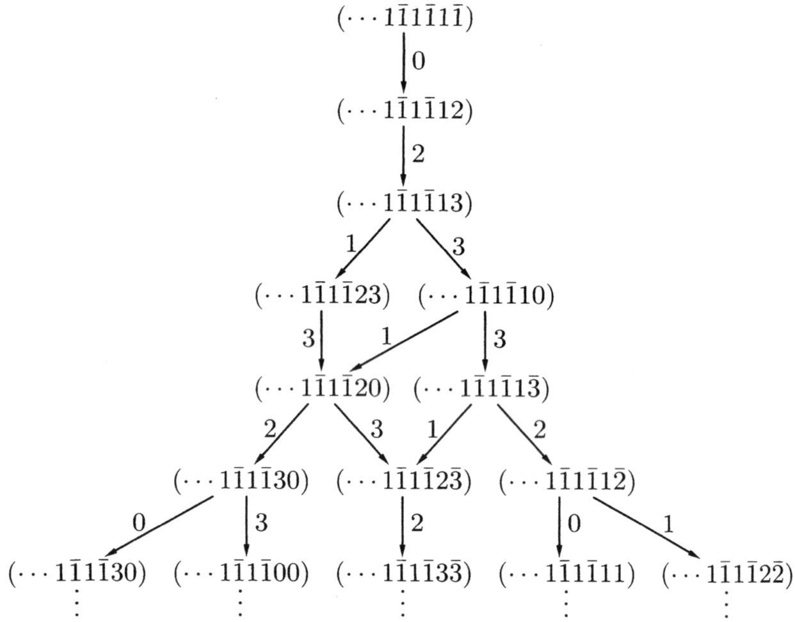

(4) Crystal graph $\mathcal{B}(\Lambda_3)$ for $B_3^{(1)}$:

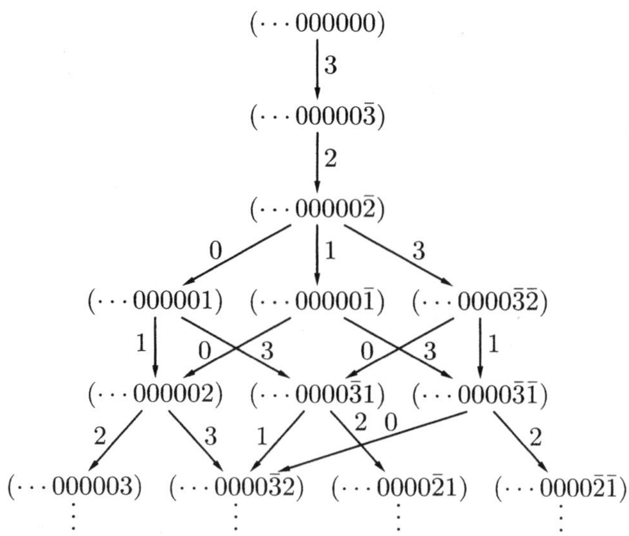

11.2. Combinatorics of Young walls

In this section, we explain the notion of *Young walls*. The Young walls are built of colored blocks of three different shapes:

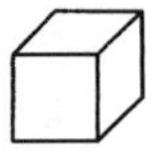 : unit width, unit height, unit thickness,

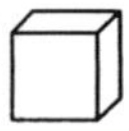 : unit width, unit height, half-unit thickness,

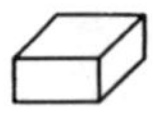 : unit width, half-unit height, unit thickness.

With these colored blocks, we will build a wall of thickness less than or equal to one unit which extends infinitely to the left, just as if we were playing with LEGO® blocks. Given a dominant integral weight λ of level 1, we fix a frame Y_λ called the ***ground-state wall*** of ***weight*** λ, and build Young walls on this frame. For each type of classical quantum affine algebras, we use different sets of colored blocks and ground-state walls, as are described below.

(1) $A_n^{(1)}$ $(n \geq 1)$

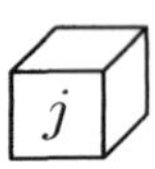

$(j = 0, 1, \ldots, n)$: unit width, unit height, unit thickness,

$Y_{\Lambda_i} =$ 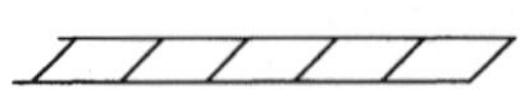 $(i = 0, 1, \ldots, n)$.

(2) $A_{2n-1}^{(2)}$ $(n \geq 3)$

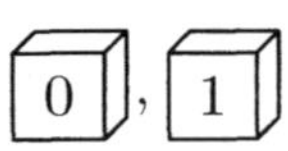

: unit width, unit height, half-unit thickness,

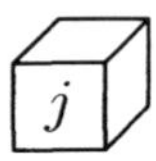

$(j = 2, \ldots, n)$: unit width, unit height, unit thickness,

$Y_{\Lambda_0} =$ 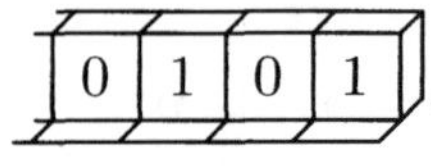

,

$Y_{\Lambda_1} =$ 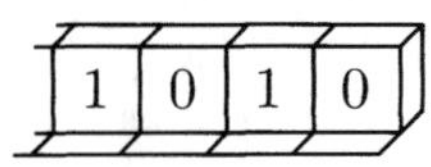

.

(3) $D_n^{(1)}$ $(n \geq 4)$

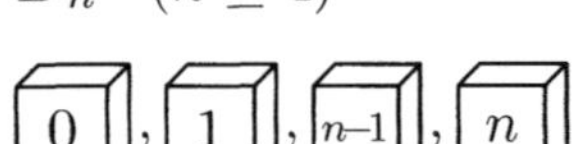

: unit width, unit height, half-unit thickness,

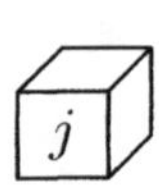

$(j = 2, \ldots, n-2)$: unit width, unit height, unit thickness,

$Y_{\Lambda_0} =$ 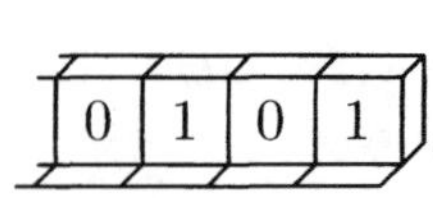

,

$Y_{\Lambda_1} =$ 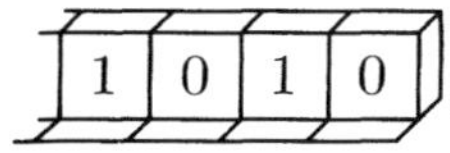

,

$Y_{\Lambda_{n-1}} =$ 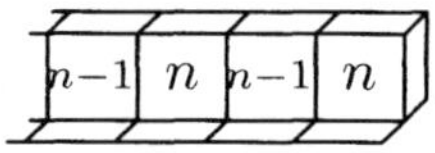

,

$Y_{\Lambda_n} =$ 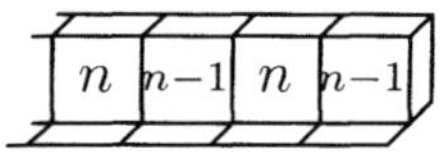

.

(4) $A_{2n}^{(2)}$ $(n \geq 2)$.

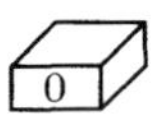

: unit width, half-unit height, unit thickness,

$(j = 1, \ldots, n)$: unit width, unit height, unit thickness,

$Y_{\Lambda_0} =$ 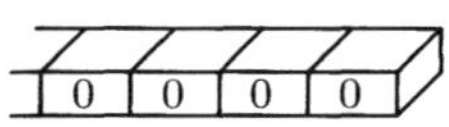

.

(5) $D_{n+1}^{(2)}$ $(n \geq 2)$.

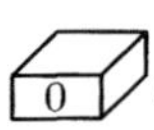

, 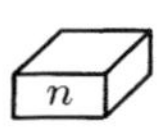

: unit width, half-unit height, unit thickness,

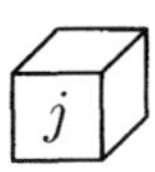

$(j = 1, \ldots, n-1)$: unit width, unit height, unit thickness,

$Y_{\Lambda_0} =$ 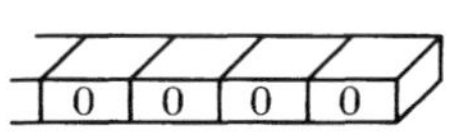

,

$Y_{\Lambda_n} =$ 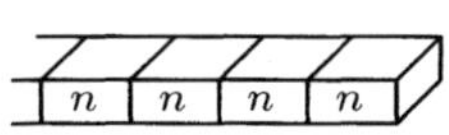

.

(6) $B_n^{(1)}$ $(n \geq 3)$

, 

: unit width, unit height, half-unit thickness,

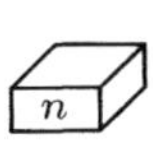

: unit width, half-unit height, unit thickness,

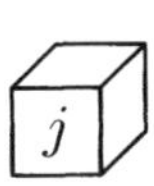

$(j = 2, \ldots, n-1)$: unit width, unit height, unit thickness,

$Y_{\Lambda_0} =$ 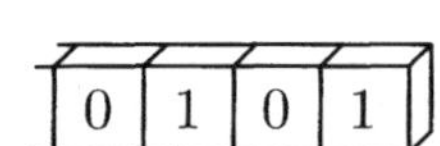

,

$Y_{\Lambda_1} =$ 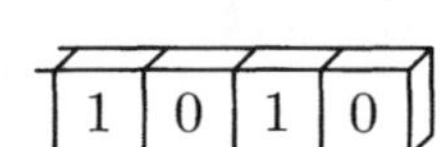

,

$Y_{\Lambda_n} =$ 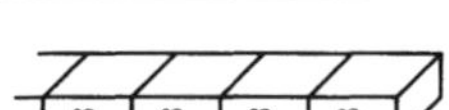

.

For convenience, we will use the following notations:

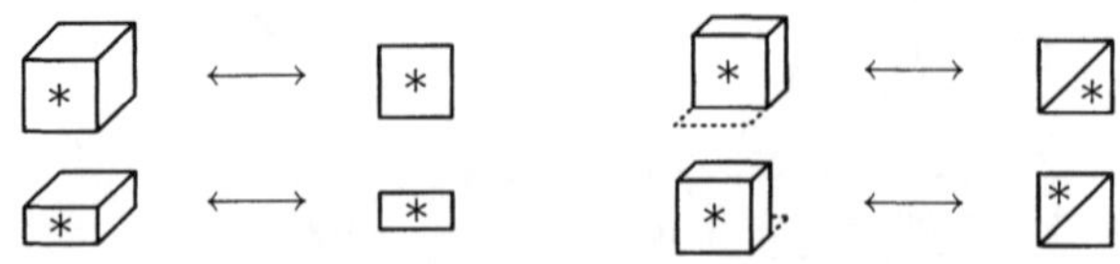

For example, we have

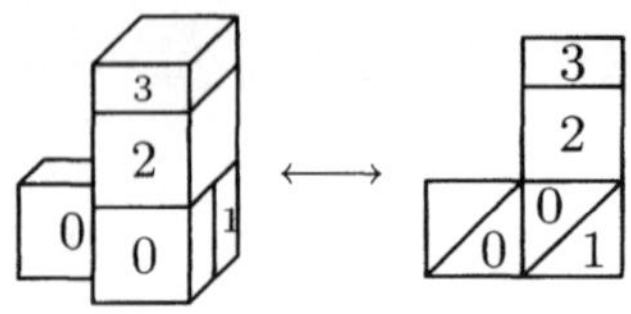

When using this notation on a wall, we will usually shade the blocks constituting the ground-state wall and discard the blocks in the ground-state wall extending infinitely to the left.

The rules for building the walls are given as follows:

(1) The walls must be built on top of one of the ground-state walls.

(2) The colored blocks should be stacked in the patterns given below for each affine type and ground-state wall.

(3) No block can be placed on top of a column of half-unit thickness.

(4) Except for the right-most column, there should be no free space to the right of any block.

Rule (4) implies that the height of the columns are weakly decreasing as we go from right to left. We now give the pattern mentioned in rule (2).

(a) $A_n^{(1)}$ $(n \geq 1)$

On Y_{Λ_i} :

					0	1
					n	0
					$n-1$	n
					$\vdots$	$\vdots$
0	1	2	3	$\cdots$	i	$i+1$
n	0	1	2	$\cdots$	$i-1$	i

(b) $A_{2n-1}^{(2)}$ $(n \geq 3)$

On Y_{Λ_0} :

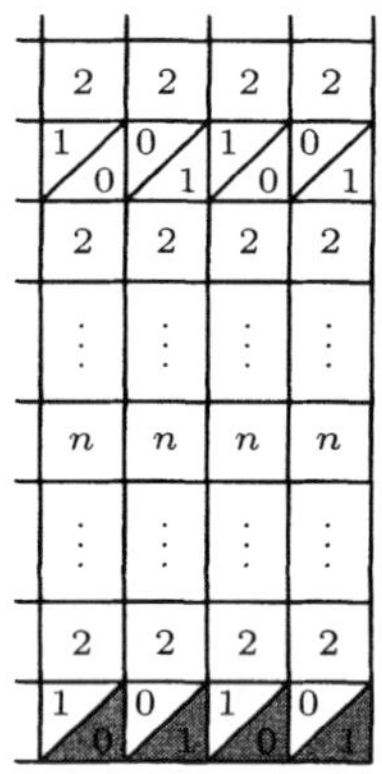

On Y_{Λ_1} :

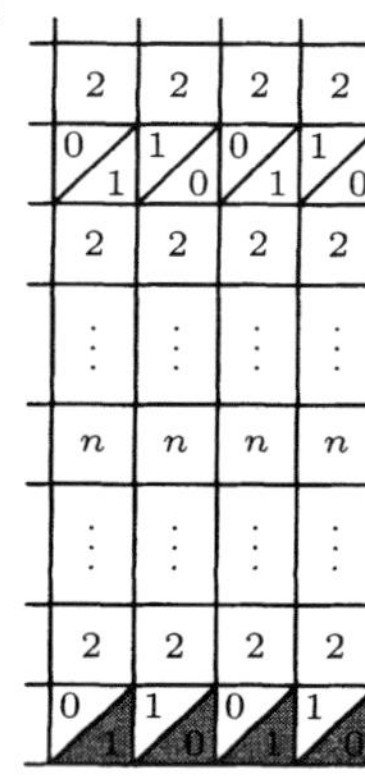

(c) $D_n^{(1)}$ $(n \geq 4)$

On Y_{Λ_0} :

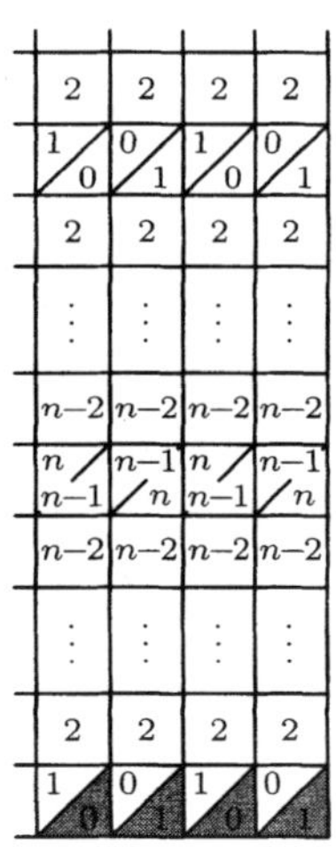

On Y_{Λ_1} :

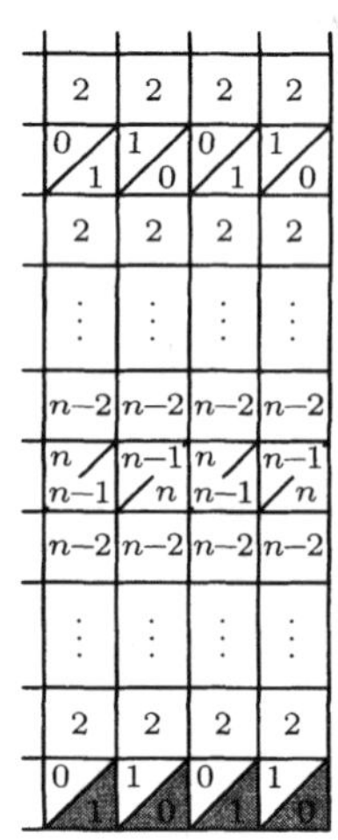

On $Y_{\Lambda_{n-1}}$:

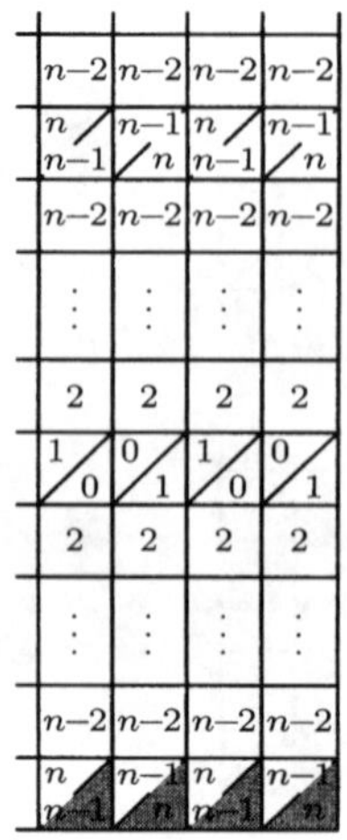

On Y_{Λ_n} :

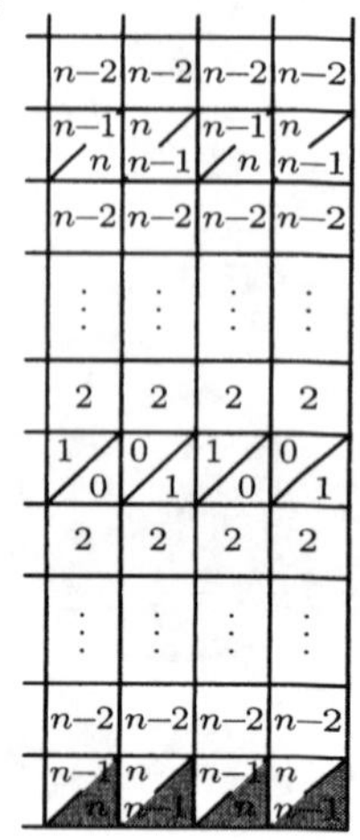

(d) $A_{2n}^{(2)}$ $(n \geq 2)$

On Y_{Λ_0} :

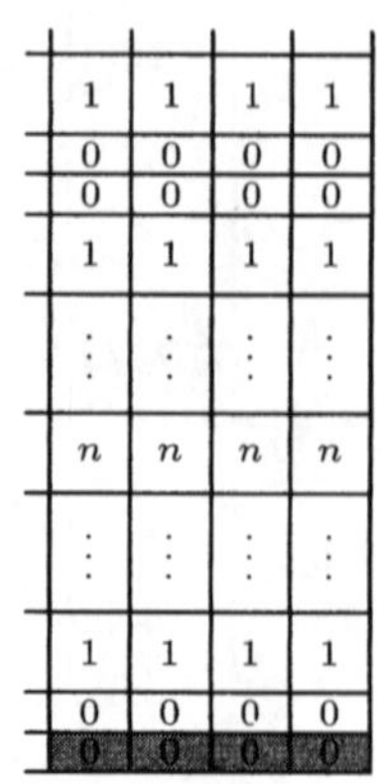

(e) $D_{n+1}^{(2)}$ $(n \geq 2)$

On Y_{Λ_0} :

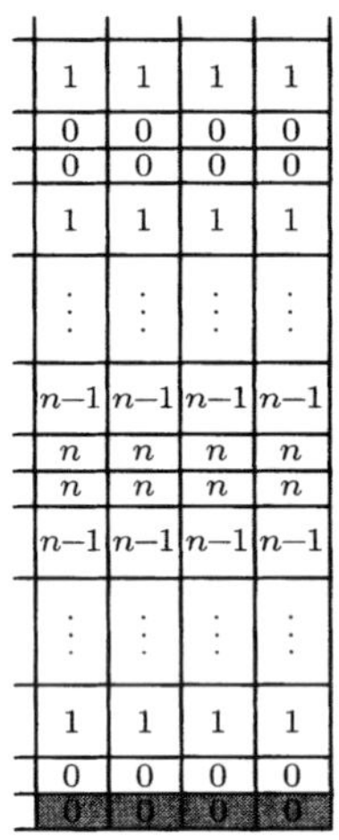

On Y_{Λ_n} :

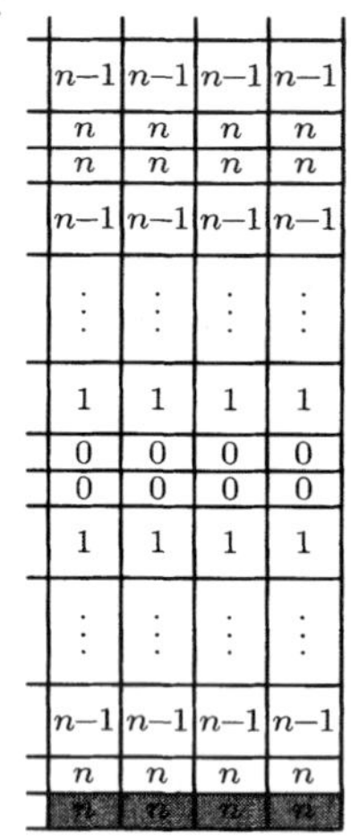

(f) $B_n^{(1)}$ $(n \geq 3)$

On Y_{Λ_0} :

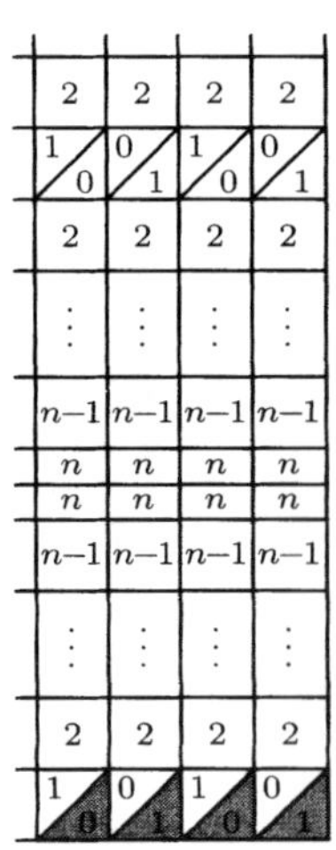

On Y_{Λ_1} :

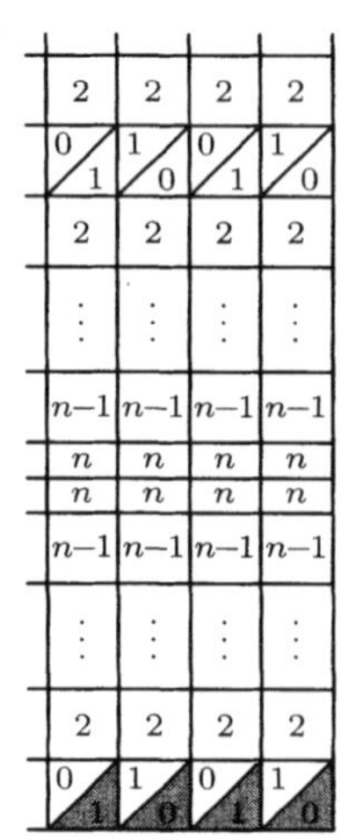

On Y_{Λ_n} :

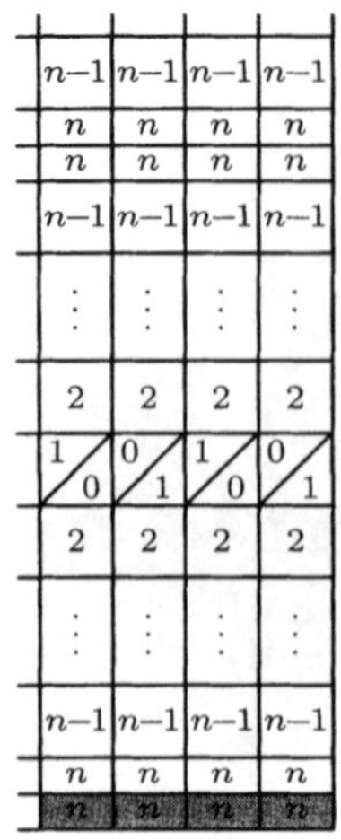

Definition 11.2.1.

(1) A wall built on the ground-state wall Y_λ following the rules listed above is called a ***Young wall*** of ***ground-state*** λ.

(2) A column in a Young wall is called a ***full column*** if its height is a multiple of the unit length and its top is of unit thickness.

(3) For the quantum classical affine algebras of type $A_{2n-1}^{(2)}$ $(n \geq 3)$, $D_n^{(1)}$ $(n \geq 4)$, $A_{2n}^{(2)}$ $(n \geq 1)$, $D_{n+1}^{(2)}$ $(n \geq 2)$, and $B_n^{(1)}$ $(n \geq 3)$, a Young wall is said to be ***proper*** if none of the full columns have the same height.

(4) For the quantum affine algebra of type $A_n^{(1)}$ $(n \geq 1)$, every Young wall is defined to be ***proper***.

Let us explain these concepts with an example.

Example 11.2.2. Consider the Young wall

$$Y = (y_k)_{k=0}^{\infty} = (\dots, y_k, \dots, y_1, y_0)$$

for $B_3^{(1)}$ built on the ground-state wall Y_{Λ_0}:

Y =

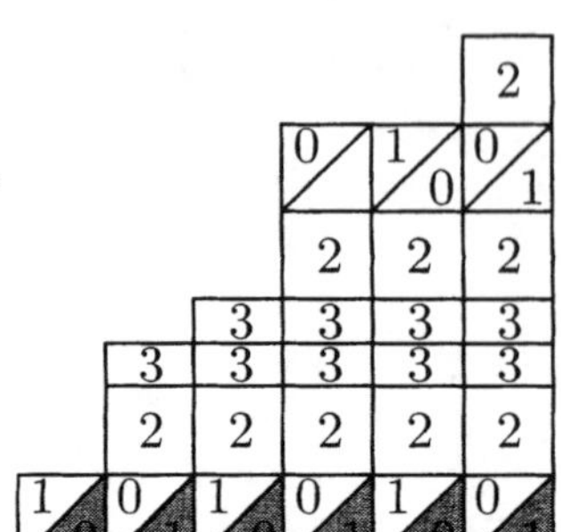

Recall that the shaded blocks make up the ground-state and that the ground-state wall extending infinitely to the left has been truncated in this drawing. The columns y_0, y_1, y_3, y_5 are full columns and y_2, y_4 are not full. Hence Y is a proper Young wall.

On the other hand,

$Y' =$ 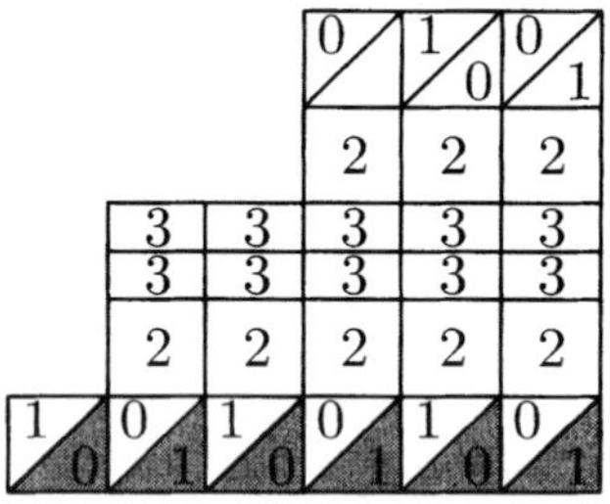

is not a proper Young wall because the full columns y'_0 and y'_1 (also y'_3 and y'_4) have the same height.

11.3. The crystal structure

Let λ be a dominant integral weight of level 1 and let $\mathcal{F}(\lambda)$ denote the set of all proper Young walls built on the ground-state wall Y_λ. In this section, we define an affine crystal structure on $\mathcal{F}(\lambda)$.

The action of Kashiwara operators is defined using the i-signature of proper Young walls in a similar way to playing the Tetris® game. Our *gravity* works *downward* as usual, but will also work *toward the right* as soon as the block touches the wall. We should shift a falling block either to the left or to the right and settle it in a secure and *correct* place on the wall. In the following, we will explain how to find the *correct* position on the wall.

Definition 11.3.1.

(1) A block of color i in a proper Young wall is called a ***removable i-block*** if the wall remains a proper Young wall after removing the block. A column in a proper Young wall is said to be ***i-removable*** if the top of that column is a removable i-block.

(2) A place in a proper Young wall where one may add an i-block to obtain another proper Young wall is called an ***i-admissible slot***. A column in a proper Young wall is said to be ***i-admissible*** if the top of that column is an i-admissible slot.

Example 11.3.2. In the following figure, we take a proper Young wall for $B_3^{(1)}$ built on the ground-state wall Y_{Λ_0} and indicate all the removable blocks and admissible slots.

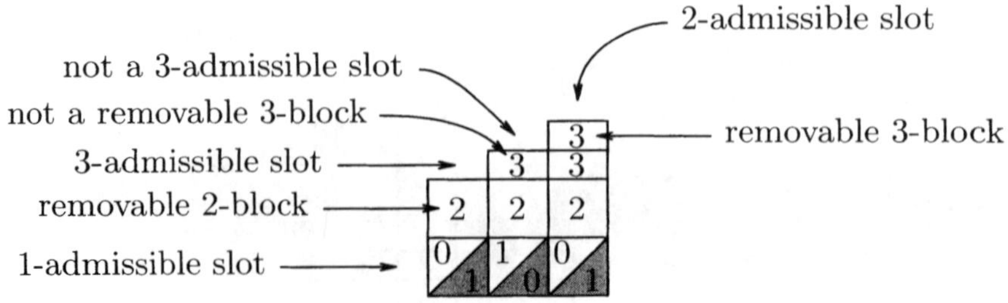

We now define the action of Kashiwara operators $\tilde{e}_i$, $\tilde{f}_i$ $(i \in I)$ on $\mathcal{F}(\lambda)$. Fix $i \in I$ and let $Y = (y_k)_{k=0}^{\infty} \in \mathcal{F}(\lambda)$ be a proper Young wall.

(1) To each column y_k of Y, we assign its i-signature as follows:
   (a) we assign $--$ if the column y_k is twice i-removable (the i-block will be of half-unit height in this case);
   (b) we assign $-$ if the column is once i-removable, but not i-admissible (the i-block may be of unit height or of half-unit height);
   (c) we assign $-+$ if the column is once i-removable and once i-admissible (the i-block will be of half-unit height in this case);
   (d) we assign $+$ if the column is once i-admissible, but not i-removable (the i-block may be of unit height or of half-unit height);
   (e) we assign $++$ if the column is twice i-admissible (the i-block will be of half-unit height in this case).

(2) From the (infinite) sequence of $+$'s and $-$'s, cancel out every $(+, -)$ pair to obtain a finite sequence of $-$'s followed by $+$'s, reading from left to right. This sequence is called the ***i-signature*** of the proper Young wall Y.

(3) We define $\tilde{e}_i Y$ to be the proper Young wall obtained from Y by removing the i-block corresponding to the right-most $-$ in the i-signature of Y. We define $\tilde{e}_i Y = 0$ if there exists no $-$ in the i-signature of Y.

(4) We define $\tilde{f}_i Y$ to be the proper Young wall obtained from Y by adding an i-block to the column corresponding to the left-most $+$ in the i-signature of Y. We define $\tilde{f}_i Y = 0$ if there exists no $+$ in the i-signature of Y.

We also define the maps

$$\mathrm{wt}: \mathcal{F}(\lambda) \to P, \qquad \varepsilon_i : \mathcal{F}(\lambda) \to \mathbf{Z}, \qquad \varphi_i : \mathcal{F}(\lambda) \to \mathbf{Z}$$

by

$$\mathrm{wt}(Y) = \lambda - \sum_{i \in I} k_i \alpha_i, \tag{11.2}$$

$$\varepsilon_i(Y) = \text{the number of } - \text{ in the } i\text{-signature of } Y, \tag{11.3}$$

$$\varphi_i(Y) = \text{the number of } + \text{ in the } i\text{-signature of } Y, \tag{11.4}$$

where k_i is the number of i-blocks in Y that have been added to the ground-state wall Y_λ.

Then it is straightforward to verify that the following theorem holds (Exercise 11.2).

Theorem 11.3.3. *The maps* $\mathrm{wt} : \mathcal{F}(\lambda) \to P$, $\tilde{e}_i, \tilde{f}_i : \mathcal{F}(\lambda) \to \mathcal{F}(\lambda) \cup \{0\}$, $\varepsilon_i, \varphi_i : \mathcal{F}(\lambda) \to \mathbf{Z}$ *define a* $U_q(\mathfrak{g})$*-crystal structure on the set* $\mathcal{F}(\lambda)$ *of all proper Young walls.*

Let δ be the null root for the quantum affine algebra $U_q(\mathfrak{g})$ and write

$$\begin{cases} \delta = a_0\alpha_0 + a_1\alpha_1 + \cdots + a_n\alpha_n & \text{for } \mathfrak{g} = A_n^{(1)}, \ldots, B_n^{(1)}, \\ 2\delta = a_0\alpha_0 + a_1\alpha_1 + \cdots + a_n\alpha_n & \text{for } \mathfrak{g} = D_{n+1}^{(2)} \end{cases}$$

The part of a column consisting of a_0-many 0-blocks, a_1-many 1-blocks, ..., a_n-many n-blocks in some cyclic order is called a ***δ-column***.

Example 11.3.4.

(1) The following are δ-columns for $B_3^{(1)}$.

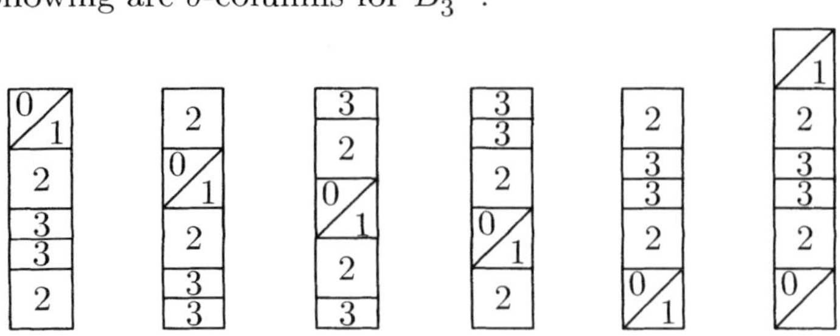

(2) The following are δ-columns for $D_3^{(2)}$.

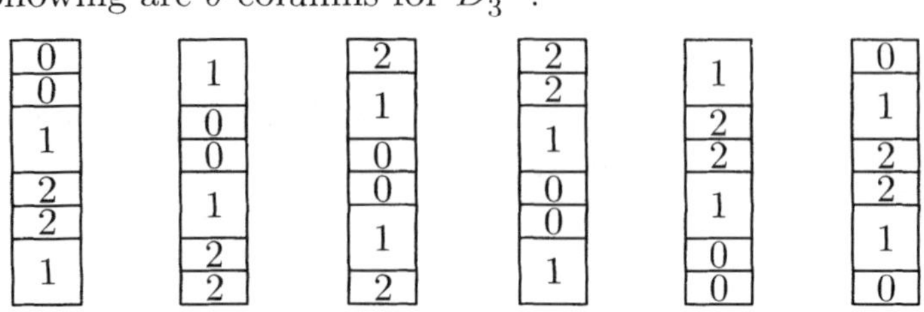

Definition 11.3.5.

(1) A column in a proper Young wall is said to contain a ***removable*** δ if we may remove a δ-column from Y and still obtain a proper Young wall.

(2) A proper Young wall is said to be ***reduced*** if none of its columns contain a removable δ.

Example 11.3.6. For the quantum affine algebra $U_q(B_3^{(1)})$, the first Young wall drawn below is reduced, but the second one is not reduced.

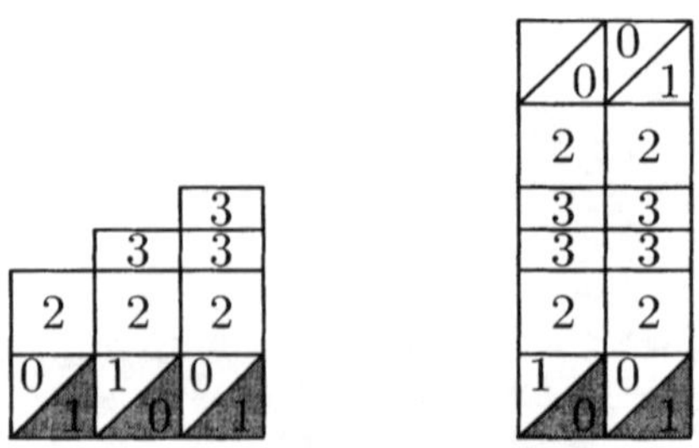

Let $\mathcal{Y}(\lambda)$ be the set of all reduced proper Young walls built on the ground-state wall Y_λ.

Proposition 11.3.7. *For any $Y \in \mathcal{Y}(\lambda)$, we have*

$$\tilde{e}_i Y \in \mathcal{Y}(\lambda) \cup \{0\} \quad \textit{and} \quad \tilde{f}_i Y \in \mathcal{Y}(\lambda) \cup \{0\}.$$

Hence the set $\mathcal{Y}(\lambda)$ has an affine crystal structure for the quantum affine algebra $U_q(\mathfrak{g})$.

Proof. Let $Y \in \mathcal{Y}(\lambda)$ and $i \in I$. If $\tilde{e}_i Y \neq 0$, by definition of removable block, $\tilde{e}_i Y$ is a proper Young wall. Suppose that $\tilde{e}_i Y$ is not reduced. This means that removing an i-block for Y has created a removable δ (in $\tilde{e}_i Y$). Hence, except for the $A_n^{(1)}$ case, which can be dealt with similarly, Y must have a column of the form:

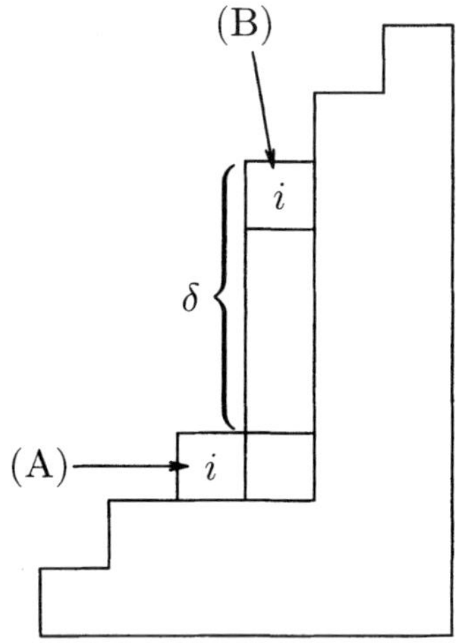

The i-block at (A) is removed to give $\tilde{e}_i Y$. Then, in this case, the Tetris® rules for the Kashiwara operators tell us that the operator $\tilde{e}_i$ would remove the i-block at (B), not the one at (A), which is a contradiction. Hence $\tilde{e}_i Y$ is reduced.

Similarly, if $\tilde{f}_i Y \neq 0$, then $\tilde{f}_i Y$ is a reduced proper Young wall (Exercise 11.3). □

11.4. Crystal graphs for basic representations

In this section, we will prove that the crystal graph $\mathcal{B}(\lambda)$ for the basic representation $V(\lambda)$ of the quantum affine algebra $U_q(\mathfrak{g})$ is isomorphic to the affine crystal $\mathcal{Y}(\lambda)$ consisting of all reduced proper Young walls built on the ground-state wall Y_λ. Thus, the crystal $\mathcal{Y}(\lambda)$ would be the connected component of $\mathcal{F}(\lambda)$ containing the ground-state wall Y_λ.

Theorem 11.4.1. *There is an isomorphism of $U_q(\mathfrak{g})$-crystals*

$$\mathcal{Y}(\lambda) \xrightarrow{\sim} \mathcal{B}(\lambda) \tag{11.5}$$

sending $Y_\lambda \mapsto u_\lambda$, where u_λ is the highest weight vector in $\mathcal{B}(\lambda)$.

The rest of this section is devoted to proving this theorem. It suffices to show that $\mathcal{Y}(\lambda) \cong \mathcal{P}(\lambda)$ as $U_q'(\mathfrak{g})$-crystals. Let us define a map $\Psi : \mathcal{Y}(\lambda) \to \mathcal{P}(\lambda)$ as follows: we read the top parts of each column in a reduced proper Young wall and write down a sequence of elements from $\mathcal{B}$, the perfect crystal of level 1, to obtain a path. The rules for obtaining the path are given below for each affine type in question.

- $A_n^{(1)}$ $(n \geq 1)$

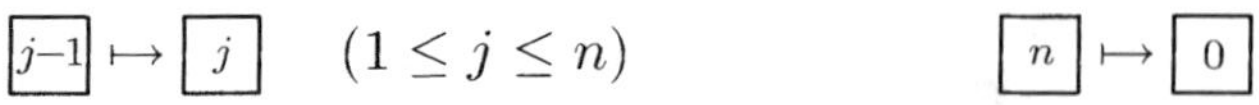

- $A_{2n-1}^{(2)}$ $(n \geq 3)$

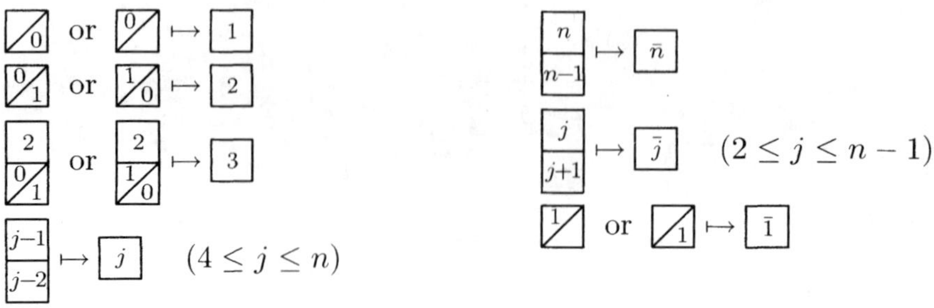

- $D_n^{(1)}$ $(n \geq 4)$

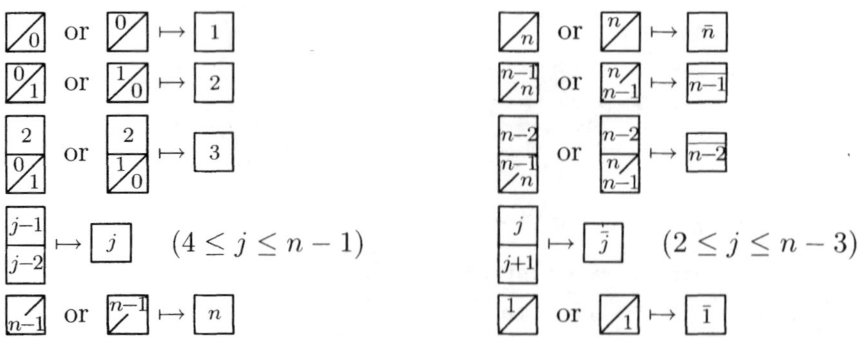

- $A_{2n}^{(2)}$ $(n \geq 2)$

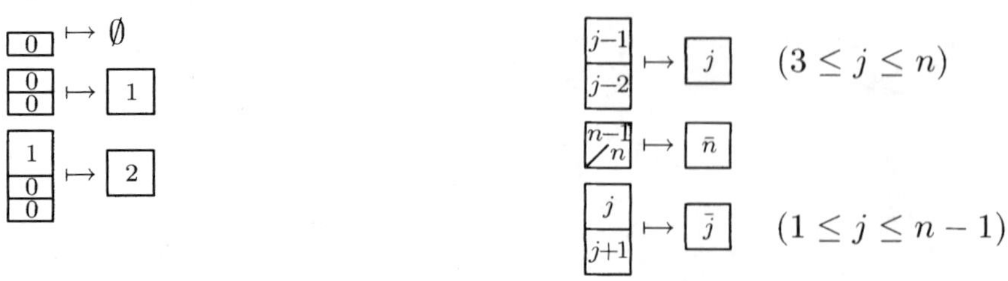

- $D_{n+1}^{(2)}$ $(n \geq 2)$

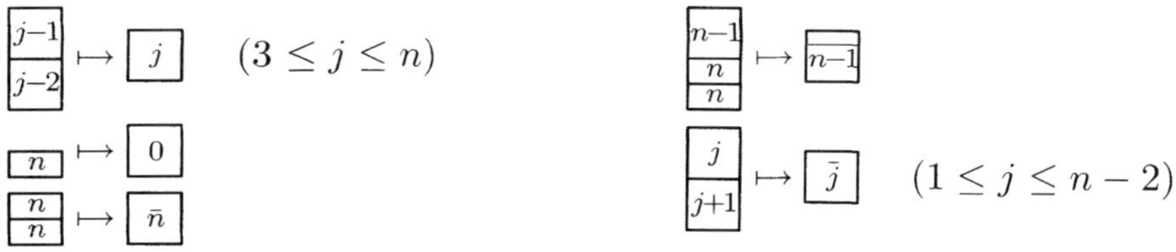

• $B_n^{(1)}$ $(n \geq 3)$

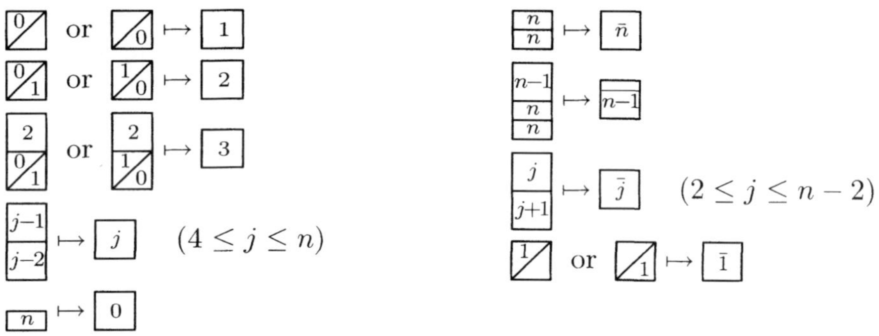

It is clear that this map sends the *ground-state walls* to the appropriate *ground-state paths* and that the image does indeed lie in the set of λ-paths. The surjectivity of Ψ follows immediately from the definition, and the injectivity of Ψ follows from the fact that the proper Young walls in $\mathcal{Y}(\lambda)$ are *reduced* (Exercise 11.4). Hence we have only to show that the map Ψ commutes with the Kashiwara operators. We give a proof of our claim only for the case of $B_n^{(1)}$. Other cases may be proved in similar manners and are less complicated.

Let $Y = (y_k)_{k=0}^{\infty} \in \mathcal{Y}(\lambda)$, where y_k denotes the kth column of Y (counting from right to left) and let $\mathbf{p} = (\mathbf{p}(k))_{k=0}^{\infty} = \Psi(Y)$ be the image of Y. Recall that the action of the Kashiwara operators is determined by the *i-signature* of Y and $\mathbf{p}$. Fix $i \in I$ and consider the Kashiwara operator $\tilde{f}_i$.

(1) $i = 0$: Suppose that the top of the kth column y_k of Y is the cube $\boxed{\diagup_1}$. The column y_k is certainly not 0-removable. If it is 0-admissible, we would assign a $+$ to the column, which is what we would also do with the corresponding element $\mathbf{p}(k) = \boxed{\bar{1}}$. If it is not 0-admissible, then the top of the column y_{k-1} must be either $\boxed{\diagup_0}$ or $\boxed{{}^1\!\diagup_0}$. In each of the two cases, y_{k-1} is neither 0-admissible nor 0-removable, so we would assign nothing to y_k and y_{k-1}. Now, consider the corresponding terms $\mathbf{p}(k)$ and $\mathbf{p}(k-1)$ of the path $\mathbf{p} = (\mathbf{p}(k))_{k=0}^{\infty} = \Psi(Y)$. We have $\mathbf{p}(k) = \boxed{\bar{1}}$ and $\mathbf{p}(k-1) = \boxed{1}$ or $\boxed{2}$. We put down $+$ for $\mathbf{p}(k) = \boxed{\bar{1}}$ and $-$ for $\mathbf{p}(k-1) = \boxed{1}$ or $\boxed{2}$, which cancel out to give nothing. Thus the resulting 0-signatures coincide with each other.

Similarly, we can verify that we would assign the same 0-signatures to y_k and $\mathbf{p}(k)$ for the other cases of y_k. In Table 11.1, we list all the possible nontrivial cases for $i = 0$.

(2) $i = 1$**:** This case is quite similar to the $i = 0$ case.

(3) $i = 2$**:** We list all the possible nontrivial cases in Table 11.2.

(4) $3 \le i \le n-2$**:** We list all the nontrivial possibilities in Table 11.3.

(5) $i = n-1$**:** We list all the nontrivial possibilities in Table 11.4.

(6) $i = n$**:** Observe that the top parts of the columns of Y that have a nontrivial contribution to the n-signature of Y have the following form.

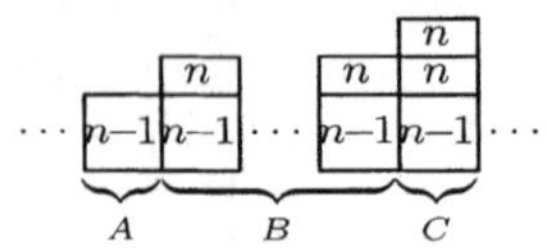

Here, we do not exclude the possibility of some of the parts A, B, or C missing in this drawing. The ground-state wall of weight Λ_n could also be considered as a degenerate case of the above. It suffices to verify that the n-signature for this segment and that of the corresponding part in the path are the same.

In the table given below, we list all the possible combinations for the parts A, B, and C, and write down their n-signatures.

presence of			signature			total
A	B	C	A	B	C	signature
yes	yes	yes	$+$		$-$	
yes	yes	no	$+$	$+$		$++$
yes	no	yes	$+$		$-$	
yes	no	no	$++$			$++$
no	yes	yes		$-$	$-$	$--$
no	yes	no		$-+$		$-+$
no	no	yes			$--$	$--$

Now one can easily verify that the n-signature for the above segment of Y coincides with that of the corresponding part in the path $\mathbf{p}$.

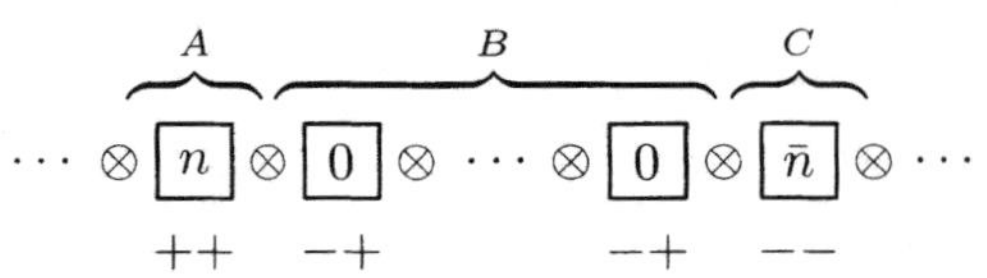

Table 11.1. $i = 0$

column			signature			path element			signature		
y_{k+1}	y_k	y_{k-1}	y_{k+1}	y_k	y_{k-1}	$\mathbf{p}(k+1)$	$\mathbf{p}(k)$	$\mathbf{p}(k-1)$	$\mathbf{p}(k+1)$	$\mathbf{p}(k)$	$\mathbf{p}(k-1)$
$\boxed{/_{1}}$	$\boxed{/_{0}}$					$\boxed{\bar{1}}$	$\boxed{1}$		$+$	$-$	
$\boxed{^{1}/}$	$\boxed{^{0}/}$					$\boxed{\bar{1}}$	$\boxed{1}$		$+$	$-$	
dashed $\boxed{/}$ over $\boxed{2}$	$\boxed{/_{0}}$ over $\boxed{2}$ or $\boxed{^{0}/}$ over $\boxed{2}$					$\boxed{\bar{2}}$	$\boxed{1}$		$+$	$-$	
other	$\boxed{/_{0}}$ or $\boxed{^{0}/}$			$-$			$\boxed{1}$			$-$	
$\boxed{^{1}/}$	$\boxed{^{0}/_{1}}$					$\boxed{\bar{1}}$	$\boxed{2}$		$+$	$-$	
$\boxed{/_{1}}$	$\boxed{^{1}/_{0}}$					$\boxed{\bar{1}}$	$\boxed{2}$		$+$	$-$	
other	$\boxed{^{0}/_{1}}$ or $\boxed{^{1}/_{0}}$			$-$			$\boxed{2}$			$-$	
	dashed $\boxed{/}$ over $\boxed{2}$	$\boxed{/_{0}}$ over $\boxed{2}$ or $\boxed{^{0}/}$ over $\boxed{2}$					$\boxed{\bar{2}}$	$\boxed{1}$		$+$	$-$
	dashed $\boxed{/}$ over $\boxed{2}$	other		$+$			$\boxed{\bar{2}}$			$+$	
	$\boxed{/_{1}}$	$\boxed{/_{0}}$					$\boxed{\bar{1}}$	$\boxed{1}$		$+$	$-$
	$\boxed{^{1}/}$	$\boxed{^{0}/}$					$\boxed{\bar{1}}$	$\boxed{1}$		$+$	$-$
	$\boxed{/_{1}}$	$\boxed{^{1}/_{0}}$					$\boxed{\bar{1}}$	$\boxed{2}$		$+$	$-$
	$\boxed{^{1}/}$	$\boxed{^{0}/_{1}}$					$\boxed{\bar{1}}$	$\boxed{2}$		$+$	$-$
	$\boxed{/_{1}}$ or $\boxed{^{1}/}$	other		$+$			$\boxed{\bar{1}}$			$+$	

Table 11.2. $i = 2$

column			signature			path element			signature		
y_{k+1}	y_k	y_{k-1}	y_{k+1}	y_k	y_{k-1}	$\mathbf{p}(k+1)$	$\mathbf{p}(k)$	$\mathbf{p}(k-1)$	$\mathbf{p}(k+1)$	$\mathbf{p}(k)$	$\mathbf{p}(k-1)$
	$\boxed{{}^0\!\diagup_1}$ or $\boxed{{}^1\!\diagup_0}$	$\boxed{2}$ / $\boxed{\diagup}$					$\boxed{2}$	$\boxed{3}$		$+$	$-$
	$\boxed{{}^0\!\diagup_1}$ or $\boxed{{}^1\!\diagup_0}$	other		$+$			$\boxed{2}$			$+$	
$\boxed{{}^0\!\diagup_1}$ or $\boxed{{}^1\!\diagup_0}$	$\boxed{2}$ / $\boxed{\diagup}$					$\boxed{2}$	$\boxed{3}$		$+$	$-$	
other	$\boxed{2}$ / $\boxed{\diagup}$			$-$			$\boxed{3}$			$-$	
	$\boxed{2}$ (dashed) / $\boxed{3}$	$\boxed{2}$ / $\boxed{3}$					$\boxed{\bar{3}}$	$\boxed{\bar{2}}$		$+$	$-$
	$\boxed{2}$ (dashed) / $\boxed{3}$	other		$+$			$\boxed{\bar{3}}$			$+$	
$\boxed{2}$ (dashed) / $\boxed{3}$	$\boxed{2}$ / $\boxed{3}$					$\boxed{\bar{3}}$	$\boxed{\bar{2}}$		$+$	$-$	
other	$\boxed{2}$ / $\boxed{3}$			$-$			$\boxed{\bar{2}}$			$-$	

Table 11.3. $3 \leq i \leq n-2$

column			signature			path element			signature		
y_{k+1}	y_k	y_{k-1}	y_{k+1}	y_k	y_{k-1}	$\mathbf{p}(k+1)$	$\mathbf{p}(k)$	$\mathbf{p}(k-1)$	$\mathbf{p}(k+1)$	$\mathbf{p}(k)$	$\mathbf{p}(k-1)$
	i (dashed) / $i-1$	i / $i-1$					i	$i+1$		$+$	$-$
	i (dashed) / $i-1$	other		$+$			i			$+$	
i (dashed) / $i-1$	i / $i-1$					i	$i+1$		$+$	$-$	
other	i / $i-1$			$-$			$i+1$			$-$	
	i (dashed) / $i+1$	i / $i+1$					$\overline{i+1}$	$\bar{i}$		$+$	$-$
	i (dashed) / $i+1$	other		$+$			$\bar{i}$			$+$	
i (dashed) / $i+1$	i / $i+1$					$\overline{i+1}$	$\bar{i}$		$+$	$-$	
other	i / $i+1$			$-$			$\bar{i}$			$-$	

Table 11.4. $i = n - 1$

column			signature			path element			signature		
y_{k+1}	y_k	y_{k-1}	y_{k+1}	y_k	y_{k-1}	$\mathbf{p}(k+1)$	$\mathbf{p}(k)$	$\mathbf{p}(k-1)$	$\mathbf{p}(k+1)$	$\mathbf{p}(k)$	$\mathbf{p}(k-1)$
	$n-1$ (dashed) / $n-2$	$n-1$ / $n-2$					$\boxed{n-1}$	$\boxed{n}$		$+$	$-$
	$n-1$ (dashed) / $n-2$	other		$+$			$\boxed{n-1}$			$+$	
$n-1$ (dashed) / $n-2$	$n-1$ / $n-2$					$\boxed{n-1}$	$\boxed{n}$		$+$	$-$	
other	$n-1$ / $n-2$			$-$			$\boxed{n}$			$-$	
	n / n	$n-1$ / n / n					$\boxed{\bar{n}}$	$\boxed{\overline{n-1}}$		$+$	$-$
	n / n	other		$+$			$\boxed{\bar{n}}$			$+$	
n / n	$n-1$ / n / n					$\boxed{\bar{n}}$	$\boxed{\overline{n-1}}$		$+$	$-$	
other	$n-1$ / n / n			$-$			$\boxed{\overline{n-1}}$			$-$	

Thus, we have shown that for all indices $i \in I$, the i-signature of the reduced proper Young wall Y is identical to that of the path $\mathbf{p} = \Psi(Y)$. It is now straightforward to verify that the action of the Kashiwara operators is compatible with the identification of Y with $\mathbf{p}$. Here are some examples.

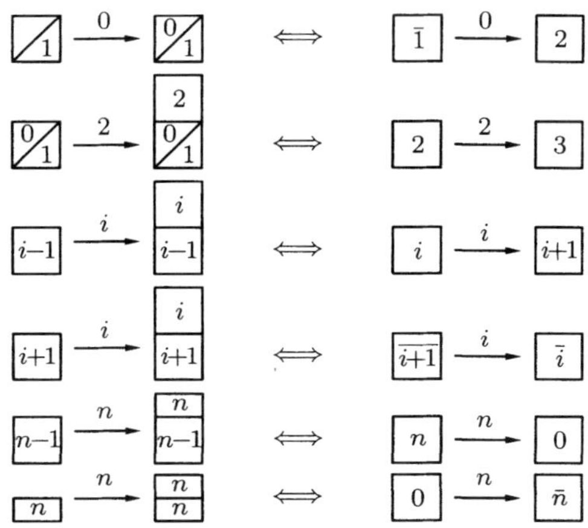

Therefore, we obtain the desired crystal isomorphism $\Psi : \mathcal{Y}(Y) \xrightarrow{\sim} \mathcal{P}(\lambda)$.

Example 11.4.2. In this example, we illustrate the top parts of the affine crystal $\mathcal{Y}(Y)$ consisting of reduced proper Young walls. Compare them with the path realization of crystal graphs given in Section 11.1.

(1) The crystal $\mathcal{Y}(\Lambda_0)$ for $A_2^{(1)}$.

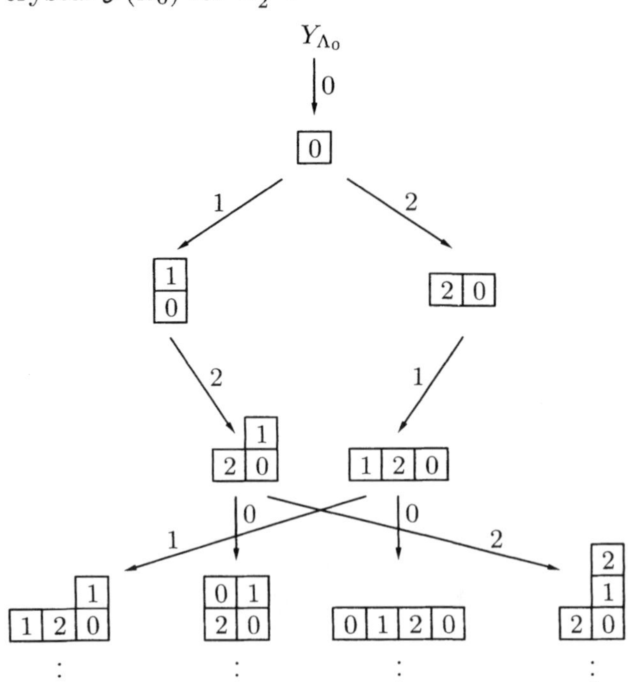

(2) The crystal $\mathcal{Y}(\Lambda_0)$ for $A_5^{(2)}$.

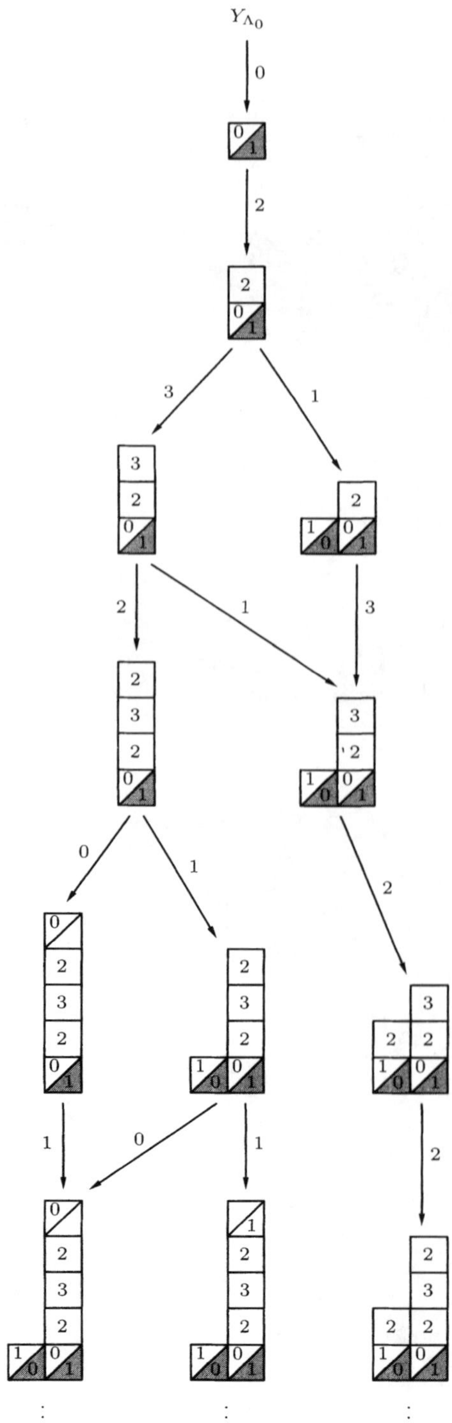

(3) The crystal $\mathcal{Y}(\Lambda_0)$ for $D_4^{(1)}$.

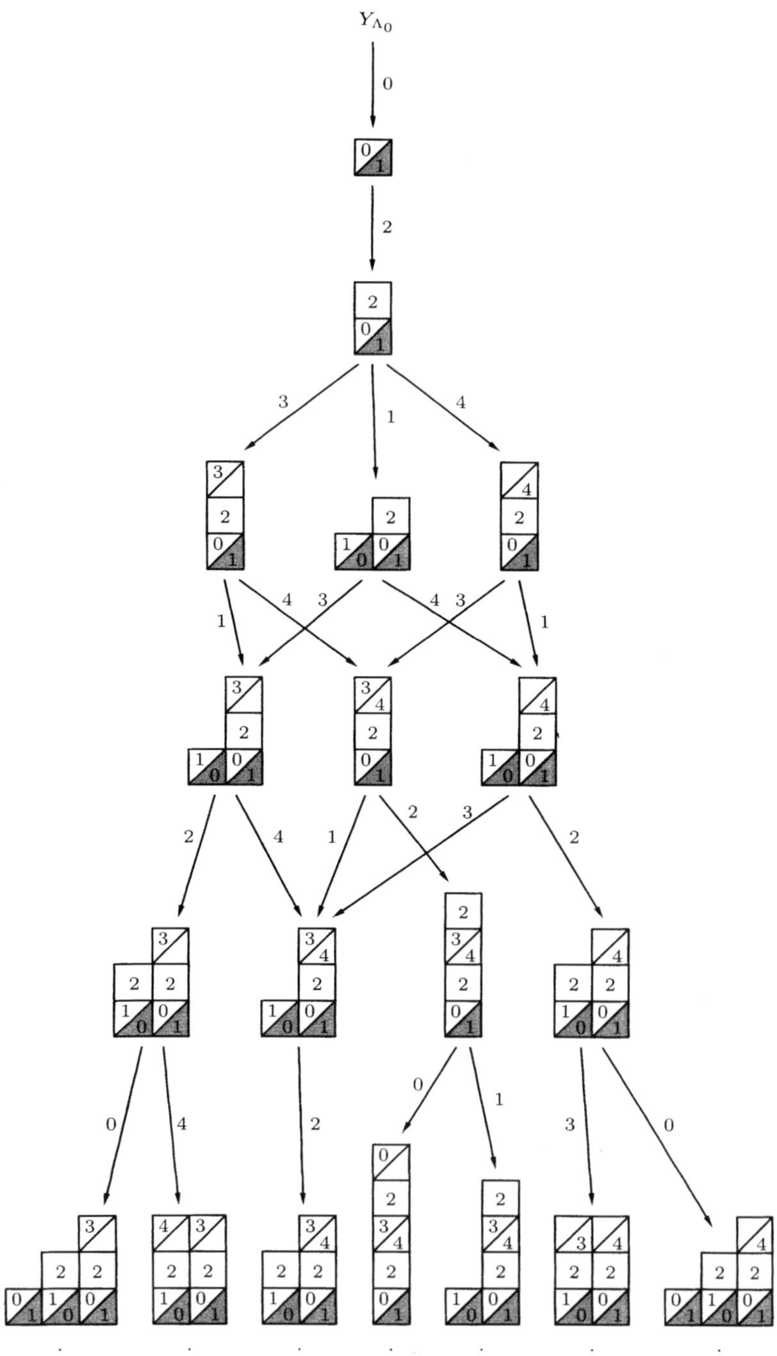

(4) The crystal $\mathcal{Y}(\Lambda_0)$ for $A_4^{(2)}$.

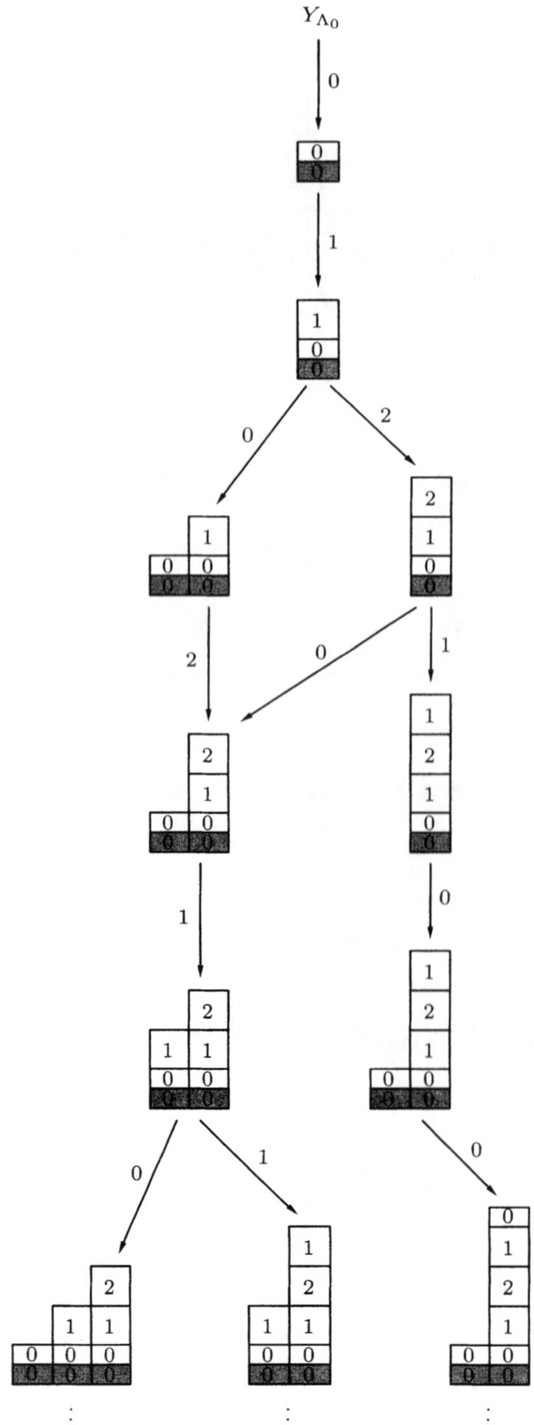

(5) The crystal $\mathcal{Y}(\Lambda_0)$ for $D_3^{(2)}$.

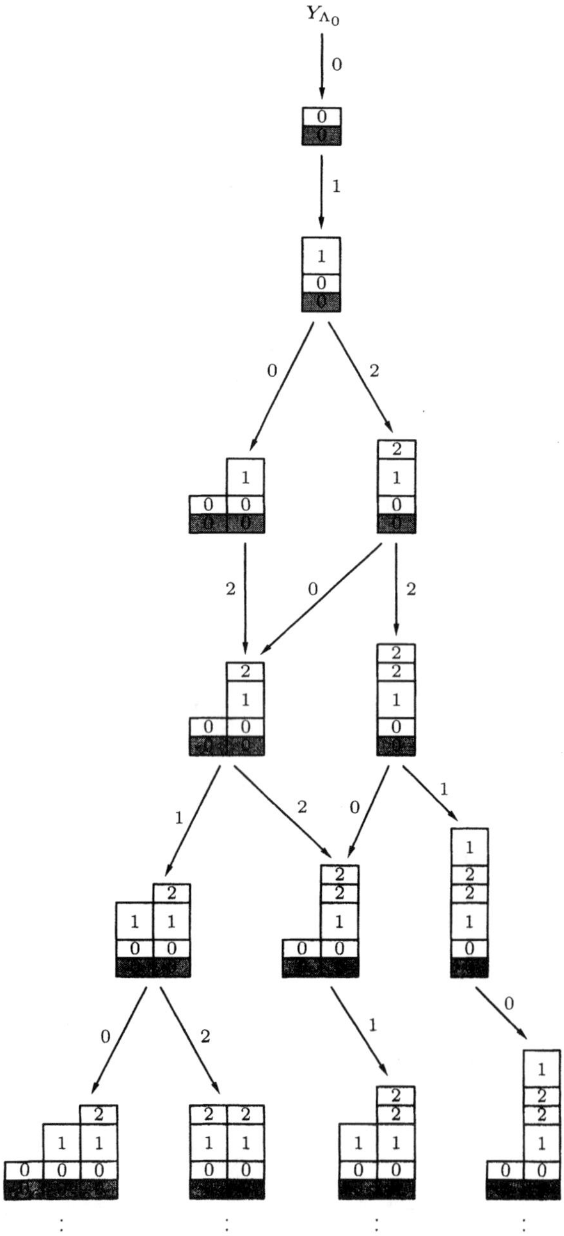

(6) The crystal $\mathcal{Y}(\Lambda_0)$ for $B_3^{(1)}$.

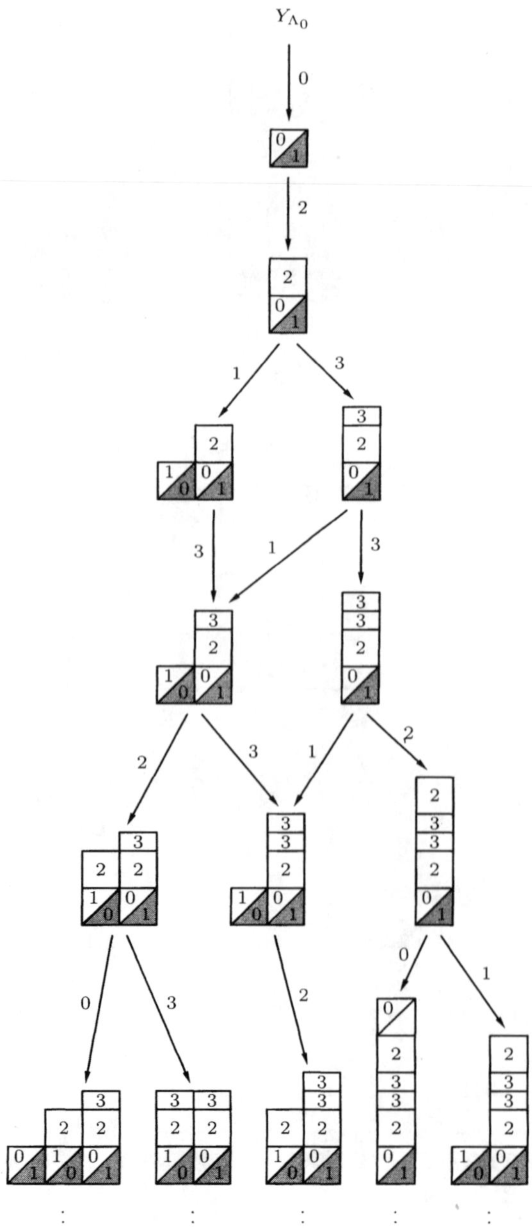

(7) The crystal $\mathcal{Y}(\Lambda_3)$ for $B_3^{(1)}$.

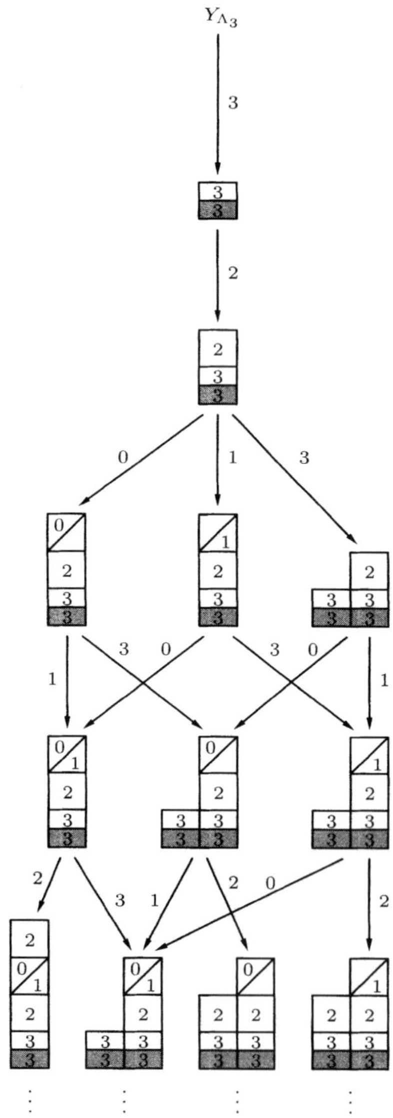

Exercises

11.1. Verify the path realization of crystal graphs for basic representations given in Examples 11.1.1–11.1.6.

11.2. Prove Theorem 11.3.3.

11.3. Complete the proof of Proposition 11.3.7.

11.4. Prove that the map given in Theorem 11.4.1 is a bijection.

11.5. Fill in the parts of the proof for Theorem 11.4.1 which we have omitted.

11.6. Verify the crystal graphs given in Example 11.4.2.

11.7. A perfect crystal for $U_q(C_2^{(1)})$ is given in Example 10.5.2 (6). Using this perfect crystal, realize the crystals for the basic representations of $U_q(C_2^{(1)})$ as the set of reduced proper Young walls (see [**16**]).

Bibliography

1. E. Abe, *Hopf Algebras*, Cambridge University Press, Cambridge, 1980.
2. G. E. Andrews, *The Theory of Partitions*, Encyclopedia of Mathematics and its Applications, Vol. 2, Addison-Wesley, Mass.-London-Amsterdam, 1976.
3. R. J. Baxter, *Exactly Solved Models in Statistical Mechanics*, Academic Press Inc., London, 1989.
4. G. Benkart, S.-J. Kang, and K. C. Misra, *Graded Lie algebras of Kac-Moody type*, Adv. Math. **97** (1993), no. 2, 154–190.
5. S. Berman and R. V. Moody, *Lie algebra multiplicities*, Proc. Amer. Math. Soc. **76** (1979), no. 2, 223–228.
6. V. Chari and A. Pressley, *Quantum affine algebras*, Comm. Math. Phys. **142** (1991), no. 2, 261–283.
7. ———, *A Guide to Quantum Groups*, Cambridge University Press, Cambridge, 1994.
8. C. W. Curtis and I. Reiner, *Methods of Representation Theory, Vol. I*, John Wiley & Sons Inc., New York, 1981.
9. E. Date, M. Jimbo, and M. Okado, *Crystal base and q-vertex operators*, Comm. Math. Phys. **155** (1993), no. 1, 47–69.
10. V. G. Drinfel′d, *Hopf algebras and the quantum Yang-Baxter equation*, Soviet Math. Dokl. **32** (1985), no. 1, 254–258.
11. ———, *Quantum groups*, Proceedings of the International Congress of Mathematicians (Berkeley, 1986), American Mathematical Society, Providence, 1987, pp. 798–820.
12. A. J. Feingold and I. B. Frenkel, *A hyperbolic Kac-Moody algebra and the theory of Siegel modular forms of genus* 2, Math. Ann. **263** (1983), no. 1, 87–144.
13. A. J. Feingold and J. Lepowsky, *The Weyl-Kac character formula and power series identities*, Adv. in Math. **29** (1978), no. 3, 271–309.
14. I. B. Frenkel and V. G. Kac, *Basic representations of affine Lie algebras and dual resonance models*, Invent. Math. **62** (1981), no. 1, 23–66.
15. W. Fulton and J. Harris, *Representation Theory: a First Course*, Springer-Verlag, New York, 1991.

16. J. Hong and S.-J. Kang, *Crystal graphs for basic representations of the quantum affine algebra* $U_q(C_2^{(1)})$, Representations and Quantizations (Shanghai, 1998), China Higher Education Press and Springer-Verlag, Beijing, 2000, pp. 213–227.

17. J. E. Humphreys, *Introduction to Lie Algebras and Representation Theory*, 2nd ed., Springer-Verlag, New York, 1978.

18. T. W. Hungerford, *Algebra*, 5th ed., Springer-Verlag, New York, 1989.

19. N. Jacobson, *Lie Algebras*, 2nd ed., Dover Publications Inc., New York, 1979.

20. ———, *Basic Algebra II*, W. H. Freeman and Company, San Francisco, 1980.

21. J. C. Jantzen, *Lectures on Quantum Groups*, Graduate Studies in Mathematics, vol. 6, American Mathematical Society, 1996.

22. K. Jeong, *Crystal bases for Kac-Moody superalgebras*, J. Algebra **237** (2001), no. 2, 562–590.

23. M. Jimbo, *A q-difference analogue of* $U(\mathfrak{g})$ *and the Yang-Baxter equation*, Lett. Math. Phys. **10** (1985), no. 1, 63–69.

24. M. Jimbo and T. Miwa, *Algebraic Analysis of Solvable Lattice Models*, Regional Conference Series in Mathematics, vol. 85, American Mathematical Society, 1995.

25. A. Joseph, *Quantum Groups and Their Primitive Ideals*, Springer-Verlag, Berlin, 1995.

26. V. G. Kac, *Infinite-dimensional Lie algebras, and the Dedekind* η*-function*, Functional Anal. Appl. **8** (1974), no. 1, 68–70.

27. ———, *Infinite-dimensional algebras, Dedekind's* η*-function, classical Möbius fuctions and the very strange formula*, Adv. Math. **30** (1978), 85–136.

28. ———, *Infinite Dimensional Lie algebras*, 3rd ed., Cambridge University Press, Cambridge, 1990.

29. V. G. Kac, R. V. Moody, and M. Wakimoto, *On* E_{10}, Differential Geometrical Methods in Theoretical Physics (Como, 1987), Kluwer Acad. Publ., Dordrecht, 1988, pp. 109–128.

30. S.-J. Kang, *Kac-Moody Lie algebras, spectral sequences, and the Witt formula*, Trans. Amer. Math. Soc. **339** (1993), no. 2, 463–493.

31. ———, *Generalized Kac-Moody algebras and the modular function* j, Math. Ann. **298** (1994), 373–384.

32. ———, *Root multiplicities of Kac-Moody algebras*, Duke Math. J. **74** (1994), no. 3, 635–666.

33. ———, *Graded Lie superalgebras and the superdimension formula*, J. Algebra **204** (1998), no. 2, 597–655.

34. ———, *Crystal bases for quantum affine algebras and combinatorics of Young walls*, RIM-GARC preprint **00-2** (2000).

35. S.-J. Kang, M. Kashiwara, and K. C. Misra, *Crystal bases of Verma modules for quantum affine Lie algebras*, Compositio Math. **92** (1994), no. 3, 299–325.

36. S.-J. Kang, M. Kashiwara, K. C. Misra, T. Miwa, T. Nakashima, and A. Nakayashiki, *Affine crystals and vertex models*, Infinite Analysis, Part A, B (Kyoto, 1991), World Scientific Publishing Co. Inc., River Edge, NJ, 1992, pp. 449–484.

37. ———, *Perfect crystals of quantum affine Lie algebras*, Duke Math. J. **68** (1992), no. 3, 499–607.

38. M. Kashiwara, *Crystalizing the q-analogue of universal enveloping algebras*, Comm. Math. Phys. **133** (1990), 249–260.

39. ______, *On crystal bases of the q-analogue of universal enveloping algebras*, Duke Math. J. **63** (1991), 465–516.

40. ______, *Similarity of crystal bases*, Lie Algebras and Their Representations (Seoul, 1995), S.-J. Kang, M.-H. Kim, I.-S. Lee (eds.), Contemporary Mathematics **194**, American Mathematical Society, Providence, RI, 1996, pp. 177–186.

41. M. Kashiwara and T. Nakashima, *Crystal graphs for representations of the q-analogue of classical Lie algebras*, J. Algebra **165** (1994), no. 2, 295–345.

42. W. L. Li, *Classification of generalized Cartan matrices of hyperbolic type*, Chinese Ann. Math. Ser. B **9** (1988), no. 1, 68–77.

43. P. Littelmann, *A Littlewood-Richardson rule for symmetrizable Kac-Moody algebras*, Invent. Math. **116** (1994), no. 1-3, 329–346.

44. ______, *Crystal graphs and Young tableaux*, J. Algebra **175** (1995), no. 1, 65–87.

45. ______, *Paths and root operators in representation theory*, Ann. of Math. **(2) 142** (1995), no. 3, 499–525.

46. D. E. Littlewood and A. R. Richardson, *Group characters and algebra*, Philos. Trans. Roy. Soc. London, Ser. A **233** (1934), 99–142.

47. G. Lusztig, *Quantum deformation of certain simple modules over enveloping algebras*, Adv. Math. **70** (1988), 237–249.

48. ______, *Canonical bases arising from quantized enveloping algebras*, J. Amer. Math. Soc. **3** (1990), no. 2, 447–498.

49. R. V. Moody and A. Pianzola, *Lie Algebras with Triangular Decompositions*, John Wiley & Sons Inc., New York, 1995.

50. T. Nakashima, *Crystal base and a generalization of the Littlewood-Richardson rule for the classical Lie algebras*, Commun. Math. Phys. **154** (1993), no. 2, 215–243.

51. D. H. Peterson, *Freudenthal-type formulas for root and weight multiplicities*, preprint (1983, unpublished).

52. D. Robinson, *On representations of the symmetric group*, Amer. J. Math. **60** (1938), 745–760.

53. M. E. Sweedler, *Hopf Algebras*, W. A. Benjamin, Inc., New York, 1969.

54. Z.-X. Wan, *Introduction to Kac-Moody Algebra*, World Scientific Publishing Co. Inc., Teaneck, NJ, 1991.

55. H. Weyl, *Classical Groups*, 2nd ed., Princeton University Press, 1946.

56. S. Yamane, *Perfect crystals of $U_q(G_2^{(1)})$*, J. Algebra **210** (1998), no. 2, 440–486.

57. A. Young, *On quantitative substitutional analysis II*, Proc. Lond. Math. Soc. **(1) 34** (1902), 361–397.

58. ______, *On quantitative substitutional analysis III*, Proc. Lond. Math. Soc. **(2) 28** (1927), 255–291.

59. ______, *On quantitative substitutional analysis IV*, Proc. Lond. Math. Soc. **(2) 31** (1929), 253–272.

Index of symbols

Index